Liquid Glass Transition

Liquid Glass Transition
A Unified Theory from the Two Band Model

Toyoyuki Kitamura

Nagasaki Institute of Applied Science
Nagasaki
Japan

ELSEVIER

AMSTERDAM • BOSTON • HEIDELBERG • LONDON • NEW YORK
OXFORD • PARIS • SAN DIEGO • SAN FRANCISCO • SINGAPORE
SYDNEY • TOKYO

Elsevier
32 Jamestown Road, London NW1 7BY
225 Wyman Street, Waltham, MA 02451, USA

First edition 2013

British Library Cataloguing-in-Publication Data
A catalogue record for this book is available from the British Library

Library of Congress Cataloging-in-Publication Data
A catalog record for this book is available from the Library of Congress

ISBN: 978-0-124-07177-3

For information on all Elsevier publications
visit our website at store.elsevier.com

This book has been manufactured using Print On Demand technology. Each copy is produced to order and is limited to black ink. The online version of this book will show color figures where appropriate.

Contents

Preface

Near the liquid-glass transition, a particle locates in a randomly distributed well-localized potential in a quantum state. A particle makes up and down transitions between two levels for a while and hops to a neighboring vacancy. This picture yields the two band model in the random system. In this book, we develop the unified theory of the liquid-glass transition on the basis of the two band model by statistical quantum field theory associated with the temperature Green's function method. In order to develop in a self-contained manner, we explain all the tools we use in this book in Chapters 3–5. The concept of free physical fields plays a key role in statistical quantum field theory. The Ward-Takahashi relations associated with the spontaneous breakdown of the symmetry also play an important role in establishing the Hamiltonian, the Nambu-Goldstone theorem and the approximation procedures among the Green's functions, the self-energy parts and the vertex parts.

The dynamical and dissipative processes in random systems are treated at two stages: at the first stage, we consider the configurationally averaged Hamiltonian, by which we treat the dynamical processes. At the second stage, we take into account the randomness by introducing the scattering processes of particles by the random potentials due to random frequencies and random hopping amplitudes. The configuration average of the simultaneously scattering processes of two particles by the same random potentials constrains the dynamical processes, which yields the relaxation and dissipative processes.

In the two band systems, bubbles of intra-band and inter-band particle-hole pairs are created. The sum of an infinite series of interacting bubbles of intra-band and inter-band particle-hole pairs yield sound and phonons respectively. The extra bubbles of intra-band and inter-band particle-hole pairs not constituting the sound and phonons make the intra-band and inter-band fluctuation entropies, which compensate with diffusion and viscosity, respectively, and then the system reaches the thermal equilibrium. The liquid and glassy states have a kind of the long range order; the pair distribution function looks the same to each observer at his own particle in a sense of the configuration average. In the glassy state, the pair distribution function holds the same every time so that it breaks the symmetry of translation, which leads to the Nambu-Goldstone bosons, phonons; the up and down transitions of particles between the two levels yield phonons. On the other hand, in the liquid state, the pair distribution function changes with time so that the hopping of particles is essential, which yields the density fluctuation mode, sound. Sound is not the Nambu-Goldstone boson.

Conventional theories have not discriminated between sound and phonons clearly. Sometimes, sound is considered to be a longitudinal phonon in the long wavelength regime. This is not correct. Despite of the limitations of the macroscopic and microscopic classical theories for sound and elastic waves (phonons), macroscopic classical theories in the long wavelength regime have discriminated between sound and elastic waves; the classical fluid mechanics has dealt with sound and the classical elastic theory with elastic waves. The microscopic classical theory can also discriminate between them, where elastic waves are valid over the long and short wavelength regimes, while sound is limited in the long wavelength regime. The macroscopic and microscopic classical theories for sound and elastic waves and their limitations are presented in Chapter 2.

Near the liquid-glass transition, since the energy bands of particles are so narrow that there appear different situations from the classical regime: First, the sound keeps alive over the long and short wavelength regimes. The sound disappears at a disappearing temperature and a reciprocal particle distance. This fact relates with the Kauzmann entropy crisis. Second, in a supercooling process in a liquid, enormous bubbles of intra-band particle-hole pairs not constituting sound are created so that the extra bubbles of intra-band particle-hole pairs not constituting the sound form the intra-band fluctuation entropy, which mixes with the sound entropy. This mixing entropy with the wave vectors near the reciprocal particle distance yields just the Kauzmann entropy, which has the form of the Curie law with the sound disappearing temperature as the critical temperature and a negative sign. The hopping probability of a particle is proportional to the exponent of the Kauzmann entropy per particle from the Einstein relation on probability and entropy. Thus the hopping amplitude (the exponent of the Kauzmann entropy per particle) manifests the Vogel-Tamman-Fulture (VTF) law. The Kauzmann entropy crisis seems to occur as the temperature decreases toward the sound disappearing temperature, but the inter-band fluctuation entropy increases as the temperature decreases. Eventually, the inter-band fluctuation entropy crosses over the Kauzmann entropy, where the glass transition occurs above the sound disappearing temperature so that it prevents the entropy crisis.

This scenario in the supercooling process naturally elucidates the universal features in the liquid-glass transition such as the Kauzmann entropy on the gap of the specific heat at the glass transition temperature and the VTF law on the relaxation and transport coefficients. This is why we call this scenario the unified theory of the liquid-glass transition. Conventionally, the Adam-Gibbs formula on the relaxation and dissipative processes has been used to arrange the experimental results. The Adam-Gibbs formula is a phenomenological one. The formula describes the hopping amplitude as the exponent of the inverse configuration entropy. Since the configuration entropy corresponding to the Kauzmann entropy is involved in the exponent of the hopping amplitude in the reverse order, the configuration entropy cannot explain the Kauzmann entropy crisis and the gap of the specific heat at the glass transition temperature. We develop the unified theory of the liquid-glass transition in Chapters 6–10.

The two level systems are ubiquitous. The two band model plays an essential role in the two level systems, where the constituent matter hops. If we identify the matter and the inter-matter interaction potentials in a two level system, with a particle and inter-particle interaction potentials in a liquid, we can extend the unified theory of the liquid-glass transition to the matter system. We call the Kauzmann entropy crisis in the system, a panic. Thus panics are also ubiquitous as the two level systems. Panics occur in financial market, psychology, ... , and even in nature. In this context, the liquid-glass transition is regarded as one of panics in nature. We develop the financial panics as the extension of the two band model in Chapter 11.

I would like to dedicate this book in the memory of Dr. Shozo Takeno.

This page intentionally left blank

1 Introduction

A glass is a disordered material like a viscous liquid and behaves mechanically as a solid [1–6]. A glass is normally formed by supercooling the viscous liquid fast enough to avoid crystallization. The liquid-glass transition occurs in diverse manner depending on materials, supercooling processes, history of materials, and so on. Among the diversity, there are universal features such as:

1. The Kauzmann paradox on the entropy crisis [7]:
   As the temperature decreases toward the glass transition temperature, the entropy of the system seems to drop to a negative value.
2. The Vogel-Tamman-Fulcher (VTF) law [8–10] on the relaxation and the transport processes:
   The relaxation time and the viscosity drastically increase, while approaching the liquid-glass transition. The ratio of the values of the relaxation time and the viscosity at the glass transition to those at higher temperatures becomes 10^{17}.
3. The jump of specific heat at the glass transition temperature:
   There is a gap in the values of the specific heats between just below and just above the glass transition temperature.
4. The boson peak [11]:
   There appears a peak of the density states at the harmonic frequency of phonons below the glass transition temperature.

In the crystal forming process, cooling a liquid in an equilibrium state, the specific volume decreases with the temperature decreasing till the melting temperature T_m, at which the specific volume decreases with the temperature keeping constant till the completion of crystallization, where the first order phase transition accompanied with the emission of the latent heat takes place, and then the specific heat decreases with the temperature decreasing as schematically illustrated in Figure 1.1.

In the glass forming process to avoid the crystallization, supercooling the liquid, the specific volume decreases beyond T_m upward concavely with the temperature decreasing till the glass transition temperature T_g, where the liquid-glass transition takes place, and then the specific volume decreases at first upward concavely, then linearly with the temperature decreasing as schematically illustrated in Figure 1.1. T_g depends on the supercooling processes.

In order to establish a unified theory to elucidate the universal features in such complex systems, at the first step we must start with a simple model in a simple liquid composed with a kind of isotropic particles and develop the theory in a most tractable way; we construct the two band model as a simple model and make the mean

Liquid Glass Transition. DOI: http://dx.doi.org/10.1016/B978-0-12-407177-3.00001-5

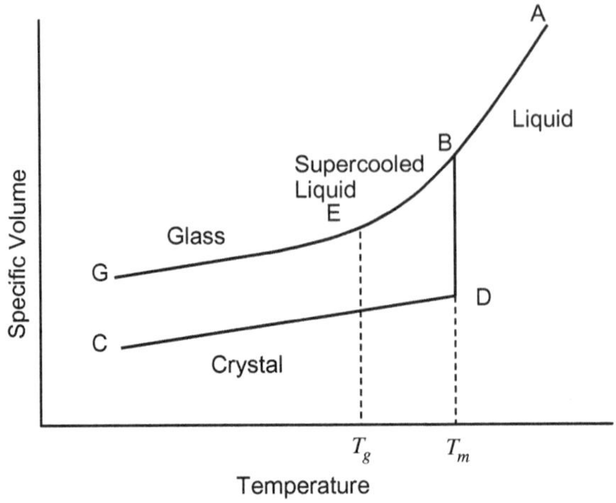

Figure 1.1 Changes of specific volume with the temperature decreasing in the glass forming and the crystal forming processes. In the crystal forming process, cooling a liquid in an equilibrium state, the specific volume decreases with the temperature decreasing A till B (at the melting temperature T_m), at which the specific volume decreases B till D (the completion of crystallization) with the temperature constant, where the first order transition accompanied with the emission of the latent heat takes place and then the specific volume decreasing D to C with the temperature decreasing. In the glass forming process to avoid crystallization, supercooling the liquid, the specific volume decreases A to B in the liquid state, B to E (at the glass transition temperature T_g) in the supercooled liquid state and then E to G in the glassy state. The specific volume changes upward concavely around the glass transition temperature.

field approximation (the random phase approximation) as a tractable way. Despite the simple model and the crude approximation, once we have established a unified theory, it plays a guiding role in establishing more rigorous theories, rigorously simulating the dynamical equations including nonlinear interactions, extending to more complex systems, arranging experimental results, and so on.

In order to construct the two band model, we must clarify the structure of condensed states. The structure of a system relates with the long range order of the system, which means that the same order holds over the whole system. The long range order spontaneously breaks the translational symmetry; the symmetry broken state is associated with the Nambu-Goldstone (NG) bosons (phonons) [12–15]. In the crystalline state, the long range order is represented by the periodic structure; the pair distribution function looks exactly the same to each observer at his own particle [16]. A particle locates in a periodically distributed localized potential making up and down transitions between exactly the same two levels for a long time. The modes of the propagation of the up and down transitions through the interactions are a longitudinal and two transverse phonons. In the liquid and glassy states, although particles are randomly distributed, there is a kind of the long range order; the pair distribution function looks the same to each observer at his own particle in a sense of the configuration average [17]. A particle locates in a randomly distributed localized potential making up and down transitions between the two levels and then hops to a neighboring vacancy. The hopping yields the two levels bandwidths. The modes of the propagation of the up and down transition through the interactions are also phonons. In the glassy state, since the hopping seldom occurs, the long range order holds; the pair distribution function holds the same every time so that phonons are essential, while in the liquid state, since the hopping always occurs, the long range order does not holds; the pair distribution function changes every time and the hopping causes the density fluctuations so that the sound mode is essential. The sound is not the NG boson.

The sound plays an essential role in the liquid-glass transition. In the two band model, the sum of an infinite series of interacting bubbles of *intra-band* particle-hole pairs constitutes sound mode, while the sum of an infinite series of interacting bubbles of *inter-band* particle-hole pairs constitutes phonon modes. The sound disappears at the sound disappearing temperature T_0 and a reciprocal particle distance K. In the supercooling process, enormous bubbles of intra-band particle-hole pairs are created so that extra bubbles of intra-band particle-hole pairs not constituting the sound form the fluctuation entropy, which mixes with the entropy of sound around the wave number $\sim K$. This mixing entropy is just the Kauzmann entropy, which is of the form of the Curie law with a negative sign and the critical temperature T_0. The glass transition occurs at the glass transition temperature T_g, where the Kauzmann entropy crosses over the fluctuation entropy due to the bubbles of inter-band particle-hole pairs. The probability corresponding to the Kauzmann entropy per particle from the Einstein relation manifests the VTF law; the probability is proportional to the exponent of the Kauzmann entropy per particle.

In Introduction, we briefly survey the structure of the condensed states, the two band model, the collective modes (sound and phonons), and the mechanism of the liquid-glass transition.

1.1 The Structure of the Condensed States and the Quantum Regime

A pair distribution function and an inter-particle interaction potential yield a well-localized potential of a particle, which can be approximated as a harmonic potential. The structure of the pair distribution function determines the structure of phonons. On the other hand, when a particle has a vacancy at the nearest neighbor, the potential of the particle has another minimum potential in the direction to the vacancy so that the particle can hop to the position of the vacancy.

The long range order spontaneously breaks down the translational symmetry, which yields the NG bosons, a longitudinal and two transverse phonons. In the glassy state, phonons are essential, while in the liquid state, sound is essential due to the hopping so that the sound is not NG bosons.

In this section, we also verify the liquid-glass transition occurs in the quantum regime.

1.1.1 *The Structure of Condensed States, the Pair Distribution Function, and Hopping of Particles*

In both the liquid and glassy states, particles are randomly distributed, but both the states have almost the same pair distribution function $g(\boldsymbol{R}_m, \boldsymbol{R}_n)$ which is defined as two particle density at the positions $\boldsymbol{R}_m$ and $\boldsymbol{R}_n$ under the configuration average of the other particles. For a simple liquid, the pair distribution function depends only on the separation, $R = |\boldsymbol{R}_m - \boldsymbol{R}_n|$, so it is simply written as $g(R)$ called the radial distribution function [17]. $g(R)$ has a peak at a mean inter-particle distance a and with the distance R increasing it converges oscillatingly to the average density ρ. Randomly distributed particles and the pair distribution function $g(R)$ is schematically illustrated in Figure 1.2.

The potential between particles has a hard core and a valley with a minimum near a mean inter-particle distance a illustrated in Figure 1.3a. A particle at a site $\boldsymbol{R}_m$ feels a well-localized potential made up by the surrounding particles. In order to see how the potential well is made up by the surrounding particles, we consider a one-dimensional system with the x-axis under the nearest neighbor approximation as schematically illustrated in Figure 1.3b. The potentials of a particle at the 0-site, made by the particles at the $-a$ and a-sites, are illustrated by the two dotted lines. The total potential made up by the surrounding particles, $\Phi(x)$, is well localized as illustrated as thick line. We can regard the well-localized potential as a harmonic potential. Figure 1.3c shows the case for the existence of a vacancy at a-site shown by a dotted circle. The potential which the particle at the 0-site feels, $\Phi(x)$, has double minimums so that the particle at the 0-site can hop to the vacancy at the a-site.

Thus we can regard a well-localized potential as a harmonic potential. The particle in the harmonic potential at a random position $\boldsymbol{R}_m$ has a random harmonic frequencies, ω_m. In a case of the existence of a vacancy at $\boldsymbol{R}_n$ in the x_i-direction of a particle at $\boldsymbol{R}_m$ as illustrated in Figure 1.4, since the m-particle misses the negative potential at $\boldsymbol{R}_m$ and the wall which should be made by the n-particle, the potential of the m-particle has a

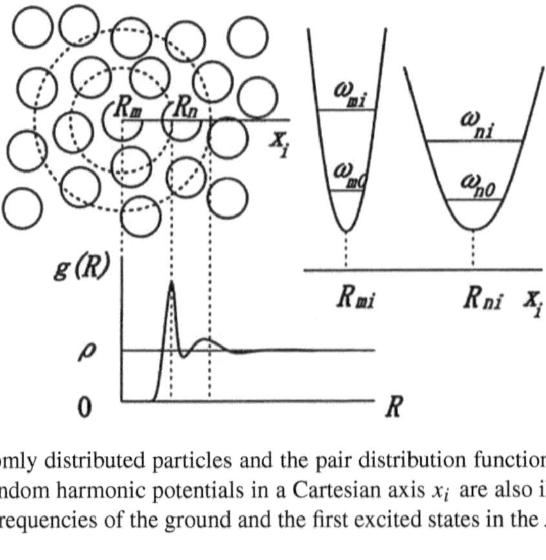

Figure 1.2 Randomly distributed particles and the pair distribution function are schematically illustrated. The random harmonic potentials in a Cartesian axis x_i are also illustrated. ω_{n0} and ω_{ni} are the eigenfrequencies of the ground and the first excited states in the x_i-component.

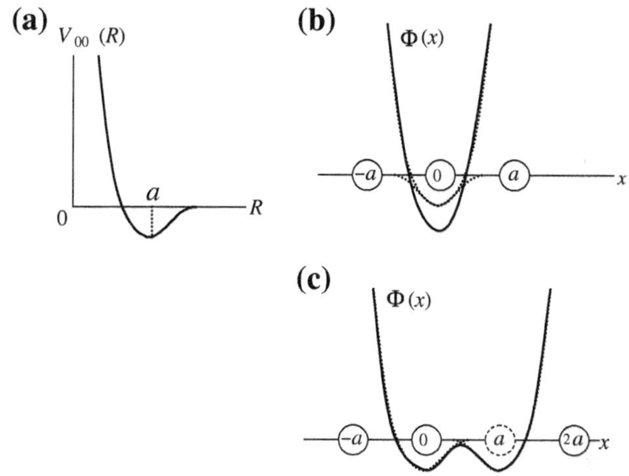

Figure 1.3 The potential between particles and the potentials made up by the surrounding particles in a one-dimensional system with the x-axis. (a) The potential between particles $V_{00}(R)$ is schematically illustrated. It has a hard core and a valley with a minimum near a mean inter-atomic distance a. (b) The total potential which a particle at the 0-site feels, made up by the surrounding particles at $-a$ and a sites, $\Phi(x)$, is a well localized as schematically illustrated by a thick line. The potential made by each particle is illustrated by its own dotted line. (c) For the case of the existence of a vacancy at a site, $\Phi(x)$, has double minimums as schematically illustrated by a thick line.

saddle point upward convex in the direction to the vacancy x_i and downward convex in the direction perpendicular to x_i. The m-particle can hop to the vacancy with a hopping magnitude, $J_{m\mu n\mu}$ as illustrated in Figure 1.4.

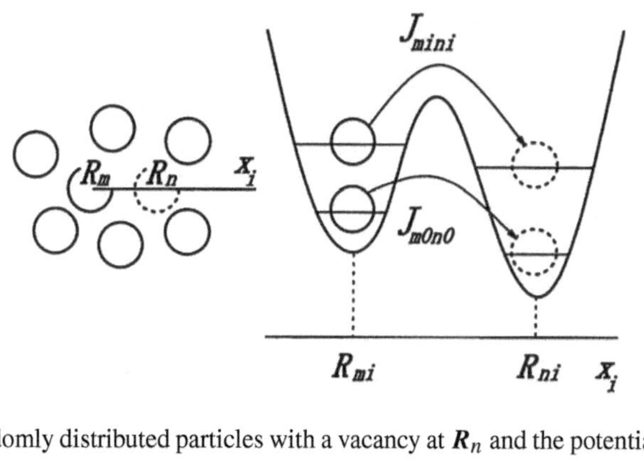

Figure 1.4 Randomly distributed particles with a vacancy at $\boldsymbol{R}_n$ and the potential of a particle at $\boldsymbol{R}_m$ in the x_i-direction. The potential at a saddle point is upward convex in the x_i-direction and downward convex in the perpendicular to x_i-axis. J_{m0n0} and J_{mini} are the hopping magnitude between the ground states and that between the excited states, respectively.

In the liquid state, a particle stays temporarily in a harmonic potential at $\boldsymbol{R}_m$ making the harmonic oscillation with the harmonic frequency ω_m and then hops to a surrounding vacancy. The hopping of a particle changes a pair distribution function to another distribution function, but the configurationally averaged distribution $g(R)$ holds the same over space. In the glassy state, since all the particles are freezed in their own harmonic potentials at random positions, they scarcely hop, so the configurationally averaged pair distribution function $g(R)$ holds the same over space and time. Thus, hopping is essential in the liquid state, while the rigidity of $g(R)$ is essential in the glassy state.

1.1.2 The Nambu-Goldstone Bosons, Phonons, and Sound

In the condensed states, the liquid, glassy, and crystalline ones, the Hamiltonians are of the same form invariant under the spatial translation. However, the long range orders in the glassy and crystalline states spontaneously break down the symmetry of the spatially translational invariance. The spontaneous breakdown of the symmetries is associated with the Nambu-Goldstone bosons [12–15]. The Nambu-Goldstone bosons are a longitudinal and two transverse phonons. The Nambu-Goldstone bosons play a role in recovering the symmetries.

In the crystalline state, the long range order is represented by a periodic structure. The pair distribution function $g(\boldsymbol{R})$ has delta function type peaks at any integral multiple of the primitive vectors; $g(\boldsymbol{R})$ looks exactly the same to each observer at his own particle [16]. The liquid and glassy states also have a kind of the structure, the long range order in a sense of the configuration average. The pair distribution function $g(R)$ has the structure starting with a peak at an inter-particle distance and oscillatorily damping toward the mean particle density in distance R as illustrated in Figure 1.2; $g(R)$ looks the same to each observer at his own particle. According to the structure of the pair distribution function $g(\boldsymbol{R})$ in a crystal and $g(R)$ in the glass, the symmetries of the spatially translational invariance spontaneous break down. This spontaneously breakdown of the symmetry is associated with the Nambu-Goldstone bosons, which are a longitudinal and two transverse phonons.

The structure of condensed states determine their structure of phonons. The phonon dispersion curves are essentially obtained by the nearest neighbor particles. The dispersion curves of phonons in the crystalline state are periodic and clear because of exactly periodic long range order, while in the glassy state phonon dispersion curves are vague, because the long range order is only guaranteed by the configurationally averaged pair distribution function, $g(R)$. The vagueness is substantial in a liquid due to the hopping of particles.

The hopping of particles plays an essential role in the liquid state. The hopping of particles causes density fluctuations. The density fluctuation waves are just sound. Conventionally, phonons and sound have not been discriminated clearly. Sometimes sound has been confused with a longitudinal phonon in the long wavelength regions. We must discriminate sound and phonons clearly. Sound is not a Nambu-Goldstone boson. Sound is substantial in fluid systems such as gaseous and liquid states. But the liquid state is a condensed one so that there remains the pair distribution function $g(R)$ in a sense of the configuration average, which yields vague phonons. Sound plays an essential role in the liquid state. The hopping also causes the relaxation of sound and phonons. Diffusivity appears according to the density gradient; it is associated with the relaxation of sound. Viscosity appears according to the flow gradient; it is associated with the relaxation of phonons.

1.1.3 The Liquid-Glass Transition Occurs in the Quantum Regime

The liquid-glass transition conventionally has been treated from the classical theory. The classical regime is given by the criterion that the de Broglie thermal wavelength denoted by λ_B is very much less than the distribution width of a particle trapped in a potential well a:

$$\lambda_B \ll a. \tag{1.1.1}$$

λ_B is estimated by using the equipartition law, $\hbar\epsilon_p = (\hbar p)^2/2M = k_B T/2$, and the de Broglie wavelength, $\lambda_B = 2\pi/p$, as

$$\lambda_B = \frac{h}{\sqrt{Mk_B T}}, \tag{1.1.2}$$

where M means the mass of a particle. This criterion (1.1.1) means that we can see an object with a scale a classically by use of light with wavelengths $\sim \lambda_B$. The criterion is also derived by using the Boltzmann distribution function under the condition $e^{\frac{\hbar(\mu-\epsilon_p)}{k_B T}} \ll 1$ for any ϵ_p [18]. Since $e^{\frac{\hbar(\mu-\epsilon_p)}{k_B T}} < e^{\frac{\hbar(\mu-\epsilon_0)}{k_B T}} = e^{\frac{\hbar\mu}{k_B T}}$, we have

$$e^{\frac{\hbar\mu}{k_B T}} = \frac{N}{V}\frac{\lambda_B^3}{(2\pi)^{\frac{3}{2}}} \ll 1, \quad N = \frac{V}{(2\pi)^3}\int e^{\frac{\hbar(\mu-\epsilon_p)}{k_B T}}\, d^3p = e^{\frac{\hbar\mu}{k_B T}}\, V\frac{(2\pi)^{\frac{3}{2}}}{\lambda_B^3}, \tag{1.1.3}$$

where $\hbar\mu$ is the chemical potential, N the number of particles, and V the volume of the system. In this book, energy and momentum are measured by frequency and wave vector, respectively.

Near the liquid-glass transition temperature, a particle stays for a long time in a harmonic potential making the harmonic oscillation with the averaged frequency ω and seldom hops to a surrounding vacancy. Under the circumstances, a particle is actually trapped in a harmonic potential. Since the glass transition temperature T_g is estimated to be the same order of the melting temperature T_m, which is the same order of the harmonic frequency ω as shown in Section 7.4 [19,20]. Thus we obtain

$$\hbar\omega \sim k_B T_g. \tag{1.1.4}$$

To estimate λ_B near the glass transition, if we put $\hbar\omega \sim k_B T_g$ in Eq. (1.1.2), we obtain

$$\lambda_B \sim \sqrt{\frac{h}{M\omega}} \quad \text{near } T_g, \tag{1.1.5}$$

which is the same order as the width of the zero point motion for the harmonic oscillator λ_0,

$$\lambda_0 \sim \sqrt{\frac{h}{M\omega}}. \tag{1.1.6}$$

From the harmonic potential illustrated in Figure 1.2, we can estimate

$$\lambda_0 \sim a, \tag{1.1.7}$$

so we have

$$\lambda_B \sim a \quad \text{near } T_g, \tag{1.1.8}$$

the criterion (1.1.1) is not satisfied. Thus we must treat the system quantum statistically. This is also why we must deal with phonons quantum statistically.

1.2 The Two Band Model for the Liquid-Glass Transition

In this section, we draw the framework of the unified theory of the liquid-glass transition. To treat the dynamical processes in a random system such as a liquid and a glass, we divide the calculation processes of the random system into two stages. At the first stage, we investigate the dynamical motions in the configurationally averaged system. At the next stage, we take the randomness of the system into account as scattering processes by random potentials.

As the first stage, we construct the two band model and develop the dynamical processes of sound and phonons in the configurationally averaged system. The sound and phonons are represented by the sum of an infinite series of bubbles of intra-band and inter-band particle-hole pairs, respectively.

As the second stage, we obtain the relaxation time by taking into account the configuration average of the simultaneously scattering processes of two particles by the same random potentials due to random harmonic frequencies and random hopping magnitudes. We obtain the dynamical equation in the dissipative processes by introducing the relaxation times and then we obtain the diffusivity and the viscosity.

Finally we investigate the supercooling process. The enormous bubbles of intra-band particle-hole pairs not constituting the sound create the fluctuation entropy, which mixes with the sound entropy with wave numbers $\sim K$. The mixing entropy leads to the Kauzmann entropy, which yields the VTF law and the gap of specific heat.

1.2.1 The Two Band Model

At the first stage, we investigate the dynamical motions in the configurationally averaged system.

Near the glass transition temperature, a particle stays for a long time in a harmonic potential and seldom hops to a surrounding vacancy. The particle is actually trapped in a harmonic potential; the energy of the particle is quantized as $\tilde{\omega}_n = \left(n + \frac{3}{2}\right)\omega$. Since $\lambda_0 \sim a$, we consider the two levels; the ground state, $\tilde{\omega}_0 = \frac{3}{2}\omega$ and the first excited states, $\tilde{\omega}_1 = \frac{5}{2}\omega$. The first excited states are degenerated; $\tilde{\omega}_i = \tilde{\omega}_1$, where $i = x, y, z$

and $\tilde{\omega}_i$ means that ith component is excited, $\frac{3}{2}\omega$ and otherwise, $\frac{1}{2}\omega$. So we denote eigenfrequencies as $\tilde{\omega}_\mu$, where $\mu = 0, i$. A particle staying in a harmonic potential make up and down transitions between the two levels, $\tilde{\omega}_0$ and $\tilde{\omega}_1$. The propagation of up and down transitions of particles yields phonons.

On the other hand, the hopping yields energy bands. We introduce the hopping of μth particles from $\boldsymbol{R}_m$ to $\boldsymbol{R}_n$ as $J_\mu(\boldsymbol{R}_m - \boldsymbol{R}_n) \equiv \langle J_\mu(\boldsymbol{R}_m, \boldsymbol{R}_n)\rangle_c$, where $< \cdots >_c$ means the configuration average. If we denote the Fourier transformation of $J_\mu(\boldsymbol{R}_m - \boldsymbol{R}_n)$ as $J_\mu(\boldsymbol{p})$, we obtain two bands, energy dispersion curves of particles illustrated in Figure 1.5:

$$\epsilon_{\mu \boldsymbol{p}} = \tilde{\omega}_\mu + J_\mu(\boldsymbol{p}), \tag{1.2.1}$$

Thus we reach the two band model. Using this two band model, we can establish a unified theory of the liquid-glass transition to elucidate the universal features of the liquid-glass transition [20,21].

The two band model is ubiquitous. It can be applied to soft matter such as liquids, colloids, polymers, and a number of biological materials. A particle in a colloid can also be seen to locate in a shallow local potential with a low harmonic frequency and hops to another place frequently, which follows the two band model so that the jamming transition in colloids can be treated in a similar manner to the liquid-glass transition.

1.2.2 Sound

The elementary excitations in the two band model can be decomposed into intra-band and inter-band transitions as illustrated in Figure 1.5. Intra-band transitions occur when particles hop to vacancies so that they are associated with particle density fluctuation waves. The propagator of an elementary excitation, $\epsilon_{0\boldsymbol{p}} - \epsilon_{0\boldsymbol{p}-\boldsymbol{q}}$, is a bubble of particle-hole pairs in the Feynman diagram. The bubble of intra-band particle-hole pairs is

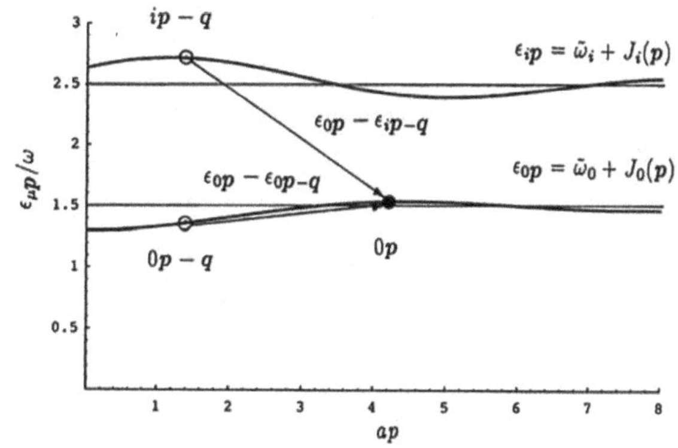

Figure 1.5 Energy dispersion curves normalized by the configurationally averaged frequency ω as a function of ap. Here we put $g(R) = \rho\delta(R - a)$. $J_0(\boldsymbol{p})$ is isotropic, while $J_i(\boldsymbol{p})$ depends on the direction, so we have shown the averaged value over the directions. The minimum gap is near a reciprocal inter-particle distance. An intra-band elementary excitation, $\epsilon_{0\boldsymbol{p}} - \epsilon_{0\boldsymbol{p}-\boldsymbol{q}}$, and an inter-band elementary excitation, $\epsilon_{0\boldsymbol{p}} - \epsilon_{i\boldsymbol{p}-\boldsymbol{q}}$, are also shown.

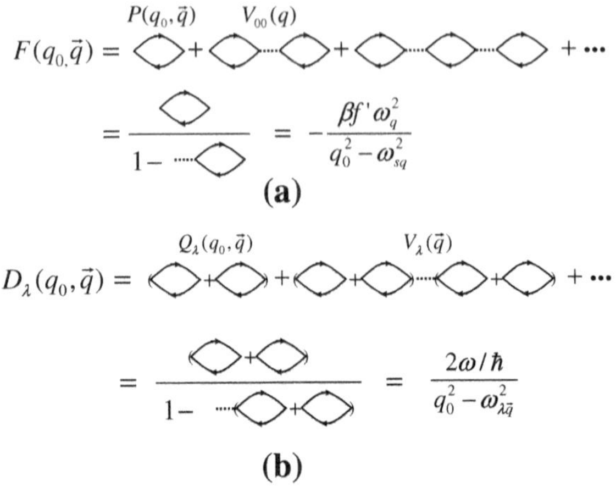

Figure 1.6 How to construct sound and phonons with the respective bubbles are illustrated. (a) $F(q_0, \boldsymbol{q})$: the propagator of sound. $P(q_0, q)$ represents the bubble of intra-band particle-hole pairs. A dotted line indicates the interaction potential $V_{00}(\boldsymbol{q})$. (b) $D_\lambda(q_0, \boldsymbol{q})$: the propagator of phonons with a mode λ. $Q(q_0, \boldsymbol{q})$ represents the pair of bubbles of inter-band particle-hole pairs. A dotted line indicates $V_\lambda(\boldsymbol{q})$.

defined by the average of the sum of the bubbles of particle-hole pairs with respect to $\boldsymbol{p}$ under the condition $\beta\hbar J_0(\boldsymbol{p}) \ll 1$, and is given by a propagator of a frequency ω_q:

$$P(q_0, \boldsymbol{q}) = \frac{1}{N\hbar}\sum_{\boldsymbol{p}} \frac{f(\epsilon_{0\boldsymbol{p}-\boldsymbol{q}}) - f(\epsilon_{0\boldsymbol{p}})}{q_0 - (\epsilon_{0\boldsymbol{p}} - \epsilon_{0\boldsymbol{p}-\boldsymbol{q}})} \cong -\frac{\beta f' {\omega_q}^2}{{q_0}^2 - {\omega_q}^2},$$

as illustrated in Figure 1.6a, where q_0 is a frequency, f is a distribution function, f' is its derivative and we have considered the inversion symmetry with respect to $\boldsymbol{q}$ under the summation of $\boldsymbol{p}$; $\omega_{\boldsymbol{q}} = \omega_q$ and $P(q_0, \boldsymbol{q}) = P(q_0, q)$. The second equation is obtained under the condition $\beta\hbar J_0(\boldsymbol{p}) \ll 1$; the band of $\epsilon_{0\boldsymbol{p}}$ is very narrow, $\beta\hbar\epsilon_{0\boldsymbol{p}} \ll 1$. The frequency of the bubble of the particle-hole pairs is given by

$${\omega_q}^2 \cong \frac{{v_p}^2 q^2}{3},$$

where $\omega_q \cong v_p q/\sqrt{3}$, where v_p is an average velocity of particles.

In order to derive sound, we introduce the interaction potential between particle-hole pairs $V_{00}(\boldsymbol{R}_m - \boldsymbol{R}_n)$ and its Fourier transformed one, $V_{00}(\boldsymbol{q}) = V_{00}(q)$. Sound is the sum of an infinite series of interacting bubbles of intra-band particle-hole pairs through the interaction $V_{00}(\boldsymbol{q})$:

$$\begin{aligned} F(q_0, \boldsymbol{q}) &= P(q_0, \boldsymbol{q}) + P(q_0, \boldsymbol{q}) V_{00}(q) F(q_0, \boldsymbol{q}) \\ &= \frac{1}{1/P(q_0, \boldsymbol{q}) - V_{00}(q)} = -\frac{\beta f' \omega_q^2}{q_0^2 - \omega_{sq}^2}, \end{aligned}$$

the sound dispersion curves are given by

$$\omega_{sq}{}^2 = (1 - \beta f' V_{00}(q))\omega_q{}^2 = \frac{\omega_q{}^2}{S(q)}, \tag{1.2.2}$$

where $S(q)$ is the static structure factor; $S(q) = \frac{1}{1-\beta f' V_{00}(q)}$, $\beta = \frac{1}{k_B T}$ and f' is a derivative of the distribution function of the particle. If we employ the Boltzmann, the Bose, or the Fermi distribution, we obtain $f' \cong -1, -2$, or 0, respectively. Sound velocity in the long wavelength region, c_T, is given by

$$c_T^2 = \frac{v_p^2}{3S(0)},$$

where c_T corresponds to the isothermal sound velocity in the classical theory.

The interaction potential $V_{00}(q)$ has the negatively minimum value at a reciprocal particle distance $K \sim \frac{2\pi}{a}$, so sound disappears at the sound disappearing temperature T_0:

$$1 - \beta_0 f' V_{00}(K) = 0, \tag{1.2.3}$$

where $\beta_0 = \frac{1}{k_B T_0}$. The sound disappearing was first investigated by Schneider et al. [22].

1.2.3 Phonons and Boson Peak

The inter-band transitions illustrated in Figure 1.5 occur when a particle in a harmonic potential makes up and down transitions between the two bands so that they are associated with phonons. The first excited states correspond to p-states; the spread of wave functions in the direction of x, y, and z corresponds to the displacement of particles to the respective directions from the ground state of the particle. The propagator of elementary excitations, $\epsilon_{1\boldsymbol{p}} - \epsilon_{0\boldsymbol{p}-\boldsymbol{q}} \sim -\omega$ and $\epsilon_{0\boldsymbol{p}} - \epsilon_{1\boldsymbol{p}-\boldsymbol{q}} \sim \omega$, are bubbles of inter-band particle-hole pairs in the Feynman diagrams. The bubble of inter-band particle-hole pairs is defined by the average of the sum of bubbles of particle-hole pairs with respect to $\boldsymbol{p}$ is given by a propagator of frequencies, $-\omega$ and ω:

$$Q(q_0, \boldsymbol{q}) = \frac{1}{N\hbar} \sum_{\boldsymbol{p}} \left\{ \frac{1}{q_0 - (\epsilon_{0\boldsymbol{p}} - \epsilon_{i\,\boldsymbol{p}-\boldsymbol{q}})} - \frac{1}{q_0 + (\epsilon_{0\boldsymbol{p}} - \epsilon_{i\,\boldsymbol{p}-\boldsymbol{q}})} \right\}$$
$$\times \left\{ f(\epsilon_{i\,\boldsymbol{p}-\boldsymbol{q}}) - f(\epsilon_{0\boldsymbol{p}}) \right\} \cong \frac{2\omega/\hbar}{q_0{}^2 - \omega^2}$$

as illustrated in Figure 1.6b.

The phonons are the sum of an infinite series of interacting bubbles of inter-band particle-hole pair through the interaction potential $V_{ij}(\boldsymbol{q})$, which is the Fourier transformation of $\zeta^2 V_{ij}(\boldsymbol{R}_m - \boldsymbol{R}_n)$, where $V_{ij}(\boldsymbol{R})$ is written as $V_{ij}(\boldsymbol{R}) = \frac{\partial^2 V_{00}(\boldsymbol{R})}{\partial R_i \partial R_j}$ and $\zeta^2 = \frac{\hbar}{2\omega M}$. If we put $\boldsymbol{q} = q\boldsymbol{e}_z$ ($\boldsymbol{e}_z$ is the unit vector of the z-axis), we can diagonalize $V_{ij}(\boldsymbol{q})$ as $V_\lambda(\boldsymbol{q}) = V_{ii}(\boldsymbol{q})$, where λ is longitudinal for $\lambda = z$ and transverse for

$\lambda = x, y$. Therefore we obtain the phonon propagators:

$$D_\lambda(q_0, \boldsymbol{q}) = Q(q_0, \boldsymbol{q}) + Q(q_0, \boldsymbol{q})V_\lambda(\boldsymbol{q})D(q_0, \boldsymbol{q}) = \frac{1}{1/Q(q_0, \boldsymbol{q}) - V_\lambda(\boldsymbol{q})}$$
$$= \frac{2\omega/\hbar}{{q_0}^2 - {\omega_{\lambda \boldsymbol{q}}}^2}$$

as illustrated in Figure 1.6b.

Phonon dispersion curves, $\omega_{\lambda \boldsymbol{q}}$, are given by

$$\begin{aligned} {\omega_{\lambda \boldsymbol{q}}}^2 &= \frac{2\omega}{\hbar}[V_\lambda(\boldsymbol{q}) - V_\lambda(0)] = \frac{1}{M}\zeta^{-2}[V_\lambda(\boldsymbol{q}) - V_\lambda(0)] \\ &= \frac{1}{M}\int d^3R\, g(R)(e^{i\boldsymbol{q}\cdot\boldsymbol{R}} - 1)V_\lambda(\boldsymbol{R}) \end{aligned} \quad (1.2.4)$$

as illustrated in Figure 1.7.

The phonons dispersion curves are rather clearly compared with sound one, and have the following features [20]:

1. In the long wavelength regime, ${\omega_{\lambda \boldsymbol{q}}}^2 \cong c_\lambda^2 q^2$, where c_λ is a phonon velocity.
2. Near the reciprocal inter-particle distance, the longitudinal mode has a roton type minimum, which is a vestige of the periodic structure in the crystalline state [23, 24].
3. In the short wavelength regime, they converge to the average harmonic frequency ω; $\omega^2 = \frac{1}{M}\int d^3R\, g(R)V_\lambda(\boldsymbol{R})$. These flat phonon dispersion curves mean the localization of phonons and makes a peak of the density states. This is the boson peak. It should be noted that the density of states is inversely proportional to the q-derivative of the phonon dispersion curves. This frequency is related with the boson peak [11].

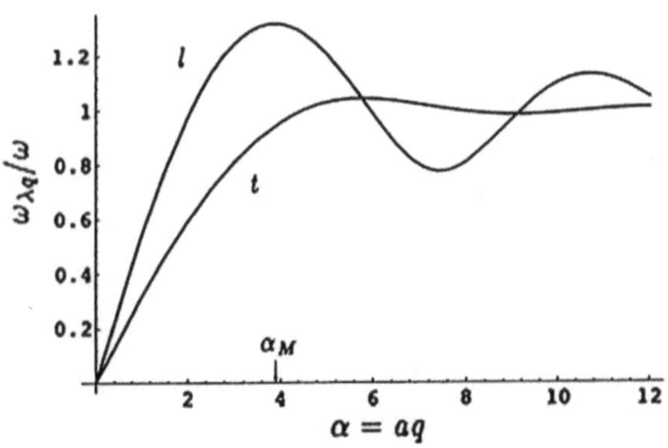

Figure 1.7 A longitudinal and two transverse phonon dispersion curves $\omega_{\lambda \boldsymbol{q}}$ normalized by the configurationally averaged harmonic frequency ω are illustrated as functions of aq. α_M is the Debye cut off. In the short wavelength limit, $\lim_{q \longrightarrow \infty} \omega_{\lambda \boldsymbol{q}} = \omega$: this frequency corresponds to boson peak.

1.2.4 The Dissipative and Relaxation Processes

At the next stage, we take the randomness of the system into account as the scattering processes of particles by random potentials.

The dissipative and relaxation processes originate from the simultaneously scattering processes of two particles by the same random potential. Random potentials come from random harmonic frequencies, $\{\omega_m\}$, and random hopping magnitudes, $\{J_{mn}\}$. The sum of the configurationally averaged simultaneously scattering processes over all sites yields the random scattering potentials in which two particles participate:

$$U_J \sim \langle \hat{J}_{mn} \hat{J}_{mn} \rangle_c \quad \text{due to the random hopping magnitudes}$$

and

$$U_\omega \sim \langle \hat{\omega}_m \hat{\omega}_m \rangle_c \quad \text{due to the random harmonic frequencies,}$$

where $\wedge$, means the deviation from the average value and $\langle \cdots \rangle_c$ means the configuration average. The statistical average on a random system constrains the dynamical processes, which lead to dissipation. Since two particles simultaneously participate in the same random scattering potentials $U_J + U_\omega$, the relaxation time for sound, τ_0, is given by

$$\frac{1}{\tau_0} \propto \sqrt{(U_J + U_\omega)}. \tag{1.2.5}$$

The relaxation times for phonons, τ_i, are also given by

$$\frac{1}{\tau_i} = \frac{1}{\tau_M} \propto \frac{1}{\tau_0}, \tag{1.2.6}$$

where we have taken $\boldsymbol{q} = q\boldsymbol{e}_z$ and τ_M is called the Maxwell relaxation time [25].

The dynamical processes including the relaxation times are given by

$$q_0^2 + \mathrm{i}\frac{1}{\tau_\mu} q_0 - \omega_{\mu\boldsymbol{q}}{}^2 = 0, \tag{1.2.7}$$

where q_0 is a frequency. $(q_0, \boldsymbol{q})$ corresponds to $(\mathrm{i}\frac{\partial}{\partial t}, -\mathrm{i}\nabla)$. In the short wavelength $aq \gg 1$ and high frequencies $\tau_\mu q_0 \gg 1$ regimes they yield

$$q_0^2 - \omega_{\mu\boldsymbol{q}}{}^2 = 0,$$

while in the long wavelength $aq \ll 1$ and low frequency $\tau_\mu q_0 \ll 1$ regimes yield

$$\begin{aligned} q_0^2 + \mathrm{i}\frac{1}{\tau_\mu} q_0 - \omega_{\mu\boldsymbol{q}}{}^2 &\cong \mathrm{i}\frac{1}{\tau_\mu} q_0 - \omega_{\mu\boldsymbol{q}}{}^2 \\ &= \frac{1}{\tau_\mu}(\mathrm{i}q_0 - \tau_\mu v_\mu^2 q^2) \equiv \frac{1}{\tau_\mu}(\mathrm{i}q_0 - D_\mu q^2) = 0. \end{aligned}$$

1. For $\mu = 0$: sound, $\omega_{0\boldsymbol{q}}{}^2 \equiv \omega_{sq}^2 = c_T^2 q^2 (v_0 \equiv c_T)$ and diffusivity $D \equiv D_0 = \tau_0 c_T^2$.

2. For $\mu = i = \lambda$: phonons, ${\omega_{\lambda q}}^2 \cong c_\lambda^2 q^2$ and viscosity, $\eta_\lambda \equiv D_\lambda = \tau_M c_\lambda^2 (\tau_i \equiv \tau_M)$.

Thus we obtain Diffusivity D and Viscosity η_λ:

$$D = {c_T}^2 \tau_0, \tag{1.2.8}$$

$$\eta_\lambda = {c_\lambda}^2 \tau_M, \tag{1.2.9}$$

where the diffusivity is essential in liquids, while the viscosity is essential in glasses.

1.2.5 The Kauzmann Entropy and the Kauzmann Entropy Crisis

Finally, we investigate the supercooling process. In the supercooling process, enormous bubbles of intra-band particle-hole pairs not constituting the sound, are created.

Sound ω_{sq} is given from the sum of an infinite series of interacting bubbles of intra-band particle-hole pairs, ω_q, through $V_{00}(q)$. The extra bubbles of particle-hole pairs not constituting sound are fluctuations of the sound system, which lowers the entropy of the system. We call the entropy of the extra bubbles of intra-band particle-hole pairs not constituting sound, the fluctuation entropy of sound. In the equilibrium state, the dissipation due to the diffusion compensates the negative fluctuation entropy of sound so that the system becomes a local equilibrium. The extra bubbles of intra-band particle-hole pairs not constituting sound play an essential role in the supercooling process, where the fluctuation entropy mixes with the sound entropy. In order to obtain the entropy of the sound and the fluctuations in the supercooling process, since ω_{sq} and ω_q depend on the temperature, we must start with the Free energy;

$$\begin{aligned} F_{\text{sound}} + F_{\text{fluct.}} &= \frac{1}{2\beta}\sum_{q} \ln(1 - e^{-\beta\omega_{sq}}) - \frac{1}{\beta}\sum_{q} \ln(1 - e^{-\beta\omega_q}) \cong \frac{1}{\beta}\sum_{q} \ln\frac{\omega_{sq}}{\omega_q} \\ &= \frac{1}{2\beta}\sum_{q} \ln(1 - \beta f' V_{00}(q)) \cong \frac{1}{2\beta}\sum_{q\in N_0} \ln(1 - \beta f' V_{00}(K)), \\ &= \frac{1}{2\beta}\sum_{q\in N_0} \ln\left(1 - \frac{\beta}{\beta_0}\right) = \frac{1}{2\beta}\sum_{q\in N_0} \ln\left(1 - \frac{T_0}{T}\right), \end{aligned}$$

where we have used the conditions: $\beta\hbar\omega_{sq}, \beta\hbar\omega_q \ll 1$ and considered the dominant contribution comes from the regions $q \sim K$, which we denote as $\boldsymbol{q} \in N_0$. N_0 means the number of states of wave vectors $\boldsymbol{q}$ around $q \sim K$. Thus we obtain the entropy, $S_K = \{S_{\text{sound}} + S_{\text{fluct.}}\}|_{\boldsymbol{q}\in N_0}$:

$$S_K \cong -\frac{N_0 k_B}{2}\frac{T_0}{T - T_0}. \tag{1.2.10}$$

We call S_K the Kauzmann entropy. The fluctuation entropy $S_{\text{fluct}}|_{\boldsymbol{q}\notin N_0}$ is compensated with the entropy of the diffusion so that S_K remains. As the temperature decreases toward T_0, the magnitude of $|S_K|$ overwhelms the entropy of the liquid state at the Kauzmann temperature T_K and then the entropy of the system seems to become a negative infinity as illustrated in Figure 1.8. This situation is the Kauzmann entropy crisis (the Kauzmann paradox). However, the phonon fluctuation entropy S_Ω, which is discussed below, prevents the Kauzmann entropy crisis at the glass transition temperature T_g as illustrated in Figure 1.8. There is an inequality: $T_0 < T_K < T_g < T_m$.

Now we show the Kauzmann's original presentation of the entropy crisis [7] in Figure 1.9, where the temperature dependence of the entropy difference between several supercooled liquids and their crystals, ΔS normalized by that at the melting temperature ΔS_m. If the entropy ΔS is extrapolated toward the lower temperature below the glass transition temperature T_g, ΔS becomes zero at the Kauzmann temperature T_K and then negatively diverges at T_0. Thus we can identify ΔS to the Kauzmann entropy S_K.

1.2.6 The Gap of the Specific Heat at the Glass Transition Temperature T_g

The Kauzmann entropy S_K reflects the specific heat C_K due to the Kauzmann entropy:

$$C_K = \begin{cases} \frac{N_0 k_B T}{2} \frac{T_0}{(T-T_0)^2} & \text{for } T_g < T, \\ 0 & \text{for } T < T_g. \end{cases} \tag{1.2.11}$$

The specific heat C_K explain the gap of the specific heat at the glass transition temperature T_g. The Kauzmann entropy S_K and the specific heat C_K due to S_K normalized by Nk_B are illustrated in $x = k_B T/\hbar\omega$ for $x_0 = 1/4$ and $1/10$ in Figure 1.10. Some experimental results are shown in Figure 1.12.

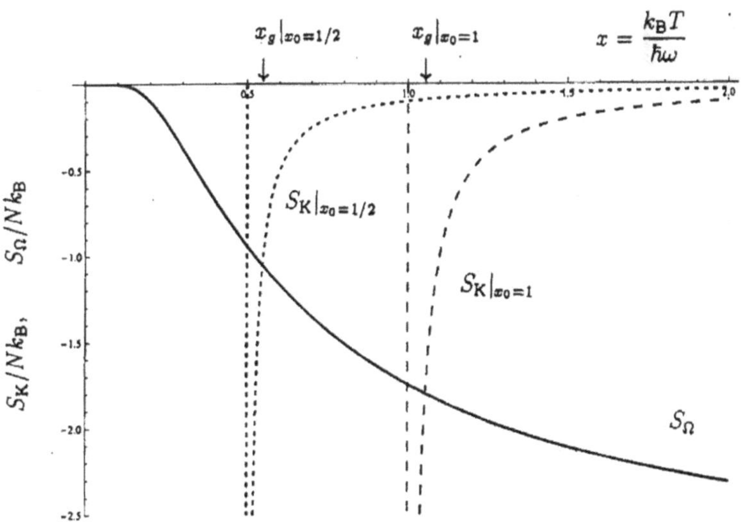

Figure 1.8 The Kauzmann entropy S_K and the phonon fluctuation entropy S_Ω normalized by Nk_B are illustrated in $x = k_B T/\hbar\omega$ for $N_0/N = 1/5$ for a fragile $x_0 = k_B T_0/\hbar\omega$ and a more fragile $x_0 = 1$. The Kauzmann entropy S_K diverges at the sound disappearing temperature x_0, but S_Ω crosses over S_K so that S_Ω prevents the Kauzmann entropy crisis at the glass transition temperature $x_g = k_B T/\hbar\omega$.

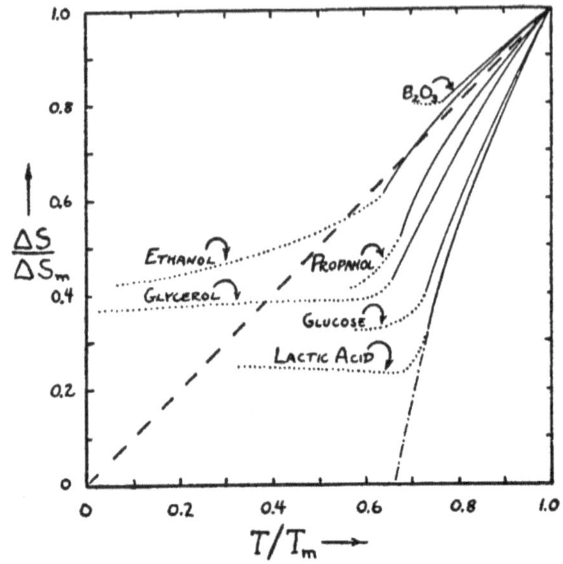

Figure 1.9 The temperature dependence of the entropy difference between several supercooled liquids and their crystals, ΔS, normalized by that at the melting temperature ΔS_m. The temperature at which the extrapolated ΔS crosses the horizontal axis is the Kauzmann temperature T_K. (Reproduced from Kauzmann's original presentation of the entropy crisis in [7].)

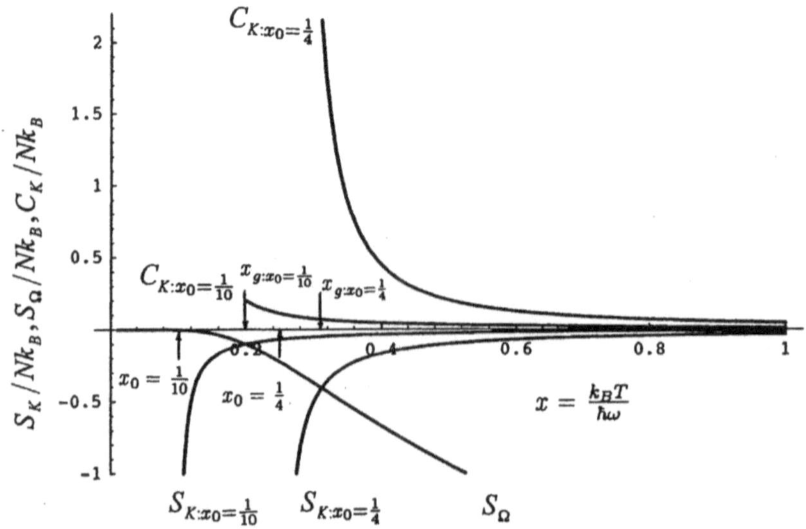

Figure 1.10 The Kauzmann entropy S_K, phonon fluctuation entropy S_Ω, and the specific heat C_K due to S_K normalized by Nk_B are illustrated in $x = k_BT/\hbar\omega$ for $x_0 = 1/4$ and $1/10$.

1.2.7 The Hopping Amplitude and the VTF Law

A hopping of a particle from a site to a vacancy corresponds to a jump from a deep valley to another deep valley in the multi-dimensional configuration space in the energy landscape model (ELM) as the α-relaxation process [28–31] as illustrated in Figure 1.11. The successive hopping constitutes a configuration space. The hopping probability is proportional to the configuration number, which is e^{S_K/k_B} from the Einstein relation. The hopping probability of a particle is proportional to the hopping amplitude:

$$J = e^{\frac{zS_K}{Nk_B}} = e^{-\frac{E}{T-T_0}}, \tag{1.2.12}$$

where $E = \frac{zN_0T_0}{2N}$ and z is the average number of the surrounding particles. This is just the VTF law.

Near the glass transition temperature, it should be noted that $\epsilon_{\mu p}$, ω_{sq}, S_K, and J are determined self-consistently. We start with the particle energy $\hbar\epsilon_{\mu p} \propto \hbar J \ll k_B T$; $v_p \propto J$. Then we obtain ${\omega_{sq}}^2 = (1 + \beta f' V(q)){\omega_q}^2 = \frac{{\omega_q}^2}{S(q)} = \frac{v_p^2 q^2}{3S(q)} \propto J^2$. There disappears the sound at K and T_0. Then we obtain the Kauzmann entropy $S_K \cong -\frac{N_0 k_B}{2}\frac{T_0}{T-T_0}$. From the Einstein relation, we can determine the hopping amplitude $J = e^{\frac{zS_K}{Nk_B}} = e^{-\frac{E}{T-T_0}}$, which satisfies $\hbar J \ll k_B T$. Thus we can determine $\omega_{\mu p}$, ω_{sq}, S_K, and J, self-consistently.

Above T_g, the term U_J dominates U_Ω so that

$$\tau_0^{-1}, \tau_M^{-1} \propto \sqrt{U_J} \propto J,$$

while

$$v_p, c_T \propto J, \quad \text{and} \quad c_\lambda = \text{constant},$$

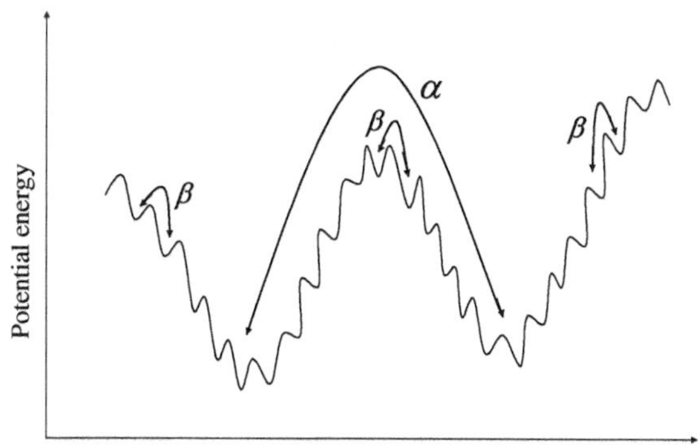

Figure 1.11 Potential energy and the multi-dimensional configuration space, and the α-and β-relaxation processes in the energy landscape model are schematically illustrated. The α-relaxation process associated with the hopping of particles corresponds to the jumps between deep valleys, while the β-relaxation associated with the up and down transitions of particles corresponds to the jumps between shallow valleys.

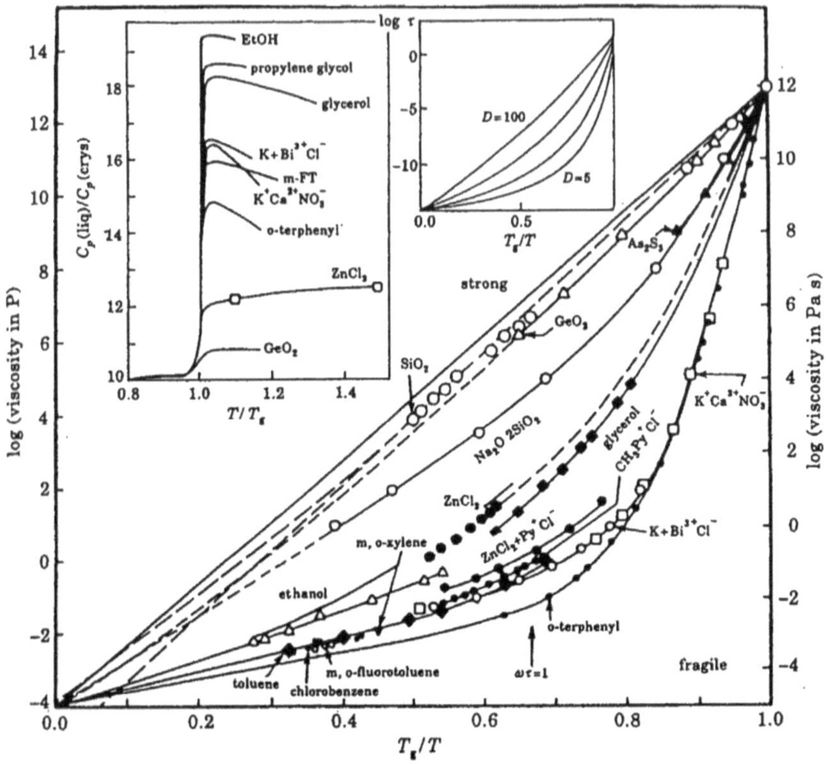

Figure 1.12 Viscosity of different glass-formers as a function of $\frac{T_g}{T}$. Left inset: the temperature dependent specific heat normalized by that of crystals for different glass-formers. (From Ref. [1], by permission.)

so that the diffusivity $D = c_T{}^2\tau_0 \propto J$, the viscosity $\eta_\lambda = c_\lambda{}^2\tau_M \propto J^{-1}$. Thus the VTF law governs as follows:

$$v_p, c_T, c_\lambda, \tau_0^{-1}, \tau_M^{-1}, D, \eta_\lambda{}^{-1} \propto J. \tag{1.2.13}$$

The relation, $D\eta_\lambda = \text{constant}$, satisfies the Stokes' law.

Experimental results on viscosity η of different glass-formers as a function of $\frac{T_g}{T}$ obeying the VTF law; $\ln \eta \propto -\frac{1}{T-T_0}$ are shown in Figure 1.12 [2]. If we consider η as a function of

$$t \equiv \frac{T_g}{T}, \quad t_0 = \frac{T_g}{T_0},$$

$\eta(t)$ is given by $\eta(t) = \eta(0)e^{\frac{E}{T_0}\frac{t}{t_0-t}}$;

$$\ln \frac{\eta(t)}{\eta(0)} = F\frac{t}{t_0 - t}, \quad F = \frac{E}{T_0} = \frac{zN_0}{2N}.$$

According to Figure 1.12, since $\log \frac{\eta(1)}{\eta(0)} \sim 17$,

$$t_0 = 1 + \frac{F}{\ln \frac{\eta(1)}{\eta(0)}} = 1 + \frac{F}{2.303 \log \frac{\eta(1)}{\eta(0)}} \cong 1 + \frac{F}{39.1}.$$

The smaller F is, i.e. the smaller t_0 is, the more fragility it is. We can fit the parameters t_0 and F from the experimental results. It should be noted that D in right inset is identified with F in this book, where D is used for diffusivity.

1.2.8 Phonon Fluctuation Entropy and the Liquid-Glass Transition

Phonon $\omega_{\lambda \boldsymbol{q}}$ are created by an infinite series of interacting bubbles of inter-band particle-hole pairs through $V_\lambda(q)$. The extra bubbles of inter-band particle-hole pairs not constituting the phonons are the fluctuations of the phonon system, which lowers the entropy of the system. We call the extra bubbles of inter-band particle-hole pairs not constituting phonons the fluctuation entropy of phonons. In the equilibrium state, the dissipation due to the viscosity compensates the negative fluctuation entropy so that the system becomes a local equilibrium. In the supercooling process, phonon fluctuation entropy plays an essential role in determining the glass transition temperature. Phonon fluctuation entropy is given by

$$S_\Omega \cong -3Nk_B \frac{\beta\hbar\omega}{e^{\beta\hbar\omega} - 1}. \tag{1.2.14}$$

An up and down transition of a particle at a site corresponds to a jump of a shallow valley in the multi-dimensional configuration space in the energy landscape model (ELM) as the β-relaxation process as illustrated in Figure 1.11. The successive up and down transitions constitute a configuration space. The probability of the magnitude of the randomness of harmonic frequencies is proportional to the configuration number e^{S_Ω/k_B}. The probability of the magnitude of randomness of harmonic frequencies per particle is proportional to

$$\Omega = e^{zS_\Omega/Nk_B} = \exp\left\{\frac{-3z\beta\hbar\omega}{e^{\beta\hbar\omega} - 1}\right\}, \tag{1.2.15}$$

where z is the average number of the surrounding particles.

As the temperature T decreases toward T_0, the Kauzmann entropy S_K rapidly drops to a negative infinity so that the system seems to occur the entropy crisis as shown in Figure 1.8. But as the temperature decreases, the negative phonon fluctuation entropy S_Ω increases and at last crosses over S_K above T_0, so that S_Ω prevents the crisis. The Kauzmann entropy S_K dominates above the crossover temperature, but the phonon fluctuation entropy S_Ω dominates below the crossover temperature. Thus we can identify the crossover temperature with the liquid-glass transition temperature T_g:

$$S_K = S_\Omega|_{T=T_g}. \tag{1.2.16}$$

The sound is a density fluctuation wave essential in liquids, while the phonons are elastic waves essential in glasses. The glass transition is a sort of dynamical transition.

1.2.9 Panic; Supercooling Process Associated with the Kauzmann Entropy Crisis and the VTF Law

The liquid-glass transition occurs in diverse manner depending on materials, supercooling processes, history of materials, and so on. Among the diversity, there are universal features such as the Kauzmann entropy crisis and the VTF law on the relaxation and transport processes. Yet the supercooling processes associated with the universal features such as the Kauzmann paradox and the VTF law are ubiquitous, not only in nature but also in human society. We call this ubiquitous nature of the supercooling processes associated with the universal features, panic. The financial and the psychological panics are well known. Regarding the constituent matters in a panic as the constituent particle in a liquid, we can elucidate the mechanism of the panic in a parallel way to the supercooling process associated with the Kauzmann entropy crisis and the VTF law in the liquid-glass transition. The bubbles of matter-antimatter pairs play an essential role in the panic. The sum of an infinite series of the bubbles of matter-antimatter particles yields sound-like mode.

A panic is characterized as follows:

1. *Cause:* A panic occurs on the occasion of the supercooling process to avoid completing matter-antimatter pair cycles under the circumstances: extremely fast cooling causes enormous bubbles of matter-antimatter pairs not constituting the sound-like mode, which form the Kauzmann entropy crisis and trigger the panic avoiding the ordered state.
2. *Dynamical process:* As the system cools down, the dynamical processes of the system are governed by the universal features such as the Kauzmann entropy $\widetilde{S}_{\mathrm{K}} = -\frac{\widetilde{N}_0 k_{\mathrm{B}}}{2} \frac{\widetilde{T}_0}{\widetilde{T} - \widetilde{T}_0}$ and the VTF law $\widetilde{J} = \exp\left[-\frac{\widetilde{E}}{\widetilde{T} - \widetilde{T}_0}\right]$. The system eventually freezes at the freezing temperature T_f ($T_f > \widetilde{T}_0$), where the Kauzmann entropy $\widetilde{S}_{\mathrm{K}}$ and phonon fluctuation entropy $\widetilde{S}_\Omega$ crossover.

Here we regard the activity of the matter as the temperature $\widetilde{T}$. We denote the quantities in the matter system by putting the tilde on the corresponding quantities in the liquid-glass transition. Thus we can elucidate the mechanism of panic from the two band model.

1.2.10 Mode Coupling Theory and the Adam-Gibbs Formula

The dynamical equation, Eq. (1.2.7) is rewritten as

$$q_0 - \frac{{\omega_{\mu \boldsymbol{q}}}^2}{q_0 + \frac{\mathrm{i}}{\tau_\mu}} = 0. \tag{1.2.17}$$

Equation (1.2.17) for $\mu = 0$ corresponds to that in the Mode Coupling Theory (MCT) [32–38]. The terms, ${\omega_{sq}}^2$ and i/τ_0 correspond to a local frequency of a particle in a cage, $\Omega_q^2 = \frac{q^2 k_{\mathrm{B}} T}{M S(q)}$ and the Memory function, $M(q_0, \boldsymbol{q})$ in MCT, respectively. In our theory, ${\omega_{sq}}^2 = \omega_q^2 / S(q) = \frac{v_p^2 q^2}{3S(q)}$ and i/τ_0 are determined self-consistently. While in MCT, starting with Ω_q^2 which is equal to ${\omega_{sq}}^2 = \frac{q^2 k_B T}{M S(q)}$ in the classical limit, $M(q_0, \boldsymbol{q})$ is determined by the dynamical process due to a nonlinear interaction.

The VTF law is constructed from the Einstein relation on entropy and probability so that the VTF law includes naturally the Kauzmann entropy in exponent; $1/\tau_0 \propto J = \exp\{z S_K / N k_B\}$. The Kauzmann entropy explains the Kauzmann paradox, the jump of the specific heat and the VTF law on the relaxation times and the transport coefficients so that the present theory elucidate the universal features self-consistently. On the other hand, the well-known Adam-Gibbs formula for the VTF law [26] includes inversely the configuration entropy, S_c corresponding to the Kauzmann entropy, in exponent; $1/\tau_{AG} \propto \exp\{-A/(T S_c)\}$, where τ_{AG} is the relaxation time in the Adam-Gibbs formula and A is a constant. The Adam-Gibbs formula has been usually used in arranging experimental data; the transport coefficient and the relaxation times. However, the configuration entropy cannot explain the Kauzmann paradox and the jump of the specific heat well so that the Adam-Gibbs formula faces some conflicting points.

Recently, some of relaxation time data have questioned a dynamical divergence of the VTF law just above the glass transition [39,40]. In this book, as the first step we develop a unified theory of the liquid-glass transition in the mean field approximation in simple liquids. When we calculate the VTF law in multi-component liquids, we regard a molecule as a particle. However, to elucidate those relaxation time data, we should take into account the role of the subordinate particles. In calculating relaxation times and transport coefficients, we should take into account each hopping amplitude due to its own Kauzmann entropy of the subordinate particle.

1.3 Perspective of This Book

In the previous section, we have briefly surveyed the unified theory of the liquid-glass transition: (1) At the first stage, we have investigated the dynamical processes in the configurationally averaged system. (2) At the second stage, we have investigated the relaxation and dissipative in the random system. (3) At the final state, we have investigated the dynamical and dissipative processes in the supercooling system. In this book, we develop the unified theory of the liquid-glass transition in detail. Before developing, we first review the macroscopic classical theory on sound and elastic waves (phonons), and try to develop the microscopic classical theory along the unified theory of the liquid-glass transition surveyed in the previous section. Next we present all the tools which we use in developing the unified theory.

This book is composed of four parts:

1. *Introductory part:* Introduction and the classical theory of sound and elastic waves, and the liquid-glass transition (Chapters 1 and 2).
2. *Preliminary part:* Quantum field theory for many particle systems at finite temperature (Chapters 3–5).

3. *Core part:* The unified theory of the liquid-glass transition on the basis of the two band model (Chapters 6–10).
4. *Extensional part:* Excitations in a Bose-condensed liquid and the panic, the polymerization of RNA and the financial panic (Chapter 11).

1.3.1 Introductary Part

Conventionally, sound and phonons (elastic waves) have not been discriminated clearly. Sometimes sound has been considered to be a longitudinal phonon in the long wavelength regime. Phonons are the NG bosons, but sound is not the NG boson. In Chapter 2, we show that the classical theories have discriminated between sound and phonons (elastic waves) clearly: in the fluid mechanics, sound is described as adiabatic density fluctuation waves [41], while in the elastic theory, phonons are described as elastic waves [42]. Furthermore, in order to clearly show the discrimination between sound and elastic waves, we develop the microscopic classical theory in the condensed matter: sound originates from the dynamical motion of particles located at the bottoms of the potential wells, while elastic waves from that of excited particles in the potential wells (displaced particles from the bottoms). In this context, we also develop the classical theories of the liquid-glass transition, comparing with the unified theory outlined in the previous section.

1.3.2 Preliminary Part

In developing the unified theory, we employ the quantum field theory for many particle systems at finite temperature, i.e. Matsubara finite temperature Green's function method [43–45]. In order to make this book self-contained, we include the quantum field theory for many particle systems at finite temperature in the preliminary part. All the tools indispensable to establish the unified theory are contained in the preliminary part; the unified theory is an application of the quantum field theory for many particles systems at finite temperature. If one has learned the quantum field theoretical method, one can skip these chapters. However, there are some important aspects to understand the unified theory.

First, in the equilibrium state, the Hamiltonian of a system is essentially described by free physical particles; in the glassy state, free particles, and phonons, while in the liquid state, free particles, sound, and unstable phonons. The quantization of these free physical particles are classified according to the type of their equations of motion of the free physical particles. This classification is very important in quantizing free physical particles, constructing the corresponding Green's functions, and developing the perturbation theory by using the Wick's theorem.

Second, the Ward-Takahashi (WT) relations play an important role in constructing the Hamiltonian, free physical particles, and calculating the Green's functions [46–49]. The WT relations show the relations that Green's functions, the self-energy parts, and the vertex parts must satisfy, so that in the approximation procedures we must take these terms such that they satisfy the WT relations. Therefore, the WT relations guarantee the self-consistent approximations. The WT relation also yields the Nambu-Goldstone (NG) theorem associated with the spontaneous breakdown of the symmetry.

Third, by analytically continuing the Matsubara frequency in a Matsubara temperature Green's function to the real frequency from above or below the real axis, the Matsubara temperature Green's function leads to the retarded or the advanced Green's function. The retarded or advanced Green's function method is equivalent to Mori's memory function approach [50–53]. Thus the temperature Green's function method, the retarded or advanced method, and memory function method are all equivalent. Furthermore, the temperature Green's function method is very tractable. Therefore, developing more rigorous calculation beyond the mean field approximation, we can obtain the same results as the mode coupling theory including the nonlinear dynamics.

1.3.3 Core Part

The unified theory of the liquid-glass transition on the basis of the two band model are developed in Chapters 6–10 [20,21].

In Chapter 6, we first show the WT relations at finite temperature. Using the WT relations, we establish the Hamiltonian, the two band model in the condensed states. We also derive the Nambu-Goldstone (NG) theorem associated with the spontaneous breakdown of the spatially translational invariance. The NG bosons are phonons in the glassy state. The structure of phonons reflect the structure of the glassy state. Therefore, we show the structure of phonons, the dispersion curves, and the life-times of phonons. Boson peak reflects the rigidity of the glassy state.

Chapter 7 is the most fundamental for the unified theory of the liquid-glass transition. Since a particle can be considered as a simple particle or a molecule, the unified theory in Chapter 7 essentially covers the liquid-glass transitions. The Kauzmann entropy associated with the extra bubbles of particle-hole pairs not constituting sound plays a core role in the unified theory. Understanding the self-consistent structure of the unified theory, we can rearrange the experimental results and develop more rigorous theories.

Chapter 8 has the two aims: the construction of the nonlinear interaction potentials of phonons and the derivation of the dynamical equations of phonon operators with the nonlinear potentials. The nonlinear interaction potentials of phonons play an important role in deriving thermal conductivity, since thermal conductivity relates with nonlinear interaction potentials. On the other hand, we derive the dynamical equations of phonon operators with the nonlinear potentials. We verify that the dynamical equations of phonon operators with the nonlinear potentials reduce to those in the classical theories by applying the dynamical equations of phonon operators to solitons and localized modes. The dynamical equations of phonon operators yield not only the conventional results, but also finite temperature dynamics of solitons and localized modes, and quantum bound states of solitons.

In Chapter 9, using the nonlinear interaction potential for phonons, we derive phonon entropy fluctuation modes and thermal conductivity for phonons. Since phonons relate with the frequencies of the modes, phonon density fluctuations relate with the phonon entropy fluctuations, which relate with heat fluctuations due to phonons. Thus the correlations of phonon density fluctuations yield the phonon entropy fluctuation modes and their corresponding phonon thermal conductivities, where the relaxation time of phonon entropy fluctuations is derived from the simultaneously scattering processes of two phonon density fluctuation modes by random scattering potentials due to the random frequencies and the random hopping magnitudes. The relaxation time of phonon entropy fluctuations is closely related to the phonon relaxation time τ_M. In a similar manner to constructing the nonlinear interaction potential of phonons, we construct a nonlinear interaction potentials of sound and we derive the sound entropy modes and thermal conductivity for sound.

In Chapter 10, we extend the theory of the liquid-glass transition in one-component liquids to that in multi-component liquids. A particle in a one-component liquid can be a particle or a molecule. But in actual case of a molecule, subordinate particles constituting the molecule have important roles: the subordinate particles not only form the structure and the polarity of the molecule, but they also have the individual freedoms. Therefore, we should develop a theory in multi-component liquids. Including the electric polarization, we show phonons and the liquid-glass transitions in multi-component liquids.

1.3.4 Extensional Part

In Chapter 11, we extend the two band model to the phase symmetry breaking system and the panics.

First, we extend the WTR to the spontaneously breaking system of phase symmetry, a Bose-condensed liquid [54–56]. Combining particle and sound excitations through the dielectric method, we obtain excitations in the Bose-condensed liquid. Second, we extend the unified theory of the liquid-glass transition to the panics, the polymerization of the RNA [57], and the financial panic [58]. Finally, we show the criteria of the panic, the supercooling processes associated with the Kauzmann entropy crisis, and the VTF law.

References

[1] C.A. Angell, P.H. Poole, J. Shao, Nuovo Cim. D 16 (1994) 993.
[2] C.A. Angell, Science 267 (1995) 1924.
[3] C.A. Angell, in: M Rubí, C. Pérez-Vicente (Eds.), Complex Behaviour of Glassy Systems, Springer, Berlin, 1997.
[4] N.E. Cusack, The Physics of Structurally Disordered Matter, Adam Hilger, Bristol, 1987.
[5] S.R. Elliott, Physics of Amorphous Materials, Longmann Scientific & Technical, New York, 1990.
[6] R. Zallen, The Physics of Amorphous Solids, John Wiley and Sons, New York, 1998.
[7] W. Kauzmann, Chem. Rev. 43 (1948) 219.
[8] M. Vogel, Phys. Z. 22 (1921) 645.
[9] G. Tamman, W. Hesse, Z. Anorg, Allg. Chem. 156 (1926) 245.
[10] G.S. Fulcher, I. Am, Ceram. Soc. 8 (1925) 339.
[11] B. Frick, D. Richter, Science 267 (1995) 1939.
[12] Y. Nambu, G. Jona-Lasinio, Phys. Rev. 122 (1961) 345.
[13] Y. Nambu, G. Jona-Lasinio, Phys. Rev. 124 (1961) 246.
[14] J. Goldstone, Nuovo Cim. 19 (1961) 154.

[15] J. Goldstone, A. Salam, S. Weinberg, Phys. Rev. 127 (1962) 965.
[16] C. Kittel, Introduction to Solid State Physics, fifth ed., John Wiley and Sons, New York, 1976.
[17] J.P. Hansen, I.R. MacDonald, Theory of Simple Liquids, second ed., Academic Press, London, 1990.
[18] R. Kubo, Statistical Mechanics, North-Holland, 1965.
[19] T. Kitamura, Il Nuovo Cim. D 11 (1989) 1441.
[20] T. Kitamura, Phys. Rep. 383 (2003) 1.
[21] T. Kitamura, AIP Conf. Proc. 708 (2004) 631.
[22] T. Schneider, R. Brout, H. Thomas, J. Feder, Phys. Rev. Lett. 25 (1970) 1423.
[23] S. Takeno, Prog. Theor. Phys. Suppl. 45 (1970) 137.
[24] S. Takeno, M. Goda, Prog. Theor. Phys. 45 (1971) 331.
[25] M.C. Phillips, A.J. Barlow, J. Lamb, Proc. Roy. Soc. London A 329 (1972) 193.
[26] G. Adam, J.H. Gibbs, J. Chem. Phys. 43 (1965) 139.
[27] M. Goldstein, J. Chem. Phys. 51 (1969) 3728.
[28] F.H. Stillinger, Science 267 (1995) 1935.
[29] T. Keyes, Phys. Rev. E 59 (1999) 3207.
[30] P.G. Debenedetti, F.H. Stillinger, Nature 410 (2001) 259.
[31] L.-M. Martinez, C.A. Angell, Nature 410 (2001) 663.
[32] K. Kawasaki, Ann. Phys. NY 61 (1970) 1.
[33] T. Geszti, J. Phys. C 16 (1983) 5805.
[34] E. Leutheusser, Phys. Rev. A 29 (1984) 2765.
[35] U. Bengtzelius, W. Götze, A. Sjölander, J. Phys. C 17 (1984) 5915.
[36] W. Götze, L. Sjögren, Rep. Prog. Phys. 55 (1992) 241.
[37] W. Götze, L. Sjögren, Transp. Theory Stat. Phys. 24 (1995) 801.
[38] J.-L. Barrat, in: P. Jund, R. Jullien (Eds.), Physics of Glasses, AIP, NY, 1999.
[39] T. Hechsher, A.I. Nielsen, N.B. Olsen, J.C. Dyre, Nat. Phys. 4 (2008) 737.
[40] G. Floudas, M. Paluch, A. Grzybowski, K. Nagai, in: F. Kremer (Ed.), Molecular Dynamics of Glass-Forming Systems: Effects of Pressure, Advances in Dielectrics, Springer-Verlag, Berlin, Heidelberg, 2011.
[41] L.D. Landau, E.M. Lifshitz, Fluid Mechanics, Pergamon Press, London, 1963.
[42] L.D. Landau, E.M. Lifshitz, Theory of Elasticity, Pergamon Press, London, 1970.
[43] T. Matsubara, Prog. Theor. Phys. 14 (1955) 351.
[44] A.A. Abricosov, L.P. Gorkov, I.E. Dzyaloshinskii, Method of Quantum Field Theory in Statistical Physics, Prentice-Hall, Englewood Cliffs, NJ, 1963.
[45] A.L. Fetter, I.D. Walecka, Quantum Theory of Many Particle System, McGraw-Hill, New York, 1971.
[46] H. Umezawa, H. Matsumoto, M. Tachiki, Thermo Field Dynamics and Condensed States, North-Holland, Amsterdam, 1982.
[47] J.C. Ward, Phys. Rev. 78 (1950) 182.
[48] Y. Takahashi, Nuovo Cim. 6 (1957) 370.
[49] Y. Takahashi, Phys. Rev. D 15 (1977) 1589.
[50] R. Zwanzig, in: W.E. Britton, B.W. Downs, J. Downs (Eds.), Lectures in Theoretical Physics, vol. 3, Wiley Interscience, New York, 1961, p. 135.
[51] H. Mori, Prog. Theor. Phys. 33 (1965) 423.
[52] H. Mori, Prog. Theor. Phys. 34 (1965) 399.
[53] M. Ichiyanagi, J. Phys. Soc. Jpn. 32 (1972) 604.
[54] T. Kitamura, Physica A 290 (2001) 39.

[55] A. Griffin, Excitations in a Bose-Condensed Liquid, Cambridge University Press, London, 1993.
[56] H.R. Glyde, Excitations in Liquid and Solid Helium, Oxford University Press, New York, 1994.
[57] T. Kitamura, M. Peyrard, S.C. Lopez, Phys. Biol. 2 (2005) 200.
[58] T. Kitamura, Physica A 389 (2010) 2586.

2 Sound and Elastic Waves in the Classical Theory

Conventionally, the discrimination between sound and phonons (elastic waves) has not sufficiently been clarified. Sometimes, sound has been considered to be a longitudinal phonon in the long wavelength regime. This is not correct. The condensed states, the glassy and liquid states have a kind of the long range order, the pair distribution function, in a sense of the configuration average. In the glassy state, the pair distribution function holds the same every time, so elastic waves (phonons) are essential; elastic waves are the NG bosons. While in the liquid state, the pair distribution function changes with time; the hopping of particles causes the density fluctuations, so density fluctuation waves (sound) are essential; sound is not the NG boson.

In this chapter, we clarify the discrimination between sound and phonons (elastic waves) from both the macroscopic and microscopic classical theories. From the macroscopic classical theory, sound is treated in the fluid mechanics, while elastic waves are treated in the elastic theory; the sound relates with the dynamics of density fluctuations in fluids, while the elastic waves relate with the displacements in glasses. On the other hand, from the microscopic classical theory, sound is treated by the dynamical motion of particles located at the bottom of the potential wells, while elastic waves by that of excited particles in the potential wells (displaced particles from the bottoms).

The macroscopic classical theory only deal with the long wavelength regime, while the microscopic classical theory can deal with both the long and short wavelength regimes near the liquid–glass transition temperature, where the bandwidths of particle energy are very narrow; the systems are in the quantum statistical regime. Since we have known the unified theory of the liquid–glass transition in Introduction, we can identify the potential of a particle made up by the surrounding particles, with the squared frequency of the bubbles of intra-band particle-hole pairs. Thus we try to approach to the liquid–glass transition from the microscopic classical theory.

2.1 Sound in the Classical Fluid Mechanics

In the classical theory, sound is the density fluctuation mode. The density fluctuation mode depends on the thermodynamical processes. There are isothermal and adiabatic sounds corresponding to the isothermal and adiabatic processes, respectively. Sound in normal fluids is adiabatic not isothermal. But the isothermal sound plays an essential role in the supercooling process associated with the Kauzmann entropy crisis and the

Liquid Glass Transition. DOI: http://dx.doi.org/10.1016/B978-0-12-407177-3.00002-7

VTF law in the liquid–glass transition. The adiabatic and isothermal sounds relate with the adiabatic and isothermal compressibilities, respectively. Here we clarify the adiabatic and isothermal sounds from the macroscopic classical theory, the classical fluid mechanics.

2.1.1 Sound in the Isothermal and Adiabatic Processes

First, we survey sound waves in the classical fluid mechanics [1,2]. An ideal fluid is governed by the Euler's equation:

$$M\rho\frac{d\boldsymbol{v}}{dt} = M\rho\left[\frac{\partial \boldsymbol{v}}{\partial t} + (\boldsymbol{v}\cdot\nabla)\boldsymbol{v}\right] = -\nabla p, \tag{2.1.1}$$

and the equation of continuity:

$$\frac{\partial \rho}{\partial t} + \nabla\cdot(\rho\boldsymbol{v}) = 0, \tag{2.1.2}$$

where $\boldsymbol{v}$ is the velocity of fluid, ρ the fluid density, and p the pressure. Since sound waves oscillate with small amplitudes, the velocity $\boldsymbol{v}$ is small so that the term $(\boldsymbol{v}\cdot\nabla)\boldsymbol{v}$ in the Euler's equation (2.1.1) can be neglected. For the same reason, the relative changes in the fluid density and the pressure are small, so we put

$$p = p_0 + p', \quad \rho = \rho_0 + \rho', \tag{2.1.3}$$

where ρ_0 and p_0 are the constant equilibrium density and pressure, and ρ' and p' are their variations in the sound wave; $\rho' \ll \rho_0$, $p' \ll p_0$. Substituting Eq. (2.1.3) into Eqs. (2.1.1) and (2.1.2) and considering quantities of the first order of ρ', p', and $\boldsymbol{v}$, we obtain

$$M\rho_0\frac{\partial \boldsymbol{v}}{\partial t} + \nabla p' = 0 \tag{2.1.4}$$

and

$$\frac{\partial \rho'}{\partial t} + \rho_0\nabla\cdot\boldsymbol{v} = 0. \tag{2.1.5}$$

Equations (2.1.4)and (2.1.5) have the unknown functions $\boldsymbol{v}$, p', and ρ'. To eliminate one of these, we consider the relation between p' and ρ' : $p' = \frac{\partial p}{\partial \rho_0}\rho'$. The term $\frac{\partial p}{\partial \rho_0}$ relates with the compressibility χ as

$$\chi = -\frac{1}{V}\frac{\partial V}{\partial p} = \frac{1}{\rho}\frac{\partial \rho}{\partial p}.$$

The compressibility takes different values in the isothermal and adiabatic processes. So first we investigate the isothermal process, where the temperature is constant. Therefore we consider the following relation:

$$p' = \left(\frac{\partial p}{\partial \rho_0}\right)_T \rho'. \tag{2.1.6}$$

Combining Eq. (2.1.5) with Eq. (2.1.6), we obtain

$$\frac{\partial p'}{\partial t} + \rho_0 \left(\frac{\partial p}{\partial \rho_0}\right)_T \nabla \cdot \boldsymbol{v} = 0. \tag{2.1.7}$$

Since we are concerned with such a system that the velocity circulation around any closed contour C is zero:

$$\int_C \boldsymbol{v} \cdot d\mathbf{s} = 0; \quad \nabla \times \boldsymbol{v} = 0, \tag{2.1.8}$$

the velocity can be expressed as the gradient of some scalar. This scalar is called the velocity potential; we denote it as ϕ:

$$\boldsymbol{v} = \nabla \phi. \tag{2.1.9}$$

Equations (2.1.4) and (2.1.7) with Eq. (2.1.9) lead to

$$\frac{\partial^2 \phi}{\partial t^2} - c_T{}^2 \nabla^2 \phi = 0 \tag{2.1.10}$$

and the sound velocity in the isothermal process:

$$c_T{}^2 = \frac{1}{M}\left(\frac{\partial p}{\partial \rho}\right)_T, \tag{2.1.11}$$

where for brevity we have omitted the subscript in p_0 and ρ_0.

In the adiabatic process, where the entropy S is constant, the relation between p' and ρ' is given by

$$p' = \left(\frac{\partial p}{\partial \rho_0}\right)_S \rho'. \tag{2.1.12}$$

In a similar manner to the isothermal process, we obtain

$$\frac{\partial^2 \phi}{\partial t^2} - c_S{}^2 \nabla^2 \phi = 0 \tag{2.1.13}$$

and the sound velocity in the adiabatic process:

$$c_S{}^2 = \frac{1}{M}\left(\frac{\partial p}{\partial \rho}\right)_S. \tag{2.1.14}$$

Next we investigate the Free energy for the isothermal sound and the internal energy for the adiabatic sound.

2.1.2 Free Energy of Sound in the Classical Fluid Mechanics

In order to derive thermodynamical quantities in the isothermal process, we must deal with the Free energy of the system, because the thermodynamical variables of the Free energy are temperature T and Volume V:

$$dF = S\,dT - p\,dV. \tag{2.1.15}$$

Here we consider the Free energy in unit volume of the classical fluid $f = \frac{F}{V}$ and the thermodynamical variables T and density $\rho = \frac{N}{V}$. Therefore from Eq. (2.1.15): $V\,df - fV\frac{d\rho}{\rho} = S\,dT + Vp\frac{d\rho}{\rho}$, we obtain

$$df = \frac{S}{V}dT + \frac{f+p}{\rho}d\rho = \frac{S}{V}dT + \frac{F+pV}{N}d\rho = \frac{S}{V}dT + \frac{G}{N}d\rho, \tag{2.1.16}$$

where $G = F + pV$ which is the Gibbs Free energy:

$$dG = S\,dT + V\,dp \tag{2.1.17}$$

Now we investigate the Free energy of sound waves in the isothermal process. The Free energy in a unit volume of the fluid with sound waves is written by

$$f_s = f + \frac{M}{2}\rho v^2, \tag{2.1.18}$$

where f is the Free energy in unit volume and v is the fluid velocity. We substitute $\rho = \rho_0 + \rho'$, $f = f_0 + f'$ where the primed quantities denote the deviations of the respective quantities from their values when the fluid rests. Considering only the terms up to the second order of the deviations and velocity, and that the term $\frac{M}{2}\rho' v^2$ is of the third order, we have

$$f_s = f_0 + \rho'\frac{\partial f}{\partial \rho} + \frac{1}{2}\rho'^2\frac{\partial^2 f}{\partial \rho^2} + \frac{M}{2}\rho_0 v^2. \tag{2.1.19}$$

From Eq. (2.1.16), we obtain

$$\left(\frac{\partial f}{\partial \rho}\right)_T = \frac{G}{N}. \tag{2.1.20}$$

The second derivative is

$$\left(\frac{\partial^2 f}{\partial \rho^2}\right)_T = \frac{1}{N}\left(\frac{\partial G}{\partial \rho}\right)_T = \frac{1}{\rho}\left(\frac{\partial p}{\partial \rho}\right)_T = Mc_T^2/\rho. \tag{2.1.21}$$

Thus the Free energy in unit volume of the fluid in the isothermal process is given by

$$f_s = f_0 + \frac{G}{N}\rho' + \frac{M}{2}c_T^2\rho'^2/\rho + \frac{M}{2}\rho_0 v^2.$$

The first term f_0 is the Free energy in unit volume when the fluid is at rest, so it does not relate to the sound waves. The second term $\frac{G}{N}\rho'$ is the change in energy due to the change in the mass of fluid in unit volume, so this term disappears in the total energy, since the total mass of fluid is unchanged; $\int \rho'\,dV = 0$. Thus the total change in the Free energy of the fluid in the isothermal process caused by the sound wave is given by:

$$f_s = \frac{M}{2}\rho_0 v^2 + \frac{M}{2}c_T^2\rho'^2/\rho_0. \tag{2.1.22}$$

We can regard f_s as the density of sound Free energy in the isothermal process. It should be noted that the square form of $\frac{M}{2}c_T^2\rho'^2/\rho_0$ in f_s can be written in the isothermal process. The square form of $\frac{M}{2}c_S^2\rho'^2/\rho_0$ can be written in the density of sound energy in the adiabatic process discussed below. When we construct the Hamiltonian of isothermal sound, we reconfirm the second term in f_s.

2.1.3 Energy of Sound in the Classical Fluid Mechanics

Next we investigate the adiabatic process. In order to derive thermodynamical quantities in the adiabatic process, we must deal with the Energy (internal energy) of the system, because the thermodynamical variables of the Energy are entropy S and volume V:

$$dE = T\,dS - p\,dV. \tag{2.1.23}$$

Here we consider the Energy per unit volume in the classical fluid $e = \frac{E}{V}$ and the thermodynamical variables S and density $\rho = \frac{N}{V}$. Therefore, from Eq. (2.1.23): $V\,de - eV\frac{d\rho}{\rho} = T\,dS + Vp\frac{d\rho}{\rho}$, we obtain

$$de = \frac{T}{V}dS + \frac{e+p}{\rho}d\rho = \frac{T}{V}dS + \frac{H}{N}d\rho, \tag{2.1.24}$$

where $H = E + pV$ which is the entropy

$$dH = T\,dS + V\,dp. \tag{2.1.25}$$

Here we investigate the energy of sound waves in a classical fluid [1]. The energy in unit volume of the fluid e_s is given by

$$e_s = e + \frac{M}{2}\rho v^2, \tag{2.1.26}$$

where e is the internal energy and v is the fluid velocity. We substitute $\rho = \rho_0 + \rho'$, $e = e_0 + e'$, where the primed quantities denote the deviations of the respective quantities from their values when the fluid rests. Considering only the terms up to the second order of the deviations and velocity, and that the term $\frac{M}{2}\rho' v^2$ is of the third order, we have

$$e_s = e_0 + \rho'\frac{\partial e}{\partial \rho} + \frac{1}{2}\rho'^2\frac{\partial^2 e}{\partial \rho^2} + \frac{M}{2}\rho_0 v^2. \tag{2.1.27}$$

From Eq. (2.1.24), we obtain

$$\left(\frac{\partial e}{\partial \rho}\right)_S = \frac{H}{N} \tag{2.1.28}$$

The second derivative is

$$\left(\frac{\partial^2 e}{\partial \rho^2}\right)_S = \frac{1}{N}\left(\frac{\partial H}{\partial \rho}\right)_S = \frac{1}{\rho}\left(\frac{\partial p}{\partial \rho}\right)_S = Mc_S^2/\rho. \tag{2.1.29}$$

Thus the energy in a unit volume of the fluid in the adiabatic process e_s is given by

$$e_s = e_0 + \frac{H}{N}\rho' + \frac{M}{2}c_s^2\rho'^2/\rho + \frac{M}{2}\rho_0 v^2.$$

The first term e_0 is the Energy in a unit volume when the fluid is at rest, so it does not relate to the sound wave. The second term $(H/N)\rho'$ is the change in energy due to the change in the mass of fluid in unit volume, so this term disappears in the total energy, since the total mass of fluid is unchanged; $\int \rho' dV = 0$. Thus the total change in the energy of the fluid in the adiabatic process caused by the sound wave is given by:

$$e_s = \frac{M}{2}\rho_0 v^2 + \frac{M}{2}c_S^2\rho'^2/\rho_0. \tag{2.1.30}$$

We can regard e_s as the density of sound energy in the adiabatic process.

2.1.4 The Relation Between the Isothermal and Adiabatic Sounds

In order to relate the isothermal sound velocity c_T with the adiabatic sound velocity c_S, we must investigate how the fluctuations of entropy S depend on the fluctuations of pressure p, density $\rho = N/V$, and temperature T. First, we rewrite the sound velocity in the adiabatic process. If we consider entropy $S(p, \rho)$ depending on p and ρ, we obtain

$$dS(p,\rho) = \left(\frac{\partial S}{\partial p}\right)_\rho dp + \left(\frac{\partial S}{\partial \rho}\right)_p d\rho. \tag{2.1.31}$$

Putting $dS(p, \rho) = 0$ in Eq. (2.1.31) in the adiabatic process, we have

$$\left(\frac{\partial p}{\partial \rho}\right)_S = -\frac{\left(\frac{\partial S}{\partial \rho}\right)_p}{\left(\frac{\partial S}{\partial p}\right)_\rho}. \tag{2.1.32}$$

Thus we obtain

$$c_S^2 = \frac{1}{M}\left(\frac{\partial p}{\partial \rho}\right)_S = -\frac{1}{M}\frac{\left(\frac{\partial S}{\partial \rho}\right)_p}{\left(\frac{\partial S}{\partial p}\right)_\rho}. \tag{2.1.33}$$

Next, we rewrite the last equation. If we consider entropy $S(T(p, \rho), p)$ depending on $T(p, \rho)$ and p, we obtain

$$\begin{aligned} dS(T(p,\rho),p) &= \left(\frac{\partial S}{\partial T}\right)_p dT + \left(\frac{\partial S}{\partial p}\right)_T dp \\ &= \left(\frac{\partial S}{\partial T}\right)_p \left[\left(\frac{\partial T}{\partial p}\right)_\rho dp + \left(\frac{\partial T}{\partial \rho}\right)_p d\rho\right] \\ &\quad + \left(\frac{\partial S}{\partial p}\right)_T dp. \end{aligned} \tag{2.1.34}$$

So we have

$$\left(\frac{\partial S}{\partial \rho}\right)_p = \left(\frac{\partial S}{\partial T}\right)_p \left(\frac{\partial T}{\partial \rho}\right)_p = \frac{C_p}{T}\left(\frac{\partial T}{\partial \rho}\right)_p, \tag{2.1.35}$$

where C_p is specific heat at constant pressure. Similarly, if we consider entropy $S(T(p,\rho),\rho)$ depending on $T(p,\rho)$ and ρ, we obtain

$$\left(\frac{\partial S}{\partial p}\right)_\rho = \left(\frac{\partial S}{\partial T}\right)_\rho \left(\frac{\partial T}{\partial p}\right)_\rho = \frac{C_V}{T}\left(\frac{\partial T}{\partial p}\right)_\rho, \tag{2.1.36}$$

where C_V is specific heat at constant volume. Substituting Eqs. (2.1.35) and (2.1.36) in Eq. (2.1.33), we have

$$c_S^2 = -\frac{1}{M}\frac{C_p}{C_V}\frac{\left(\frac{\partial T}{\partial \rho}\right)_p}{\left(\frac{\partial T}{\partial p}\right)_\rho}. \tag{2.1.37}$$

In a similar manner to the derivation of Eq. (2.1.32), using the equation

$$dT = \left(\frac{\partial T}{\partial \rho}\right)_p d\rho + \left(\frac{\partial T}{\partial p}\right)_\rho dp = 0,$$

we obtain

$$\left(\frac{\partial p}{\partial \rho}\right)_T = -\frac{\left(\frac{\partial T}{\partial \rho}\right)_p}{\left(\frac{\partial T}{\partial p}\right)_\rho}. \tag{2.1.38}$$

Thus the substitution of Eq. (2.1.38) into Eq. (2.1.37) leads to

$$c_S^2 = \frac{C_p}{C_V}\frac{1}{M}\left(\frac{\partial p}{\partial \rho}\right)_T = \gamma c_T^2, \tag{2.1.39}$$

where

$$\gamma = C_p/C_V. \tag{2.1.40}$$

The relations between sound velocity and compressibility in the isothermal and adiabatic processes are given by

$$\chi_T = -\frac{1}{V}\left(\frac{\partial V}{\partial p}\right)_T = \frac{1}{\rho}\left(\frac{\partial \rho}{\partial p}\right)_T = \frac{1}{M\rho c_T^2},$$
$$\chi_S = -\frac{1}{V}\left(\frac{\partial V}{\partial p}\right)_S = \frac{1}{\rho}\left(\frac{\partial \rho}{\partial p}\right)_S = \frac{1}{M\rho c_S^2}, \tag{2.1.41}$$

where

$$\chi_S = \frac{1}{\gamma}\chi_T. \tag{2.1.42}$$

In actual fluids, the adiabatic process plays a role in forming sound, because sound velocity is faster than heat propagation. But the isothermal process plays an essential role in the supercooling process associated with the Kauzmann entropy crisis and the VTF laws.

2.2 Elastic Waves in the Classical Elastic Theory

Elastic waves are generated in solids, glasses, and crystals. Conventionally, the terms elastic waves and phonons are used as synonym; phonons are quantized particles of classical elastic waves. Since the classical elastic theory only deals with the macroscopic behaviors, the classical elastic waves are valid only in the long wavelength regime, where both the dispersion curves of classical elastic waves and phonons yield the same. In the short wavelength regime, the elastic waves miss the microscopic structure which phonons possess: the periodic structure in crystals and the roton type minimum near the reciprocal particle distance and boson peak in glasses, which are explained by the microscopic classical theory in the next section. But it is worthy to recognize that the elastic waves originate from the dynamical motion of the displacements of matter in solids, compared with sound waves from that of the density fluctuations in fluids.

2.2.1 *Strain Tensor and Stress Tensor*

In order to derive the fundamental equation of motion in continuum solid [3], first we consider the deformation in a solid under action of force. Under deformation a position $\boldsymbol{x} = (x_1, x_2, x_3)$ moves to a point $\boldsymbol{x}' = (x_1', x_2', x_3')$, then the displacements $\boldsymbol{u}(\boldsymbol{x})$ are given by

$$\boldsymbol{u}(\boldsymbol{x}) = \boldsymbol{x}'(\boldsymbol{x}) - \boldsymbol{x}, \tag{2.2.1}$$

where $\boldsymbol{x}'(\boldsymbol{x})$ and $\boldsymbol{u}(\boldsymbol{x})$ are functions of $\boldsymbol{x}$. Under the deformation, the relative distance $d\boldsymbol{x}$ becomes $d\boldsymbol{x}' = d\boldsymbol{x} + d\boldsymbol{u}$ and the distance $dl^2 = d\boldsymbol{x}^2$ becomes $dl'^2 = d\boldsymbol{x}'^2 = (d\boldsymbol{x} + d\boldsymbol{u})^2$. Since $du_i = \sum_j \frac{\partial u_i}{\partial x_j} dx_j$, we obtain

$$\begin{aligned} dl'^2 &= \sum_i (dx_i + du_i)^2 \\ &= dl^2 + 2\sum_{ij} \frac{\partial u_i}{\partial x_j} dx_i\, dx_j + \sum_{ijk} \frac{\partial u_i}{\partial x_j} \frac{\partial u_i}{\partial x_k} dx_j\, dx_k \\ &= dl^2 + \sum_{ij} \left(\frac{\partial u_i}{\partial x_j} + \frac{\partial u_j}{\partial x_i} \right) dx_i\, dx_j + \sum_{ijk} \frac{\partial u_k}{\partial x_i} \frac{\partial u_k}{\partial x_j} dx_i\, dx_j, \end{aligned}$$

where from the second line to the third line we have appropriately changed the subscripts i, j, k. Then we can write down as

$$dl'^2 = dl^2 + \sum_{ij} 2u_{ij}\, dx_i\, dx_j, \tag{2.2.2}$$

where strain tensor u_{ij} is defined as

$$u_{ij} = \frac{1}{2}\left(\frac{\partial u_i}{\partial x_j} + \frac{\partial u_j}{\partial x_i} + \sum_k \frac{\partial u_k}{\partial x_i}\frac{\partial u_k}{\partial x_j}\right). \tag{2.2.3}$$

Under an infinitesimal deformation we can neglect the second term as the infinitesimal small value of the second order, so we can put

$$u_{ij} = \frac{1}{2}\left(\frac{\partial u_i}{\partial x_j} + \frac{\partial u_j}{\partial x_i}\right). \tag{2.2.4}$$

Next we consider internal stress in the solid. If $\boldsymbol{F}$ is force per unit volume, the force acting on a volume dV is $\boldsymbol{F}\,dV$. The ith component force acting on the solid is written as

$$\int F_i\,dV.$$

Here we introduce stress tensor σ_{ij} as

$$F_i = \nabla \cdot \boldsymbol{\sigma}_i; \quad F_i = \sum_j \frac{\partial \sigma_{ij}}{\partial x_j}. \tag{2.2.5}$$

Using the Gauss's theorem in vector analysis we obtain:

$$\int_V \nabla \cdot \boldsymbol{\sigma}_i\,dV = \int_S \boldsymbol{\sigma}_i \cdot d\boldsymbol{S}; \quad \int_V \sum_j \frac{\partial \sigma_{ij}}{\partial x_j} dV = \int_S \sum_j \sigma_{ij}\,dS_j, \tag{2.2.6}$$

where V and S on the integrals mean the closed volume and the closed surface, respectively. Thus we have

$$\int_V F_i\,dV = \int_S \sum_j \sigma_{ij}\,dS_j. \tag{2.2.7}$$

This equation states that the stress tensor σ_{ij} is the ith component force acting on the unit surface perpendicular to j-axis. Since the continuum is rotational invariance, σ_{ij} is symmetry:

$$\sigma_{ij} = \sigma_{ji} \tag{2.2.8}$$

If the substance is under the uniform pressure p from the surroundings, the force acting on dS_i is $-p\,dS_i = -\sum_j p\delta_{ij}\,dS_j$, where δ_{ij} is the Kronecker delta. Then we obtain the stress tensor under the uniform pressure p from the surroundings:

$$\sigma_{ij} = -p\delta_{ij}. \tag{2.2.9}$$

2.2.2 Thermodynamics in Deformation

Now we investigate thermodynamics in case of deformation. First we derive the work which the inner stress $F_i = \sum_j \frac{\partial \sigma_{ij}}{\partial x_j}$ carries out through the deformation δu_i. The work of the inner stress per unit volume is $\delta w = F_i \delta u_i = \sum_j \frac{\partial \sigma_{ij}}{\partial x_j} \delta u_i$. The work which the substance has done is given by

$$
\begin{aligned}
\int_V \delta w \, dV &= \int_V \sum_{ij} \frac{\partial \sigma_{ij}}{\partial x_j} \delta u_i \, dV \\
&= \int_V \sum_{ij} \frac{\partial \sigma_{ij} \delta u_i}{\partial x_j} dV - \sum_{ij} \int_V \sigma_{ij} \frac{\partial \delta u_i}{\partial x_j} dV \\
&= \int_S \sum_{ij} \sigma_{ij} \, \delta u_i \, dS_j - \int_V \sum_{ij} \sigma_{ij} \frac{\partial \delta u_i}{\partial x_j} dV \\
&= -\frac{1}{2} \int_V \sum_{ij} \sigma_{ij} \left(\frac{\partial \delta u_i}{\partial x_j} + \frac{\partial \delta u_j}{\partial x_i} \right) dV = - \int_V \sum_{ij} \sigma_{ij} \, \delta u_{ij} \, dV,
\end{aligned}
$$

where from the second line to the third line we have used the Gauss's theorem, from the third line to the fourth line we have used $\sigma_{ij} = 0$ on the surface S and the symmetry of σ_{ij} in Eq. (2.2.8). Thus we obtain

$$
\delta w = -\sum_{ij} \sigma_{ij} \, \delta u_{ij}. \tag{2.2.10}
$$

Here we denote the inner energy and entropy per unit volume as $e = \frac{E}{V}$ and $s = \frac{S}{V}$, respectively. From the thermodynamic equation, $de = T\,ds - \delta w$, we obtain

$$
de = T\,ds + \sum_{ij} \sigma_{ij} \, du_{ij}. \tag{2.2.11}
$$

In case of uniform pressure p, from Eq. (2.2.9) we obtain

$$
\sum_{ij} \sigma_{ij} \, du_{ij} = -\sum_{ij} p \delta_{ij} \, \delta u_{ij} = -p \sum_i u_{ii} = -p \, dV.
$$

The last equation is obtained as follows: in the principal axis, where u_{ij} is diagonalized as $u_{ij} = \delta_{ij} u^{(i)}$, the volume $dx_1 \, dx_2 \, dx_3$ is transformed as $dx'_1 \, dx'_2 \, dx'_3 = dx_1 \, dx_2 \, dx_3 (1 + u^{(1)})(1 + u^{(2)})(1 + u^{(3)}) \cong dx_1 \, dx_2 \, dx_3 (1 + \sum_i u^{(i)})$. So the volume change per unit volume $dV = \sum_i u^{(i)} = \sum_i u_{ii}$, where the last equation comes from $\mathrm{Tr}(u_{ij})$ = invariant in any coordinates. Thus we have

$$
de = T\,ds - p\,dV. \tag{2.2.12}
$$

In case of general work of inner stress, Eq. (2.2.10), if we introduce Free energy per unit volume $f = e - Ts$, we obtain

$$
df = -s\,dT + \sum_{ij} \sigma_{ij} \, du_{ij}. \tag{2.2.13}
$$

From Eqs. (2.2.11) and (2.2.13), we obtain

$$(\sigma_{ij})_S = \left(\frac{\partial e}{\partial u_{ij}}\right)_S; \quad (\sigma_{ij})_T = \left(\frac{\partial f}{\partial u_{ij}}\right)_T, \tag{2.2.14}$$

where $(\sigma_{ij})_S$ and $(\sigma_{ij})_T$ are the adiabatic and isothermal stress tensors, respectively.

2.2.3 The Fick's Law in the Isothermal Process

In order to derive stress tensors σ_{ij} in Eq. (2.2.14), we have to determine the Free energy f in the isothermal process and the Energy e in the adiabatic process as the function of strain tensors u_{ij}. First we investigate the isothermal process. Since we need elastic waves, we develop f up to the second order of u_{ij}. From the equilibrium condition, if $u_{ij} = 0$, then $\sigma_{ij} = \frac{\partial f}{\partial u_{ij}} = 0$. This fact says that f does not have the linear term of u_{ij}. Thus we adopt the following form:

$$f = f_0 + \frac{\lambda}{2}\left(\sum_k u_{kk}\right)^2 + \mu \sum_{ij} u_{ij}^2, \tag{2.2.15}$$

where we call λ and μ Lamè coefficients. f_0 is the Free energy without deformation. So we do not consider this term. To divide u_{ij} into shearing strain keeping the constant volume and bulk strain keeping it uniform, we rewrite as follows:

$$u_{ij} = \left(u_{ij} - \frac{1}{3}\delta_{ij}\sum_k u_{kk}\right) + \frac{1}{3}\delta_{ij}\sum_k u_{kk}. \tag{2.2.16}$$

Substituting Eq. (2.2.16) into Eq. (2.2.15), we obtain

$$f = \mu \sum_{ij}\left(u_{ij} - \frac{1}{3}\delta_{ij}\sum_k u_{kk}\right)^2 + \frac{K_T}{2}\left(\sum_k u_{kk}\right)^2, \quad K_T = \lambda + \frac{2}{3}\mu, \tag{2.2.17}$$

where K_T is called bulk modulus in the isothermal process and μ shear modulus. Since $\sum_{ij}\left(u_{ij} - \frac{1}{3}\delta_{ij}\sum_k u_{kk}\right)\delta_{ij} = 0$ and $d\sum_k u_{kk} = d\sum_{ij}\delta_{ij}u_{ij}$, we have

$$\begin{aligned} df &= 2\mu \sum_{ij}\left(u_{ij} - \frac{1}{3}\delta_{ij}\sum_k u_{kk}\right) d\left(u_{ij} - \frac{1}{3}\delta_{ij}\sum_k u_{kk}\right) \\ &\quad + K_T\left(\sum_k u_{kk}\right) d\left(\sum_k u_{kk}\right) \\ &= \sum_{ij}\left[2\mu\left(u_{ij} - \frac{1}{3}\delta_{ij}\sum_k u_{kk}\right) + K_T\delta_{ij}\sum_k u_{kk}\right] du_{ij}. \end{aligned}$$

Thus we obtain the stress tensor

$$(\sigma_{ij})_T = 2\mu\left(u_{ij} - \frac{1}{3}\delta_{ij}\sum_k u_{kk}\right) + K_T\delta_{ij}\sum_k u_{kk}. \tag{2.2.18}$$

If we trace both sides of Eq. (2.2.18), we obtain

$$\sum_i (\sigma_{ii})_T = 3K_T \sum_i u_{ii}. \tag{2.2.19}$$

Substituting Eq. (2.2.19) into Eq. (2.2.18), we obtain Fick's law (the linear relation between u_{ij} and σ_{ij}):

$$u_{ij} = \frac{1}{9K_T}\delta_{ij}\sum_k (\sigma_{kk})_T + \frac{1}{2\mu}\left[(\sigma_{ij})_T - \frac{1}{3}\delta_{ij}\sum_k (\sigma_{kk})_T\right]. \tag{2.2.20}$$

Under uniform pressure p, $(\sigma_{ij})_T = -p\delta_{ij}$, Eq. (2.2.19) leads to

$$\sum_i u_{ii} = -\frac{p}{K_T}. \tag{2.2.21}$$

Since $\sum_i u_{ii}$ is infinitesimally small, so is p, we can write

$$\frac{\sum_i u_{ii}}{p} = \frac{1}{V}\frac{dV}{p} = \frac{1}{V}\left(\frac{\partial V}{\partial p}\right)_T = -\chi_T,$$

so that we obtain

$$\frac{1}{K_T} = -\frac{1}{V}\left(\frac{\partial V}{\partial p}\right)_T = \chi_T, \tag{2.2.22}$$

where K_T is called isothermal bulk modulus and χ_T is isothermal compressibility in Eq. (2.1.40).

2.2.4 The Fick's Law in the Adiabatic Process

In the adiabatic process, since S is constant, we must take thermodynamical valuables s and u_{ij} from Eq. (2.2.11), so we use internal energy e in Eq. (2.2.11) instead of f in Eq. (2.2.13). Then we can write

$$e = \mu \sum_{ij}\left(u_{ij} - \frac{1}{3}\delta_{ij}\sum_k u_{kk}\right)^2 + \frac{K_S}{2}\left(\sum_k u_{kk}\right)^2, \tag{2.2.23}$$

where K_S is bulk modulus in the adiabatic process. From $(\sigma_{ij})_S = (\frac{\partial e}{\partial u_{ij}})_S$ in Eq. (2.2.14), the stress tensor is given by

$$(\sigma_{ij})_S = 2\mu\left(u_{ij} - \frac{1}{3}\delta_{ij}\sum_k u_{kk}\right) + K_S\delta_{ij}\sum_k u_{kk}. \tag{2.2.24}$$

If we trace both sides of Eq. (2.2.24), we obtain

$$\sum_i (\sigma_{ii})_S = 3K_S \sum_i u_{ii}. \tag{2.2.25}$$

Substituting Eq. (2.2.25) into Eq. (2.2.24), we obtain Fick's law (the linear relation between u_{ij} and σ_{ij}):

$$u_{ij} = \frac{1}{9K_S}\delta_{ij}\sum_k(\sigma_{kk})_S + \frac{1}{2\mu}\left[(\sigma_{ij})_S - \frac{1}{3}\delta_{ij}\sum_k(\sigma_{kk})_S\right]. \tag{2.2.26}$$

Under uniform pressure p, $(\sigma_{ij})_S = -p\delta_{ij}$, Eq. (2.2.25) leads to

$$\sum_i u_{ii} = -\frac{p}{K_S}. \tag{2.2.27}$$

In a similar manner to K_T, we obtain

$$\frac{1}{K_S} = -\frac{1}{V}\left(\frac{\partial V}{\partial p}\right)_S = \chi_S, \tag{2.2.28}$$

where K_S is called adiabatic bulk modulus and χ_S adiabatic compressibility. From Eq. (2.1.41), we have

$$K_S = \gamma K_T. \tag{2.2.29}$$

Shear modulus μ is the same in both the isothermal and adiabatic processes.

2.2.5 Elastic Waves

We start with the equation of motion:

$$M\rho\ddot{u}_i = \sum_j \frac{\partial\sigma_{ij}}{\partial x_j}, \tag{2.2.30}$$

where

$$\begin{aligned}\sum_j \frac{\partial\sigma_{ij}}{\partial x_j} &= \sum_j \frac{\partial}{\partial x_j}\left[2\mu\left(u_{ij} - \frac{1}{3}\delta_{ij}\sum_k u_{kk}\right) + K\delta_{ij}\sum_k u_{kk}\right] \\ &= K\frac{\partial}{\partial x_i}\sum_k u_{kk} + 2\mu\left[\sum_j \frac{\partial}{\partial x_j}\frac{1}{2}\left(\frac{\partial u_i}{\partial x_j} + \frac{\partial u_j}{\partial x_i}\right) - \frac{1}{3}\frac{\partial}{\partial x_i}\sum_k u_{kk}\right] \\ &= K\frac{\partial}{\partial x_i}\nabla\boldsymbol{u} + 2\mu\left[\frac{1}{2}\Delta u_i + \frac{1}{2}\frac{\partial}{\partial x_i}\nabla\boldsymbol{u} - \frac{1}{3}\frac{\partial}{\partial x_i}\nabla\boldsymbol{u}\right] \\ &= \left(K + \frac{1}{3}\mu\right)\frac{\partial}{\partial x_i}\nabla\boldsymbol{u} + \mu\Delta u_i,\end{aligned} \tag{2.2.31}$$

and $\nabla\cdot\nabla = \Delta$. Here we have considered both the isothermal and adiabatic processes omitting subscripts T and S. Then we obtain

$$M\rho\ddot{\boldsymbol{u}} = \left(K + \frac{1}{3}\mu\right)\nabla\nabla\boldsymbol{u} + \mu\Delta\boldsymbol{u}. \tag{2.2.32}$$

Now we consider elastic plane waves propagating to x-axis: $\boldsymbol{u}(x, 0, 0; t)$. Then we obtain

$$M\rho \begin{pmatrix} \ddot{u}_x \\ \ddot{u}_y \\ \ddot{u}_z \end{pmatrix} = \left(K + \frac{1}{3}\mu\right) \begin{pmatrix} \frac{\partial^2 u_x}{\partial x^2} \\ 0 \\ 0 \end{pmatrix} + \mu \begin{pmatrix} \frac{\partial^2 u_x}{\partial x^2} \\ \frac{\partial^2 u_y}{\partial x^2} \\ \frac{\partial^2 u_z}{\partial x^2} \end{pmatrix} = \begin{pmatrix} (K + \frac{4}{3}\mu)\frac{\partial^2 u_x}{\partial x^2} \\ \mu \frac{\partial^2 u_y}{\partial x^2} \\ \mu \frac{\partial^2 u_z}{\partial x^2} \end{pmatrix}. \tag{2.2.33}$$

Thus we obtain a longitudinal mode and two transverse modes:

$$\frac{\partial u_x}{\partial t^2} - c_l^2 \frac{\partial^2 u_x}{\partial x^2} = 0, \quad c_l = \sqrt{\frac{3K + 4\mu}{M\rho}} = \sqrt{\frac{\lambda + 2\mu}{M\rho}}, \tag{2.2.34}$$

$$\frac{\partial u_{y,z}}{\partial t^2} - c_t^2 \frac{\partial^2 u_{y,z}}{\partial x^2} = 0, \quad c_t = \sqrt{\frac{\mu}{M\rho}}, \tag{2.2.35}$$

where c_l and c_t are velocities of the longitudinal and transverse modes, respectively.

Conventionally, the discrimination between sound and elastic waves have not sufficiently been distinguished. Classical arguments in Sections 2.1 and 2.2 show that sound is density fluctuation mode essential in liquids and elastic waves are kinetic displacement modes essential in solids as follows:

1. Dynamical equation for sound are derived by the Euler's equation (2.1.1) and the number conservation law (2.1.2) with respect to number density, while that for elastic waves are derived by the equation of motion Eq. (2.2.30) or Eq. (2.2.32) with respect to displacements under deformation in continuum.
2. Free energy and Energy for sound are constructed by density fluctuations in Eqs. (2.1.22) and (2.1.30), while those for elastic waves are constructed by displacements in Eq. (2.2.17) or (2.2.23).
3. Compressibility related with sound velocity is derived from the relation between density fluctuations and pressure in Eq. (2.1.6), while that related with elastic waves is derived from the relation between displacements and the pressure in Eq. (2.2.21).

2.3 Sound and Phonons in the Classical Microscopic Theory

In the previous sections, we have developed the macroscopic classical theory of sound and elastic waves in continuum; sound and elastic waves are density fluctuation waves and kinetic modes of particle displacements, respectively. On the other hand, the microscopic classical theories of sound and phonons are developed in a system composed of particles in well-localized potentials. The dynamical motions of particles are governed through the interaction potentials between the particles. Sound is the density fluctuation mode of well-localized potentials, while phonons are the kinetic modes of displacement particles from the bottoms of the well-localized potentials.

As we have seen the unified theory of the liquid–glass transition under the supercooling process associated with the Kauzmann entropy and the VTF law in the liquid–glass transition in Introduction, bubbles of intra-band particle-hole pairs play an essential role in forming the sound, while the extra bubbles of particle-hole pairs not constituting the sound mode play an essential role in constructing the Kauzmann entropy; the extra bubbles play a role in the fluctuation entropy. Therefore, in order to investigate the supercooling process, we must find out the terms corresponding to the bubbles of intra-band density particle-hole pairs. We identify the potential of a particle made up by the surrounding particles, with the terms corresponding to the bubble diagram of intra-band particle-hole pairs. Thus we can develop a theory of the liquid–glass transition in a parallel way to the unified theory. Developing the microscopic classical theory for sound and phonons, we clarify the validity and the limitation of the microscopic classical theory.

2.3.1 The Hamiltonians and the Equations of Motion for Sound and Phonons

Now we investigate the equation of motion for particles. To do so, we first construct the potential energy for particles. In our model, an m-particle is located in a random harmonic potential at a random site $\boldsymbol{R}_m$. In the harmonic potential, the m-particle can do a harmonic motion with the harmonic frequency ω_m associated with the displacement $\boldsymbol{u}_m$ from the assigned position $\boldsymbol{R}_m$. Therefore, the position of the m-particle is $\boldsymbol{R}_m + \boldsymbol{u}_m$. The potential energy of the system can be written as the summation of the interaction potentials between the particle positions, $\{\boldsymbol{R}_m + \boldsymbol{u}_m\}$. Thus we start with the following potential energy:

$$\begin{aligned}\Phi &= \frac{1}{2}\sum_{m\neq n} V(\{\boldsymbol{R}_m + \boldsymbol{u}_m\} - \{\boldsymbol{R}_n + \boldsymbol{u}_m\}) \\ &\cong \frac{1}{2}\sum_{m\neq n}\left[V(\boldsymbol{R}_m - \boldsymbol{R}_n) + \frac{1}{2}\sum_{ij} V_{ij}(\boldsymbol{R}_m - \boldsymbol{R}_n)(u_{mi} - u_{ni})(u_{mj} - u_{nj})\right] \\ &= \frac{1}{2}\sum_{m\neq n}\left[V(\boldsymbol{R}_m - \boldsymbol{R}_n) + \sum_{ij} V_{ij}(\boldsymbol{R}_m - \boldsymbol{R}_n)(u_{mi}u_{mj} - u_{mi}u_{nj})\right], \quad (2.3.1)\end{aligned}$$

where $V_{ij}(\boldsymbol{R}) = \frac{\partial^2 V(\boldsymbol{R})}{\partial R_i \partial R_j}$ and we have neglected the first derivative terms, because we can expect that the configuration average of the terms vanishes. In transforming the second equation to the third one, we have used the symmetries of the potential.

The first potential term,

$$\frac{1}{2}\sum_{m\neq n} V(\boldsymbol{R}_m - \boldsymbol{R}_n),$$

relates to sound [1,2,4]. Sound plays an essential role in liquids; sound is the density fluctuation modes of randomly localized particles assigned at the positions $\{\boldsymbol{R}_m\}$.

The second potential term,

$$\frac{1}{2}\sum_{m\neq n,ij} V_{ij}(\boldsymbol{R}_m - \boldsymbol{R}_n)(u_{mi}u_{mj} - u_{mi}u_{nj}),$$

relates to phonons [3,5–7]. Phonons play an essential role in solids; phonons are kinetic modes of particles with the displacements $\{\boldsymbol{u}_m\}$. The periodic structure in a crystal or the pair distribution function in a glass breaks the spatially translational symmetry. Phonons are the Nambu–Goldstone bosons associated with the spontaneous breakdown of the spatially translational symmetry [8–11]. In a liquid, although phonons relax soon associated with viscosity, the phonon fluctuations play an important role in the liquid–glass transition.

First we derive the Hamiltonian of the sound waves from the first potential term in Eq. (2.3.1). In order to construct the translational invariant Hamiltonian for sound, we add the constant term defined by

$$V(0) \equiv -\sum_{n'(\neq m)} V(\boldsymbol{R}_m - \boldsymbol{R}_{n'}) \tag{2.3.2}$$

to the first potential term and subtract it in the first potential. Thus we obtain the Hamiltonian for sound:

$$H_{\text{sound}} = \frac{1}{2}\sum_{m} M\dot{\boldsymbol{R}}_m^2 + \frac{1}{2}\sum_{mn}[-\delta_{mn}V(0) + V(\boldsymbol{R}_m - \boldsymbol{R}_n)]\,. \tag{2.3.3}$$

From the Hamiltonian (2.3.3), the equation of motion is given by

$$M\ddot{\boldsymbol{R}}_m = -\sum_{n}\nabla_{\boldsymbol{R}_m} V(\boldsymbol{R}_m - \boldsymbol{R}_n). \tag{2.3.4}$$

Next we investigate the second term of the potential energy in Eq. (2.3.1). In order to construct the translational invariant Hamiltonian, we define

$$V_{ij}(0) = -\sum_{n'(\neq m)} V_{ij}(\boldsymbol{R}_m - \boldsymbol{R}_{n'}) \tag{2.3.5}$$

In the harmonic approximation, we can put

$$-V_{ij}(0) = M\omega^2\delta_{ij}, \tag{2.3.6}$$

where ω is the configurationally averaged harmonic frequency, $\langle \omega_m \rangle_c$, and $M\omega^2$ is the spring constant. The second term of the potential energy is written as

$$\frac{1}{2}\sum_{m\neq n,ij} V_{ij}(\boldsymbol{R}_m-\boldsymbol{R}_n)(u_{mi}u_{mj}-u_{mi}u_{nj})$$
$$=\frac{1}{2}\sum_{mij}\left[\sum_{n'(\neq m)} V_{ij}(\boldsymbol{R}_m-\boldsymbol{R}_{n'})u_{mi}u_{mj}-\sum_{n(\neq m)} V_{ij}(\boldsymbol{R}_m-\boldsymbol{R}_n)u_{mi}u_{nj}\right]$$
$$=\frac{1}{2}\sum_{mij}\left[-V_{ij}(0)u_{mi}u_{mj}-\sum_{n(\neq m)} V_{ij}(\boldsymbol{R}_m-\boldsymbol{R}_n)u_{mi}u_{nj}\right]$$
$$=-\frac{1}{2}\sum_{mnij} V_{ij}(\boldsymbol{R}_m-\boldsymbol{R}_n)u_{mi}u_{nj}.$$

Thus we obtain the Hamiltonian for phonons:

$$H_{\text{phonon}}=\frac{1}{2}\sum_{mi} M\dot{u}_{mi}^2-\frac{1}{2}\sum_{mnij} V_{ij}(\boldsymbol{R}_m-\boldsymbol{R}_n)u_{mi}u_{nj}$$
$$=\frac{1}{2}\sum_{mi} M\dot{u}_{mi}^2+\frac{1}{2}\sum_{mij}\left[\sum_{n'(\neq m)} V_{ij}(\boldsymbol{R}_m-\boldsymbol{R}_{n'})u_{mi}u_{mj}\right.$$
$$\left.-\sum_{n(\neq m)} V_{ij}(\boldsymbol{R}_m-\boldsymbol{R}_n)u_{mi}u_{nj}\right]. \qquad (2.3.7)$$

From the Hamiltonian (2.3.7), the equation of motion is given by

$$M\ddot{u}_{mi}=\sum_{nj} V_{ij}(\boldsymbol{R}_m-\boldsymbol{R}_n)u_{nj}$$
$$=-\left[\sum_{n'(\neq m)j} V_{ij}(\boldsymbol{R}_m-\boldsymbol{R}_{n'})u_{mj}-\sum_{n(\neq m)j} V_{ij}(\boldsymbol{R}_m-\boldsymbol{R}_n)u_{nj}\right]. \qquad (2.3.8)$$

We use the Hamiltonian (2.3.3) and the equation of motion (2.3.4) in constructing sound mode, and Eqs. (2.3.7) and (2.3.8) in constructing phonon modes.

2.3.1.1 Sound and diffusivity

In order to investigate sound waves, we consider the local particle density

$$\rho(\boldsymbol{R})=\sum_{m=1}^{N}\delta(\boldsymbol{R}-\boldsymbol{R}_m), \qquad (2.3.9)$$

we introduce the following density fluctuations as the Fourier transformation of Eq. (2.3.9):

$$\rho_q = \int d\boldsymbol{R}\rho(\boldsymbol{R})e^{-\mathrm{i}\boldsymbol{q}\cdot\boldsymbol{R}} = \sum_m e^{-\mathrm{i}\boldsymbol{q}\cdot\boldsymbol{R}_m}. \tag{2.3.10}$$

We investigate the equations of motion for sound waves in the two ways:

1. The equations of motion for the density fluctuations.
2. The Canonical equations of motion.

When we investigate these equations we make use of the following

1. In Fourier transforming $V(\boldsymbol{R}_m)$ to $V(\boldsymbol{q})$, we first transform $V(\boldsymbol{R}_m)$ continuous function $V(\boldsymbol{R})$ of continuous variable $\boldsymbol{R}$ by multiplying the factor $\frac{V}{N}$. Then we obtain

$$V(\boldsymbol{R}_m) = \frac{V}{N}\frac{1}{(2\pi)^3}\int d\boldsymbol{q}\; V(\boldsymbol{q})e^{-\mathrm{i}\boldsymbol{q}\cdot\boldsymbol{R}_m} \equiv \frac{1}{N}\sum_{\boldsymbol{q}} V(\boldsymbol{q})e^{-\mathrm{i}\boldsymbol{q}\cdot\boldsymbol{R}_m},$$

$$\delta_{mn} = \frac{V}{N}\frac{1}{(2\pi)^3}\int d\boldsymbol{q}e^{\mathrm{i}\boldsymbol{q}\cdot(\boldsymbol{R}_m-\boldsymbol{R}_n)} = \frac{1}{N}\sum_{\boldsymbol{q}} e^{\mathrm{i}\boldsymbol{q}\cdot(\boldsymbol{R}_m-\boldsymbol{R}_n)}.$$

It should be noted this procedure is valid under the condition $V \to \infty$.
2. In calculating the equations of motion for sound waves, we put

$$\sum_{\boldsymbol{q}} e^{\mathrm{i}\boldsymbol{q}\cdot(\boldsymbol{R}_m-\boldsymbol{R}_n)} \cong N\delta_{\boldsymbol{R}_m,\boldsymbol{R}_n},$$

$$\sum_m e^{\mathrm{i}(\boldsymbol{q}-\boldsymbol{q}')\cdot\boldsymbol{R}_m} \cong N\delta_{\boldsymbol{q},\boldsymbol{q}'}.$$

These approximations are called random phase approximation.

Now we investigate the equations of motion for sound waves:

2.3.2 The Equations of Motion for Density Fluctuations

From the equation of motion (2.3.4), we obtain

$$\ddot{\rho}_q = \sum_m\left[(-\mathrm{i}\boldsymbol{q}\cdot\dot{\boldsymbol{R}}_m)^2 e^{-\mathrm{i}\boldsymbol{q}\cdot\boldsymbol{R}_m} + \mathrm{i}\boldsymbol{q}\cdot\sum_n\frac{1}{M}\{\nabla_{\boldsymbol{R}_m}V(\boldsymbol{R}_m-\boldsymbol{R}_n)\}e^{-\mathrm{i}\boldsymbol{q}\cdot\boldsymbol{R}_m}\right]. \tag{2.3.11}$$

In the first term, we take the configuration average of $(\boldsymbol{q}\cdot\dot{\boldsymbol{R}}_m)^2$ replaced $\dot{\boldsymbol{R}}_m$ by the average particle velocity $\boldsymbol{v}_p$, and then use the equipartition law:

$$\langle(\boldsymbol{q}\cdot\dot{\boldsymbol{R}}_m)^2\rangle_c = \langle(\boldsymbol{q}\cdot\boldsymbol{v}_p)^2\rangle_c = \frac{q^2{v_p}^2}{3} \equiv \omega_q^2 = \frac{k_{\mathrm{B}}T}{M}q^2, \tag{2.3.12}$$

In the second term, making the random phase approximation, we have

$$\sum_{mn}\{\nabla_{\boldsymbol{R}_m} V(\boldsymbol{R}_m - \boldsymbol{R}_n)\}e^{-\mathrm{i}\boldsymbol{q}\cdot\boldsymbol{R}_m} = \frac{1}{N}\sum_{mn\boldsymbol{q}'} \mathrm{i}\boldsymbol{q}' V(\boldsymbol{q}')e^{\mathrm{i}\boldsymbol{q}'\cdot(\boldsymbol{R}_m-\boldsymbol{R}_n)}e^{-\mathrm{i}\boldsymbol{q}\cdot\boldsymbol{R}_m}$$
$$\cong \sum_n \mathrm{i}\boldsymbol{q} V(\boldsymbol{q})e^{-\mathrm{i}\boldsymbol{q}\cdot\boldsymbol{R}_n} = \mathrm{i}\boldsymbol{q} V(\boldsymbol{q})\rho_{\boldsymbol{q}}. \quad (2.3.13)$$

Thus Eq. (2.3.11) leads to

$$\ddot{\rho}_{\boldsymbol{q}} \cong -q^2\frac{k_{\mathrm{B}}T}{M}\rho_{\boldsymbol{q}} - \frac{1}{M}q^2 V(q)\rho_{\boldsymbol{q}}$$
$$= -{\omega_{sq}}^2\rho_{\boldsymbol{q}}, \quad (2.3.14)$$

where

$${\omega_{sq}}^2 = \frac{1}{M}\{k_{\mathrm{B}}T + V(q)\}q^2$$
$$= \frac{k_{\mathrm{B}}T}{M}\{1 + \beta V(q)\}q^2 = \{1 + \beta V(q)\}{\omega_q}^2 = \frac{{\omega_q}^2}{S(q)}$$
$$= c_T^2 q^2, \quad (2.3.15a)$$
$$1/S(q) = 1 + \beta V(q), \quad (2.3.15b)$$
$$c_T^2 = (k_{\mathrm{B}}T + V(q))/M, \quad (2.3.15c)$$
$${\omega_q}^2 = v_p^2 q^2/3 = k_{\mathrm{B}}Tq^2/M, \quad (2.3.15d)$$

where we have identified with $V(\boldsymbol{R})$ with $V_{00}(\boldsymbol{R})$. In this section we use $f' = -1$ for the Boltzmann distribution function. $S(q)$ is the static structure factor [2,4,12]. There is the thermodynamical relations [13–16], $S(0) = \rho k_{\mathrm{B}}T\chi_T$. We clarify the properties of $S(q)$ in the latter chapter. It should be note that our system is in the isothermal process, so the sound velocity c_T is the isothermal one.

2.3.3 The Canonical Equations of Motion for Sound

Now we rewrite the Hamiltonian for sound, H_{sound}, in Eq. (2.3.3) by the density fluctuations, $\{\rho_{\boldsymbol{q}}\}$. The term $\frac{1}{2}\sum_{mn} V(\boldsymbol{R}_m - \boldsymbol{R}_n)$ can be written as

$$\frac{1}{2}\sum_{mn} V(\boldsymbol{R}_m - \boldsymbol{R}_n) = \frac{1}{2N}\sum_{mn}\sum_{\boldsymbol{q}} e^{\mathrm{i}\boldsymbol{q}\cdot(\boldsymbol{R}_m-\boldsymbol{R}_n)}V(\boldsymbol{q}) = \frac{M}{2N}\sum_{\boldsymbol{q}}\frac{V(q)}{M}\rho_{\boldsymbol{q}}^{\dagger}\rho_{\boldsymbol{q}}, \quad (2.3.16)$$

where we have put $\rho_{\boldsymbol{q}}^{\dagger} \equiv \rho_{-\boldsymbol{q}}$.

The term, $-V(0) = \sum_{n'(\neq m)} V(\boldsymbol{R}_m - \boldsymbol{R}_{n'})$ is the potential of a particle at $\boldsymbol{R}_m$, which is made up by the surrounding particles. If we consider a spherical harmonic potential, we can put

$$-V(0) = k_{\mathrm{B}}T, \quad (2.3.17)$$

which corresponds to the term $k_{\rm B}T$ in Eq. (2.3.12). The first term of the potential in Eq. (2.3.3) is written as

$$-\frac{1}{2}\sum_{mn}\delta_{mn}V(0) = \frac{1}{2N}\sum_{mnq} e^{iq\cdot(R_m-R_n)}k_{\rm B}T = \frac{M}{2N}\sum_{q}\frac{k_{\rm B}T}{M}\rho_q^\dagger\rho_q$$
$$\equiv \frac{M}{2}\sum_{q}\frac{k_{\rm B}T}{M}q^2\eta_q^\dagger\eta_q$$
$$= \frac{M}{2}\sum_{q}\omega_q^2\eta_q^\dagger\eta_q. \qquad (2.3.18)$$

Thus we obtain

$$\frac{1}{2}\sum_{mn}[-\delta_{mn}V(0) + V(R_m - R_n)] = \frac{M}{2N}\sum_{q}\frac{k_{\rm B}T + V(q)}{M}\rho_q^\dagger\rho_q$$
$$= \frac{M}{2N}\sum_{q}c_T^2\rho_q^\dagger\rho_q$$
$$= \frac{M}{2}\sum_{q}\frac{k_{\rm B}T}{M}q^2(1+\beta V(q))\eta_q^\dagger\eta_q$$
$$= \frac{M}{2}\sum_{q}\omega_q^2(1+\beta V(q))\eta_q^\dagger\eta_q$$
$$= \frac{M}{2}\sum_{q}\omega_{sq}^2\eta_q^\dagger\eta_q, \qquad (2.3.19)$$

where the second equation corresponds to the second term of the Free energy of sound waves in Eq. (2.1.22), and in the third equation we have defined the canonical coordinates as

$$\eta_q \equiv \frac{\rho_q}{q\sqrt{N}}, \quad \pi_q = M\dot{\eta}_q. \qquad (2.3.20)$$

Equation (2.3.20) represents a harmonic potential with a frequency ω_{sq}.

Next we investigate the kinetic term of Eq. (2.3.3). Since we have already known the dynamical equation for sound (2.3.14), we can expect

$$\frac{1}{2}\sum_{m}M\dot{R}_m^2 = \frac{1}{2M}\sum_{q}\pi_q^\dagger\pi_q.$$

To see this situation, we consider

$$\frac{1}{2M}\sum_{q}\pi_{q}^{\dagger}\pi_{q} = \frac{M}{2N}\sum_{q}\frac{\dot{\rho}_{q}^{\dagger}\dot{\rho}_{q}}{q^{2}} = \frac{M}{2N}\sum_{mnq}\dot{\boldsymbol{R}}_{mq}\dot{\boldsymbol{R}}_{nq}\, e^{\mathrm{i}\boldsymbol{q}\cdot(\boldsymbol{R}_m-\boldsymbol{R}_n)}$$
$$\cong \frac{M}{2N}\sum_{mnq}\dot{\boldsymbol{R}}_{m}\dot{\boldsymbol{R}}_{n}\, e^{\mathrm{i}\boldsymbol{q}\cdot(\boldsymbol{R}_m-\boldsymbol{R}_n)}$$
$$\cong \frac{M}{2}\sum_{m}\dot{\boldsymbol{R}}_{m}^{2}, \tag{2.3.21}$$

where the subscript $\boldsymbol{q}$ on $\dot{\boldsymbol{R}}_{mq}$ means the velocity $\dot{\boldsymbol{R}}_m$ has only the component parallel to $\boldsymbol{q}$. Thus we obtain

$$H_{\text{sound}} = \sum_{q}\left[\frac{1}{2M}\pi_{q}^{\dagger}\pi_{q} + \frac{M}{2}{\omega_{sq}}^{2}\eta_{q}^{\dagger}\eta_{q}\right], \tag{2.3.22}$$

where η_q is the reference coordinate and π_q is the canonical momentum, $\pi_q = \partial H_{\text{sound}}/\partial\dot{\eta}_q^{\dagger} = M\dot{\eta}_q$ satisfying Eq. (2.3.20) and the canonical equations of motion:

$$\dot{\pi}_{q} = -\frac{\partial H_{\text{sound}}}{\partial\eta_{q}^{\dagger}} = -M{\omega_{sq}}^{2}\eta_{q}, \quad \dot{\eta}_{q} = \frac{\partial H_{\text{sound}}}{\partial\pi_{q}^{\dagger}} = \frac{\pi_{q}}{M}. \tag{2.3.23}$$

Thus we obtain the dynamical equation for sound:

$$\ddot{\eta}_{q} = -\omega_{sq}^{2}\eta_{q}. \tag{2.3.24}$$

This fact shows that the Hamiltonian of sound (2.3.22) is described as a harmonic oscillator with the sound frequency ω_{sq} as the harmonic frequency.

Now we investigate the physical meaning of Eq. (2.3.17) [2]. Equations (2.3.18) and (2.3.19) say that

$$-\frac{1}{2}\sum_{mn}\delta_{mn}V(0) = \frac{M}{2}\sum_{q}{\omega_{q}}^{2}\eta_{q}^{\dagger}\eta_{q},$$
$$\frac{1}{2}\sum_{mn}[-\delta_{mn}V(0) + V(\boldsymbol{R}_m - \boldsymbol{R}_n)] = \frac{M}{2}\sum_{q}\omega_{q}^{2}(1+\beta V(q))\eta_{q}^{\dagger}\eta_{q}$$
$$= \frac{M}{2}\sum_{q}\omega_{sq}^{2}\eta_{q}^{\dagger}\eta_{q}.$$

Judging from the unified theory of the liquid–glass transition explained in Introduction, we can conclude that ω_q is an elementary excitation associated with the intra-band particle-hole pairs, η_q and $\eta_q^{\dagger}$. Thus we can regard ω_q as the frequencies of the bubbles of intra-band particle-hole pairs. As explained in Introduction, the sound wave ω_{sq} is formed as the sum of an infinite series of interacting bubbles of intra-band particle-hole pairs. The extra bubbles of intra-band particle-hole pairs not constituting the sound yield the fluctuation entropy of sound. The mixing between the sound entropy and the fluctuation entropy of sound yields the Kauzmann entropy near the glass transition temperature.

So far we have investigated the sound waves in the long wavelength regime. However, near T_g the motion of the particles rapidly slows down and finally the particles become trapped in the randomly distributed harmonic potentials made up by the surrounding particles. Thus near T_g, an m-particle stays for a long time in the harmonic potential at $\boldsymbol{R}_m$, making the harmonic oscillations associated with the displacements $\boldsymbol{u}_m$ from the assigned position $\boldsymbol{R}_m$, and seldom hops to a surrounding vacancy. The hopping magnitude of an m-particle to an n-vacancy is denoted as J_{mn}, which is proportional to the hopping amplitude J governed by the VTF law as explained in Introduction, $\hbar J \ll k_{\mathrm{B}}T$. The particle energy, $\epsilon_{\boldsymbol{p}}$ is $\epsilon_{\boldsymbol{p}} \propto J$. Since the excitations of particle-hole pairs, $(\epsilon_{\boldsymbol{p}-\boldsymbol{q}} - \epsilon_{\boldsymbol{p}}) \propto J$, the excitations of the particle-hole pairs change little with respect to $\boldsymbol{p}$ so that we can calculate the elementary excitations ω_q at any wave vector $\boldsymbol{q}$ under such circumstances. Thus near T_g, at any wave vector, the relation ${\omega_{sq}}^2 = \frac{{\omega_q}^2}{S(q)} = (1 + \beta V(q)){\omega_q}^2$ holds [2,4,12,13]. $V(q)$ has a negative minimum value at a reciprocal particle distance, K [2,12]. Thus the sound disappears at the sound disappearing temperature T_0:

$$1 + \beta_0 V(K) = 0. \tag{2.3.25}$$

It should be noted that such a regime $\hbar J \ll k_{\mathrm{B}}T$ belongs to quantum one. Therefore the precise treatments in the short wavelength regime must be done from the quantum field theory.

2.3.4 Dissipative and Relaxation Processes

Next we investigate the dissipative and relaxation processes [2]. Dissipative process originates from the simultaneously scattering processes of two particles by the same random potential. Random potentials come from random harmonic frequencies, $\{\omega_m\}$ and random hopping magnitudes, $\{J_{mn}\}$. The sum of the configurationally averaged simultaneously scattering processes over all sites yields the random scattering potentials in which two particles participate:

$$U_J \sim \langle \hat{J}_{mn} \hat{J}_{mn} \rangle_c, \quad \text{due to the random hopping magnitudes,}$$

$$U_\omega \sim \langle \hat{\omega}_m \hat{\omega}_m \rangle_c, \quad \text{due to the random harmonic frequencies,}$$

where ^, means the deviation from the average value and $\langle \cdots \rangle_c$ means the configuration average. The statistical average on a random system constrains the dynamical processes, which lead to dissipation. Since two particles simultaneously participate in the same random scattering potentials $U_J + U_\omega$, the relaxation time, τ_0, for sound is given by

$$\frac{1}{\tau_0} \propto \sqrt{(U_J + U_\omega)}. \tag{2.3.26}$$

It should be noted that such microscopically scattering processes must be treated from the quantum field theory.

If we introduce $\frac{1}{\tau_0}$ in Eq. (2.3.14), we obtain

$$\ddot{\rho}_{\boldsymbol{q}} + \frac{1}{\tau_0}\dot{\rho}_{\boldsymbol{q}} + \omega_{sq}^2 \rho_{\boldsymbol{q}} = 0. \tag{2.3.27}$$

The Fourier component $\rho_{\boldsymbol{q}}(t) = \tilde{\rho}_{\boldsymbol{q}}\, e^{-\mathrm{i}q_0 t}$ leads to

$$q_0 - \frac{\omega_{sq}^2}{q_0 + \frac{\mathrm{i}}{\tau_0}} = 0, \tag{2.3.28}$$

where q_0 is a frequency:

1. $q_0\tau_0 \gg 1$:

$$q_0^2 = \omega_{sq}^2. \tag{2.3.29}$$

Sound mode ω_{sq} is stable.

2. $q_0\tau_0 \ll 1$:

$$q_0 + \mathrm{i}\tau_0\omega_{sq} = q_0 + \mathrm{i}Dq^2 = 0, \quad D = c_s^2\tau_0. \tag{2.3.30}$$

Sound mode dissipates so that the diffusion appears. D is the diffusion coefficient.

2.3.4.1 Phonons, boson peak, and viscosity

Introducing a normal mode component $u_{mi} = u_{\boldsymbol{q}i}\, e^{\mathrm{i}\boldsymbol{q}\cdot\boldsymbol{R}_m}$ and $\boldsymbol{q} = q\hat{z}$ in the equation of motion for phonons (2.3.8):

$$\begin{aligned}
M\ddot{u}_{\lambda\boldsymbol{q}} &= \sum_n V_{ii}(\boldsymbol{R}_m - \boldsymbol{R}_n)e^{-\mathrm{i}\boldsymbol{q}\cdot(\boldsymbol{R}_m-\boldsymbol{R}_n)}u_{\lambda\boldsymbol{q}} \\
&= -\left[\sum_{n'(\neq m)} V_{ii}(\boldsymbol{R}_m - \boldsymbol{R}_{n'}) - \sum_{n(\neq m)} V_{ii}(\boldsymbol{R}_m - \boldsymbol{R}_n)e^{-\mathrm{i}\boldsymbol{q}\cdot(\boldsymbol{R}_m-\boldsymbol{R}_n)}\right]u_{\lambda\boldsymbol{q}} \\
&= -\sum_n \left[V_{ii}(\boldsymbol{R}_m - \boldsymbol{R}_n) - V_{ii}(\boldsymbol{R}_m - \boldsymbol{R}_n)e^{-\mathrm{i}\boldsymbol{q}\cdot(\boldsymbol{R}_m-\boldsymbol{R}_n)}\right]u_{\lambda\boldsymbol{q}} \\
&= -\sum_n V_{ii}(\boldsymbol{R}_m - \boldsymbol{R}_n)\left[1 - e^{-\mathrm{i}\boldsymbol{q}\cdot(\boldsymbol{R}_m-\boldsymbol{R}_n)}\right]u_{\lambda\boldsymbol{q}} \\
&= -\omega_{\lambda\boldsymbol{q}}^2 u_{\lambda\boldsymbol{q}},
\end{aligned} \tag{2.3.31}$$

where from the second to the third line we have considered that the term $V_{ii}(0)$ in the first and second terms of the third equation cancels. Here $\lambda = i$, where $i = z$ corresponds to a longitudinal mode, and $i = x, y$ to transverse modes. Taking the configuration average, we obtain

$$\begin{aligned}
{\omega_{\lambda\boldsymbol{q}}}^2 &= \frac{1}{M}\left\langle \sum_n V_{ii}(\boldsymbol{R}_m - \boldsymbol{R}_n)\left[1 - e^{-\mathrm{i}\boldsymbol{q}\cdot(\boldsymbol{R}_m-\boldsymbol{R}_n)}\right]\right\rangle_c \\
&= \frac{1}{M}\int d^3R\, g(R)\left(1 - e^{-\mathrm{i}\boldsymbol{q}\cdot\boldsymbol{R}}\right)V_{ii}(R),
\end{aligned} \tag{2.3.32}$$

and similarly, we obtain

$$\omega^2 = \frac{1}{M}\left\langle \sum_n V_{ii}(\boldsymbol{R}_m - \boldsymbol{R}_n) \right\rangle_c = \frac{1}{M}\int d^3R\, g(R) V_{ii}(R), \tag{2.3.33}$$

where

$$\begin{aligned} {\omega_{lq}}^2 &= \frac{4\pi}{M}\int R^2\, dR g(R) V''(R) l(\alpha), \\ l(\alpha) &= \frac{1}{3} - \left(\frac{\sin\alpha}{\alpha} + \frac{2\cos\alpha}{\alpha^2} - \frac{2\sin\alpha}{\alpha^3} \right), \end{aligned} \tag{2.3.34}$$

$$\begin{aligned} {\omega_{tq}}^2 &= \frac{4\pi}{M}\int R^2\, dR g(R) V''(R) t(\alpha), \\ t(\alpha) &= \frac{1}{3} - \left(\frac{-2\cos\alpha}{\alpha^2} + \frac{2\sin\alpha}{\alpha^3} \right), \end{aligned} \tag{2.3.35}$$

$$\omega^2 == \frac{4\pi}{3M}\int R^2\, dR g(R) V''(R). \tag{2.3.36}$$

Here $g(R)$ is the pair distribution function and $\alpha = qR$. $\lambda = l$ means a longitudinal component for $i = z$ and $\lambda = t$ transverse components for $i = x, y$. We have put

$$V_{ij}(R) = \frac{x_i x_j}{R^2}(V''(R) - \frac{1}{R}V'(R)) + \frac{\delta_{ij}}{R}V'(R) \cong \frac{x_i x_j}{R^2}V''(R), \tag{2.3.37}$$

where $'$ means the derivative with respect to the argument.

The phonon dispersion curves are illustrated in Figure 1.7 [2], where we have put $g(R) \cong \rho\delta(R - a)$. a is the average particle distance. The phonons are the Nambu–Goldstone bosons associated with the structure which is represented by the pair distribution function in a glass. The longitudinal dispersion curve has a roton type minimum near the reciprocal particle distance [2, 17, 18], which reflects the pair distribution function. The roton type minimum is a vestige of the periodic phonon dispersion curves in the crystal with a periodic distribution function.

Now we investigate the two limiting cases:

1. $\alpha \ll 1$:

$$l(\alpha) \longrightarrow \frac{\alpha^2}{10} + \cdots, \quad t(\alpha) \longrightarrow \frac{\alpha^2}{30} + \cdots \tag{2.3.38}$$

In the long wavelength regime, phonon velocity c_λ is defined as $\omega_{\lambda q} \equiv c_\lambda q$.

2. $\alpha \gg 1$:

$$\omega_{l,tq} \longrightarrow \omega. \tag{2.3.39}$$

In the short wavelength regime, the phonon dispersion curves tend to the frequency ω. These flat phonon dispersion curves mean the localization of phonons and make a peak of the density states. This is the boson peak. It should be noted that the density of states is inversely proportional to the q-derivative of the phonon dispersion curves.

Next we investigate the Hamiltonian equation. If we put $u_{mi} = \sum_q u_{qi}\, e^{iq\cdot R_m}$ in Eq. (2.3.7), we obtain

$$H_{\text{phonon}} = \sum_{\lambda q}\left[\frac{1}{2}M\dot{u}^{\dagger}_{q\lambda}\dot{u}_{q\lambda} + \frac{1}{2}M\omega_{\lambda q}{}^{2}u^{\dagger}_{q\lambda}u_{q\lambda}\right]$$
$$= \sum_{\lambda q}\left[\frac{1}{2M}p^{\dagger}_{q\lambda}p_{q\lambda} + \frac{1}{2}M\omega_{\lambda q}{}^{2}u^{\dagger}_{q\lambda}u_{q\lambda}\right]. \tag{2.3.40}$$

This is the well-known Hamiltonian for phonons, where we have taken $u_{q\lambda}$ as a reference coordinate and its momentum $p_{q\lambda} = \partial H_{\text{phonon}}/\partial \dot{u}^{\dagger}_{q\lambda} = M\dot{u}_{q\lambda}$.

Here we investigate the physical meaning of Eq. (2.3.6) [2]. The first potential in Eq. (2.3.7) is written as

$$-V_{ij}(0)u_{mi}u_{mi} = \sum_{n'(\neq m)} V_{ij}(\boldsymbol{R}_m - \boldsymbol{R}_{n'})u_{mi}u_{mj} = M\omega^2 u^{\dagger}_{q\lambda}u_{q\lambda}. \tag{2.3.41}$$

Judging from the unified theory of the liquid–glass transition as explained in Introduction, we conclude that the frequency ω is an elementary excitations associated with the inter-band particle-hole pairs, $u_{q\lambda}$ and $u^{\dagger}_{q\lambda}$. Thus we can regard ω as the frequency of the bubbles of inter-band particle-hole pairs. As explained in Introduction, the phonon $\omega_{\lambda q}$ is formed as the sum of an infinite series of the interacting bubbles of inter-band particle-hole pairs. The extra bubbles of inter-band particle-hole pairs not constituting the phonons yield the fluctuation entropy of phonons.

The dissipation and the relaxation processes occur in the similar manner to the sound. We introduce the Maxwell relaxation time, τ_M [2,19], to Eq. (2.3.31) as

$$\ddot{u}_{q\lambda} + \frac{1}{\tau_M}\dot{u}_{q\lambda} + \omega^2_{\lambda q}u_{q\lambda} = 0, \quad \frac{1}{\tau_M} \propto \frac{1}{\tau_0}. \tag{2.3.42}$$

The Fourier component $u_{q\lambda}(t) = \tilde{u}_{q\lambda}e^{-iq_0 t}$ leads to

$$q_0 - \frac{\omega_{\lambda q}}{q_0 + \frac{i}{\tau_M}} = 0, \tag{2.3.43}$$

1. $q_0\tau_M \gg 1$:

$$q_0^2 = \omega_{\lambda q}{}^2. \tag{2.3.44}$$

Phonon modes are stable.

2. $q_0\tau_M \ll 1$:

$$q_0 + i\tau_M\omega_{\lambda q} = q_0 + i\eta_\lambda q^2 = 0, \quad \eta_\lambda = \tau_M c_\lambda^2, \tag{2.3.45}$$

where we are concerned with the long wavelength regime. Phonon modes dissipate so that the viscosity appears.

2.4 The Kauzmann Entropy, the Vogel–Tamman–Fulcher Law and Specific Heat

Now we investigate the entropies for dynamical and dissipative system [2]. First, we consider the sound system with the Hamiltonian (2.3.22). As explained in Introduction, the sound wave, ω_{sq} is formed as the sum of an infinite series of interacting bubbles of intra-band particle-hole pairs with elementary excitations, ω_q. The extra bubbles not constituting the sound yield the fluctuation entropy of the sound, which lowers the entropy of the system. However, the dissipation due to the diffusion compensates the negative fluctuation entropy so that the system becomes a local equilibrium. Thus when we investigate the entropy for the sound system, we must consider three entropies: sound, sound fluctuations, and dissipation. In the local equilibrium, the dissipation entropy compensates the sound fluctuation entropy. But in the supercooling process, since enormous bubbles are created so that the sound fluctuation entropy plays an essential role in the liquid–glass transition.

Since sound depends on temperatures, to take into account temperature dependence correctly, we must start with the Free energy. The classical partition function for sound is given by

$$Z_{\text{sound}}^{\text{cl}} = \prod_q \left[\int e^{-\beta H_{\text{sound}}} d\sqrt{\pi_q^\dagger \pi_q} d\sqrt{\eta_q^\dagger \eta_q} \frac{1}{\hbar} \right] = \prod_q \frac{1}{\beta\hbar\omega_{sq}}. \tag{2.3.46}$$

Since the classical condition, $\beta\hbar\omega_{sq} \ll 1$ is satisfied, the quantum partition function reduces to the classical one:

$$Z_{\text{sound}}^{\text{qn}} = \prod_q (1 - e^{-\beta\hbar\omega_{sq}})^{-1} \longrightarrow Z_{\text{sound}}^{\text{cl}} = \prod_q \frac{1}{\beta\hbar\omega_{sq}}, \quad \text{for } \beta\hbar\omega_{sq} \ll 1.$$

The classical Free energy is given by

$$F_{\text{sound}} = k_B T \sum_q \ln[\beta\hbar\omega_{sq}]. \tag{2.3.47}$$

Similarly, we can obtain the Free energy for the sound fluctuations:

$$F_{\text{fluctuation}} = -k_B T \sum_q \ln[\beta\hbar\omega_q]. \tag{2.3.48}$$

Thus we obtain the Free energy of the sound system denoted as F_{intra}:

$$\begin{aligned} F_{\text{intra}} &= F_{\text{sound}} + F_{\text{fluctuation}} + F_{\text{dissipation}} \\ &= k_B T \sum_q \ln[\beta\hbar\omega_{sq}] - k_B T \sum_q \ln[\beta\hbar\omega_q] + F_{\text{dissipation}}. \end{aligned} \tag{2.3.49}$$

From the sound disappearing condition Eq. (2.3.25), the dominant contribution of the temperature dependence to the Free energy of the first and second ones comes from the regions $q \sim K$, where

$$\ln[\beta\hbar\omega_{sq}] - \ln[\beta\hbar\omega_q] = \frac{1}{2} \ln[1 + \beta V(q)].$$

If we denote the number of $\boldsymbol{q}$-states near K by N_0, we obtain

$$F_{\text{intra}} \cong \frac{N_0 k_B T}{2} \ln\left[1 - \frac{T_0}{T}\right] + k_B T \sum_{q \notin N_0} \ln\left[\beta\hbar\omega_{sq}\right], \tag{2.3.50}$$

where $\boldsymbol{q} \notin N_0$ means the exclusion of the N_0 states near $q \sim K$; the number of the $\boldsymbol{q}$-states in $\boldsymbol{q} \notin N_0$ is about $N - N_0$. Note that $F_{\text{dissipation}}$ compensates the remaining $F_{\text{fluctuation}}$ with the states $N - N_0$. Thus we obtain the entropy for the intra-system denoted by S_{intra}

$$S_{\text{intra}} = S_K + S'_{\text{sound}}, \tag{2.3.51}$$

$$S_K \cong \frac{N_0 k_B}{2} \frac{T_0}{T - T_0}, \tag{2.3.52}$$

$$S'_{\text{sound}} \cong -k_B \sum_{q \notin N_0} \ln[\beta\hbar\omega_{sq}] + k_B[N - N_0]. \tag{2.3.53}$$

We call S_K the Kauzmann entropy.

A state of N-particles distributed randomly in space corresponds to a minimum of N-particle potential in a configuration space in the energy landscape model (ELM) [20–23]. A hopping of a particle from a site to a vacancy corresponds to a jump from a deep valley to another deep valley in the multi-dimensional configuration space in ELM as the α-relaxation process. The successive hopping constitutes a configuration space. A hopping generates the elementary excitations and the successive hopping yields the Kauzmann entropy S_K. The hopping probability is proportional to the configuration number, which is e^{S_K/k_B} from the Einstein relation [15,16]. The hopping probability per particle is given by the hopping amplitude:

$$J = e^{zS_K/Nk_B} = e^{-E/(T-T_0)}, \tag{2.3.54}$$

where $E = zN_0T_0/2N$ and z is of the order of the number of the surrounding particles. This equation manifests the VTF law. Thus we can put

$$v_p, c_s, \sqrt{U_J} \propto J. \tag{2.3.55}$$

It should be noted that the hopping J generates the elementary excitations, the sound ω_{sq}, and the Kauzmann entropy S_K, and S_K determines J in Eq. (2.3.54). Thus ω_{sq} is self-consistently determined.

Next we investigate the system with the Hamiltonian (2.3.40). As surveyed in Introduction, the phonons $\omega_{\lambda q}$ are formed as the sum of an infinite series of interacting bubbles of inter-band particle-hole pairs with elementary excitations, ω in Eq. (2.3.36). The extra bubbles not constituting the phonons yield the entropy of phonon fluctuations, which lowers the entropy of the system. However, the dissipation due to the viscosity compensates the negative fluctuation entropy so that the system becomes to a local equilibrium. Thus we must consider three entropies for the entropy of the phonon system denoted as S_{inter}: phonons, phonon fluctuations, and dissipation. In the local equilibrium, the dissipation entropy compensates the phonon fluctuation entropy.

In calculating the entropy of phonon system, near the liquid–glass transition we should note the condition, $\beta\hbar\omega > 1$ so that we cannot use the classical partition function Z^{cl}_{phonon}, which leads to the constant entropy:

$$Z^{cl}_{phonon} = \prod_{q}[\beta\hbar\omega_{\lambda q}] \longrightarrow S^{cl}_{phonon} = 3Nk_B.$$

We must calculate the entropy for the system quantum statistically. Thus we obtain

$$S_{inter} = S_{phonon} + S_{\Omega} + S_{dissipation}. \tag{2.3.56}$$

$$S_{phonon} = -k_B \sum_{\lambda q} \ln(1 - e^{-\beta\hbar\omega_{\lambda q}}) + k_B \sum_{\lambda q} \frac{\beta\hbar\omega_{\lambda q}}{e^{\beta\hbar\omega_{\lambda q}} - 1}, \tag{2.3.57}$$

$$S_{\Omega} = 3Nk_B \ln(1 - e^{-\beta\hbar\omega}) - 3Nk_B \frac{\beta\hbar\omega}{e^{\beta\hbar\omega} - 1}, \tag{2.3.58}$$

where S_{phonon}, S_{Ω}, and $S_{dissipation}$ are the entropies of phonon, phonon fluctuations, and phonon dissipation, respectively.

An oscillation at a site corresponds to a jump of a shallow valley in the multi-dimensional configuration space in the ELM as the β-relaxation process. The successive up and down transitions constitute another configuration space different from hoppings. The probability of the magnitude of the randomness of harmonic frequencies is proportional to the configuration number e^{S_{Ω}/k_B}. The probability of the magnitude of randomness of harmonic frequencies per particle is proportional to

$$\Omega = e^{zS_{\Omega}/Nk_B} = \exp\left[\frac{-3z\beta\hbar\omega}{e^{\beta\hbar\omega} - 1}\right]. \tag{2.3.59}$$

Thus we obtain

$$\sqrt{U_{\omega}} \propto \Omega. \tag{2.3.60}$$

Under the supercooling processes, the fluctuations are alive. The temperature dependence of S_K and S_{Ω} is illustrated in Figure 1.8.

As the temperature decreases, the Kauzmann entropy S_K rapidly drops to negative infinity near T_0 so that the system seems to occur the entropy crisis. But as the temperature decreases, the negative phonon fluctuation entropy S_{Ω} increases and at last crosses over S_K above T_0 and prevents the entropy crisis. The Kauzmann entropy S_K dominates above the crossover temperature, but the phonon fluctuation entropy S_{Ω} dominates below the crossover temperature. Thus we can identify the crossover temperature with the liquid–glass transition temperature T_g:

$$S_K = S_{\Omega}|_{T=T_g}. \tag{2.3.61}$$

The sound is a density fluctuation wave essential in liquids, while the phonons are elastic waves essential in solids. The glass transition is a sort of dynamical transition.

Next we investigate the specific heat. We obtain the specific heats, C_{K} for the Kauzmann entropy S_{K}, C_{sound} for the sound entropy S'_{K} and C_{phonon} for the phonon entropy S_{phonon}:

$$C_{\mathrm{K}} = \begin{cases} \frac{N_0 k_{\mathrm{B}}}{2} \frac{T T_0}{(T-T_0)^2} & \text{for } T_g < T, \\ 0 & \text{for } T < T_g. \end{cases} \tag{2.3.62}$$

$$C_{\mathrm{sound}} \cong k_{\mathrm{B}}(N - N_0), \tag{2.3.63}$$

$$C_{\mathrm{phonon}} = \frac{k_{\mathrm{B}}}{4} \sum_{\lambda \boldsymbol{q}} \frac{(\beta\hbar\omega_{\lambda \boldsymbol{q}})^2}{\sinh^2(\beta\hbar\omega_{\lambda \boldsymbol{q}}/2)}. \tag{2.3.64}$$

Note that the other fluctuation entropies but the Kauzmann entropy, are compensated by the respective dissipative entropies in the local equilibrium.

Next we summarize the VTF law for the relaxation times and transport coefficients:

1. $T_0 < T$:

$$v_p, c_T \propto J, \quad c_\lambda \cong \text{constant}. \tag{2.3.65}$$

2. $T_0 < T_g < T$: The term $\sqrt{U_J} \propto J$ dominates; the α-relaxation. We obtain

$$\tau_0^{-1}, \tau_M^{-1}, D^{-1}, \eta_\lambda \propto J. \tag{2.3.66}$$

   Note that $c_T \propto J$, while $c_\lambda \cong$ constant. c_T plays an essential role in the diffusivity D. $\eta_\lambda D =$ constant satisfies the Stokes law.

3. $T_0 < T < T_g$: The term $\sqrt{U_\omega} \propto \Omega$ dominates; the β-relaxation. Thus we obtain

$$\tau_0^{-1}, \tau_M^{-1} \propto \Omega, \quad D^{-1} \propto J^2, \quad \eta_\lambda \propto \Omega^{-1}. \tag{2.3.67}$$

These results are the same in the unified theory of the liquid–glass transition.

2.4.1 The Limitations of the Microscopic Classical Theory and the Roles of Quantum Statistics

We have approached to the unified theory of the liquid–glass transition from the microscopic classical theory. Since the liquid–glass transition are prepared in the liquid state, sound, and the extra bubbles of intra-band and inter-band particle-hole pairs not constituting sound and phonons play an essential role in the liquid–glass transition. However, the microscopic classical theory of sound is only valid at higher temperatures $\beta\hbar\omega < 1$ and in the long wavelength regime. Near the liquid–glass transition temperature, $\beta\hbar\omega \geqslant 1$, the classical theory of sound inevitably faces some flaws, where phonons, which must be treated quantum statistically, appear. Here we point out the limitations of the microscopic classical theory in the liquid–glass transition:

1. Near the liquid–glass transition T_g, $\beta\hbar\omega \geqslant 1$, the system dynamically slows down so that particles are trapped in the localized potentials, where the energy bands of particles become narrower as the temperature decreases. This narrowing bands allow us to extend the dynamical equation of sound to the short wavelength regime. But this narrowing process of bands must be treated quantum statistically.

2. Sound and phonons are collective excitations constructed from the sum of an infinite series of interacting bubbles of intra-band and inter-band particle-hole pairs, respectively. The fluctuation entropies of sound and phonons originate from the extra bubbles of intra-band and inter-band particle-hole pairs not constituting the sound and phonons, respectively. The microscopic classical theory cannot calculate such bubbles of intra-band and inter-band particle-hole pairs. In having driven the fluctuation entropies of sound and phonons, since we have known the unified theory of the liquid–glass transition, we could identify $-\frac{1}{2}\sum_{mn}\delta_{mn}V(0)$ for sound and $-V_{ii}(0)u_{mi}u_{mi}$ for phonons with the frequencies of the bubbles of the intra-band and inter-band particle-hole pairs, respectively. The actual frequencies must be calculated by quantum statistically.
3. The relaxation and dissipative processes have been treated phenomenologically by introducing the respective relaxation times to the dynamical equations for sound and for phonons. The more precise treatment of the relaxation times and the dynamical dissipative equations requires the method of the quantum field theory.
4. The phonon fluctuation entropy plays an important role in determining the liquid–glass transition temperature T_g. The frequency of the phonon fluctuations is the harmonic frequency ω. Near T_g, since $\beta\hbar\omega \geqslant 1$, we cannot apply the classical statistical mechanics to the liquid–glass transition.

References

[1] L.D. Landau, E.M. Lifshitz, Fluid Mechanics, Pergamon Press, London, 1963.
[2] T. Kitamura, Phys. Rep. 383 (2003) 1.
[3] L.D. Landau, E.M. Lifshitz, Theory of Elasticity, Pergamon Press, London, 1970.
[4] K.S. Singwi, K. Sköld, M.P. Tosi, Phys. Rev. A 1 (1970) 454.
[5] M. Born, K. Haung, Dynamical Theory of Crystal Lattices, Oxford University Press, Oxford, 1954.
[6] A.A. Maradudin, E.W. Montroll, G.H. Weiss, I.P. Ipatova, Solid State, Physics Supplement, vol. 3, second ed., 1971.
[7] G.K. Horton, A.A. Maradudin (Eds.), Dynamical Properties of Solids, vol. 1, North-Holland, New York, 1974.
[8] Y. Nambu, G. Jona-Lasinio, Phys. Rev. 122 (1961) 345.
[9] Y. Nambu, G. Jona-Lasinio, Phys. Rev. 124 (1961) 246.
[10] J. Goldstone, Nuovo Cim. 19 (1961) 154.
[11] J. Goldstone, A. Salam, S. Weinberg, Phys. Rev. 127 (1962) 965.
[12] T. Schneider, R. Brout, H. Thomas, J. Feder, Phys. Rev. Lett. 25 (1970) 1423.
[13] J.P. Hansen, I.R. MacDonald, Theory of Simple Liquids, second ed., Academic Press, London, 1990.
[14] R. Kubo, Statistical Mechanics, North-Holland, 1965.
[15] L.D. Landau, Statistical Physics, Pergamon Press, London, 1958.
[16] D. Kondepudi, I. Prigogine, Modern Thermodynamics, John Wiley and Sons, New York, 1998.
[17] S. Takeno, Prog. Theor. Phys. Suppl. 45 (1970) 137.
[18] S. Takeno, M. Goda, Prog. Theor. Phys. 45 (1971) 331.
[19] M.C. Phillips, A.J. Barlow, J. Lamb, Proc. Roy. Soc. London A 329 (1972) 193.

[20] F.H. Stillinger, Science 267 (1995) 1935.
[21] T. Keyes, Phys. Rev. E 59 (1999) 3207.
[22] P.G. Debenedetti, F.H. Stillinger, Nature 410 (2001) 259.
[23] L.-M. Martinez, C.A. Angell, Nature 410 (2001) 663.

3 Fundamentals of Quantum Field Theory

Conventionally, theories of the liquid-glass transition have been developed from the classical statistical mechanics. The classical theory have not substantially clarified the supercooling process associated with the universal features such as the Kauzmann entropy crisis and the VTF law. This is because the classical theory does not cover the quantum regime, i.e. the short wavelength and the low temperature regimes; the classical theory cannot explain the microscopic mechanism such as the bubbles of particle-hole pairs and the scattering processes, and the low temperature statistical mechanics. On the other hand, the two band model based on the quantum field theory at finite temperature has provided us with the unified theory of the liquid-glass transition. The quantum field theory at finite temperature plays a substantial role in the unified theory of the liquid-glass transition.

In this book, using the method of the quantum field theory at finite temperature, we develop the unified theory of the liquid-glass transition. In order to understand the unified theory, the following two things are required: (1) to understand the fundamentals of the quantum field theory, and (2) to master the technique of the finite temperature Green's function method. In this chapter we explain the fundamentals of quantum field theory [1–3]. Especially, it is essential to understand the concept of free physical fields. In the next chapter, we develop the finite temperature Green's function method for quantum statistics of many particle systems as the tools to construct the unified theory of the liquid-glass transition.

3.1 The Number Representation and the Fock Space

Quantum field theory is constructed by *states* of particles in a space and *operators* acting on the states. The operators are governed by the Heisenberg equation. A state is described by the number of particles at the state. An operator is described by annihilating and creating a particle at a state. Therefore we first construct the states in the number representation of particles and, the annihilation and creation operators of particles.

3.1.1 The Number Representation

We start with particles in a one component system. The state of a single particle is represented by a wave function $\psi(x)$, which is normalizable

$$\int d^3x|\psi(x)|^2 < \infty. \tag{3.1.1}$$

Liquid Glass Transition. DOI: http://dx.doi.org/10.1016/B978-0-12-407177-3.00003-9

In this book, the argument x is represented by $x \equiv (t, \boldsymbol{x})$. The normalizable wave function $\psi(x)$ is a wave packet not a plane wave. The wave packet means some spatially localized wave function.

By using an orthonormalized complete set $\{\varphi_i(\boldsymbol{x})\}$, any normalizable wave function $\psi(x)$ can be expanded as

$$\psi(x) = \sum_i c_i \varphi_i(\boldsymbol{x}), \tag{3.1.2}$$

where

$$\int d^3x \varphi_i^*(\boldsymbol{x}) \varphi_j(\boldsymbol{x}) = \delta_{ij}, \tag{3.1.3}$$

and if the particle has spin, the index i can have spin states. c_i can depend on time t. A well-known example of an orthonormalized complete set $\{\varphi_i(\boldsymbol{x})\}$ is given by the wave functions of a harmonic oscillator.

Now we consider the state of a many particle system. The state is represented by a function of the states of the constituent particles:

$$\psi(q_a, q_b, q_c, \ldots),$$

where q_a is the state of the first particle, q_b the second particle, and so on. Since all particles are the same, we cannot distinguish individual particles with the same state, so we transform $\psi(q_a, q_b, q_c, \ldots)$ to another function of the number of particles occupying the ith state, n_i:

$$\psi(n_1, n_2, n_3, \ldots).$$

We denote this state as

$$\psi(n_1, n_2.n_3, \ldots) \equiv |n_1, n_2, n_3, \ldots\rangle. \tag{3.1.4}$$

Assembling all of these states, we construct the set $\{|n_1, n_2, n_3, \ldots\rangle\}$. According to the symmetry of the wave functions, if a wave function is unchanged under the exchange of the states, the particles are called bosons. Then n_i can take any natural number, $n_i = 0, 1, 2, \ldots$. If a wave function is changed to that with a negative sign, the particles are called fermions. Then n_i can only take $n_i = 0$ or 1.

We investigate the features of the set $\{|n_1, n_2, \ldots, n_i, \ldots\rangle\}$.

1. First, this set is non-countable. To see this non-countability, we consider a fermion system, where the particle number takes n_i=0 or 1. We apply this fermion system to the binary number system. We consider that a state $|n_1, n_2, \ldots, n_i, \ldots\rangle$ one-to-one corresponds to a real number $0.n_1 n_2 \cdots n_i \cdots$ between 0 and 1:

$$|n_1, n_2, \ldots, n_i, \ldots\rangle \longleftrightarrow 0.n_1 n_2 \ldots n_i \ldots. \tag{3.1.5}$$

Then, the number $0.n_1 n_2 \ldots n_i \ldots$ one-to-one corresponds to a point in the interval $[0, 1]$ on the real number axis. The number of points in the interval $[0, 1]$ is not countable. Thus we conclude that the set $\{|n_1, n_2, \ldots, n_i, \ldots\rangle\}$ is not countable. It is obvious that this conclusion also holds true in the case of a boson system. The ordinary quantum mechanics is constructed in the separable Hilbert space. A separable Hilbert space means that the Hilbert space contains a countable basis $\{\xi_i\}$, by the linear combination of which any vector in the Hilbert space, ξ, can be approximated to any accuracy, i.e. for any ϵ there exists a sequence $\{c_i\}$ such that

$$|\xi - \sum_i c_i \xi_i| < \epsilon \qquad \text{for arbitrary } \epsilon. \tag{3.1.6}$$

In quantum field theory the set $\{|n_1, n_2, \ldots, n_i, \ldots\rangle\}$ is not countable, To construct a separable Hilbert space by means of members of the set $\{|n_1, n_2, \ldots, n_i, \ldots\rangle\}$, we must select a countable subset for the base of the Hilbert space. However, there are infinitely many ways of selecting countable subsets.

2. Two representations in different subsets are unitarily inequivalent to each other. If two different subsets can be used as base of representations for the operators; $\{a_i, a_i^\dagger\}$ and $\{\alpha_i, \alpha_i^\dagger\}$ in the set $\{|n_1, n_2, \ldots, n_i, \ldots\rangle\}$, these two representations are unitarily inequivalent to each other, which means that a vector of one representation for the operators $\{a_i, a_i^\dagger\}$ is not expressed by a superposition of base vectors of another representation for the operators $\{\alpha_i, \alpha_i^\dagger\}$. In quantum mechanics, we can express a vector of one representation by a superposition of base vectors of another representation, because all possible representations are unitary equivalent. This situation disappears in the case of quantum field theory. We show an example in the next section.

3.1.2 The Fock Space

In order to construct quantum field theory countable, we consider the following subset denoted by [0]-set with a finite number of particles:

$$[0]\text{-set} = \{|n_1, n_2, \ldots\rangle; \sum_i n_i = \text{finite}\}. \tag{3.1.7}$$

In actual experiments, only a finite number of quanta is excited, although the number can be arbitrally large. Therefore [0]-set is sufficient for actual physical processes.

The [0]-set does not contain infinite states which had been in the original set $\{|n_1, n_2, \ldots, n_i, \ldots\rangle\}$. On the other hand, this set contains the state of no particle; $n_i = 0$ for all i. This state is called the vacuum denoted by $|0\rangle$:

$$|0>= |0, 0, \ldots\rangle. \tag{3.1.8}$$

Now we show that the [0]-set is countable. Since $\sum_i n_i =$ finite in any member of the [0]-set, we can find an integer m such that $n_i = 0$ for $i > m$ and $n_m \neq 0$. For a given m, we can assign a second s such that $\sum_{i \leq m} = s$. For example in a boson system, for $m = 3, s = 1 : |100\rangle, |010\rangle, |001\rangle$. $s = 2 : |200\rangle.|110\rangle, |101\rangle, |020\rangle, |011\rangle, |002\rangle$. $s = 3 : |300\rangle, |210\rangle, \cdots$ and so on, where we have omitted 0s more than third row in the ket vector; for example, $|001\rangle$ means $|001000\ldots\rangle$. From Eq. (3.1.7) for the [0]-set, s is finite so that we can count the[0]-set and we can order the [0]-set represented as

$$\xi_i \quad i = 1, 2, \ldots. \tag{3.1.9}$$

For above example, $i = 1$: $m = 0$. $i = 2$: $m = 1, s = 1, i = 3$: $m = 1, s = 2$, and so on.

First we determine the measure of the amplitudes of vectors, $\{|n_1, n_2, \ldots\rangle\}$. To do this we introduce the conjugate of vectors, $\{\langle n_1, n_2, \ldots|\}$ and define the inner product as

$$\langle n_1', n_2', \ldots|n_1, n_2, \ldots\rangle = \prod_i \delta_{n_i n_i'}, \tag{3.1.10}$$

The conjugate vector of the vacuum is defined as

$$\langle 0|0\rangle = 1. \tag{3.1.11}$$

If we use the notation ξ_i in Eq. (3.1.9), the orthonormalization relation (3.1.10) reads simply as

$$(\xi_i \cdot \xi_j) = \delta_{ij}. \tag{3.1.12}$$

If we expand any two vectors using the orthonormalized bases $\{\xi_i\}$ as

$$\xi = \sum_i c_i \xi_i, \qquad \xi' = \sum_i c_i' \xi_i, \tag{3.1.13}$$

we obtain the inner product as

$$\begin{aligned}(\xi \cdot \xi') &= \sum_{ij} c_i^* c_j (\xi_i \cdot \xi_j) \\ &= \sum_i c_i^* c_i'.\end{aligned} \tag{3.1.14}$$

The norm of a vector ξ is defined by $(\xi \cdot \xi)^{1/2}$ denoted as $|\xi|$.

Now we define the Fock space. The Fock space is defined by the separable Hilbert space span by an orthonormalized bases $\{\xi_i\}$ as

$$\mathcal{H}[a] = \left\{ \xi = \sum_{i=1}^{\infty} c_i \xi_i; \ \sum_{i=1}^{\infty} |c_i|^2 = \text{finite} \right\}, \tag{3.1.15}$$

where we define $\xi = \sum_{i=1}^{\infty} c_i \xi_i$ as the limit, $N \to \infty$ of the vector series $\xi^N = \sum_{i=1}^{N} c_i \xi_i$, which is a Cauchy sequence under the condition $\sum_{i=1}^{N} |c_i|^2 = \text{finite}$. This separable Hilbert space is dence. Dence means that the norm $|\xi^N - \xi^{N'}|$ can be made smaller than an arbitrary small ϵ by choosing suitable large N and N', i.e. for any ϵ there are large enough numbers N and N':

$$|\xi^N - \xi^{N'}| < \epsilon \qquad \text{for arbitrary } \epsilon. \tag{3.1.16}$$

Intuitively dence means that the Fock space is almost covered by the set $\{\xi^N\}$ for a large enough N. For example, the set of rational number is dense in the set of real numbers.

3.1.3 Creation and Annihilation Operators; the Commutation Relations

Now we construct the vectors $\{|n_1, n_2, \dots\rangle\}$ with the operators, we introduce commutation and anti-commutation relations for the annihilation operators, α_i and creation operators $\alpha_j^\dagger$:

$$[\alpha_i, \alpha_j^\dagger]_\mp = \delta_{ij}, \tag{3.1.17}$$

$$[\alpha_i, \alpha_j]_\mp = [\alpha_i^\dagger, \alpha_j^\dagger]_\mp = 0, \tag{3.1.18}$$

where commutation and anti-commutation relations are defined by

$$[A, B]_\mp = AB \mp BA. \tag{3.1.19}$$

The commutation relations are associated with boson systems, while the anti-commutation relations with fermion systems.

3.1.3.1 Bosons

We investigate boson systems. We first note that since the properties of the commutation relations, the order of the operator for different indices are changeable:

$$F(\alpha_i, \alpha_i^\dagger)G(\alpha_j, \alpha_j^\dagger) = G(\alpha_j, \alpha_j^\dagger)F(\alpha_i, \alpha_i^\dagger).$$

We introduce the number operator:

$$N_i = \alpha_i^\dagger \alpha_i. \tag{3.1.20}$$

We can see that the operator N_i is the number operator as follows:
Since N_i is an hermitian, we have an eigenvalue and eigenvector as

$$N_i|n_1, \dots, n_i, \dots\rangle = n_i|n_1, \dots, n_i, \dots\rangle. \tag{3.1.21}$$

The following relations:

$$[N_i, \alpha_i] = -\alpha_i, \tag{3.1.22}$$

$$[N_i, \alpha_i^\dagger] = \alpha_i^\dagger, \tag{3.1.23}$$

show α_i is the lowering operator and $\alpha_i^\dagger$ raising operator such that

$$\begin{aligned} N_i\alpha_i|n_1, \cdots, n_i, \dots\rangle &= (\alpha_i N_i - \alpha_i)|n_1, \dots, n_i, \dots\rangle \\ &= (n_i - 1)\alpha_i|n_1, \dots, n_i, \dots\rangle, \end{aligned} \tag{3.1.24}$$

$$\begin{aligned} N_i\alpha_i^\dagger|n_1, \dots, n_i, \cdots\rangle &= (\alpha_i^\dagger N_i + \alpha_i^\dagger|n_1, \dots, n_i, \dots\rangle \\ &= (n_i + 1)\alpha_i^\dagger|n_1, \dots, n_i, \dots\rangle. \end{aligned} \tag{3.1.25}$$

Repeating the above process, we obtain

$$N_i \underbrace{\alpha_i \cdots \alpha_i}_{m_i} |n_1, \cdots, n_i, \dots\rangle = (n_i - m_i) \underbrace{\alpha_i \cdots \alpha_i}_{m_i} |n_1, \cdots, n_i, \dots\rangle. \tag{3.1.26}$$

$$N_i \underbrace{\alpha_i^\dagger \dots \alpha_i^\dagger}_{m_i} |n_1, \dots, n_i, \dots\rangle = (n_i + m_i) \underbrace{\alpha_i^\dagger \dots \alpha_i^\dagger}_{m_i} |n_1, \dots, n_i, \dots\rangle. \tag{3.1.27}$$

Since N_i is an hermitian, $n_i - m_i \geq 0$. We choose the maximum natural number m_0 to satisfy $n_i \geq m_0 > n_i - 1$;

$$\underbrace{\alpha_i \cdots \alpha_i}_{m_0} |n_1, \ldots, n_i, \ldots\rangle \neq 0,$$

$$\underbrace{\alpha_i \cdots \alpha_i}_{m_0+1} |n_1, \ldots, n_i, \ldots\rangle = 0.$$

Then $n_i = m_i$ is satisfied. This is because if $n_i \neq m_i$, N_i can have a negative eigenvalue $n_i - (m_0 + 1)$. Therefore we have

$$\alpha_i |n_1, \ldots, n_i = 0, \ldots\rangle = 0, \tag{3.1.28}$$

$$N_i |n_1, \ldots, n_i = 0, \ldots\rangle = 0. \tag{3.1.29}$$

Thus we conclude that N_i is the number operator.

From Eq. (3.1.27) we can put

$$|n_1, \ldots, n_i, \ldots\rangle = c_i \underbrace{\alpha_i^\dagger \ldots \alpha_i^\dagger}_{n_i} |n_1, \ldots, n_i = 0, \ldots\rangle, \tag{3.1.30}$$

where c_i is a constant number. In order to determine c_i, using the relations

$$\begin{aligned}(\alpha_i)^{n_i}(\alpha_i^\dagger)^{n_i} &= (\alpha_i)^{n_i-1}[\alpha_i, \alpha_i^\dagger](\alpha_i^\dagger)^{n_i-1} + (\alpha_i)^{n_i-1}\alpha_i^\dagger[\alpha_i, \alpha_i^\dagger](\alpha_i^\dagger)^{n_i-2} \ldots \\ &\quad + (\alpha_i)^{n_i-1}(\alpha_i^\dagger)^{n_i-1}[\alpha_i, \alpha_i^\dagger] + (\alpha_i)^{n_i-1}(\alpha_i^\dagger)^{n_i}\alpha_i \\ &= n_i(\alpha_i)^{n_i-1}(\alpha_i^\dagger)^{n_i-1} + (\alpha_i)^{n_i-1}(\alpha_i^\dagger)^{n_i}\alpha_i,\end{aligned}$$

we obtain

$$\begin{aligned}&c_i^2(\alpha_i)^{n_i}(\alpha_i^\dagger)^{n_i}|n_1, \ldots, n_i = 0, \ldots\rangle \\ &\quad = c_i^2 n_i(\alpha_i)^{n_i-1}(\alpha_i^\dagger)^{n_i-1}|n_1, \ldots, n_i = 0, \ldots\rangle \\ &\quad = c_i^2 n_i! |n_1, \ldots, n_i = 0, \ldots\rangle.\end{aligned} \tag{3.1.31}$$

Using the orthonormalization relations (3.1.10) and (3.1.11), we obtain

$$c_i = \frac{1}{\sqrt{n_i!}}. \tag{3.1.32}$$

Thus we obtain

$$|n_1, \ldots, n_i, \ldots\rangle = \prod_i \frac{1}{\sqrt{n_i!}}(\alpha_i^\dagger)^{n_i}|0\rangle. \tag{3.1.33}$$

If we operate α_i or $\alpha_i^\dagger$ on both sides of Eq. (3.1.30), we obtain

$$\alpha_i |n_1, \ldots n_i, \ldots\rangle = \sqrt{n_i}|n_1, \ldots n_i - 1, \ldots >, \tag{3.1.34}$$

$$\alpha_i^\dagger |n_1, \ldots n_i, \ldots\rangle = \sqrt{n_i + 1}|n_1, \ldots n_i + 1, \ldots\rangle. \tag{3.1.35}$$

The total number operator is given by

$$N = \sum_i N_i. \tag{3.1.36}$$

3.1.3.2 Fermions

Next we investigate fermion systems. We introduce the number operator in a similar manner to the boson systems.

$$N_i = \alpha_i^\dagger \alpha_i. \tag{3.1.37}$$

In the fermion system, even for different indices the transposition of the operators yields a negative sign such as

$$\alpha_i \alpha_j = -\alpha_j \alpha_i, \qquad \alpha_i^\dagger \alpha_j = -\alpha_j \alpha_i^\dagger$$

But N_i has two operators so that N_i and α_j or $\alpha_j^\dagger$ is changeable, where $i \neq j$. N_i is a hermitian so that it has eigenstate and eigenvalue:

$$N_i|n_1, \ldots, n_i, \ldots\rangle = n_i|n_1, \ldots, n_i, \ldots\rangle. \tag{3.1.38}$$

We can show the following relations:

$$[N_i, \alpha_i] = -\alpha_i, \tag{3.1.39}$$
$$[N_i, \alpha_i^\dagger] = \alpha_i^\dagger, \tag{3.1.40}$$

which show α_i is the lowering operator and $\alpha_i^\dagger$ raising operator such that

$$N_i \alpha_i|n_1, \ldots, n_i, \ldots\rangle = (n_i - 1)\alpha_i|n_1, \ldots, n_i, \ldots\rangle, \tag{3.1.41}$$
$$N_i \alpha_i^\dagger|n_1, \ldots, n_i, \ldots\rangle = (n_i + 1)\alpha_i^\dagger|n_1, \ldots, n_i, \ldots\rangle. \tag{3.1.42}$$

However, from Eq. (3.1.18) there is the relations

$$\alpha_i \alpha_i = \alpha_i^\dagger \alpha_i^\dagger = 0, \tag{3.1.43}$$

so we obtain

$$N_i N_i = \alpha_i^\dagger \alpha_i \alpha_i^\dagger \alpha_i = \alpha_i^\dagger(-\alpha_i^\dagger \alpha_i + 1)\alpha_i = \alpha_i^\dagger \alpha_i = N_i,$$

which shows

$$N_i(N_i - 1) = 0. \tag{3.1.44}$$

This states that the eigenvalue of the number operator N_i is 0 or 1. From Eq. (3.1.39) we find

$$N_i \alpha_i|n_1, \ldots, n_i = 0, \ldots\rangle = -\alpha_i|n_1, \ldots, n_i = 0, \ldots\rangle,$$

which shows that since the eigenvalue of N_i cannot be negative. we conclude

$$\alpha_i|n_1, \ldots, n_i = 0, \ldots\rangle = 0, \tag{3.1.45}$$

and

$$N_i|n_1, \ldots, n_i = 0, \ldots\rangle = 0. \tag{3.1.46}$$

The orthonormalization relations (3.1.10) and (3.1.11) yield

$$|n_1, \ldots, n_i, \ldots\rangle = \prod_i (\alpha_i^\dagger)^{n_i}|0\rangle, \tag{3.1.47}$$

where $n_i = 0,\ 1$.

If we operate α_i or $\alpha_i^\dagger$ on both sides of Eq. (3.1.47), we obtain

$$\alpha_i|n_1, \ldots, n_i, \ldots\rangle = \begin{cases} 0 & \text{for } n_i = 0, \\ (-1)^{\sum_{j<i} n_j}|n_1, \ldots, n_i - 1, \ldots\rangle & \text{for } n_i = 1, \end{cases} \tag{3.1.48}$$

$$\alpha_i^\dagger|n_1, \ldots, n_i, \ldots\rangle = \begin{cases} 0 & \text{for } n_i = 1, \\ (-1)^{\sum_{j<i} n_j}|n_1, \ldots, n_i + 1, \ldots\rangle & \text{for } n_i = 0, \end{cases} \tag{3.1.49}$$

The phase factor $(-1)^{\sum_{j<i} n_j}$ comes from the fact that the operator, a_i or $a_i^\dagger$, must transfer over the number states, $n_1, \ldots n_{i-1}$.

3.1.4 *Creation and Annihilation Operators for Wave Packets of Particles*

Any normalizable wave function (wave packet) can be expressed by terms of a countable set of orthonormalized functions $\varphi_i(\boldsymbol{x})$:

$$(\varphi_i^*, \varphi_j) \equiv \int d^3x \varphi_i^*(\boldsymbol{x})\varphi_j(\boldsymbol{x}) = \delta_{ij}, \tag{3.1.50}$$

Since the wave packet functions are Fourier transformable, we can formulate the theory of the Fock space in the Fourier representation. Such a formulation plays an important role in the practical computations.

We write the Fourier amplitude of $\varphi_i(\boldsymbol{x})$ by $\varphi_i(\boldsymbol{k})$:

$$\varphi_i(\boldsymbol{x}) = \int \frac{d^3k}{(2\pi)^3} e^{i\boldsymbol{k}\cdot\boldsymbol{x}} \varphi_i(\boldsymbol{k}). \tag{3.1.51}$$

Then the orthonormalization Eq. (3.1.50) leads to

$$(\varphi_i^*, \varphi_j) \equiv \int \frac{d^3k}{(2\pi)^3} \varphi_i^*(\boldsymbol{k})\varphi_j(\boldsymbol{k}) = \delta_{ij}. \tag{3.1.52}$$

In the Fourier analysis, it is very convenient to use plane waves $e^{i\boldsymbol{k}\cdot\boldsymbol{x}}$ as bases in the wave packet space, but the plane waves cannot represent the spatial distribution of single particle states, because they are not normalizable and do not construct a countable set. To reconcile these two representations, we introduce the δ-function normalization:

$$[\alpha(\boldsymbol{k}), \alpha^\dagger(\boldsymbol{k}')]_\mp = \delta(\boldsymbol{k} - \boldsymbol{k}'), \tag{3.1.53}$$

$$[\alpha(\boldsymbol{k}), \alpha(\boldsymbol{k}')]_\mp = [\alpha^\dagger(\boldsymbol{k}), \alpha^\dagger(\boldsymbol{k}')]_\mp = 0. \tag{3.1.54}$$

Then introducing

$$\alpha_i = \int \frac{d^3k}{(2\pi)^{3/2}} \varphi_i(\boldsymbol{k})\alpha(\boldsymbol{k}), \tag{3.1.55}$$

we obtain Eqs. (3.1.17) and (3.1.18).

It should be noted $\alpha(\boldsymbol{k})$ and $a^\dagger(\boldsymbol{k})$ are not defined as operators in the Fock space. When they act on the Fock space, they yield infinite norms. For example, for $|\boldsymbol{k} >= \alpha^\dagger(\boldsymbol{k})|0>$, $<\boldsymbol{k}|\boldsymbol{k}>$ leads to

$$\langle \boldsymbol{k}|\boldsymbol{k}\rangle = \lim_{\boldsymbol{k}\to 0} \delta(\boldsymbol{k}). \tag{3.1.56}$$

Since the δ-function is not normalizable, the operators $\alpha(\boldsymbol{k})$ are not realized in the Fock space; their usefulness relates to the Fourier representation of α_i in Eq. (3.1.55). In order to construct operators in the Fock space, we investigate the relations between an infinite volume and a finite system with a volume V. For an infinite volume, we obtain the Dirac delta function:

$$\delta(\boldsymbol{k}-\boldsymbol{k}') = \frac{1}{(2\pi)^3}\int d^3x e^{\mathrm{i}(\boldsymbol{k}-\boldsymbol{k}')\cdot\boldsymbol{x}}, \tag{3.1.57}$$

where $\boldsymbol{k}$ are continuous, while for a finite system with a volume V, we obtain the Kronecker delta:

$$\delta_{\boldsymbol{k},\boldsymbol{k}'} = \frac{1}{V}\int d^3x e^{\mathrm{i}(\boldsymbol{k}-\boldsymbol{k}')\cdot\boldsymbol{x}}, \tag{3.1.58}$$

where $\boldsymbol{k}$ are discrete $\boldsymbol{k} = (2\pi n_1, 2\pi n_2, 2\pi n_3)/V$; n_i: integers. Note that if we take $\boldsymbol{R}_m = (am_1, am_2, am_3)$, we have

$$\begin{aligned}\delta_{\boldsymbol{k},\boldsymbol{k}'} &= \frac{a^3}{V}\sum_{\boldsymbol{R}_m} e^{\mathrm{i}(\boldsymbol{k}-\boldsymbol{k}')\cdot\boldsymbol{R}_m} \\ &= \frac{1}{N}\sum_{\boldsymbol{R}_m} e^{\mathrm{i}(\boldsymbol{k}-\boldsymbol{k}')\cdot\boldsymbol{R}_m}.\end{aligned} \tag{3.1.59}$$

Thus we obtain the following formulae:

$$\lim_{V\to\infty} \frac{(2\pi)^3}{V}\sum_{\boldsymbol{k}} = \int d\boldsymbol{k}, \tag{3.1.60}$$

$$\lim_{V\to\infty} \frac{V}{(2\pi)^3}\delta_{\boldsymbol{k},\boldsymbol{k}'} = \delta(\boldsymbol{k}-\boldsymbol{k}'). \tag{3.1.61}$$

Under such situations, annihilation operators $\alpha_{\boldsymbol{k}}$ and creation operators $\alpha_{\boldsymbol{k}}$ for the finite volume relate to those for the infinite volume as

$$\lim_{V\to\infty}\sqrt{\frac{V}{(2\pi)^3}}\alpha_{\boldsymbol{k}} = \alpha(\boldsymbol{k}), \tag{3.1.62}$$

$$\lim_{V\to\infty}\sqrt{\frac{V}{(2\pi)^3}}\alpha^\dagger_{\boldsymbol{k}} = \alpha^\dagger(\boldsymbol{k}), \tag{3.1.63}$$

so that

$$\lim_{V\to\infty} \frac{V}{(2\pi)^3}[\alpha_{\boldsymbol{k}}, \alpha^{\dagger}_{\boldsymbol{k}'}] = [\alpha(\boldsymbol{k}), \alpha^{\dagger}(\boldsymbol{k}')] = \delta(\boldsymbol{k}-\boldsymbol{k}')$$
$$= \lim_{V\to\infty} \frac{V}{(2\pi)^3}\delta_{\boldsymbol{k},\boldsymbol{k}'}. \quad (3.1.64)$$

Thus for a finite volume the commutation relations are given by

$$[\alpha_{\boldsymbol{k}}, \alpha^{\dagger}_{\boldsymbol{k}'}]_{\mp} = \delta_{\boldsymbol{k},\boldsymbol{k}'} \quad (3.1.65)$$
$$[\alpha_{\boldsymbol{k}}, \alpha_{\boldsymbol{k}'}]_{\mp} = [\alpha^{\dagger}_{\boldsymbol{k}}, \alpha^{\dagger}_{\boldsymbol{k}'}]_{\mp} = 0. \quad (3.1.66)$$

In this book, we exclusively deal with the finite volume system.

3.2 An Example of Unitarily Inequivalent Representations; The Bogoliubov Transformation of Boson Operators

The quantum fields are constructed in the countable Fock space, while the set of the number states $|n_1, n_2, \cdots, n_i, \ldots\rangle$ is non-countable. Thus there occurs the unitarily inequivalence between the different representations of operators in a system. We show an example of unitarily inequivalent representations; the Bogoliubov transformation of boson operators.

3.2.1 The Bogoliubov Transformation of Boson Operators

We first consider two sets of boson annihilation operators, $\alpha_{\boldsymbol{k}}$ and $\beta_{\boldsymbol{k}}$, in the Fock space $\mathcal{H}(\alpha, \beta)$. The vacuum $|0\rangle$ satisfies

$$\alpha_{\boldsymbol{k}}|0\rangle = 0, \qquad \beta_{\boldsymbol{k}}|0\rangle = 0. \quad (3.2.1)$$

The commutation relations for the operators are

$$[\alpha_{\boldsymbol{k}}, \alpha^{\dagger}_{\boldsymbol{k}'}] = \delta_{\boldsymbol{k},\boldsymbol{k}'}, \quad (3.2.2)$$
$$[\beta_{\boldsymbol{k}}, \beta^{\dagger}_{\boldsymbol{k}'}] = \delta_{\boldsymbol{k},\boldsymbol{k}'}. \quad (3.2.3)$$

Other commutation relations vanish.

Now we introduce the operators, $\alpha_{\boldsymbol{k}}$ and $\beta_{\boldsymbol{k}}$ as follows:

$$a_{\boldsymbol{k}} = c_k \alpha_{\boldsymbol{k}} - d_k \beta^{\dagger}_{-\boldsymbol{k}}, \quad (3.2.4)$$
$$b_{\boldsymbol{k}} = c_k \beta_{\boldsymbol{k}} - d_k \alpha^{\dagger}_{-\boldsymbol{k}}, \quad (3.2.5)$$

where c_k and d_k are real numbers and satisfy

$$c_k^2 - d_k^2 = 1. \quad (3.2.6)$$

This relation leads to

$$[a_{\boldsymbol{k}}, a^{\dagger}_{\boldsymbol{k}'}] = \delta_{\boldsymbol{k},\boldsymbol{k}'}, \tag{3.2.7}$$

$$[b_{\boldsymbol{k}}, b^{\dagger}_{\boldsymbol{k}'}] = \delta_{\boldsymbol{k},\boldsymbol{k}'}. \tag{3.2.8}$$

Other commutation relations vanish. We call the tranformation, (3.2.4) and (3.2.5) Bogoliubov transformation. For simplicity, we rewrite Eq. (3.2.6) as follows:

$$c_k = \cosh\theta_k, \qquad d_k = \sinh\theta_k, \tag{3.2.9}$$

Now we introduce a generator of the transformation:

$$G(\theta) = \exp[A(\theta)], \tag{3.2.10}$$

where

$$A(\theta) = \sum_{\boldsymbol{k}} \theta_k[\alpha_{\boldsymbol{k}}\beta_{-\boldsymbol{k}} - \beta^{\dagger}_{-\boldsymbol{k}}\alpha^{\dagger}_{\boldsymbol{k}}], \tag{3.2.11}$$

which yields

$$[\alpha_{\boldsymbol{k}}, A(\theta)] = -\theta_k\beta^{\dagger}_{-\boldsymbol{k}}, \tag{3.2.12}$$

$$[\beta^{\dagger}_{-\boldsymbol{k}}, A(\theta)] = -\theta_k\alpha_{\boldsymbol{k}}. \tag{3.2.13}$$

For $\theta_k \ll 1$,

$$G^{-1}(\theta)\alpha_{\boldsymbol{k}}G(\theta_k) \cong \alpha_{\boldsymbol{k}} - \beta^{\dagger}_{-\boldsymbol{k}}\theta_k, \tag{3.2.14}$$

$$G^{-1}(\theta)\beta_{\boldsymbol{k}}G(\theta_k) \cong \beta_{\boldsymbol{k}} - \alpha^{\dagger}_{-\boldsymbol{k}}\theta_k, \tag{3.2.15}$$

These relations show for finite θ_k:

$$G^{-1}(\theta)\alpha_{\boldsymbol{k}}G(\theta) = \alpha_{\boldsymbol{k}}\cosh\theta_k - \beta^{\dagger}_{-\boldsymbol{k}}\sinh\theta_k = a_{\boldsymbol{k}}, \tag{3.2.16}$$

$$G^{-1}(\theta)\beta_{\boldsymbol{k}}G(\theta) = \beta_{\boldsymbol{k}}\cosh\theta - \alpha^{\dagger}_{-\boldsymbol{k}}\sinh\theta_k = b_{\boldsymbol{k}}, \tag{3.2.17}$$

These relations seem to show the Bogoliubov transformation Eqs. (3.2.4) and (3.2.5) unitary. To see this point more carefully, we expand the transformed state

$$|0\gg \equiv G^{-1}(\theta)|0\rangle, \tag{3.2.18}$$

with respect to base vectors in Fock space $\mathcal{H}(a, b)$ and calculate the coefficients of the base vectors. Since $A(\theta)$ in Eq. (3.2.11) contains the pair operators such as $\alpha_{\boldsymbol{k}}\beta_{-\boldsymbol{k}}$ or $\beta^{\dagger}_{-\boldsymbol{k}}\alpha^{\dagger}_{\boldsymbol{k}}$, $|0\gg$can be expanded as

$$|0\gg = \prod_{\boldsymbol{k}}\sum_{n}\frac{1}{(n!)^2}f_n(\theta_k)[\alpha^{\dagger}_{\boldsymbol{k}}\beta^{\dagger}_{-\boldsymbol{k}}]^n|0\rangle, \tag{3.2.19}$$

where $f_n(\theta_k)$ is given by

$$f_n(\theta_k) = \langle 0|[(\alpha_{\boldsymbol{k}}\beta_{-\boldsymbol{k}})]^n G^{-1}|0\rangle, \tag{3.2.20}$$

where the factor $1/(n!)^2$ comes from the normalization factors in the terms, $\langle 0|(\alpha_{\boldsymbol{k}}\beta_{-\boldsymbol{k}})^n \frac{1}{n!}$ and $\frac{1}{n!}(\alpha^{\dagger}_{\boldsymbol{k}}\beta^{\dagger}_{-\boldsymbol{k}})^n$ in Eq. (3.1.33).

First we calculate

$$f_0(\theta) = \langle 0|G^{-1}(\theta)|0\rangle. \tag{3.2.21}$$

To calculate this, we define the functional derivative as follows: $\delta f(\theta; \boldsymbol{l})$ is defined as $f_0(\theta)$ replaced the parametar $\theta_{\boldsymbol{k}}$ by $\theta_{\boldsymbol{k}} + \epsilon\delta_{0,\boldsymbol{k}-\boldsymbol{l}}$ for any given $\boldsymbol{l}$. Then the functional derivative is defined by

$$\frac{\delta}{\delta\theta_{\boldsymbol{l}}} f_0(\theta) = \lim_{\epsilon\longrightarrow 0} \frac{1}{\epsilon}\delta f(\theta; \boldsymbol{l}).$$

Thus we obtain

$$\frac{\delta}{\delta\theta_k} f_0(\theta) = -\langle 0|\alpha_{\boldsymbol{k}}\beta_{-\boldsymbol{k}}G^{-1}(\theta)|0\rangle, \tag{3.2.22}$$

or

$$= -\langle 0|G^{-1}(\theta)\beta^{\dagger}_{-\boldsymbol{k}}\alpha^{\dagger}_{\boldsymbol{k}}|0\rangle. \tag{3.2.23}$$

Using $G^{-1}(\theta) = G(-\theta)$, we obtain

$$\begin{aligned} &\alpha_{\boldsymbol{k}}\beta_{-\boldsymbol{k}}G^{-1}(\theta) \\ &= G^{-1}(\theta)G^{-1}(-\theta)\alpha_{\boldsymbol{k}}\beta_{-\boldsymbol{k}}G(-\theta) \\ &= G^{-1}(\theta)[\alpha_{\boldsymbol{k}}\cosh\theta_k + \beta^{\dagger}_{-\boldsymbol{k}}\sinh\theta_k][\beta_{-\boldsymbol{k}}\cosh\theta_k + \alpha^{\dagger}_{\boldsymbol{k}}\sinh\theta_k], \end{aligned} \tag{3.2.24}$$

Then Eqs. (3.2.22) and (3.2.24) leads to

$$\begin{aligned} \frac{\delta}{\delta\theta_k} f_0(\theta) &= -\sinh\theta_k\cosh\theta_k f_0(\theta) - \sinh^2\theta_k < 0|G^{-1}(\theta)\beta^{\dagger}_{-\boldsymbol{k}}\alpha^{\dagger}_{\boldsymbol{k}}|0\rangle \\ &= -\sinh\theta_k\cosh\theta_k f_0(\theta) + \sinh^2\theta_k \frac{\delta}{\delta\theta_k} f_0(\theta). \end{aligned} \tag{3.2.25}$$

Thus we obtain

$$\frac{\delta}{\delta\theta_k} f_0(\theta) = -\tanh\theta_k f_0(\theta). \tag{3.2.26}$$

The solution of Eq. (3.2.26) subject to $f_0(0) = 1$ is

$$f_0(\theta) = \exp\left[-\sum_k \ln\cosh\theta_k\right]. \tag{3.2.27}$$

Next we calculate $f_n(\theta_k)$. In a similar manner to the case of $f_0(\theta)$, we obtain

$$\frac{\delta}{\delta\theta_k} f_n(\theta) = -\langle 0|[\alpha_{\boldsymbol{k}}\beta_{-\boldsymbol{k}}]^n[\alpha_{\boldsymbol{k}}\beta_{-\boldsymbol{k}} - \beta^\dagger_{-\boldsymbol{k}}\alpha^\dagger_{\boldsymbol{k}}]G^{-1}(\theta)|0\rangle$$
$$= -f_{n+1}(\theta_{\boldsymbol{k}}) + n^2 f_{n-1}(\theta_{\boldsymbol{k}}).$$

The solution is

$$f_n(\theta_k) = f_0(\theta) n! (\tanh\theta_k)^n. \tag{3.2.28}$$

Equations (3.2.27) and (3.2.28) show that $|0\gg$ in Eq. (3.2.18) can be expanded as

$$|0\gg = f_0(\theta)\exp\left[\sum_{\boldsymbol{k}} \alpha^\dagger_{\boldsymbol{k}}\beta^\dagger_{\boldsymbol{k}} \tanh\theta_k\right]|0\rangle. \tag{3.2.29}$$

In order to estimate $f_0(\theta)$ in the limit of $V \to \infty$, we consider the relation

$$\lim_{V\to\infty} \frac{(2\pi)^3}{V}\sum_{\boldsymbol{k}} = \int d\boldsymbol{k}.$$

Then we obtain

$$\lim_{V\to\infty} f_0(\theta) = \lim_{V\to\infty} \exp\left[-\frac{V}{(2\pi)^3}\int d\boldsymbol{k} \ln\cosh\theta_k\right] = 0, \tag{3.2.30}$$

where we have considered $\cosh\theta_k > 1$. This manifests that since

$$\ln\cosh\theta_k > \alpha^\dagger_{\boldsymbol{k}}\beta^\dagger_{\boldsymbol{k}} \tanh\theta_k|0\rangle,$$

when we expand $|0\gg$ in the base vectors in $\mathcal{H}[\alpha, \beta]$, every expansion coefficient vanishes so that $|0\gg$ does not belong to $\mathcal{H}[\alpha, \beta]$.

From Eqs. (3.2.16) and (3.2.17), we obtain

$$a_{\boldsymbol{k}}|0\gg = 0, \qquad b_{\boldsymbol{k}}|0\gg = 0, \tag{3.2.31}$$

which imply that $|0\gg$ is the vacuum associated with $a_{\boldsymbol{k}}$ and $b_{\boldsymbol{k}}$. The above consideration shows that the vacuum $|0\gg$ does not belong to the Fock space $\mathcal{H}[\alpha, \beta]$. Thus $\mathcal{H}[a, b]$ and $\mathcal{H}[\alpha, \beta]$ are two unitarily inequivalent representations, which means that $\mathcal{H}[\alpha, \beta]$ cannot be expanded by the base vectors in $\mathcal{H}[a, b]$.

However, this result does not mean that the Fock space $\mathcal{H}[\alpha, \beta]$ does not relates with $\mathcal{H}[a, b]$ under the relations (3.2.4) and (3.2.5). As discussed in Section 3.1, these unitarily inequivalent representations originate from the fact that the set $|n_1, n_2, \ldots\rangle$ is not countable. There are non-countable Fock spaces, which are unitarily inequivalent, but each Fock space belongs to its own physical region in the number representation space. The Fock spaces $\mathcal{H}[\alpha, \beta]$ and $\mathcal{H}[a, b]$ belong to the same physical region under the relations (3.2.4) and (3.2.5).

3.3 The Physical Particle Representation and the Dynamical Map

In quantum field theory, there are non-countably infinite representations which are unitarily inequivalent to each other. In order to choose a Fock space, we adopt the statement in quantum mechanics that the Hilbert space is constructed by all of the observable states which constitute the complete orthonormalized system. Thus we choose a Fock space to describe a physical system which consists of observable particles. There are a various kind of observable particles. We call these observable particles physical particles. We can construct a Fock space to successively creation operators of the physical particles on the physical vacuum.

3.3.1 The Physical Particles and the Heisenberg Fields in the Heisenberg Equation

In the condensed state, the constituent particles are interacting through the interaction Hamiltonian. The interaction Hamiltonian determines its own Fock space in the configuration space. In the order state, the interaction Hamiltonian determines its own long range order, which breaks the invariance of the Hamiltonian. Under the spontaneously broken symmetry, there appear the NG bosons. The Fock space is formed by the physical particles: the constituent particles and the NG bosons. In the liquid and glassy states, there is a kind of the long range order, the pair distribution function. In the glassy state, the pair distribution function holds every time so that there is a long range order; the NG bosons, phonons appear. The physical particles are the constituent particles and phonons. However, in the liquid state, the pair distribution function changes with time so that there is not a long range order; the hopping is essential so that sound appears. The sound is not the NG boson. The physical particles are the constituent particles and sound.

First we investigate the Hamiltonian for physical particles. We denote the energy of a single particle with momentum $\hbar \boldsymbol{k}$ by $\hbar\omega_{\boldsymbol{k}}$. Here we introduce the operator H_0 the eigenvalues of which are the energy of the system. The above requirement means

$$H_0 \alpha^{\dagger}_{\boldsymbol{k}_1} \alpha^{\dagger}_{\boldsymbol{k}_2} \cdots \alpha^{\dagger}_{\boldsymbol{k}_n} |0\rangle = \sum_{i=1}^{n} \hbar\omega_{\boldsymbol{k}_i} \alpha^{\dagger}_{\boldsymbol{k}_1} \alpha^{\dagger}_{\boldsymbol{k}_2} \cdots \alpha^{\dagger}_{\boldsymbol{k}_n} |0\rangle, \tag{3.3.1}$$

which leads to

$$[H_0, a^{\dagger}_{\boldsymbol{k}_1}] \alpha^{\dagger}_{\boldsymbol{k}_2} \cdots \alpha^{\dagger}_{\boldsymbol{k}_n} |0\rangle = \hbar\omega_{\boldsymbol{k}_1} \alpha^{\dagger}_{\boldsymbol{k}_1} \alpha^{\dagger}_{\boldsymbol{k}_2} \cdots \alpha^{\dagger}_{\boldsymbol{k}_n} |0\rangle. \tag{3.3.2}$$

Since the relation (3.3.2) is true for any n, we have

$$[H_0, \alpha^{\dagger}_{\boldsymbol{k}_1}] = \hbar\omega_{\boldsymbol{k}} \alpha^{\dagger}_{\boldsymbol{k}}. \tag{3.3.3}$$

Since the energy is real, H_0 is hermitian. The hermite conjugate of Eq. (3.3.3) yields

$$[H_0, \alpha_{\boldsymbol{k}}] = -\hbar\omega_{\boldsymbol{k}} \alpha_{\boldsymbol{k}}. \tag{3.3.4}$$

Thus we obtain

$$H_0 = \sum_k \hbar\omega_k \alpha_k^\dagger \alpha_k. \tag{3.3.5}$$

Since the operator H_0 does not have any interaction terms, the operator H_0 is called the free Hamiltonian of the physical particles. Note that

$$\begin{aligned} H_0 &= \lim_{V\to\infty} \sum_k \hbar\omega_k \alpha_k^\dagger \alpha_k \\ &= \lim_{V\to\infty} \frac{(2\pi)^3}{V} \sum_k \hbar\omega_k \frac{V}{(2\pi)^3} \alpha_k^\dagger \alpha_k \\ &= \int d\boldsymbol{k} \hbar\omega_{\boldsymbol{k}} \alpha^\dagger(\boldsymbol{k})\alpha(\boldsymbol{k}). \end{aligned}$$

In a similar manner as the operator H_0, we find the momentum operator $\boldsymbol{P}$:

$$\boldsymbol{P} = \sum_k \hbar\boldsymbol{k}\alpha_k^\dagger \alpha_k. \tag{3.3.6}$$

Thus we can construct the Hamiltonian of physical particles. In the liquid state, the Hamiltonian is constructed by summing up the Hamiltonians of the physical particles, the two bands, and sound, while in the glassy state, the Hamiltonian is also constructed by summing up the physical particles, the two bands, and phonons.

Although the Hamiltonian for physical particles are constructed by summing up the Hamiltonians of physical particles. These physical particles must be constructed by the interacting constituent particles through the interaction Hamiltonians. Therefore we must investigate the Heisenberg fields $\psi(x)$ in space–time $x = (\boldsymbol{x}, t)$. The dynamics of the Heisenberg field $\psi(x)$ is governed by the Heisenberg equation with the Hamiltonian H, which generates the time translation. The Heisenberg equation is given by

$$i\hbar \frac{d}{dt}\psi(x) = [\psi(x), H]. \tag{3.3.7}$$

To calculate the commutator $[\psi(x), H]$, we need the assumption that the Heisenberg fields satisfy the equal-time canonical commutation relations. To solve the Heisenberg equation of Heisenberg fields is one of the main themes of this book.

3.3.2 An Example of the Physical Particle Representation for Heisenberg Fields

Since in general the Heisenberg equation cannot be solved exactly, we solve the Heisenberg equation by realizing the Heisenberg fields in the Fock space of the physical particles; we solve the Heisenberg equation in such a way that the Heisenberg fields are expressed in terms of certain physical particles so that all of the matrix elements, $\langle a|\psi(x)|b\rangle$, are determined for the vectors $|a\rangle$ and $|b\rangle$ in the Fock space of the free physical particles. In this way, we can determine the Fock space $\mathcal{H}[\alpha]$ by the requirement that the Heisenberg fields are expressed in terms of annihilation and creation operators of certain physical particles. We will see later that this requirement leads to

$$\langle a|H|b\rangle = \langle a|H_0|b\rangle + W_0\langle a|b\rangle. \tag{3.3.8}$$

where W_0 is a c-number, and $|a\rangle$ and $|b\rangle$ are vectors in $\mathcal{H}[\alpha]$ of the physical particle representation. It should be noted Eq. (3.3.8) is not as strong as $H = H_0 + W_0$, but it requires that $H = H_0 + W_0$ is realized in the certain Fock space $\mathcal{H}[\alpha]$. The relations under the matrix elements with a specific representation are called weak relation. Equation (3.3.8) shows a weak condition.

For a given Hamiltonian H, there is no royal road to find out a suitable Fock space $\mathcal{H}[\alpha]$. Therefore we must find out the Fock space by considering the general laws such as commutation relations, the symmetry relations called the Ward-Takahashi relations. Here we only show a simple example for Eq. (3.3.8).

We take a Hamiltonian

$$H = \sum_{\boldsymbol{k}} \left[\hbar\epsilon_k(a_{\boldsymbol{k}}^{\dagger}a_{\boldsymbol{k}} + b_{\boldsymbol{k}}^{\dagger}b_{\boldsymbol{k}}) + \hbar\nu_k(a_{\boldsymbol{k}}b_{-\boldsymbol{k}} + b_{-\boldsymbol{k}}^{\dagger}a_{\boldsymbol{k}}^{\dagger})\right], \tag{3.3.9}$$

where $a_{\boldsymbol{k}}$ and $b_{\boldsymbol{k}}$ are the Heisenberg operators satisfying the boson commutation relations:

$$[a_{\boldsymbol{k}}, a_{\boldsymbol{k}'}^{\dagger}] = \delta_{\boldsymbol{k},\boldsymbol{k}'}, \qquad [b_{\boldsymbol{k}}, b_{\boldsymbol{k}'}^{\dagger}] = \delta_{\boldsymbol{k},\boldsymbol{k}'}. \tag{3.3.10}$$

Other commutation relations vanish. We assume that ϵ_k and ν_k are positive definite functions of k^2 under a condition $\epsilon_k > \nu_k$. To find out a free Hamiltonian, we introduce the Bogoliubov transformation

$$a_{\boldsymbol{k}} = c_k\alpha_{\boldsymbol{k}} - d_k\beta_{-\boldsymbol{k}}^{\dagger} \tag{3.3.11}$$

$$b_{\boldsymbol{k}} = c_k\beta_{\boldsymbol{k}} - d_k\alpha_{-\boldsymbol{k}}^{\dagger}. \tag{3.3.12}$$

If we consider the relation

$$c_k^2 - d_k^2 = 1, \tag{3.3.13}$$

the commutation relations for $\alpha_{\boldsymbol{k}}$ and $\beta_{\boldsymbol{k}}$ are given by

$$[\alpha_{\boldsymbol{k}}, \alpha_{\boldsymbol{k}'}^{\dagger}] = \delta_{\boldsymbol{k},\boldsymbol{k}'}, \qquad [\beta_{\boldsymbol{k}}, \beta_{\boldsymbol{k}'}^{\dagger}] = \delta_{\boldsymbol{k},\boldsymbol{k}'}, \tag{3.3.14}$$

and other commutation relations vanish. Thus we can put $c_k = \cosh\theta_k$ and $d_k = \sinh\theta_k$. Substituting Eqs. (3.3.11) and (3.3.12) into Eq. (3.3.9) leads to

$$
\begin{aligned}
H &= \sum_k \hbar\epsilon_k[(c_k\alpha_{\boldsymbol{k}}^\dagger - d_k\beta_{-\boldsymbol{k}})(c_k\alpha_{\boldsymbol{k}} - d_k\beta_{-\boldsymbol{k}}^\dagger) \\
&\quad +(c_k\beta_{\boldsymbol{k}}^\dagger - d_k\alpha_{-\boldsymbol{k}})(c_k\beta_{\boldsymbol{k}} - d_k\alpha_{-\boldsymbol{k}}^\dagger)] + \sum_k \hbar\nu_k \\
&\quad \times[(c_k\alpha_{\boldsymbol{k}} - d_k\beta_{-\boldsymbol{k}}^\dagger)(c_k\beta_{-\boldsymbol{k}} - d_k\alpha_{\boldsymbol{k}}^\dagger) + (c_k\beta_{-\boldsymbol{k}}^\dagger - d_k\alpha_{\boldsymbol{k}})(c_k\alpha_{\boldsymbol{k}}^\dagger - d_k\beta_{-\boldsymbol{k}})] \\
&= \sum_k [\{\hbar\epsilon_k(c_k^2 + d_k^2) + 2\hbar\nu_k c_k d_k\}(\alpha_{\boldsymbol{k}}^\dagger\alpha_{\boldsymbol{k}} + \beta_{\boldsymbol{k}}^\dagger\beta_{\boldsymbol{k}}) \\
&\quad +\{-2\hbar\epsilon_k c_k d_k + \hbar\nu_k(c_k^2 + d_k^2)\}(\alpha_{\boldsymbol{k}}\beta_{-\boldsymbol{k}} + \alpha_{\boldsymbol{k}}^\dagger\beta_{-\boldsymbol{k}}^\dagger) + 2\epsilon_k d_k^2 - 2\nu_k c_k d_k] \\
&= \sum_k [\{\hbar\epsilon_k \cosh 2\theta_k + \hbar\nu_k \sinh 2\theta_k\}(\alpha_{\boldsymbol{k}}^\dagger\alpha_{\boldsymbol{k}} + \beta_{\boldsymbol{k}}^\dagger\beta_{\boldsymbol{k}}) \\
&\quad +\{-\hbar\epsilon_k \sinh 2\theta_k + \hbar\nu_k \cosh 2\theta_k\}(\alpha_{\boldsymbol{k}}\beta_{-\boldsymbol{k}} + \alpha_{\boldsymbol{k}}^\dagger\beta_{-\boldsymbol{k}}^\dagger) \\
&\quad +\hbar\epsilon_k(\cosh 2\theta_k - 1) - \hbar\nu_k \sinh 2\theta_k],
\end{aligned}
\tag{3.3.15}
$$

where we have considered the inversion symmetry of $\boldsymbol{k}$. Choosing θ_k such that the coefficients of the terms, $\alpha_{\boldsymbol{k}}\beta_{-\boldsymbol{k}}$ and $\alpha_{\boldsymbol{k}}^\dagger\beta_{\boldsymbol{k}}^\dagger$ vanish, we obtain

$$\tanh 2\theta_k = \frac{\nu_k}{\epsilon_k}. \tag{3.3.16}$$

Considering

$$\cosh 2\theta_k = \frac{\epsilon_k}{\sqrt{\epsilon^2 - \nu_k^2}}, \qquad \sinh 2\theta_k = \frac{\nu_k}{\sqrt{\epsilon^2 - \nu_k^2}}, \tag{3.3.17}$$

we have

$$H = H_0 + W_0, \tag{3.3.18}$$

with

$$H_0 = \sum_k \hbar\omega_k[\alpha_{\boldsymbol{k}}^\dagger\alpha_{\boldsymbol{k}} + \beta_{\boldsymbol{k}}^\dagger\beta_{\boldsymbol{k}}], \tag{3.3.19}$$

$$\omega_k = \sqrt{\epsilon^2 - \nu_k^2}, \tag{3.3.20}$$

$$W_0 = \sum_k \hbar(\omega_k - \epsilon_k), \tag{3.3.21}$$

Therefore $\alpha_{\boldsymbol{k}}$ and $\beta_{\boldsymbol{k}}$ are the annihilation operators of physical particles in the Fock space $\mathcal{H}[\alpha, \beta]$. Equation (3.3.18) shows this case is a strong relation comparing to the weak relation (3.3.8). Note that

$$H_0 = \lim_{V\to\infty}\sum_k \hbar\omega_{\boldsymbol{k}}[\alpha_{\boldsymbol{k}}^\dagger\alpha_{\boldsymbol{k}} + \beta_{\boldsymbol{k}}^\dagger\beta_{\boldsymbol{k}}] = \int d\boldsymbol{k}\hbar\omega_{\boldsymbol{k}}[\alpha^\dagger(\boldsymbol{k})\alpha(\boldsymbol{k}) + \beta^\dagger(\boldsymbol{k})\beta(\boldsymbol{k})],$$

(3.3.22)

$$W_0 = \lim_{V\to\infty} \sum_k \hbar(\omega_k - \epsilon_k) = \lim_{V\to\infty} \frac{V}{(2\pi)^3} \int d\boldsymbol{k}\hbar(\omega_k - \epsilon_k). \tag{3.3.23}$$

Although W_0 has an infinite value, this numerical value does not affect on the physical state.

3.3.3 The Dynamical Map and the Normal Product

Next we investigate the time development of physical particle operators and the Heisenberg operators. For the free Hamiltonian H_0 in Eq. (3.3.5), the time development of the annihilation operators of physical particles is governed by the Heisenberg equation:

$$\mathrm{i}\hbar \frac{d}{dt}\alpha_{\boldsymbol{k}}(t) = [\alpha_{\boldsymbol{k}}(t), H_0], \tag{3.3.24}$$

which leads to

$$\begin{aligned} \alpha_{\boldsymbol{k}}(t) &= e^{\mathrm{i}H_0t/\hbar}\alpha_{\boldsymbol{k}}e^{-\mathrm{i}H_0t/\hbar} \\ &= \alpha_{\boldsymbol{k}}e^{-\mathrm{i}\omega_{\boldsymbol{k}}t}, \end{aligned} \tag{3.3.25}$$

and

$$\begin{aligned} \alpha^\dagger_{\boldsymbol{k}}(t) &= e^{\mathrm{i}H_0t/\hbar}\alpha^\dagger_{\boldsymbol{k}}e^{-\mathrm{i}H_0t/\hbar} \\ &= \alpha^\dagger_{\boldsymbol{k}}e^{\mathrm{i}\omega_{\boldsymbol{k}}t}. \end{aligned} \tag{3.3.26}$$

The time development of the Heisenberg operators $a_{\boldsymbol{k}}(t)$ is by the Heisenberg equation

$$\mathrm{i}\hbar \frac{d}{dt}a_{\boldsymbol{k}}(t) = [a_{\boldsymbol{k}}(t), H], \tag{3.3.27}$$

which leads to

$$a_{\boldsymbol{k}}(t) = e^{\mathrm{i}Ht/\hbar}a_{\boldsymbol{k}}e^{-\mathrm{i}Ht/\hbar} \tag{3.3.28}$$

In the case of a boson model with the Hamiltonian (3.3.9), the Bogoliubov transformations (3.3.11) and (3.3.12) determine the time development of the Heisenberg operators as follows:

$$a_{\boldsymbol{k}}(t) = \cosh\theta_k e^{-\mathrm{i}\omega_k t}\alpha_{\boldsymbol{k}} - \sinh\theta_k e^{\mathrm{i}\omega_k t}\beta^\dagger_{-\boldsymbol{k}}, \tag{3.3.29}$$

$$b_{\boldsymbol{k}}(t) = \cosh\theta_k e^{-\mathrm{i}\omega_k t}\beta_{\boldsymbol{k}} - \sinh\theta_k e^{\mathrm{i}\omega_k t}\alpha^\dagger_{-\boldsymbol{k}}. \tag{3.3.30}$$

Equations (3.3.29) and (3.3.30) show how the Heisenberg operators are realized in the physical particle representation. These relations express the Heisenberg operators $a_{\boldsymbol{k}}(t)$ and $b_{\boldsymbol{k}}(t)$ in terms of physical particle operators $\alpha_{\boldsymbol{k}}$ and $\beta_{\boldsymbol{k}}$. The expression of this kind is called the dynamical map. In this simple example, the dynamical map is linear and is given by the strong relation. In more complicated cases, the dynamical map involves higher order products of the physical creation and annihilation operators and it is defined only through matrix element in the weak relation.

In the higher order products of physical particle operators, these higher order products are arranged in the form of linear combinations of the so-called *normal products*, which is a product of physical particle creation and annihilation operators where all of creation operators stand on the left side of all of annihilation operators. The calculation of matrix elements of normal products does not yield the contractions due to the commutation relations of creation and annihilation operators. The matrix element of a normal product with n creation operators and m annihilation operators represents the transition of the m particles to n particles. So the normal products make the dynamical map more transparent.

3.4 Free Physical Fields for Physical Particles

There are a various kinds of physical particles; fundamental particles, renormalized particles, collective modes, bound states, and so on. In developing the dynamical and dissipative processes in quantum field theory, we need the operators of physical particles which describe the space–time behavior of physical particles. These operators are called the free physical fields. The free physical fields are classified by the eigenvalue equations of free field equations and the structure of the free physical particles are prescribed by the orthonormalized complete set of solutions. Here we show how to construct the free physical fields.

3.4.1 *Space–Time Variation of Creation and Annihilation Operators*

To construct the free physical fields, we start with the time development of the annihilation and creation operator of the physical particles for the Hamiltonian $H_0 = \sum_k \hbar\omega_k\alpha_k^\dagger\alpha_k$ in Eq. (3.3.5):

$$\alpha_k(t) = \alpha_k e^{-\mathrm{i}\omega_k t}, \tag{3.4.1}$$

$$\alpha_k^\dagger(t) = \alpha_k^\dagger e^{\,\mathrm{i}\omega_k t}, \tag{3.4.2}$$

The momentum operator $\boldsymbol{P} = \sum_k \hbar\boldsymbol{k}\alpha_k^\dagger\alpha_k$ in Eq. (3.3.6) leads to the spatial behavior of annihilation and creation operators:

$$\begin{aligned}\alpha_k(t,\boldsymbol{x}) &= e^{-\mathrm{i}\boldsymbol{P}\cdot\boldsymbol{x}/\hbar}\alpha_k(t)e^{\mathrm{i}\boldsymbol{P}\cdot\boldsymbol{x}/\hbar},\\ &= \alpha_k e^{-\mathrm{i}(\boldsymbol{k}\cdot\boldsymbol{x}-\omega_k t)}\end{aligned} \tag{3.4.3}$$

$$\begin{aligned}\alpha_k^\dagger(t,\boldsymbol{x}) &= e^{-\mathrm{i}\boldsymbol{P}\cdot\boldsymbol{x}/\hbar}\alpha_k^\dagger(t)e^{\mathrm{i}\boldsymbol{P}\cdot\boldsymbol{x}/\hbar},\\ &= \alpha_k^\dagger e^{-\mathrm{i}(\boldsymbol{k}\cdot\boldsymbol{x}-\omega_k t)}\end{aligned} \tag{3.4.4}$$

This relation shows that an annihilation operator has a negative frequency, while a creation operator has a positive frequency.

3.4.2 *The Free Field Equations of Physical Fields*

Since the physical field $\phi(x)$ is a superposition of the plane waves in Eqs. (3.4.3) and (3.4.4), it satisfies a homogeneous differential equation:

$$\Lambda(\partial)\phi(x) = 0, \tag{3.4.5}$$

where $\partial = (\frac{\partial}{\partial t}, \nabla)$ and $x = (t, \boldsymbol{x})$. If the physical particle has spin or other degree of freedom, $\phi(x)$ is a column vector:

$$\phi(x) = \begin{pmatrix} \phi_1 \\ \cdot \\ \cdot \\ \cdot \\ \phi_n \end{pmatrix}, \tag{3.4.6}$$

where Λ is a $n \times n$ matrix.

3.4.3 *The Classification of Free Field Equations*

Now we reduce free field Eqs. (3.4.5) to the eigenvalue equations and we classify the eigenvalue equations into two types:

1. Type 1.

$$\left(\mathrm{i}\frac{\partial}{\partial t} - \epsilon(\nabla)\right)\phi(x) = 0. \tag{3.4.7}$$

2. Type 2.

$$\left(\frac{\partial^2}{\partial t^2} + \omega^2(\nabla)\right)\phi(x) = 0. \tag{3.4.8}$$

Here the derivative operators are defined as

$$\epsilon(\nabla)e^{\mathrm{i}\boldsymbol{k}\cdot\boldsymbol{x}} = \epsilon(\boldsymbol{k})e^{\mathrm{i}\boldsymbol{k}\cdot\boldsymbol{x}}, \text{ etc.} \tag{3.4.9}$$

Here we note about type 1 and type 2:

1. Type 1.
   $\epsilon(\boldsymbol{k})$ can have both positive and negative, but negative case only occurs for fermion fields. Positive one is called a normal particle, while a negative one is called a hole. For $\epsilon > 0$, ϕ has only annihilation operators of particles, while for $\epsilon(\boldsymbol{k}) < 0$, ϕ has creation operators of holes.
2. Type 2.
   ϕ has both annihilation and creation operators for each momentum.

Equations (3.4.7) and (3.4.8) show that there should exist differential operators $D(\partial)$ which satisfy

$$D(\partial)\Lambda(\partial) = \mathrm{i}\frac{\partial}{\partial t} - \epsilon(\nabla), \qquad \text{for type 1,} \tag{3.4.10}$$

$$D(\partial)\Lambda(\partial) = -\left(\frac{\partial^2}{\partial t^2} + \omega^2(\nabla)\right), \qquad \text{for type 2.} \tag{3.4.11}$$

The operator $D(\partial)$ is called the divisor. Noting that the matrix on the right-hand side is essentially a unit matrix, then acting $\Lambda^{-1}(\partial)$ from the right-hand sides, $\Lambda(\partial)$ from the left-hand sides on both Eqs. (3.4.10) and (3.4.11), we obtain

$$\Lambda(\partial)D(\partial) = \mathrm{i}\frac{\partial}{\partial t} - \epsilon(\nabla), \qquad \text{for type 1,} \tag{3.4.12}$$

$$\Lambda(\partial)D(\partial) = -\left(\frac{\partial^2}{\partial t^2} + \omega^2(\nabla)\right), \qquad \text{for type 2.} \tag{3.4.13}$$

Now we introduce the Green's function denoted by the same notation $\Delta_G(x)$ for Eqs. (3.4.10) and (3.4.11):

$$\left(\mathrm{i}\frac{\partial}{\partial t} - \epsilon(\nabla)\right)\Delta_G(x) = \delta(\boldsymbol{x})\delta(t) \qquad \text{for type 1,} \tag{3.4.14}$$

$$-\left(\frac{\partial^2}{\partial t^2} + \omega^2(\nabla)\right)\Delta_G(x) = \delta(\boldsymbol{x})\delta(t) \qquad \text{for type 2.} \tag{3.4.15}$$

Then we obtain the important result:

$$\Lambda(\partial)D(\partial)\Delta_G(x) = \delta(\boldsymbol{x})\delta(t), \tag{3.4.16}$$

which implies that $D(\partial)\Delta_G(x)$ is a Green's function for Eq. (3.4.5).

3.4.4 The Herimitian Conjugate of the Free Field Equation and the Lagrangian

If we define $\Lambda(k)$ by

$$\Lambda(k)e^{\mathrm{i}(\boldsymbol{k}\cdot\boldsymbol{x}-k_0 t)} = \Lambda(\partial)e^{\mathrm{i}(\boldsymbol{k}\cdot\boldsymbol{x}-k_0 t)}, \tag{3.4.17}$$

the equation

$$\Lambda(k)u = 0$$

is the eigenvalue equation, which yields the eigenvalues $k_0 = \pm\omega(\boldsymbol{k})$ and the eigenvector $u = (u_1, \cdots, u_n)^T$, where T is a transposition operator. Since the eigenvalues $\omega(\boldsymbol{k})$ are real, $\Lambda(k)u = 0$ should be equivalent to an eigenvalue equation of a certain hermitian matrix. This means that there is a non-singular matrix η to make $\eta\Lambda$ hermitian:

$$\Lambda^\dagger(k)\eta^\dagger = \eta\Lambda(k), \tag{3.4.18}$$

which satisfies $\eta\Lambda(k)u = 0$. Considering Eq. (3.4.17), we obtain

$$\Lambda^{\dagger}(-\partial)\eta^{\dagger} = \eta\Lambda(\partial). \tag{3.4.19}$$

Since $\Lambda D = D\Lambda$ is essentially a unit matrix in Eqs. (3.4.10–3.4.13) ; $\Lambda D = \eta\Lambda D\eta^{-1}$ and $D\Lambda = D\eta\eta^{-1}\Lambda$, we obtain

$$[\eta\Lambda(\partial)][D(\partial)\eta^{-1}] = [D(\partial)\eta^{-1}][\eta\Lambda(\partial)]. \tag{3.4.20}$$

Taking hermite conjugate of the left hand side of the equation and using Eq. (3.4.19),

$$(\eta^{\dagger})^{-1}D(-\partial)^{\dagger}\Lambda(-\partial)^{\dagger}\eta^{\dagger} = (\eta^{\dagger})^{-1}D(-\partial)^{\dagger}\eta\Lambda(\partial)$$

and putting this term equal to the right hand side,

$$(\eta^{\dagger})^{-1}D(-\partial)^{\dagger}\eta\Lambda(\partial) = [D(\partial)\eta^{-1}][\eta\Lambda(\partial)],$$

we obtain an hermitian relation:

$$(\eta^{\dagger})^{-1}D^{\dagger}(-\partial) = D(\partial)\eta^{-1}, \tag{3.4.21}$$

The matrix η is called the hermitization matrix.

If we consider Eq. (3.4.19) and

$$\eta\Lambda(\partial)\phi = \Lambda^{\dagger}(-\partial)(\phi^{\dagger}\eta)^{\dagger} = (\phi^{\dagger}\eta\Lambda(-\overleftarrow{\partial}))^{\dagger},$$

we obtain

$$\bar{\phi}(x)\Lambda(-\overleftarrow{\partial}) = 0, \tag{3.4.22}$$

where $\overleftarrow{\partial}$ means that the derivative acts on the left quantities and $\bar{\phi}$ is defined by

$$\bar{\phi}(x) = \phi^{\dagger}(x)\eta. \tag{3.4.23}$$

Since $\phi(x)$ is a column vector, $\bar{\phi}(x)$ is a row one.

The Lagrangian for free physical fields are given by

$$\mathcal{L}_{\phi}^{0} = \int d^4x\bar{\phi}(x)\Lambda(\partial)\phi(x), \tag{3.4.24}$$

which are real.

3.4.5 The Innner Product of Wave Functions

Now we assume that $\Lambda(\partial)$ is a polynomial in $\frac{\partial}{\partial t}$ up to second order and put as follows:

$$\Lambda(\partial) = \lambda_0(\nabla) + i\lambda_1(\nabla)(\frac{\partial}{\partial t}) + \lambda_2(\nabla)(\frac{\partial}{\partial t})^2. \tag{3.4.25}$$

We define

$$\overleftrightarrow{\Gamma} = \lambda_1(\nabla) - \mathrm{i}\lambda_2(\nabla)\frac{\overleftrightarrow{\partial}}{\partial t}, \qquad \frac{\overleftrightarrow{\partial}}{\partial t} = \frac{\partial}{\partial t} - \frac{\overleftarrow{\partial}}{\partial t}. \tag{3.4.26}$$

Then we have

$$\begin{aligned}
&\frac{d}{dt}\int d^3x\,\bar{f}(x)\,\overleftrightarrow{\Gamma}\,g(x) \\
&= \int d^3x\,\bar{f}(x)\left(\frac{\partial}{\partial t} + \frac{\overleftarrow{\partial}}{\partial t}\right)\lambda_1(\nabla)g(x) - \mathrm{i}\int d^3x\,\bar{f}(x) \\
&\quad\times\left[\left(\frac{\partial}{\partial t}\right)^2 - \left(\frac{\overleftarrow{\partial}}{\partial t}\right)^2\right]\lambda_2(\nabla)g(x) \\
&= -\mathrm{i}\int d^3x\,\bar{f}(x)[\Lambda(\partial) - \Lambda(-\overleftarrow{\partial})]g(x),
\end{aligned} \tag{3.4.27}$$

where a partial integration in space has been used. If $f(x)$ and $g(x)$ satisfy the free field Eq. (2.3.5), Eq. (2.3.27) vanishes:

$$\int d^3x\,\bar{f}(x)\,\overleftrightarrow{\Gamma}\,g(x) : \text{independent of } t \tag{3.4.28}$$

This quantity is not necessarily positive definite. Using Eq. (3.4.28) we construct an orthonormalized complete set of solutions for the free field Eq. (3.4.5).

First we investigate the type 2 equation. Then it can have both negative and positive frequency solutions:

$$u^r_{\boldsymbol{k}}(x) = u_{\boldsymbol{k}}e^{\mathrm{i}(\boldsymbol{k}\cdot\boldsymbol{x} - \omega_{\boldsymbol{k}}t)}, \tag{3.4.29}$$

$$v^r_{\boldsymbol{k}}(x) = v^r_{\boldsymbol{k}}e^{-\mathrm{i}(\boldsymbol{k}\cdot\boldsymbol{x} - \omega_{-\boldsymbol{k}}t)}, \tag{3.4.30}$$

where subscript r refers to the spin and the other degree of freedom. Using the relation (3.4.17), we have

$$\Lambda(k)u^r_{\boldsymbol{k}} = 0, \qquad \text{for } k_0 = \omega_{\boldsymbol{k}}, \tag{3.4.31}$$

$$\Lambda(-k)v^r_{\boldsymbol{k}} = 0, \qquad \text{for } k_0 = \omega_{-\boldsymbol{k}}. \tag{3.4.32}$$

If we take $u^r_{\boldsymbol{k}}(x)$ and $v^s_{\boldsymbol{k}'}(x)$ for $f(x)$ and $g(x)$, respectively, the quantity in Eq. (3.4.28) is a superposition of $e^{\mathrm{i}(\omega_{\boldsymbol{k}}+\omega_{-\boldsymbol{k}'})t}$, although it is time independence. Therefore it must be zero. Thus we have the following orthogonality theorem:

$$\int d^3x\bar{u}^r_{\boldsymbol{k}}(x)\,\overleftrightarrow{\Gamma}\,v^s_{\boldsymbol{k}'}(x) = 0, \tag{3.4.33}$$

$$\int d^3x\bar{v}^s_{\boldsymbol{k}}(x)\,\overleftrightarrow{\Gamma}\,u^r_{\boldsymbol{k}'}(x) = 0, \tag{3.4.34}$$

We choose $u_{\boldsymbol{k}}^r(x)$ and $v_{\boldsymbol{k}}(x)$ to satisfy the following orthonormalization condition:

$$\int d^3x \bar{u}_{\boldsymbol{k}}^r(x)\, \overleftrightarrow{\Gamma}\, u_{\boldsymbol{k}'}^s(x) = \hbar\delta_{rs}\delta_{\boldsymbol{k},\boldsymbol{k}'}, \tag{3.4.35}$$

$$\int d^3x \bar{v}_{\boldsymbol{k}}^r(x)\, \overleftrightarrow{\Gamma}\, v_{\boldsymbol{k}'}^s(x) = -\rho\hbar\delta_{rs}\delta_{\boldsymbol{k},\boldsymbol{k}'}. \tag{3.4.36}$$

Since $\Lambda(\partial)$ in Eq. (3.4.5) can be allowed to take any sign, we can choose the sign for Eq. (3.4.35) to take a positive sign. Once the sign of $\Lambda(\partial)$ is chosen as Eq. (3.4.35), there is no reason that Eq. (3.4.36) should be positive. Thus we put

$$\rho = \pm \tag{3.4.37}$$

in Eq. (3.4.36).

Using the notations

$$\lambda_i(\boldsymbol{k})e^{\mathrm{i}\boldsymbol{k}\cdot\boldsymbol{x}} = \lambda_i(\nabla)e^{\mathrm{i}\boldsymbol{k}\cdot\boldsymbol{x}}, \qquad \text{for } i = 0, 1, 2, \tag{3.4.38}$$

we obtain

$$\Gamma(\boldsymbol{k}, k_0) = \lambda_1(\boldsymbol{k}) - 2k_0\lambda_2(\boldsymbol{k}). \tag{3.4.39}$$

Then the orthonormalization conditions (4.3.35) and (4.4.36) with Eq. (3.1.39) lead to

1. For the continuous space:

$$\bar{u}_{\boldsymbol{k}}^r\Gamma(\boldsymbol{k}, \omega_{\boldsymbol{k}})u_{\boldsymbol{k}}^s = \frac{\hbar}{V}\delta_{rs}, \tag{3.4.40a}$$

$$\bar{v}_{\boldsymbol{k}}^r\Gamma(-\boldsymbol{k}, -\omega_{-\boldsymbol{k}})v_{\boldsymbol{k}}^s = -\rho\frac{\hbar}{V}\delta_{rs}, \tag{3.4.41a}$$

2. For the discrete space

$$\bar{u}_{\boldsymbol{k}}^r\Gamma(\boldsymbol{k}, \omega_{\boldsymbol{k}})u_{\boldsymbol{k}}^s = \frac{\hbar}{N}\delta_{rs}. \tag{3.4.40b}$$

$$\bar{v}_{\boldsymbol{k}}^r\Gamma(-\boldsymbol{k}, -\omega_{-\boldsymbol{k}})v_{\boldsymbol{k}}^s = -\rho\frac{\hbar}{N}\delta_{rs}, \tag{3.4.41b}$$

Next we investigate the type 1 equation, $u_{\boldsymbol{k}}^r(x)$ is identified with positive $\epsilon_{\boldsymbol{k}}$, while $v_{\boldsymbol{k}}^r(x)$ appears for negative $\epsilon_{\boldsymbol{k}}$. The orthonormalization condition is also given by Eqs. (3.4.40) and (3.4.41).

We are now ready to construct the free physical field $\phi(x)$.

1. Type 1.

$$\phi(x) = \sum_r\sum_{\boldsymbol{k}}[\theta(\epsilon_{\boldsymbol{k}})u_{\boldsymbol{k}}^r\alpha_{\boldsymbol{k}}^r + \theta(-\epsilon_{-\boldsymbol{k}})v_{-\boldsymbol{k}}^r\beta_{-\boldsymbol{k}}^{r\dagger}]e^{\mathrm{i}(\boldsymbol{k}\cdot\boldsymbol{x}-\omega_{\boldsymbol{k}}t)}. \tag{3.4.42}$$

Here $\theta(x)$ is the step function: $\theta(x) = 1$ for $0 < x$ and $\theta(x) = 0$ for $x < 0$.

2. Type 2.

$$\phi(x) = \sum_r \sum_k [u^r_k \alpha^r_k e^{\mathrm{i}(k\cdot x - \omega_k t)} + v^r_k \beta^{r\dagger}_k e^{-\mathrm{i}(k\cdot x - \omega_{-k} t)}]. \tag{3.4.43}$$

Now we show examples of type 1 and type 2, which appear in this book.

1. Type 1.

$$\Lambda(\partial) = \mathrm{i}\frac{\partial}{\partial t} + \frac{\hbar}{2M}\nabla^2. \tag{3.4.44}$$

In this case,

$$\eta = -1, \qquad D(\partial) = 1, \tag{3.4.45}$$

$$\epsilon_k = \frac{\hbar}{2M}k^2, \tag{3.4.46}$$

$$\overleftrightarrow{\Gamma} = 1, \qquad \Gamma(k, k_0) = 1, \tag{3.4.47}$$

The normalization condition for continuous or discrete spaces:

$$u^\dagger_k u_k = \frac{\hbar}{V}, \quad \text{or} \quad \frac{\hbar}{N} \tag{3.4.48}$$

which give

$$u_k = \sqrt{\frac{\hbar}{V}}, \quad \text{or} \quad \sqrt{\frac{\hbar}{N}}, \tag{3.4.49}$$

respectively. Thus we obtain

$$\phi(x) = \sqrt{\frac{\hbar}{V}} \sum_k \alpha_k e^{\mathrm{i}(k\cdot x - \epsilon_k t)}, \quad \text{or} \quad \sqrt{\frac{\hbar}{N}} \sum_k \alpha_k e^{\mathrm{i}(k\cdot x - \epsilon_k t)}, \tag{3.4.50}$$

respectively.

2. Type 2.

$$-M\frac{\partial^2}{\partial t^2} - \omega^2(\nabla). \tag{3.4.51}$$

In this case, we have

$$\eta = 1, \qquad D(\partial) = 1, \tag{3.4.52}$$

$$\overleftrightarrow{\Gamma} = \mathrm{i}\frac{\overleftrightarrow{\partial}}{\partial t}, \qquad \Gamma(k, k_0) = 2Mk_0, \tag{3.4.53}$$

$$\rho = 1. \tag{3.4.54}$$

the normalization condition for the continuous and discrete spaces are

$$u_k^\dagger(2M\omega_k)u_k = \frac{\hbar}{V}, \quad \text{or} \quad \frac{\hbar}{N}, \tag{3.4.55}$$

$$v_k^\dagger(2M\omega_{-k})v_k = \frac{\hbar}{V}, \quad \text{or} \quad \frac{\hbar}{N}, \tag{3.4.56}$$

which yield

$$u_k = \sqrt{\frac{\hbar}{V}}\frac{1}{\sqrt{2M\omega_k}}, \quad \text{or} \quad \sqrt{\frac{\hbar}{N}}\frac{1}{\sqrt{2M\omega_k}}, \tag{3.4.57}$$

$$v_k = \sqrt{\frac{\hbar}{V}}\frac{1}{\sqrt{2M\omega_{-k}}}. \quad \text{or} \quad \sqrt{\frac{\hbar}{N}}\frac{1}{\sqrt{2M\omega_{-k}}}, \tag{3.4.58}$$

respectively. Thus we obtain

$$\phi(x) = \frac{1}{\sqrt{V}}\sum_k \left[\sqrt{\frac{\hbar}{2M\omega_k}}\alpha_k e^{\mathrm{i}(\boldsymbol{k}\cdot\boldsymbol{x}-\omega_k t)} + \sqrt{\frac{\hbar}{2M\omega_{-k}}}\beta_k^\dagger e^{-\mathrm{i}(\boldsymbol{k}\cdot\boldsymbol{x}-\omega_{-k}t)}\right], \tag{3.4.59}$$

or

$$\phi(x) = \frac{1}{\sqrt{N}}\sum_k \left[\sqrt{\frac{\hbar}{2M\omega_k}}\alpha_k e^{\mathrm{i}(\boldsymbol{k}\cdot\boldsymbol{x}-\omega_k t)} + \sqrt{\frac{\hbar}{2M\omega_{-k}}}\beta_k^\dagger e^{-\mathrm{i}(\boldsymbol{k}\cdot\boldsymbol{x}-\omega_{-k}t)}\right], \tag{3.4.60}$$

respectively. We will use these results later.

3.5 The Physical Particle Representation and Perturbation Theory

The system is composed of physical particles. The free particle fields are described as the creation and annihilation operators of the physical particles. In quantum field theory the physical particles are constructed by the perturbation theory starting with the physical particles. Therefore we must clarify the relations between the physical particles and the interaction Hamiltonian in the Hamiltonian. First we briefly survey the equivalent representations between the Schrödinger, the interaction, and the Heisenberg. Next we clarify the relation between the physical particles and the interaction Hamiltonian by the adiabatic method.

3.5.1 The Schrödinger, the Heisenberg, and the Interaction Representations

In the Schrödinger representation, the operators are time independent; the Hamiltonian H is time independent, and only the state changes obeying the Schrödinger equation:

$$i\hbar\frac{\partial}{\partial t}|\Psi_S(t)\rangle = H|\Psi_S(t)\rangle. \tag{3.5.1}$$

The formal solution of Eq. (3.5.1) is obtained as

$$|\Psi_S(t)\rangle = e^{-iHt/\hbar}|\Psi_S(0)\rangle. \tag{3.5.2}$$

The state $\Psi_S(t)$ is generated by the time independent Hamiltonian. For any operator O, we have

$$\langle\Psi_S(t)|O|\Psi_S(t)\rangle = \langle\Psi_S(0)|e^{iHt/\hbar}Oe^{-iHt/\hbar}|\Psi_S(0)\rangle. \tag{3.5.3}$$

Next we investigate the Heisenberg representation. In the Heisenberg representation, the state is time independent and any operator changes obeying the Heisenberg equation:

$$i\hbar\frac{\partial}{\partial t}O_H(t) = [O_H(t), H], \tag{3.5.4}$$

$$i\hbar\frac{\partial}{\partial t}|\Psi\rangle = 0. \tag{3.5.5}$$

Equation (3.5.4) leads to

$$O_H(t) = e^{iHt/\hbar}O_H(0)e^{-iHt/\hbar}, \tag{3.5.6}$$

Therefore we have

$$\langle\Psi|O_H(t)|\Psi\rangle = \langle\Psi|e^{iHt/\hbar}O_H(0)e^{-iHt/\hbar}|\Psi\rangle. \tag{3.5.7}$$

Now we investigate the interaction representation. We consider that the Hamiltonian consists of the free Hamiltonian for physical particles H_0 and the interaction Hamiltonian H_I:

$$H = H_0 + H_I, \tag{3.5.8}$$

We introduce the interaction state vector as:

$$|\Psi_I(t)\rangle = e^{iH_0t/\hbar}|\Psi_S(t)\rangle. \tag{3.5.9}$$

Then we obtain

$$i\hbar\frac{\partial}{\partial t}|\Psi_I(t)\rangle = H_I(t)|\Psi_I(t)\rangle, \tag{3.5.10}$$

$$H_I(t) = e^{iH_0t/\hbar}H_Ie^{-iH_0t/\hbar}. \tag{3.5.11}$$

For any operator O, we define

$$O_I(t) = e^{\mathrm{i}H_0 t/\hbar} O e^{-\mathrm{i}H_0 t/\hbar}. \tag{3.5.12}$$

Equation (3.5.8) shows

$$\mathrm{i}\hbar \frac{\partial}{\partial t} O_I(t) = [O_I(t), H_0]. \tag{3.5.13}$$

If we put

$$|\Psi\rangle = |\Psi_S(0)\rangle = |\Psi_I(0)\rangle, \qquad O = O_H(0) = O_I(0), \tag{3.5.14}$$

we can show the equivalent representations:

$$\langle\Psi|O_H(t)|\Psi\rangle = \langle\Psi_S(t)|O|\Psi_S(t)\rangle = \langle\Psi_I(t)|O_I(t)|\Psi_I(t)\rangle, \tag{3.5.15}$$

where there are relations

$$\Psi_I(t) = U(t)\Psi_S, \qquad O_H(t) = U^{-1} O_I(t) U(t), \tag{3.5.16}$$

where we define

$$U(t) = e^{\mathrm{i}H_0 t/\hbar} e^{-\mathrm{i}Ht/\hbar}, \tag{3.5.17}$$

which satisfies

$$\mathrm{i}\hbar \frac{d}{dt} U(t) = H_I(t) U(t). \tag{3.5.18}$$

We can solve Eq. (3.5.18) by perturbative calculation under the initial condition $U(0) = 1$.

3.5.2 The Adiabatic Method

In order to see the relations between physical particles and the interactions, we recall the scattering process: entering particles before the reaction and outgoing particles after the reaction are required to be physical particles. Therefore to satisfy the requirement, we introduce the adiabatic method; we replace the interaction Hamiltonian H_I replaced by $e^{-\epsilon|t|} H_I(t)$ denoted by $H_{I\epsilon}(t)$. By introducing the adiabatic factor, we can guarantee to carry out the unitary transformation between the Heisenberg and the interaction representation.

For any operator $O(x)$ in the Heisenberg representation, there is a corresponding operator in the interaction representation, satisfying the following unitary transformation:

$$O(x) = U_\epsilon^{-1}(t) O_I(x) U_\epsilon(t), \tag{3.5.19}$$

The unitary operator $U_\epsilon(t)$ is chosen such that the relation

$$\mathrm{i}\hbar \frac{\partial}{\partial t} O_I(x) = [O_I(x), H_0] \tag{3.5.20}$$

holds for the free Hamiltonian H_0 of the free physical field ϕ. Note that the free physical field is independent of the interaction Hamiltonian. With this condition, the interaction Hamiltonian $H_{I\epsilon}$ is defined as

$$i\hbar\frac{d}{dt}U_\epsilon(t) = H_{I\epsilon}(t)U_\epsilon(t). \tag{3.5.21a}$$

Since

$$i\hbar\frac{d}{dt}[U_\epsilon^{-1}U_\epsilon] = i\hbar\frac{d}{dt}U_\epsilon^{-1}U_\epsilon + U_\epsilon^{-1}H_{I\epsilon}U_\epsilon = 0,$$

we have

$$i\hbar\frac{d}{dt}U_\epsilon^{-1}(t) = -U_\epsilon^{-1}(t)H_{I\epsilon}. \tag{3.5.21b}$$

From Eq. (3.5.19) we obtain

$$i\hbar\frac{\partial}{\partial t}O(x) = U_\epsilon^{-1}(t)[O_I(x), H_0 + H_{I\epsilon}]U_\epsilon(t). \tag{3.5.22}$$

For the Hamiltonian H in the Heisenberg representation, we have

$$i\hbar\frac{\partial}{\partial t}O(x) = [O(x), H]. \tag{3.5.23}$$

In the limit $\epsilon \to 0$, Eq. (3.5.22) tends to Eq. (3.5.23) therefore we have

$$H = \lim_{\epsilon\to 0}\{U_\epsilon^{-1}(t)[H_0 + H_{I\epsilon} + W_0]U_\epsilon(t)\}, \tag{3.5.24}$$

where W_0 is a c-number.

Now we prove the following significant relation:

If we consider the physical particles of the system as the physical free fields

$$H_0 = \sum_k \hbar\omega_k\alpha_k^\dagger\alpha_k, \tag{3.5.25}$$

then we obtain

$$H = H_0 + W_0 \qquad \text{under the weak relations.} \tag{3.5.26}$$

In order to prove Eq. (3.5.26), if we consider

$$\frac{i}{\hbar}[H_0, H_{I\epsilon}(t)] = e^{\epsilon|t|}\frac{dH_I(t)}{dt}, \tag{3.5.27}$$

we obtain

$$\begin{aligned}\frac{d}{dt}\left\{U_\epsilon^{-1}(t)[H_0 + H_{I\epsilon}]U_\epsilon(t)\right\} &= U_\epsilon^{-1}(t)\left\{\frac{i}{\hbar}[H_0, H_{I\epsilon}(t)] + \frac{d}{dt}H_{I\epsilon}(t)\right\}U_\epsilon(t)\\ &= \left(\frac{d}{dt}e^{\epsilon|t|}\right)U_\epsilon^{-1}(t)H_IU_\epsilon(t)\end{aligned} \tag{3.5.28}$$

Since in the limit $t \to \infty$, $U_\epsilon \to 1$, and $H_{I\epsilon} \to 0$, Eq. (3.5.28) leads to

$$\left\{U_\epsilon^{-1}(t)[H_0 + H_{I\epsilon}]U_\epsilon(t)\right\}\Big|_{t=0} = H_0 + \int_\infty^0 dt' U_\epsilon^{-1}(t') H_I U_\epsilon(t') \frac{d}{dt'} e^{\epsilon t'}$$
$$= H_0 + \epsilon \int_\infty^0 dt' U_\epsilon^{-1}(t') H_{I\epsilon} U_\epsilon(t')$$
$$= H_0 + \epsilon \int_\infty^0 dt' H_{I\epsilon}(t') + \cdots . \tag{3.5.29}$$

where from the second to the last equation, we have used the relation

$$U_\epsilon(t) = 1 - \frac{\mathrm{i}}{\hbar} \int_\infty^t dt' H_{I\epsilon}(t') U_\epsilon(t'), \tag{3.5.30}$$

which is derived from Eq. (3.5.21).

Next we calculate the second term in Eq. (3.5.29). First we note the relation for $t < 0$

$$\langle W|H_{I\epsilon}(t)|W'\rangle = e^{\epsilon t}\langle W|e^{\mathrm{i}H_0 t/\hbar} H_I(0) e^{-\mathrm{i}H_0 t/\hbar}|W'\rangle$$
$$= e^{\epsilon t}\langle W|H_I(0)|W'\rangle e^{\mathrm{i}(W-W')t/\hbar}, \tag{3.5.31}$$

where we have used the eigenvalue equations such as

$$H_0|W\rangle = W|W\rangle. \tag{3.5.32}$$

Thus, we obtain

$$\lim_{\epsilon\to 0} \epsilon \langle W| \int_\infty^0 dt' H_{I\epsilon}(t')|W'\rangle = \langle W|H_I(0)|W'\rangle \lim_{\epsilon\to 0} \frac{\hbar}{\mathrm{i}} \frac{\epsilon}{W - W' - \mathrm{i}\hbar\epsilon}$$
$$= \langle W|H_I(0)|W'\rangle \delta_{W,W'}, \tag{3.5.33}$$

where we have used the relation

$$\lim_{\epsilon\to 0} \frac{\epsilon}{W - W' - \mathrm{i}\hbar\epsilon} = -\frac{1}{\mathrm{i}\hbar}\delta_{W,W'}. \tag{3.5.34}$$

The term $\delta_{W,W'}$ is not the Dirac's delta function, but the Kronecker's delta. The states W are physical particle states. Therefore the right hand side in Eq. (3.5.33) remains only for one physical particle states. One physical states contribute to the self-energy of the free Hamiltonian H_0. But since H_0 already contains the self-energy. Thus we could prove the relation (3.5.26).

The total Hamiltonian is described by the free Hamiltonian for physical particles. However, this does not mean that there exits no interaction. Once physical particles such as fundamental particles, collective modes, and the bound states are given, the Hamiltonian of the physical particles yields the total Hamiltonian of the system. So there remains the important task to find out physical particles of the system. In this book we show: in liquids, the free Hamiltonian H_0 consists of constituent particles and sound. In glasses, H_0 consists of constituent particles and phonons.

3.6 The Spectral Representations of Two-Particle Green's Functions

First we define causal n-particle Green's functions, which is expressed by the vacuum expectation values of the T-product of the Heisenberg operators:

$$G_c^{AB\ldots Z}(x_1, x_2, \ldots, x_n) = \langle 0|T[A(x_1)B(x_2)\ldots Z(x_n)]|0\rangle, \tag{3.6.1}$$

where T-product of the Heisenberg operators rearrange the operators in time order as

$$\begin{aligned} &T[A(x_1)B(x_2)\ldots Z(x_n)] \\ &\quad = (\pm)^P A'(x_1')B'(x_2')\ldots Z'(x_n') \qquad \text{for} \qquad t_1'\rangle t_2'\rangle\ldots\rangle t_n'. \end{aligned} \tag{3.6.2}$$

Hereafter, the upper and the lower signs such as $(\pm)^P$ are referred to as bosons and fermions, respectively. P means the number of the transposition in the rearrangement.

Therefore the causal two-particle Green's function is defined by

$$\begin{aligned} G_c^{AB}(x-y) &= \langle 0|T[A(x)B(y)]|0\rangle \\ &= \theta(t_x - t_y)\langle 0|A(x)B(y)|0\rangle \pm \theta(t_y - t_x)\langle 0|B(t_y)A(t_x)|0\rangle. \end{aligned} \tag{3.6.3}$$

The retarded and the advanced two-particle Green's functions are defined as

$$G_r^{AB}(x-y) = \theta(t_x - t_y)\langle 0|[A(x), B(y)]_{\mp}|0\rangle, \tag{3.6.4}$$

and

$$G_a^{AB}(x-y) = -\theta(t_y - t_x)\langle 0|[A(x), B(y)]_{\mp}|0\rangle, \tag{3.6.5}$$

respectively. In $[A, B]_{\pm}$, the commutation relation and anti-commutation are for bosons and fermions, respectively.

Using

$$O(t) = e^{\mathrm{i}Ht/\hbar} O e^{-\mathrm{i}Ht/\hbar}, \qquad O(t, \boldsymbol{x}) = e^{-\mathrm{i}\boldsymbol{P}\cdot\boldsymbol{x}} O(t) e^{\mathrm{i}\boldsymbol{P}\cdot\boldsymbol{x}},$$

we obtain

$$\begin{aligned} G_c^{AB}(x-y) &= \sum_n \{\theta(t_x - t_y)\langle 0|A(x)|n\rangle\langle n|B(y)|0\rangle \\ &\quad \pm\theta(t_y - t_x) < 0|B(y)|n\rangle\langle n|A(x)|0\rangle\} \\ &= \sum_n \Big\{\theta(t_x - t_y)\langle 0|A(0)|n\rangle\langle n|B(0)|0\rangle e^{\mathrm{i}[\boldsymbol{k}_n\cdot(\boldsymbol{x}-\boldsymbol{y}) - \omega_n(t_x - t_y)]} \\ &\quad \pm\theta(t_y - t_x)\langle 0|B(0)|n\rangle\langle n|A(0)|0\rangle e^{-\mathrm{i}[\boldsymbol{k}_n\cdot(\boldsymbol{x}-\boldsymbol{y}) - \omega_n(t_x - t_y)]}.\Big\} \end{aligned} \tag{3.6.6}$$

If we use the formula

$$\theta(t_x - t_y) = \lim_{\epsilon\to 0} \frac{\mathrm{i}}{2\pi} \int_{-\infty}^{\infty} dw \frac{1}{w + \mathrm{i}\epsilon} e^{-\mathrm{i}w(t_x - t_y)}, \tag{3.6.7}$$

and we define

$$\sigma^{AB}(\boldsymbol{k}, w) = \frac{(2\pi)^3}{\hbar} \sum_n \langle 0|A(0)|n\rangle\langle n|B(0)|0\rangle\delta(\boldsymbol{k} - \boldsymbol{k}_n)\delta(w - \omega_n), \tag{3.6.8}$$

we obtain

$$\begin{aligned} G_c^{AB}(x - y) &= \frac{\mathrm{i}\hbar}{(2\pi)^4} \int_0^\infty dw \int d^3k dw' \\ &\times \left\{ \frac{1}{w' + \mathrm{i}\epsilon} \sigma^{AB}(\boldsymbol{k}, w) e^{\mathrm{i}[\boldsymbol{k}\cdot(\boldsymbol{x}-\boldsymbol{y}) - (w+w')(t_x - t_y)]} \right. \\ &\quad \left. \pm \frac{1}{w' + \mathrm{i}\epsilon} \sigma^{BA}(\boldsymbol{k}, w) e^{-\mathrm{i}[\boldsymbol{k}\cdot(\boldsymbol{x}-\boldsymbol{y}) - (w+w')(t_x - t_y)]} \right\} \\ &= \frac{\mathrm{i}\hbar}{(2\pi)^4} \int_0^\infty dw \int d^3k dk_0 \\ &\times \left\{ \frac{\sigma^{AB}(\boldsymbol{k}, w)}{k_0 - w + \mathrm{i}\epsilon} \mp \frac{\sigma^{BA}(-\boldsymbol{k}, w)}{k_0 + w - \mathrm{i}\epsilon} \right\} e^{\mathrm{i}[\boldsymbol{k}\cdot(\boldsymbol{x}-\boldsymbol{y}) - k_0(t_x - t_y)]}, \end{aligned} \tag{3.6.9}$$

where we have changed variables in getting the final equation. If we define the Fourier transformation of the two-point Green's function as

$$G_{c,r,a}^{AB}(x - y) = \frac{\mathrm{i}\hbar}{(2\pi)^4} \int d^4k e^{\mathrm{i}k(x-y)} G_{c,r,a}^{AB}(k), \tag{3.6.10}$$

where $d^4k = d^3kdk_0$ and $kx = \boldsymbol{k}\cdot\boldsymbol{x} - k_0 t$, we obtain the causal two-point Green's function:

$$G_c^{AB}(k) = \int_0^\infty dw \left\{ \frac{\sigma^{AB}(\boldsymbol{k}, w)}{k_0 - w + \mathrm{i}\epsilon} \mp \frac{\sigma^{BA}(-\boldsymbol{k}, w)}{k_0 + w - \mathrm{i}\epsilon} \right\}, \tag{3.6.11}$$

where if we define

$$\sigma(\boldsymbol{k}, w) = \theta(w)\sigma^{AB}(\boldsymbol{k}, w) \mp \theta(-w)\sigma^{BA}(-\boldsymbol{k}, -w), \tag{3.6.12}$$

we can obtain from Eq. (3.6.11)

$$G_c^{AB}(k) = \int_{-\infty}^\infty dw \sigma(\boldsymbol{k}, w) \left\{ \frac{\theta(w)}{k_0 - w + \mathrm{i}\epsilon} + \frac{\theta(-w)}{k_0 - w - \mathrm{i}\epsilon} \right\}. \tag{3.6.13}$$

Expressions (3.6.11) and (3.6.13) are called the spectral representation of the causal two-point Green's function.

Now we consider single particle states with frequency $\omega_{\boldsymbol{k}}$ in Eq. (3.6.8), σ^{AB} has a following form:

$$\sigma^{AB}(\boldsymbol{k}, w) = Z^{AB}(\boldsymbol{k})\delta(w - \omega_{\boldsymbol{k}}) + \tilde{\sigma}^{AB}(\boldsymbol{k}, w). \tag{3.6.14}$$

Here $Z^{AB}(\boldsymbol{k})$ is the contribution of the single particle states and $\tilde{\sigma}^{AB}$ is that from the many particle states. The $Z^{AB}(\boldsymbol{k})$-term with frequency $\omega_{\boldsymbol{k}}$ contributes to the Green's function (3.6.11) as a single pole singularity with the residue $Z^{AB}(\boldsymbol{k})$ in k_0-plane, while since $\tilde{\sigma}^{AB}$ is a smooth function of w for a given $\boldsymbol{k}$, $\tilde{\sigma}^{AB}$-term contributes to the Green's function as cut singularity along the real k_0-axis. The $Z^{AB}(\boldsymbol{k})$-term has the value: $1 \geq Z^{AB}(\boldsymbol{k}) \geq 0$.

From Eq. (3.6.8) we obtain

$$\langle 0|A(x)B(y)|0\rangle = \frac{\hbar}{(2\pi)^3}\int d^3k dw \sigma^{AB}(\boldsymbol{k}, w) e^{\mathrm{i}(\boldsymbol{k}\cdot(\boldsymbol{x}-\boldsymbol{y})-wt)}, \tag{3.6.15}$$

$$\langle 0|[A, B]_{\mp}|0\rangle|_{t_x=t_y} = \frac{\hbar}{(2\pi)^3}\int d^3k e^{\mathrm{i}\boldsymbol{k}\cdot(\boldsymbol{x}-\boldsymbol{y})}\int_{-\infty}^{\infty} dw\sigma(\boldsymbol{k}, w). \tag{3.6.16}$$

If B is canonical conjugate to A, we have

$$[A(x), B(y)]_{\mp}\delta(t_x - t_y) = \hbar\delta^{(4)}(x - y). \tag{3.6.17}$$

Equations (3.6.16) and (3.6.17) lead to the sum rule

$$\int_{-\infty}^{\infty} dw\sigma(\boldsymbol{k}, w) = 1. \tag{3.6.18}$$

For $B = A^{\dagger}$, Eq. (3.6.8) leads to

$$\sigma^{AB}(\boldsymbol{k}, w) \geq 0, \tag{3.6.19}$$

Equations (3.6.14) and (3.6.18) yield

$$Z^{AB}(\boldsymbol{k}) \geq 0, \qquad \tilde{\sigma}^{AB}(\boldsymbol{k}, w) \geq 0. \tag{3.6.20}$$

Next we investigate the spectral representations of retarded and advanced Green's functions. From the definition Eq.(3.6.10), we obtain

$$G_r^{AB} = \int dw \frac{\sigma(\boldsymbol{k}, w)}{k_0 - w + i\epsilon}, \tag{3.6.21}$$

$$G_a^{AB} = \int dw \frac{\sigma(\boldsymbol{k}, w)}{k_0 - w - i\epsilon}. \tag{3.6.22}$$

Now we return to the causal Green's function. We consider a case that the physical field ϕ is a linear term of the dynamical map of A and $\bar{\phi} = \phi^{\dagger}\eta$ is the linear term of the dynamical map of B:

$$A(x) = Z^A(\nabla)\phi + \cdots, \tag{3.6.23}$$

$$B(x) = Z^B(\nabla)\bar{\phi} + \cdots, \tag{3.6.24}$$

where the dots mean the higher order normal products. We then have

$$G_c^{AB} = Z^A(\nabla_x)Z^B(\nabla_y)G_c^0(x - y) + \cdots, \tag{3.6.25}$$

the dots mean the contributions of many particle states and $G_c^0(x - y)$ is given by

$$G_c^0(x - y) = \langle 0|T[\phi(x)\bar{\phi}(y)]|0\rangle, \tag{3.6.26}$$

which is the Green's function for the physical particles.

3.7 Invariance, the Noether Current and the Ward-Takahashi Relations

The symmetry or invariance of the system plays an essential role in developing the theory. When the system holds invariant under operations such as translation, rotation, and so on, the system is called to have the symmetry or invariance of translation, rotation, and so on. The symmetry or invariance requires certain relations between Green's functions, self-energy parts, and vertex parts, which are called the Ward-Takahashi (WT) relations. Therefore, the WT relations play an important role in determining the Green's functions, the self-energy parts, and the vertex parts in the approximation procedures. The WT relations also play an important role in determining the Hamiltonian, the dynamical map, the NG theorem, and so on. First we present a general method, the Noether current in treating the invariance in quantum field theory.

3.7.1 The Noether Current and Lagrangian

We consider a system, in which the Lagrangian density $\mathcal{L}[\psi(x), \partial_\mu \psi(x)]$ governs the dynamics. Here $\psi(x)$ represents a set of Heisenberg fields $\psi_i;\ i = 1, 2, \cdots$. We use the following four-dimensional relations:

$$\begin{aligned} &x_\mu = (x_0 = t, x_1, x_2, x_3), \\ &\partial^\mu = (\partial/\partial t, \partial/\partial x_1, \partial/\partial x_2, \partial/\partial x_3), \\ &x_\mu = g_{\mu\nu} x^\nu, \quad x^\mu = g^{\mu\nu} x_\nu; \qquad \partial_\mu = g_{\mu\nu}\partial^\nu, \quad \partial^\mu = g^{\mu\nu}\partial_\nu. \\ &\qquad -g_{00} = g_{11} = g_{22} = g_{33} = 1, \\ &\qquad -g^{00} = g^{11} = g^{22} = g^{33} = 1, \qquad \text{otherwise } g_{\mu\nu} = g^{\mu\nu} = 0, \end{aligned}$$

where we have used the relations

$$A_\mu B^\mu \equiv \sum_{\mu=0}^{3} A_\mu B^\mu, \qquad A^\mu B_\mu \equiv \sum_{\mu-0}^{3} A^\mu B_\mu, \qquad A_\mu B^\mu = A^\mu B_\mu.$$

First we derive the Lagrange equation. To do so, we start with the action integral:

$$S = \int_{t_1}^{t_2} d^4x \mathcal{L}[\psi(x), \psi_\mu(x)], \tag{3.7.1}$$

where we have defined

$$\psi_\mu(x) = \partial_\mu \psi(x). \tag{3.7.2}$$

We require that the action integral S is invariant for arbitrary variation $\delta\psi(x)$ which vanishes at the end points t_1 and t_2, $\delta\psi(t_1, \boldsymbol{x}) = \delta\psi(t_2, \boldsymbol{x}) = 0$. We carry out the variation of Eq. (3.7.1) with respect to $\psi(x)$

$$\delta S = \int_{t_1}^{t_2} d^4x \left[\frac{\partial \mathcal{L}}{\partial \psi(x)} \delta\psi(x) + \frac{\partial \mathcal{L}}{\partial \psi_\mu(x)} \delta\psi_\mu(x) \right]. \tag{3.7.3}$$

Integrating the second terms by parts and considering the boundary conditions, we find

$$\delta S = \int_{t_1}^{t_2} d^4x \left[\frac{\partial \mathcal{L}}{\partial \psi(x)} - \partial_\mu \frac{\partial \mathcal{L}}{\partial \psi_\mu(x)} \right] \delta\psi(x). \tag{3.7.4}$$

Since $\delta\psi(x)$ is arbitrary, we obtain the Lagrange equation:

$$\frac{\partial \mathcal{L}}{\partial \psi(x)} - \partial_\mu \frac{\partial \mathcal{L}}{\partial \psi_\mu(x)} = 0. \tag{3.7.5}$$

In the Lagrangian density $\mathcal{L}$, the canonical conjugate of $\psi(x)$ is defined by

$$\pi(x) = \frac{\partial \mathcal{L}}{\partial \psi^0(x)}, \tag{3.7.6}$$

which satisfies

$$[\psi(x), \pi(y)]_{\mp} = \mathrm{i}\hbar\delta^{(4)}(x-y). \tag{3.7.7}$$

The Hamiltonian density $\mathcal{H}$ is defined by

$$\mathcal{H}(x) = \pi(x)\psi^0(x) - \mathcal{L}(x). \tag{3.7.8}$$

The Hamiltonian is given by

$$H = \int d^3x \mathcal{H}(x). \tag{3.7.9}$$

The Heisenberg equation for $\psi(x)$ is also given by

$$\mathrm{i}\hbar\psi^0(x) = [\psi(x), H]. \tag{3.7.10}$$

A continuous transformation

$$\psi(x) \to \psi'(x') = \psi(x) + \theta\delta\psi(x), \tag{3.7.11}$$

$$x_\mu \to x'_\mu = x_\mu + \theta\delta x_\mu, \tag{3.7.12}$$

with an infinitesimal constant θ, changes the Lagrangian as

$$d^4x'\mathcal{L}[\psi'(x'), \partial'_\mu\psi(x')] - d^4x\mathcal{L}[\psi(x), \partial_\mu\psi(x)] = \theta d^4x\delta\mathcal{L}(x). \tag{3.7.13}$$

Now we define

$$\theta\delta^0\psi(x) = \psi'(x) - \psi(x). \tag{3.7.14}$$

Since

$$\psi'(x') = \psi'(x) + \theta\delta x_\mu\partial^\mu\psi'(x) = \psi'(x) + \theta\delta x_\mu\partial^\mu\psi(x), \tag{3.7.15}$$

using Eqs. (3.7.11) and (3.7.14), we obtain

$$\begin{aligned}\theta\delta\psi(x) &= \psi'(x') - \psi(x) \\ &= \psi'(x) + \theta\delta x_\mu \partial^\mu \psi(x) - \psi(x) \\ &= \theta\delta^0 \psi(x) + \theta\delta x_\mu \partial^\mu \psi(x).\end{aligned}$$

This equation yields

$$\delta^0 \psi(x) = \delta\psi(x) - \delta x_\mu \partial^\mu \psi(x). \tag{3.7.16}$$

If we consider the relation

$$d^4x' = \frac{\partial(x_0' x_1' x_2' x_3')}{\partial(x_0 x_1 x_2 x_3)} d^4x, \tag{3.7.17}$$

$$\frac{\partial(x_0' x_1' x_2' x_3')}{\partial(x_0 x_1 x_2 x_3)} = 1 + \theta\partial^\mu \delta x_\mu, \tag{3.7.18}$$

Equation (3.7.13) yields

$$\begin{aligned}\theta\delta\mathcal{L}(x) &= \mathcal{L}[\psi'(x'), \partial_\mu'\psi'(x')] - \mathcal{L}[\psi(x), \partial_\mu\psi(x)] + \theta(\partial^\mu \delta x_\mu \mathcal{L}(x) \\ &= \frac{\partial\mathcal{L}}{\partial\psi(x)}[\psi'(x') - \psi(x)] \\ &\quad + \frac{\partial\mathcal{L}}{\partial\psi_\mu(x)}[\partial_\mu'\psi'(x') - \partial_\mu\psi(x)] + \theta(\partial^\mu\delta x_\mu \mathcal{L}(x),\end{aligned} \tag{3.7.19}$$

where in deriving Eq. (3.7.18), we have considered that the terms with off-diagonal component are of the higher order, and $\mathcal{L}(x)$ means $\mathcal{L}[\psi(x), \partial_\mu\psi(x)]$. Using Eq. (3.7.11) and the relation

$$\begin{aligned}\partial_\mu' &= \frac{\partial}{\partial x'^\mu} = \frac{\partial x^\nu}{\partial x'^\mu}\partial_\nu = \frac{\partial}{\partial x'^\mu}(x'^\nu - \theta\delta x^\nu)\partial_\nu \\ &= \partial_\mu - \theta\partial_\mu\delta x^\nu \partial_\nu = \partial_\mu - \theta\partial_\mu \delta x_\nu \partial^\nu,\end{aligned}$$

we obtain

$$\partial_\mu'\psi'(x') = \partial_\mu\psi(x) + \theta\partial_\mu\delta\psi(x) - \theta\partial^\nu\psi(x)\partial_\mu\delta x_\nu. \tag{3.7.20}$$

Inserting Eqs. (3.7.11) and (3.7.20) into Eq. (3.7.19), and then using the Lagrange Eq. (3.7.5), we obtain

$$\begin{aligned}\delta\mathcal{L}(x) &= \frac{\partial\mathcal{L}}{\partial\psi}\delta\psi(x) + \frac{\partial\mathcal{L}}{\partial\psi_\mu}\left[\partial_\mu\delta\psi(x) - \partial^\nu\psi(x)\partial_\mu\delta x_\nu\right] + \partial^\nu\delta x_\nu\mathcal{L} \\ &= \partial_\mu\left[\frac{\partial\mathcal{L}}{\partial\psi_\mu}\delta\psi(x)\right] - \frac{\partial\mathcal{L}}{\partial\psi_\mu}\partial^\nu\psi(x)\partial_\mu\delta x_\nu + \partial^\nu\delta x_\nu\mathcal{L}.\end{aligned} \tag{3.7.21}$$

Using the Lagrange Eq. (3.7.5), we have

$$\begin{aligned}\partial_\mu\left[\frac{\partial\mathcal{L}}{\partial\psi_\mu}\partial^\nu\psi(x)\right] &= \frac{\partial\mathcal{L}}{\partial\psi_\mu}\partial_\mu\partial^\nu\psi + \frac{\partial\mathcal{L}}{\partial\psi}\partial^\nu\psi \\ &= \frac{\partial\mathcal{L}}{\partial\psi_\mu}\partial^\nu\partial_\mu\psi + \frac{\partial\mathcal{L}}{\partial\psi}\partial^\nu\psi \\ &= \partial^\nu\mathcal{L}[\psi(x), \partial_\mu\psi(x)].\end{aligned} \tag{3.7.22}$$

Therefore we can rewrite Eq. (3.7.21) as

$$\begin{aligned}\delta\mathcal{L}(x) &= \partial_\mu\left\{\frac{\partial\mathcal{L}}{\partial\psi_\mu}[\delta\psi - \partial^\nu\psi\delta x_\nu] + \delta x^\mu\mathcal{L}\right\},\\ &= \partial_\mu\left\{\frac{\partial\mathcal{L}}{\partial\psi_\mu}\delta^0\psi(x) + \delta x^\mu\mathcal{L}\right\},\end{aligned} \tag{3.7.23}$$

where we have used Eq. (3.7.16). If we define the Noether current [4]:

$$N_\mu = \frac{\partial\mathcal{L}}{\partial\psi^\mu}\delta^0\psi(x) + \delta x_\mu\mathcal{L}, \tag{3.7.24}$$

Equation (3.7.23) leads to

$$\partial^\mu N_\mu = \delta\mathcal{L}. \tag{3.7.25}$$

This expresses that the divergence of the Noether current is equal to the change of the Lagrangian. If Eq. (3.7.13) is invariant under the transformations (3.7.11) and (3.7.12); $\delta\mathcal{L} = 0$, we obtain the Noether's theorem:

$$\partial^\mu N_\mu = 0. \tag{3.7.26}$$

Now we construct the generator for the transformation $\delta^0\psi(x)$. To do so, we investigate

$$\begin{aligned}N_0(x) &= \pi(x)\delta^0\psi(x) \pm \delta x_0\mathcal{L}\\ &= \pi(x)\left[\delta\psi(x) - \nabla\psi(x)\cdot\delta\boldsymbol{x} - \psi^0\delta x_0\right] + \delta x_0\mathcal{L}\\ &= \pi(x)\left[\delta\psi(x) - \nabla\psi(x)\cdot\delta\boldsymbol{x}\right] - \delta x_0\mathcal{H}(x).\end{aligned} \tag{3.7.27}$$

If we define

$$N(t) = \int d^3x N_0(x), \tag{3.7.28}$$

the canonical equal-time commutation relations (3.7.7) and the Heisenberg Eq. (3.7.10) lead to

$$\begin{aligned}[\psi(x), N(t)]\delta(t_x - t) &= \mathrm{i}\hbar\delta(t_x - t)\left[\delta\psi(x) - \nabla\psi(x)\cdot\delta\boldsymbol{x} - \psi^0\delta x_0\right]\\ &= \mathrm{i}\hbar\delta(t_x - t)\delta^0\psi(x),\end{aligned} \tag{3.7.29}$$

which yields

$$e^{\mathrm{i}\theta N/\hbar}\psi(x)e^{-\mathrm{i}\theta N/\hbar} = \psi(x) + \theta\delta^0\psi(x) + \cdots. \tag{3.7.30}$$

This equation shows that $N(x)$ is a generator of $\delta^0\psi(x)$.

Therefore if we want to transform

$$\psi(x) \longrightarrow \psi(x + \theta\delta x), \tag{3.7.31}$$

we take the following generator

$$N = \int d^3x N_0(x), \qquad N_0(x) = \pi(x)\,[\nabla\psi(x)\cdot\delta\boldsymbol{x}] - \delta x_0\mathcal{H}(x), \tag{3.7.32}$$

which leads to

$$e^{\mathrm{i}\theta N/\hbar}\psi(x)e^{-\mathrm{i}\theta N/\hbar} = \psi(x+\theta\delta x) = \psi(x) + \theta[\nabla\psi(x)\cdot\delta\boldsymbol{x} + \psi^0\delta x_0] + \cdots. \tag{3.7.33}$$

N is the generator of the spatial and time translational tranformation.

3.7.2 The Ward-Takahashi Relation

We start with the time derivative of a Green's function with $N(t)$:

$$\begin{aligned}
&\frac{\partial}{\partial t}\langle 0|T[N(t)\psi(x_1)\ldots\psi(x_n)]|0\rangle \\
&= \frac{\partial}{\partial t}\sum_{i=1}^{n}\langle 0|T\psi(x_1) \\
&\quad \ldots\{\theta(t-t_i)N(t)\psi(x_i) + \theta(t_i - t)\psi(x_i)N(t)\}\ldots\psi(x_n)|0\rangle \\
&= \sum_{i=1}^{n}\delta(t-t_i)\langle 0|T[\psi(x_1)\ldots[N(t),\psi(x_i)]\ldots\psi(x_n)|0\rangle \\
&\quad + \langle 0|T[\dot{N}(t)\psi(x_1)\ldots\psi(x_n)]|0\rangle.
\end{aligned} \tag{3.7.34}$$

Equation (3.7.25) yields

$$\int d^3x\partial^\mu N_\mu = \int d^3x\dot{N}_0(x) + \int d^3x\nabla\cdot\boldsymbol{N}(x) = \int d^3x\dot{N}_0(x) = \int d^3x\delta\mathcal{L},$$

where we have used the Gauss's theorem:

$$\int_V d^3x\nabla\cdot\boldsymbol{N}(t) = \int_S d^2x N_n(x) = 0,$$

N_n means the normal component of $\boldsymbol{N}$ on the surface S. Using Eq. (3.7.28), we obtain

$$\dot{N}(t) = \int d^3x\delta\mathcal{L}(x), \tag{3.7.35}$$

By using Eqs. (3.7.29) and (3.7.35), (3.7.34) yields

$$\begin{aligned}
&\frac{\partial}{\partial t}\langle 0|T[N(t)\psi(x_1)\ldots\psi(x_n)]|0\rangle \\
&= -\mathrm{i}\hbar\sum_{i=1}^{n}\delta(t-t_i)\langle 0|T[\psi(x_1)\cdots\delta^0\psi(x_i)\ldots\psi(x_n)|0\rangle \\
&\quad + \int d^3x\langle 0|T[\delta\mathcal{L}\psi(x_1)\ldots\psi(x_n)]|0\rangle.
\end{aligned} \tag{3.7.36}$$

Carrying out the t-integration from $-\infty$ to ∞ on both sides of Eq. (3.7.36), we obtain

$$i\hbar\sum_{i=1}^{n}\langle 0|T[\psi(x_1)\cdots\delta^0\psi(x_i)\ldots\psi(x_n)|0\rangle$$
$$=\int d^4x\langle 0|T[\delta\mathcal{L}\psi(x_1)\ldots\psi(x_n)]|0\rangle, \tag{3.7.37}$$

where we have considered $N(\pm\infty)=0$. This is the Ward-Takahashi (WT) relation associated with the transformation, Eqs. (3.7.11) and (3.7.12) [5–7]. The WT relations restrict the relations among the Green's functions so that the WT relations play an important role in calculating the Green's functions self-consistently. In the condensed state, the Lagrangian or the Hamiltonian is invariant under the spatial translation and rotation, but the symmetry is spontaneously broken; the pair distribution function appears. In constructing the WT relations, we add an infinitesimal term $\mathcal{L}_\epsilon$ to the invariant Lagrangian to make the condensed state and at the end of the calculation we carry out $\lim \mathcal{L}_\epsilon \longrightarrow 0$. In the WT relation for the boson fields inducing the symmetry breaking, we can obtain the Nambu–Goldstone theorem [8–11].

References

[1] H. Umezawa, H. Matsumoto, M. Tachiki, Thermo Field Dynamics and Condenced States, North-Holland, Amsterdum, 1982.
[2] J. von Neumann, Mathematische Grundlagen der Quantunmechanik, Springer, Belrin, 1932.
[3] P.A.M. Dirac, Principles of Quantum Mechanics, Oxford, 1930.
[4] E. Noether, Goett. Nachr. (1918)235.
[5] J.C. Ward, Phys. Rev, 78 (1950) 182.
[6] Y. Takahashi, Nuovo Cimento 6 (1957) 370.
[7] Y. Takahashi, Phys. Rev. D15 (1977) 1589.
[8] Y. Nambu, G. Jona-Lasinio, Phys. Rev. 122 (1961) 345.
[9] Y. Nambu, G. Jona-Lasinio, Phys. Rev. 124 (1961) 246.
[10] J. Goldstone, Nuovo Cim. 19 (1961) 154.
[11] J. Goldstone, A. Salam, S. Weinberg, Phys. Rev. 127 (1962) 965.

4 Temperature Green's Functions

The quantum field theory, especially the Green's function method associated with the Feynman diagrams, has played an essential role in the theory of elementary particles. However, the theory of elementary particles almost deals with few particle systems, so the systems are not thermodynamical and statistical ensembles. Therefore, the systems are limited to the zero temperature.

On the other hand, since actual condensed systems consist of enormous number of particles, the aspect of thermodynamical and statistical ensembles appears; we must treat the condensed systems with thermodynamical and statistical theory. So we must extend the Green's function method at zero temperature to that in the grand canonical ensemble, the Matsubara temperature Green's function method [1–6] in the equilibrium state. The Feynman diagrammatic rules at finite temperature is almost the same as those at the zero temperature. The Wick's theorem according to the periodic nature of the temperature Green's functions with respect to the imaginary Matsubara temperature frequencies helps us treat the temperature Green's function method tractable.

Even in the non-equilibrium state, we must approach from thermodynamical and statistical theory, because there is no general theory approaching non-equilibrium systems. By analytically continuing the imaginary Matsubara temperature frequency to the real frequency in the temperature Green's function from the upper or the lower plane, the temperature Green's function reduces to the advanced or retarded Green's function, respectively, the method of which is equivalent to the Mori's memory function method [7–10]. Therefore we can deal with the dynamical and dissipative processes from the temperature Green's function method associated with the Feynman diagrams. Since the Feynman diagrammatic method is very tractable, the temperature Green's function method associated with the Feynman diagrams plays an essential role in the theory of the condensed matter.

4.1 Definition of the Temperature Green's Functions

The Matsubara temperature Green's function method is developed in the grand canonical ensemble, where thermodynamics is governed by the thermodynamical function $\Omega(T, V, \mu)$, the thermodynamical valuables of which are temperature T, the Volume V, and chemical potential μ. We first determine the statistical operator in the grand canonical ensemble relating to the thermodynamical function. Using the statistical operator, we define the temperature Green's function.

Liquid Glass Transition. DOI: http://dx.doi.org/10.1016/B978-0-12-407177-3.00004-0

4.1.1 Thermodynamical Functions

First we briefly review the thermodynamical equations [11]. The fundamental thermodynamical equation is

$$dE = T\,dS - p\,dV + \mu\,dN, \tag{4.1.1}$$

so that the internal energy $E(S, V, N)$ is a function of S, V, N and

$$T = \left(\frac{\partial E}{\partial S}\right)_{VN}, \quad p = -\left(\frac{\partial E}{\partial V}\right)_{SN}, \quad \mu = \left(\frac{\partial E}{\partial N}\right)_{SV}. \tag{4.1.2}$$

The Legendre transformation between the variables $(T, V, N), (T, p, N)$, and (T, V, μ), leads to the Helmholtz free energy $F(T, V, N)$, the Gibbs free energy $G(T, p, N)$, and the thermodynamical potential $\Omega(T, V, \mu)$ defined by

$$F(T, V, N) = E - TS, \tag{4.1.3a}$$

$$G(T, p, N) = E - TS + pV, \tag{4.1.3b}$$

$$\Omega(T, V, \mu) = E - TS - \mu N. \tag{4.1.3c}$$

Equations (4.1.1) and (4.1.3) yield

$$dF = -S\,dT - p\,dV + \mu\,dN, \tag{4.1.4a}$$

$$dG = -S\,dT + V\,dp + \mu\,dN, \tag{4.1.4b}$$

$$d\Omega = -S\,dT - p\,dV - N\,d\mu, \tag{4.1.4c}$$

where

$$S = -\left(\frac{\partial F}{\partial T}\right)_{VN}, \quad p = -\left(\frac{\partial F}{\partial V}\right)_{TN}, \quad \mu = \left(\frac{\partial F}{\partial N}\right)_{TV}, \tag{4.1.5a}$$

$$S = -\left(\frac{\partial G}{\partial T}\right)_{p\mu}, \quad V = \left(\frac{\partial G}{\partial V}\right)_{TN}, \quad \mu = \left(\frac{\partial G}{\partial \mu}\right)_{Tp}, \tag{4.1.5b}$$

$$S = -\left(\frac{\partial \Omega}{\partial T}\right)_{V\mu}, \quad p = -\left(\frac{\partial \Omega}{\partial V}\right)_{T\mu}, \quad N = -\left(\frac{\partial \Omega}{\partial \mu}\right)_{TV}. \tag{4.1.5c}$$

Since $E(S, V, N)$ and the quantities S, V, N are extended quantities, there is the relation:

$$\lambda E = E(\lambda S, \lambda V, \lambda N). \tag{4.1.6}$$

Differentiating both the sides in λ and putting $\lambda = 1$, we obtain

$$E = S\left(\frac{\partial E}{\partial S}\right)_{VN} + V\left(\frac{\partial E}{\partial V}\right)_{SN} + N\left(\frac{\partial E}{\partial S}\right)_{SV} = ST - Vp + N\mu. \tag{4.1.7}$$

This equation yields the Gibbs-Duhems relation:

$$S\,dT - V\,dp + N\,d\mu = 0. \tag{4.1.8}$$

Thus we obtain

$$F = -pV + \mu N, \tag{4.1.9a}$$
$$G = \mu N, \tag{4.1.9b}$$
$$\Omega = -pV. \tag{4.1.9c}$$

Here we investigate the thermodynamical potential $\Omega(T, V, \mu)$. The grand partition function is given by

$$\begin{aligned} Z_G &= \mathrm{Tr} e^{-\beta(H-\mu N)} \\ &\equiv \sum_N \sum_i \langle N E_i | e^{-\beta(H-\mu N)} | N E_i \rangle \\ &= \sum_N \sum_i e^{-\beta(E_i - \mu N)} \\ &= \sum_N e^{\beta\mu N} \int dE e^{-\beta E} W(E, V, N) \\ &= \sum_N \int dE \exp\left[-\frac{1}{k_B} \left\{ -\frac{\mu}{T} N + \frac{1}{T} E - S(E, V, N) \right\} \right], \end{aligned} \tag{4.1.10}$$

where $\beta = 1/k_B T$, $W(E, V, N)$ is the number of states and

$$S(E, V, N) = k_B \ln W(E, V, N).$$

The term $\exp\left[-\frac{\mu}{T} N + \frac{1}{T} E - S(E, V, N) \right]$ has a sharp maximum at E^* and N^*, where

$$\frac{1}{T} = \frac{\partial S}{\partial E}, \quad \frac{\mu}{T} = -\frac{\partial S}{\partial N}.$$

Here we expand around E^* and N^* as

$$\begin{aligned} &\exp\left[-\frac{1}{k_B} \left\{ -\frac{\mu}{T} N + \frac{1}{T} E - S(E, V, N) \right\} \right] \\ &= \exp\left[-\frac{1}{k_B} \left\{ -\frac{\mu}{T} N^* + \frac{1}{T} E^* - S(E^*, V, N^*) \right. \right. \\ &\quad \left. \left. -\frac{1}{2} \left(\frac{\partial^2 S}{\partial E^2} (\Delta E)^2 + 2 \frac{\partial^2 S}{\partial E \partial N} \Delta E \Delta N + \frac{\partial^2 S}{\partial N^2} (\Delta N)^2 \right) + \cdots \right\} \right]. \end{aligned}$$

We substitute this equation into Eq. (4.1.9b) and put $\sum_N \sim \int dN$. We can expect that the term $\exp\left[\frac{1}{2k_B} \left\{ \frac{\partial^2 S}{\partial E^2} (\Delta E)^2 + 2 \frac{\partial^2 S}{\partial E \partial N} \Delta E \Delta N + \frac{\partial^2 S}{\partial N^2} (\Delta N)^2 \right\} \right]$ shows the Gaussian distribution in $\Delta E = E - E^*$ and $\Delta N = N - N^*$, for example $\frac{\partial^2 S}{\partial E^2} = \frac{\partial(\frac{1}{T})}{\partial E} = -\frac{C}{T^2}$, where $C = \frac{\partial E}{\partial T}$ is a heat capacity. Carrying out the Gaussian integration in N and E, we obtain

$$\begin{aligned} Z_G &\approx \exp\left[-\frac{1}{k_\mathrm{B}}\left\{-\frac{\mu}{T}N^* + \frac{1}{T}E^* - S(E^*, V, N^*)\right\}\right] \\ &\quad\times \int dN\, dE \exp\left[\frac{1}{2k_\mathrm{B}}\left\{\frac{\partial^2 S}{\partial E^2}(\Delta E)^2 + 2\frac{\partial^2 S}{\partial E \partial N}\Delta E \Delta N \right.\right. \\ &\qquad\qquad \left.\left. + \frac{\partial^2 S}{\partial N^2}(\Delta N)^2\right\} + \cdots\right] \\ &\approx \exp\left[-\frac{1}{k_\mathrm{B}}\left\{-\frac{\mu}{T}N^* + \frac{1}{T}E^* - S(E^*, V, N^*)\right\}\right] \frac{2\pi k_\mathrm{B}}{\sqrt{\frac{\partial^2 S}{\partial E^2}\frac{\partial^2 S}{\partial N^2} - \left\{\frac{\partial^2 S}{\partial E \partial N}\right\}^2}}. \end{aligned} \tag{4.1.11}$$

Since $E \sim O(N)$ and $S \sim O(N)$, we can expect the term

$$\frac{2\pi k_\mathrm{B}}{\sqrt{\frac{\partial^2 S}{\partial E^2}\frac{\partial^2 S}{\partial N^2} - \left\{\frac{\partial^2 S}{\partial E \partial N}\right\}^2}} \sim O(N),$$

while the term

$$\begin{aligned} &\exp\left[-\frac{1}{k_\mathrm{B}}\left\{-\frac{\mu}{T}N^* + \frac{1}{T}E^* - S(E^*, V, N^*)\right\}\right] \\ &\quad = \exp\left[-\frac{1}{k_\mathrm{B}T}\{E^* - TS^* - \mu N^*\}\right] \sim O(e^N). \end{aligned}$$

Thus we can put

$$\Omega(T, V, \mu) = -\frac{1}{\beta}\ln Z_G, \tag{4.1.12}$$

where we have put E^*, S^*, N^* as E, S, N.

Here we introduce the following definition:

$$\mathcal{H} \equiv H - \mu N. \tag{4.1.13}$$

The statistical operator ρ_G corresponding to Eq. (4.1.10) is given by

$$\rho_G = \frac{e^{-\beta\mathcal{H}}}{Z_G}, \tag{4.1.14}$$

which is rewritten as from Eq. (4.1.12):

$$\rho_G = e^{\beta(\Omega - \mathcal{H})}. \tag{4.1.15}$$

For any operator O, the ensemble average $\langle O \rangle$ is given by

$$\begin{aligned}\langle O \rangle &= \mathrm{Tr}(\rho_G O) \\ &= \mathrm{Tr}(e^{\beta(\Omega - \mathcal{H})} O) \\ &= \frac{\mathrm{Tr}(e^{-\beta\mathcal{H}} O)}{\mathrm{Tr} e^{-\beta\mathcal{H}}},\end{aligned} \tag{4.1.16}$$

where the ensemble average is carried out by the statistical operator ρ_G.

4.1.2 The Definition of the Single Particle Temperature Green's Function

In order to define the temperature Green's function, we introduce the modified Heisenberg operator for any operator O

$$O(\boldsymbol{x}\tau) \equiv e^{\mathcal{H}\tau/\hbar} O(\boldsymbol{x}) e^{-\mathcal{H}\tau/\hbar}, \tag{4.1.17}$$

where τ is real. It should be noted that the Heisenberg operator O_H is expressed as the operator O analytically continued to a pure imaginal value $\tau = \mathrm{i}t$, apart from the substitution of H for $\mathcal{H}$:

$$O_H(\boldsymbol{x}, t) = O(\boldsymbol{x}, \tau = \mathrm{i}t) = e^{\mathrm{i}Ht/\hbar} O e^{-\mathrm{i}Ht/\hbar}. \tag{4.1.18}$$

The Heisenberg operator $O_H(\boldsymbol{x}, t)$ is a self-adjoint operator:

$$O_H^\dagger(\boldsymbol{x}, t) = O_H(\boldsymbol{x}, t). \tag{4.1.19}$$

However, $O(\boldsymbol{x}, \tau)$ is not a self-adjoint operator so that the adjoint of $O(\boldsymbol{x}, \tau)$, $O^\dagger(\boldsymbol{x}, \tau)$ must be defined as

$$O^\dagger(\boldsymbol{x}, \tau) \equiv e^{\mathcal{H}\tau/\hbar} O^\dagger(\boldsymbol{x}) e^{-\mathcal{H}\tau/\hbar}. \tag{4.1.20}$$

τ is an imaginary time so that $O(\boldsymbol{x}, \tau)$, $O^\dagger(\boldsymbol{x}, \tau)$ are called imaginary time operators. Hereafter, we represent $x = (\boldsymbol{x}, \tau)$.

The field operators are given by

$$\psi_\alpha(x) = e^{\mathcal{H}\tau/\hbar} \psi_\alpha(\boldsymbol{x}) e^{-\mathcal{H}\tau/\hbar}, \tag{4.1.21}$$

$$\psi_\alpha^\dagger(x) = e^{\mathcal{H}\tau/\hbar} \psi_\alpha^\dagger(\boldsymbol{x}) e^{-\mathcal{H}\tau/\hbar}. \tag{4.1.22}$$

The single particle temperature Green's function is defined as

$$\mathcal{G}_{\alpha\beta}(x, x') \equiv -\mathrm{Tr}\left\{\rho_G T_\tau \left[\psi_\alpha(x)\psi_\beta(x')\right]\right\}. \tag{4.1.23}$$

The ordering operator T_τ arranges the operators from left to right in order of decreasing τ. T_τ also includes the signature factor $(\pm 1)^P$, where P is the number of permutations of the operators from the original ordering. The upper and lower signs refer to the bosons and fermions. For operators, $A(\tau)$ and $B(\tau')$, we can write

$$T_\tau\left[A(\tau)B(\tau')\right] = \theta(\tau - \tau')A(\tau)B(\tau') \pm \theta(\tau' - \tau)B(\tau')A(\tau), \tag{4.1.24}$$

where $\theta(\tau)$ is the theta function; $\theta(\tau) = 1$ for $\tau > 0$ and $\theta(\tau) = 0$ for $\tau < 0$.

4.1.3 The Free Particle Green's Functions

First we investigate the free particle Green's functions. The free fields are given by

$$\psi_\sigma(\boldsymbol{x}) = \frac{1}{\sqrt{V}} \sum_{p,\sigma} e^{\mathrm{i}\boldsymbol{p}\cdot\boldsymbol{x}} \eta_\sigma a_{\boldsymbol{p}\sigma}, \tag{4.1.25}$$

$$\psi_\sigma^\dagger(\boldsymbol{x}) = \frac{1}{\sqrt{V}} \sum_{p,\sigma} e^{-\mathrm{i}\boldsymbol{p}\cdot\boldsymbol{x}} \eta_\sigma a_{\boldsymbol{p}\sigma}^\dagger, \tag{4.1.26}$$

where σ is the spin indices and η_σ is the spin states for fermions. For bosons we discard the spin indices and the spin states. Since we are considering the ensemble average, we do not need the discrimination between the particles and holes. The Heisenberg field operators are given by

$$\begin{aligned}\psi_\sigma(\boldsymbol{x}\tau) &= \frac{1}{\sqrt{V}} \sum_{p,\sigma} e^{\mathrm{i}\boldsymbol{p}\cdot\boldsymbol{x}} \eta_\sigma e^{\mathcal{H}_0\tau/\hbar} a_{\boldsymbol{p}\sigma} e^{-\mathcal{H}_0\tau/\hbar} \\ &= \frac{1}{\sqrt{V}} \sum_{p,\sigma} e^{\mathrm{i}\boldsymbol{p}\cdot\boldsymbol{x}} \eta_\sigma a_{\boldsymbol{p}\sigma}(\tau),\end{aligned} \tag{4.1.27}$$

$$\begin{aligned}\psi_\sigma^\dagger(\boldsymbol{x}\tau) &= \frac{1}{\sqrt{V}} \sum_{p,\sigma} e^{-\mathrm{i}\boldsymbol{p}\cdot\boldsymbol{x}} \eta_\sigma e^{\mathcal{H}_0\tau/\hbar} a_{\boldsymbol{p}\sigma}^\dagger e^{-\mathcal{H}_0\tau/\hbar} \\ &= \frac{1}{\sqrt{V}} \sum_{p,\sigma} e^{-\mathrm{i}\boldsymbol{p}\cdot\boldsymbol{x}} \eta_\sigma a_{\boldsymbol{p}\sigma}^\dagger(\tau),\end{aligned} \tag{4.1.28}$$

The differentiation of the operators on the right in τ leads to

$$\begin{aligned}\hbar\frac{\partial}{\partial\tau} a_{\boldsymbol{p}\sigma}(\tau) &= e^{\mathcal{H}_0\tau/\hbar}\left[\mathcal{H}_0, a_{\boldsymbol{p}\sigma}\right] e^{-\mathcal{H}_0\tau/\hbar} \\ &= -\hbar\epsilon_p^0 a_{\boldsymbol{p}\sigma}(\tau),\end{aligned} \tag{4.1.29}$$

where the $\mathcal{H}_0$ and ϵ_p^0 are given by

$$\mathcal{H}_0 = \sum_{p\sigma} \hbar\epsilon_{\boldsymbol{p}}^0 a_{\boldsymbol{p}\sigma}^\dagger a_{\boldsymbol{p}\sigma}, \quad \epsilon_{\boldsymbol{p}}^0 = (\hbar^2 p^2/2M - \mu)/\hbar. \tag{4.1.30}$$

The integration of Eq. (4.1.29) yields

$$a_{\boldsymbol{k}\sigma}(\tau) = a_{\boldsymbol{p}\sigma} e^{-\epsilon_{\boldsymbol{p}}^0\tau}, \tag{4.1.31}$$

and similarly

$$a_{\boldsymbol{k}\sigma}^\dagger(\tau) = a_{\boldsymbol{p}\sigma}^\dagger e^{\epsilon_{\boldsymbol{p}}^0\tau}. \tag{4.1.32}$$

Note that Eq. (4.1.32) is not the adjoint of Eq. (4.1.31).

The free particle temperature Green's function is given by

$$\mathcal{G}^0_{\alpha\beta}(x,x') = -e^{\beta\Omega^0}\mathrm{Tr}\left\{e^{-\beta\mathcal{H}_0}T_\tau\left[\psi_\alpha(x)\psi^\dagger_\beta(x')\right]\right\}. \tag{4.1.33}$$

First we consider the case $\tau' < \tau$:

$$\begin{aligned}\mathcal{G}^0_{\alpha\beta}(x,x') &= -\frac{1}{V}\sum_{\boldsymbol{p}\boldsymbol{p}'}\sum_{\sigma\sigma'} e^{i\boldsymbol{p}\cdot\boldsymbol{x}-\boldsymbol{p}'\cdot\boldsymbol{x}'}(\eta_\sigma)_\alpha(\eta_{\sigma'})_\beta e^{-\epsilon^0_{\boldsymbol{p}}\tau+\epsilon^0_{\boldsymbol{p}'}\tau'}\left\langle a_{\boldsymbol{p}\sigma}a^\dagger_{\boldsymbol{p}'\sigma'}\right\rangle_0 \\ &= -\frac{1}{V}\delta_{\alpha\beta}\sum_{\boldsymbol{p}} e^{\mathrm{i}\boldsymbol{p}\cdot(\boldsymbol{x}-\boldsymbol{x}')}e^{-\epsilon^0_{\boldsymbol{p}}(\tau-\tau')}\left\langle a_{\boldsymbol{p}\sigma}a^\dagger_{\boldsymbol{p}\sigma}\right\rangle_0,\end{aligned} \tag{4.1.34}$$

where

$$\begin{aligned}\langle\cdots\rangle_0 &= \mathrm{Tr}(\rho^0_G\cdots) = e^{\beta\Omega^0}\mathrm{Tr}(e^{-\beta\mathcal{H}^0}\cdots) \\ &= e^{\beta\Omega^0}\sum_{n_{\boldsymbol{p}_1}}\cdots\sum_{n_{\boldsymbol{p}_n}}\cdots\langle n_{\boldsymbol{p}_1}\cdots n_{\boldsymbol{p}_n}\cdots|e^{-\beta\mathcal{H}^0}\cdots|n_{\boldsymbol{p}_1}\cdots n_{\boldsymbol{p}_n}\cdots\rangle \\ &= e^{\beta\Omega^0}\prod_{\boldsymbol{p}_n}\sum_{n_{\boldsymbol{p}_n}}\left[e^{-\beta\hbar\epsilon^0_{\boldsymbol{p}_n}n_{\boldsymbol{p}_n}}\cdots\right],\end{aligned} \tag{4.1.35}$$

$$\begin{aligned}e^{-\beta\Omega^0} &= \mathrm{Tr}e^{-\beta\mathcal{H}^0} = \prod_{\boldsymbol{p}_n}\sum_{n_{\boldsymbol{p}_n}} e^{-\beta\hbar\epsilon^0_{\boldsymbol{p}_n}n_{\boldsymbol{p}_n}} \\ &= \prod_{\boldsymbol{p}_n}\left[1\mp e^{-\beta\hbar\epsilon^0_{\boldsymbol{p}_n}}\right]^{\mp},\end{aligned} \tag{4.1.36}$$

where the upper and the lower signs refer to the bosons and fermions, respectively, and in summing up $\sum_{n_{\boldsymbol{p}_n}} e^{-\beta\hbar\epsilon^0_{\boldsymbol{p}_n}n_{\boldsymbol{p}_n}}$ we have taken into account

$$\sum\nolimits_{n_{\boldsymbol{p}_n}=0}^{\infty} e^{-\beta\hbar\epsilon^0_{\boldsymbol{p}_n}n_{\boldsymbol{p}_n}} = \frac{1}{1-e^{-\beta\hbar\epsilon^0_{\boldsymbol{p}_n}}} \quad \text{for bosons,} \tag{4.1.37}$$

$$\sum\nolimits_{n_{\boldsymbol{p}_n}=0}^{1} e^{-\beta\hbar\epsilon^0_{\boldsymbol{p}_n}n_{\boldsymbol{p}_n}} = 1+e^{-\beta\hbar\epsilon^0_{\boldsymbol{p}_n}} \quad \text{for fermions.} \tag{4.1.38}$$

Since

$$\begin{aligned}\langle a^\dagger_{\boldsymbol{p}\sigma}a_{\boldsymbol{p}\sigma}\rangle_0 &= e^{\beta\Omega^0}\prod_{\boldsymbol{p}_n\neq\boldsymbol{p}}\left[1\mp e^{-\beta\hbar\epsilon^0_{\boldsymbol{p}_n}}\right]^{\mp}\sum_{n_{\boldsymbol{p}}=0}^{\infty} e^{-\beta\hbar\epsilon^0_{\boldsymbol{p}}n_{\boldsymbol{p}}}n_{\boldsymbol{p}} \\ &= \left[1\mp e^{-\beta\hbar\epsilon^0_{\boldsymbol{p}}}\right]^{\pm}\times\left\{\begin{array}{l}\frac{e^{-\beta\hbar\epsilon^0_{\boldsymbol{p}}}}{(1-e^{-\beta\hbar\epsilon^0_{\boldsymbol{p}}})^2} \\ e^{-\beta\hbar\epsilon^0_{\boldsymbol{p}}}\end{array}\right. \\ &= \frac{1}{e^{\beta\hbar\epsilon^0_{\boldsymbol{p}}}\mp 1},\end{aligned} \tag{4.1.39}$$

we obtain

$$f(\epsilon^0_{\boldsymbol{p}}) = \langle a^\dagger_{\boldsymbol{p}\sigma}a_{\boldsymbol{p}\sigma}\rangle_0 = \frac{1}{e^{\beta\hbar\epsilon^0_{\boldsymbol{p}}}\mp 1}, \tag{4.1.40}$$

and

$$\langle a_{p\sigma}a^{\dagger}_{p\sigma}\rangle_0 = 1 \mp \langle a^{\dagger}_{p\sigma}a_{p\sigma}\rangle_0 = 1 \mp f(\epsilon^0_p). \tag{4.1.41}$$

Thus we obtain

$$\mathcal{G}^0_{\alpha\beta}(x, x') = -\delta_{\alpha\beta}\frac{1}{V}\sum_p e^{\mathrm{i}\boldsymbol{p}\cdot(\boldsymbol{x}-\boldsymbol{x}')}e^{-\epsilon^0_p(\tau-\tau')}(1 \pm f(\epsilon^0_p)) \quad \text{for } \tau' < \tau. \tag{4.1.42}$$

A similar calculation for $\tau < \tau'$ yields

$$\mathcal{G}^0_{\alpha\beta}(x, x') = -\delta_{\alpha\beta}\frac{1}{V}\sum_p e^{\mathrm{i}\boldsymbol{p}\cdot(\boldsymbol{x}-\boldsymbol{x}')}e^{-\epsilon^0_p(\tau-\tau')}f(\epsilon^0_p) \quad \text{for } \tau < \tau'. \tag{4.1.43}$$

4.2 Perturbation Theory and the Wick's Theorem at Finite Temperature

In order to calculate the finite temperature Green's functions, we employ the perturbation theory of the interaction Hamiltonian in the interaction representation. The periodic nature of the finite temperature Green's function originated from the property of the trace operation in the ensemble average yields the Wick's theorem, which prescribes the contractions of the expanded perturbation terms. The contraction processes make the calculation procedure very easy.

4.2.1 *The Interaction Representation*

For any operator O, we formally define the interaction representation $O_I(\tau)$ and the Heisenberg representation $O_H(\tau)$ as:

$$O_I(\tau) \equiv e^{\mathcal{H}_0\tau/\hbar}Oe^{-\mathcal{H}_0\tau/\hbar}, \tag{4.2.1}$$

$$O_H(\tau) \equiv e^{\mathcal{H}\tau/\hbar}Oe^{-\mathcal{H}\tau/\hbar}, \tag{4.2.2}$$

where $\mathcal{H}_0 = H_0 - \mu N$, $\mathcal{H} = H - \mu N$, and $H = H_0 + H_I$. There is a relation between the two representations:

$$\begin{aligned} O_H(\tau) &= e^{\mathcal{H}\tau/\hbar}e^{-\mathcal{H}_0\tau/\hbar}O_I(\tau)e^{\mathcal{H}_0\tau/\hbar}e^{-\mathcal{H}\tau/\hbar} \\ &= \mathcal{U}^{-1}(\tau)O_I(\tau)\mathcal{U}(\tau), \end{aligned} \tag{4.2.3}$$

where

$$\mathcal{U}(\tau) = e^{\mathcal{H}_0\tau/\hbar}e^{-\mathcal{H}\tau/\hbar}. \tag{4.2.4}$$

Differentiating Eq. (4.2.4), we obtain

$$\hbar\frac{\partial}{\partial\tau}\mathcal{U}(\tau) = -\mathcal{H}_I(\tau)\mathcal{U}(\tau), \tag{4.2.5}$$

where

$$\mathcal{H}_I(\tau) = e^{\mathcal{H}_0\tau/\hbar} H_I e^{-\mathcal{H}_0\tau/\hbar}. \tag{4.2.6}$$

Equation (4.2.5) under the condition $\mathcal{U}(0) = 1$ yields an integral equation:

$$\mathcal{U}(\tau) = 1 + \left(-\frac{1}{\hbar}\right) \int_0^\tau d\tau' \mathcal{H}_I(\tau')\mathcal{U}(\tau'). \tag{4.2.7}$$

The solution of Eq. (4.2.7) can be obtained by iteration; the first approximation putting $\mathcal{U}(\tau') = 1$ in the integrand on the right-hand side of Eq. (4.2.7) yields the solution, $\mathcal{U}_1(\tau) = 1 + (-\frac{1}{\hbar}) \int_0^\tau d\tau' \mathcal{H}_I(\tau')$, the second approximation putting $\mathcal{U}(\tau') = \mathcal{U}_1(\tau')$ yields the solution $\mathcal{U}_2(\tau), \ldots$ Thus we obtain

$$\begin{aligned}\mathcal{U}(\tau) = 1 &+ \left(-\frac{1}{\hbar}\right) \int_0^\tau d\tau' \mathcal{H}_I(\tau') \\ &+ \left(-\frac{1}{\hbar}\right)^2 \int_0^\tau d\tau' \int_0^{\tau'} d\tau'' \mathcal{H}_I(\tau')\mathcal{H}_I(\tau'') + \cdots .\end{aligned} \tag{4.2.8}$$

The third term can be rewritten as

$$\begin{aligned}&\int_0^\tau d\tau' \int_0^{\tau'} d\tau'' \mathcal{H}_I(\tau')\mathcal{H}_I(\tau'') \\ &= \frac{1}{2} \int_0^\tau d\tau' \int_0^{\tau'} d\tau'' \mathcal{H}_I(\tau')\mathcal{H}_I(\tau'') + \frac{1}{2} \int_0^\tau d\tau'' \int_{\tau''}^\tau d\tau' \mathcal{H}_I(\tau')\mathcal{H}_I(\tau'') \\ &= \frac{1}{2} \int_0^\tau d\tau' \int_0^{\tau'} d\tau'' \mathcal{H}_I(\tau')\mathcal{H}_I(\tau'') + \frac{1}{2} \int_0^\tau d\tau' \int_{\tau'}^\tau d\tau'' \mathcal{H}_I(\tau'')\mathcal{H}_I(\tau') \\ &= \frac{1}{2} \int_0^\tau d\tau' \int_0^\tau d\tau'' \left[\theta(\tau' - \tau'')\mathcal{H}_I(\tau')\mathcal{H}_I(\tau'') + \theta(\tau'' - \tau')\mathcal{H}_I(\tau'')\mathcal{H}_I(\tau')\right] \\ &= \frac{1}{2!} \int_0^\tau d\tau' \int_0^\tau d\tau'' T_\tau \left[\mathcal{H}_I(\tau')\mathcal{H}_I(\tau'')\right],\end{aligned} \tag{4.2.9}$$

where we have considered the following:

1. From the first to the second equation: $\int_0^\tau d\tau' \int_0^{\tau'} d\tau'' \mathcal{H}_I(\tau')\mathcal{H}_I(\tau'') = \int_0^\tau d\tau'' \int_{\tau''}^\tau d\tau' \mathcal{H}_I(\tau')\mathcal{H}_I(\tau'')$, in the same integral regions.
2. From the second to the third equation: we have interchanged the variables τ' and τ'' in the second term.
3. From the third to the fourth equation: we have used the property of the θ-function.
4. From the fourth to the last equation: we have used the property of the T_τ-ordering operator.

Thus considering the same procedure as in Eq. (4.2.9), we obtain

$$\begin{aligned}\mathcal{U}(\tau) &= \sum_{n=0}^\infty \frac{1}{n!}(-\frac{1}{\hbar})^n \int_0^\tau d\tau_1 \cdots \int_0^\tau d\tau_n T_\tau \left[\mathcal{H}_I(\tau_1) \cdots \mathcal{H}_I(\tau_n)\right] \\ &= T_\tau \exp\left[-\frac{1}{\hbar} \int_0^\tau d\tau' \mathcal{H}_I(\tau')\right].\end{aligned} \tag{4.2.10}$$

Here we consider another matrix $\mathcal{U}(\tau_1, \tau_2)$, where $\tau_2 < \tau_1$ is defined by

$$\mathcal{U}(\tau_1, \tau_2) = T_\tau \exp\left[-\frac{1}{\hbar}\int_{\tau_2}^{\tau_1} d\tau' \mathcal{H}_I(\tau')\right], \tag{4.2.11}$$

where

$$\mathcal{U}(\tau) = \mathcal{U}(\tau, 0). \tag{4.2.12}$$

$\mathcal{U}(\tau_1, \tau_2)$ has the properties

$$\mathcal{U}(\tau_1, \tau_3) = \mathcal{U}(\tau_1, \tau_2)\mathcal{U}(\tau_2, \tau_3) \quad \text{for } \tau_1 > \tau_2 > \tau_3, \tag{4.2.13}$$

$$\mathcal{U}(\tau_1, \tau_2) = \mathcal{U}(\tau_1, 0)\mathcal{U}(0, \tau_2) = \mathcal{U}(\tau_1)\mathcal{U}^{-1}(\tau_2) \quad \text{for } \tau_1 > \tau_2. \tag{4.2.14}$$

From Eqs. (4.2.4) and (4.2.12), we can write

$$e^{-\mathcal{H}\tau/\hbar} = e^{-\mathcal{H}_0\tau/\hbar}\mathcal{U}(\tau, 0), \tag{4.2.15}$$

$$\begin{aligned} e^{-\beta\Omega} &= \mathrm{Tr} e^{-\beta\mathcal{H}} \\ &= \mathrm{Tr}[e^{-\beta\mathcal{H}_0}\mathcal{U}(\beta\hbar, 0)]. \end{aligned} \tag{4.2.16}$$

The exact temperature Green's function can be rewritten in the interaction representation. For $\tau > \tau'$, we have

$$\begin{aligned} \mathcal{G}_{\alpha\beta}(x, x') &= -e^{\beta\Omega}\mathrm{Tr}[e^{-\beta\mathcal{H}}\psi_\alpha(x)\psi_\beta^\dagger(x')] \\ &= -e^{\beta\Omega}\mathrm{Tr}\big\{[e^{-\beta\mathcal{H}_0}\mathcal{U}(\beta\hbar, 0)][\mathcal{U}(0, \tau)\psi_{I\alpha}(x)\mathcal{U}(\tau, 0)] \\ &\qquad [\mathcal{U}(0, \tau')\psi_{I\beta}(x')\mathcal{U}(\tau', 0)]\big\} \\ &= -\frac{\mathrm{Tr}[e^{-\beta\mathcal{H}_0}\mathcal{U}(\beta\hbar, \tau)\psi_{I\alpha}(x)\mathcal{U}(\tau, \tau')\psi_{I\beta}^\dagger(x')\mathcal{U}(\tau', 0)]}{\mathrm{Tr}[e^{-\beta\mathcal{H}_0}\mathcal{U}(\beta\hbar, 0)]}. \end{aligned} \tag{4.2.17}$$

Similarly, for $\tau < \tau'$ we obtain

$$\mathcal{G}_{\alpha\beta}(x, x') = \mp\frac{\mathrm{Tr}[e^{-\beta\mathcal{H}_0}\mathcal{U}(\beta\hbar, \tau')\psi_{I\alpha}(x')\mathcal{U}(\tau', \tau)\psi_{I\beta}^\dagger(x)\mathcal{U}(\tau, 0)]}{\mathrm{Tr}[e^{-\beta\mathcal{H}_0}\mathcal{U}(\beta\hbar, 0)]}. \tag{4.2.18}$$

Expressions (4.2.17) and (4.2.18) can be combined into a single formula:

$$\begin{aligned} &\mathcal{G}_{\alpha\beta}(x, x') \\ &= -\frac{\mathrm{Tr}\{e^{-\beta\mathcal{H}_0}T_\tau[\psi_{I\alpha}(x)\psi_{I\beta}^\dagger(x')\mathcal{U}(\beta\hbar)]\}}{\mathrm{Tr}[e^{-\beta\mathcal{H}_0}\mathcal{U}(\beta\hbar)]} \\ &= -\frac{\langle T_\tau[\psi_{I\alpha}(x)\psi_{I\beta}^\dagger(x')\mathcal{U}(\beta\hbar)]\rangle_0}{\langle\mathcal{U}(\beta\hbar)\rangle_0} \\ &= -\frac{\sum_{n=0}^{\infty}\frac{1}{n!}(-\frac{1}{\hbar})^n\int_0^{\beta\hbar}d\tau_1\cdots\int_0^{\beta\hbar}d\tau_n\langle T_\tau[\mathcal{H}_I(\tau_1)\cdots\mathcal{H}_I(\tau_n)\psi_{I\alpha}(x)\psi_{I\beta}^\dagger(x')]\rangle_0}{\sum_{n=0}^{\infty}\frac{1}{n!}(-\frac{1}{\hbar})^n\int_0^{\beta\hbar}d\tau_1\cdots\int_0^{\beta\hbar}d\tau_n\langle T_\tau[\mathcal{H}_I(\tau_1)\cdots\mathcal{H}_I(\tau_n)]\rangle_0}, \end{aligned} \tag{4.2.19}$$

where $\langle\cdots\rangle_0 = \mathrm{Tr}\{\rho_{G_0}\cdots\} = \mathrm{Tr}\{e^{\beta(\Omega_0-\mathcal{H}_0)}\ldots\}$.

4.2.2 The Periodic Nature of the Temperature Green's Function

In the temperature Green's function $\mathcal{G}(x, x') = \mathcal{G}(\boldsymbol{x}\tau, \boldsymbol{x}'\tau')$, τ and τ' have a domain $0 \leqslant \tau, \tau' \leqslant \beta\hbar$ so that the difference $\tau - \tau'$ stays as $-\beta\hbar \leqslant \tau - \tau' \leqslant \beta\hbar$. In this domain, the temperature Green's function has a periodic or antiperiodic nature. To show this, we first start with the definition of the temperature Green's function:

$$\begin{aligned}\mathcal{G}_{\alpha\beta}(\boldsymbol{x}\tau, \boldsymbol{x}'\tau') &= -e^{\beta\Omega}\mathrm{Tr}\{e^{-\beta\mathcal{H}}[\psi_\alpha(x)\psi^\dagger_\beta(x')]\}\\ &= -e^{\beta\Omega}\mathrm{Tr}\{e^{-\beta\mathcal{H}}[e^{\mathcal{H}(\tau-\tau')/\hbar}\psi_\alpha(\boldsymbol{x})\\ &\quad \times e^{-\mathcal{H}(\tau-\tau')/\hbar}\psi^\dagger_\beta(\boldsymbol{x}')]\} \quad \text{for } \tau > \tau',\end{aligned} \tag{4.2.20}$$

$$\begin{aligned}\mathcal{G}_{\alpha\beta}(\boldsymbol{x}\tau, \boldsymbol{x}'\tau') &= \mp e^{\beta\Omega}\mathrm{Tr}\{e^{-\beta\mathcal{H}}[\psi^\dagger_\beta(x')\psi_\alpha(x)]\}\\ &= \mp e^{\beta\Omega}\mathrm{Tr}\{e^{-\beta\mathcal{H}}[e^{-\mathcal{H}(\tau-\tau')/\hbar}\psi^\dagger_\beta(\boldsymbol{x}')\\ &\quad \times e^{\mathcal{H}(\tau-\tau')/\hbar}\psi_\alpha(\boldsymbol{x})]\} \quad \text{for } \tau < \tau'.\end{aligned} \tag{4.2.21}$$

Next we show a periodic or antiperiodic nature of the temperature Green's function. To see this, we put

$$\mathcal{G}(x, x') = \mathcal{G}(\boldsymbol{x}, \boldsymbol{x}', \tau - \tau') \tag{4.2.22}$$

and consider $\mathcal{G}(\boldsymbol{x}, \boldsymbol{x}', \tau - \tau' < 0)$ in Eq. (4.2.22). Carrying out a cyclic permutation of the operators, we obtain

$$\begin{aligned}&\mathcal{G}(\boldsymbol{x}, \boldsymbol{x}', \tau - \tau' < 0)\\ &\quad = \mp e^{\beta\Omega}\mathrm{Tr}\{e^{\mathcal{H}(\tau-\tau')/\hbar}\psi_\alpha(\boldsymbol{x})e^{-\mathcal{H}(\tau-\tau'+\beta\hbar)/\hbar}\psi^\dagger_\beta(\boldsymbol{x}')\}\\ &\quad = \mp e^{\beta\Omega}\mathrm{Tr}\{e^{-\beta\mathcal{H}}e^{\mathcal{H}(\tau-\tau'+\beta\hbar)/\hbar}\psi_\alpha(\boldsymbol{x})e^{-\mathcal{H}(\tau-\tau'+\beta\hbar)/\hbar}\psi^\dagger_\beta(\boldsymbol{x}')\}.\end{aligned} \tag{4.2.23}$$

Since $\tau - \tau' + \beta\hbar > 0$, using Eq. (4.2.20) for the last equation, we obtain

$$\mathcal{G}_{\alpha\beta}(\boldsymbol{x}, \boldsymbol{x}', \tau - \tau' < 0) = \pm\mathcal{G}_{\alpha\beta}(\boldsymbol{x}, \boldsymbol{x}', \tau - \tau' + \beta\hbar). \tag{4.2.24}$$

This means that $\mathcal{G}(\tau)$ is periodic for bosons and antiperiodic for fermions in the domain $-\beta\hbar < \tau < \beta\hbar$.

4.2.3 The Wick's Theorem

In the perturbation theory, to obtain the temperature Green's function $\mathcal{G}_{\alpha\beta}(x, x')$ we have to calculate the terms as follows:

$$\langle T_\tau[\mathcal{H}_I(\tau_1)\cdots\mathcal{H}_I(\tau_n)\psi_{I\alpha}(x)\psi^\dagger_{I\beta}(x')]\rangle_0.$$

$\mathcal{H}_I$ consists of a multiple product of ψ_I and $\psi^\dagger_I$. Therefore the general terms in the perturbation expansion are

$$\langle T_\tau[\psi_1\psi_2\psi_3\cdots\psi_{2n}]\rangle_0, \tag{4.2.25}$$

where $\psi_1\psi_2\psi_3\cdots\psi_{2n}$ are free field operators in the interaction representation:

$$\psi_l = \psi_{I\alpha} \text{ or } \psi_{I\alpha}^{\dagger}. \tag{4.2.26}$$

Here we consider the free fields as

$$\psi(\boldsymbol{x}) = \sum_i \phi_i(\boldsymbol{x}) a_i,$$
$$\psi^{\dagger}(\boldsymbol{x}) = \sum_i \phi_i{}^*(\boldsymbol{x}) a_i^{\dagger}, \tag{4.2.27}$$

where the eigenfunctions satisfy

$$H_0\phi_i(\boldsymbol{x}) = \hbar\epsilon_i\phi_i(\boldsymbol{x}). \tag{4.2.28}$$

If we consider Eqs. (4.1.23) and (4.1.24) for $\psi(\boldsymbol{x})$ and $\psi^{\dagger}(\boldsymbol{x})$, we can put $\phi_i(\boldsymbol{x}) = \frac{1}{\sqrt{V}}e^{\mathbf{i}\boldsymbol{k}\cdot\boldsymbol{x}}\eta_\sigma$ for $i = \boldsymbol{k}\sigma$, $\phi_i^{\dagger}(\boldsymbol{x}) = \frac{1}{\sqrt{V}}e^{-\mathbf{i}\boldsymbol{k}\cdot\boldsymbol{x}}\eta_\sigma$, and $\epsilon_i = \epsilon_{\boldsymbol{k}}^0$ for $i = \boldsymbol{k}\sigma$.

The interaction representation yields

$$\psi_I(\boldsymbol{x}\tau) = \sum_i \phi_i(\boldsymbol{x}) a_i e^{-e_i\tau},$$
$$\psi_I^{\dagger}(\boldsymbol{x}\tau) = \sum_i \phi_i{}^*(\boldsymbol{x}) a_i e^{e_i\tau}, \tag{4.2.29}$$

where we abbreviate as

$$e_i = \epsilon_i - \mu. \tag{4.2.30}$$

The free Green's function is given by

$$\mathcal{G}^0(x, x') = \mathcal{G}^0(\boldsymbol{x}\tau, \boldsymbol{x}'\tau')$$
$$= -\sum_i \phi_i(\boldsymbol{x})\phi_i{}^*(\boldsymbol{x}')e^{-e_i(\tau-\tau')} \times \begin{cases} 1 \pm f(e_i) & \tau' < \tau \\ \pm f(e_i) & \tau < \tau' \end{cases}, \tag{4.2.31}$$

where

$$f(e_i) = e^{\beta\Omega_0}\mathrm{Tr}[e^{-\beta\mathcal{H}_0}a_i^{\dagger}a_i]$$
$$= \frac{1}{e^{\beta\hbar e_i} \mp 1}. \tag{4.2.32}$$

Note that the argument of the distribution function is $e_i = \epsilon_i - \mu$.

Here we define a contraction as

$$\dot{\psi_1}\dot{\psi_2} = \langle T_\tau[\psi_1\psi_2]\rangle_0 = \mathrm{Tr}\{\rho_{G_0}T_\tau[\psi_1\psi_2]\}. \tag{4.2.33}$$

In our free Green's function, we obtain

$$\dot{\psi_I(x)}\dot{\psi_I^{\dagger}(x')} = -\mathcal{G}^0(x, x'). \tag{4.2.34}$$

The Wick's theorem at finite temperature asserts:

The Wick's theorem:

The ensemble average of the time ordered free fields is expressed as the sum over all possible fully contractions:

$$\langle T_\tau \psi_1\psi_2\psi_3\cdots\psi_{2n}\rangle_0 = [\psi_1^{\cdot}\psi_2^{\cdot}\psi_3^{\cdot}\cdots\psi_{2n}^{\cdot}] + [\psi_1^{\cdot}\psi_2^{\cdot}\psi_3^{\cdot}\cdots\psi_{2n}^{\cdot}] + \cdots. \quad (4.2.35)$$

Proof:

In proving the theorem, we consider the case that the operators are already in the proper time order, under the time ordering operator T_τ. Therefore, we prove the algebraic identity

$$\langle \psi_1\psi_2\psi_3\cdots\psi_{2n}\rangle_0 = [\psi_1^{\cdot}\psi_2^{\cdot}\psi_3^{\cdot}\cdots\psi_{2n}^{\cdot}] + [\psi_1^{\cdot}\psi_2^{\cdot}\psi_3^{\cdot}\cdots\psi_{2n}^{\cdot}] + \cdots, \quad (4.2.36)$$

subject to the restriction $\tau_1 > \tau_2 > \tau_3 > \cdots > \tau_{2n}$.

In order to prove Eq. (4.2.36), we introduce a notation

$$\psi_l(x) = \sum_i \chi_{i_l}(x)\alpha_{i_l}, \quad (4.2.37)$$

where

$$\alpha_i = a_i \text{ or } a_i^\dagger, \quad (4.2.38)$$

$$\chi_i(x) = \phi_i(\boldsymbol{x})e^{-e_i\tau} \text{ or } \phi_i^*(\boldsymbol{x})e^{e_i\tau}. \quad (4.2.39)$$

Then the left-hand side of Eq. (4.2.36) yields

$$\langle \psi_1\psi_2\psi_3\cdots\psi_{2n}\rangle_0 = \sum_{i_1,i_2,i_3,\cdots,i_{2n}} \chi_{i_1}\chi_{i_2}\chi_{i_3}\cdots\chi_{i_{2n}}\mathrm{Tr}\rho_G^0\alpha_{i_1}\alpha_{i_2}\alpha_{i_3}\cdots\alpha_{i_{2n}}. \quad (4.2.40)$$

Commuting α_{i_1} successively to the right, we obtain

$$\begin{aligned}\mathrm{Tr}\{(\rho_G^0\alpha_{i_1}\alpha_{i_2}\alpha_{i_3}\cdots\alpha_{i_{2n}})\\ = \mathrm{Tr}\rho_G^0\{[\alpha_{i_1},\alpha_{i_2}]_\mp\alpha_{i_3}\cdots\alpha_{i_{2n}}\}\pm\mathrm{Tr}\rho_G^0\{\alpha_{i_2}[\alpha_{i_1},\alpha_{i_3}]_\mp\cdots\alpha_{i_{2n}}\}+\cdots\\ + \mathrm{Tr}\rho_G^0\{\alpha_{i_2}\alpha_{i_3}\cdots[\alpha_{i_1},\alpha_{i_{2n}}]_\mp\}\pm\mathrm{Tr}\rho_G^0(\alpha_{i_2}\alpha_{i_3}\cdots\alpha_{i_{2n}}\alpha_{i_1}).\end{aligned} \quad (4.2.41)$$

Using the cyclic property of Tr and the relation

$$\alpha_{i_1}\rho_G^0 = \rho_G^0(\rho_G^0)^{-1}\alpha_{i_1}\rho_G^0 = \rho_G^0 e^{\beta\mathcal{H}_0}\alpha_{i_1}e^{-\beta\mathcal{H}_0} = \rho_G^0\alpha_{i_1}e^{\lambda_{i_1}\beta\hbar e_{i_1}}, \quad (4.2.42)$$

we obtain

$$\begin{aligned}\pm\mathrm{Tr}\rho_G^0(\alpha_{i_2}\alpha_{i_3}\cdots\alpha_{i_{2n}}\alpha_{i_1}) &= \pm e^{\lambda_{i_1}\beta\hbar e_{i_1}}\mathrm{Tr}(\alpha_{i_2}\alpha_{i_3}\cdots\alpha_{i_{2n}}\rho_G^0\alpha_{i_1})\\ &= \pm e^{\lambda_{i_1}\beta\hbar e_{i_1}}\mathrm{Tr}\{(\rho_G^0\alpha_{i_1}\alpha_{i_2}\alpha_{i_3}\cdots\alpha_{i_{2n}}),\end{aligned} \quad (4.2.43)$$

where $\lambda_{i_1} = 1$ for $\alpha_{i_1} = a_{i_1}^\dagger$ and $\lambda_{i_1} = -1$ for $\alpha_{i_1} = a_{i_1}$. Equation (4.2.41) leads to

$$\begin{aligned}
&\mathrm{Tr}(\rho_G^0 \alpha_{i_1}\alpha_{i_2}\alpha_{i_3}\cdots\alpha_{i_{2n}}) \\
&= \frac{[\alpha_{i_1},\alpha_{i_2}]_\mp}{1 \mp e^{\lambda_{i_1}\beta\hbar e_{i_1}}}\mathrm{Tr}(\rho_G^0\alpha_{i_3}\cdots\alpha_{i_{2n}}) \pm \frac{[\alpha_{i_1},\alpha_{i_3}]_\mp}{1 \mp e^{\lambda_{i_1}\beta\hbar e_{i_1}}}\mathrm{Tr}(\rho_G^0\alpha_{i_2}\cdots\alpha_{i_{2n}}) \\
&\quad + \cdots + \frac{[\alpha_{i_1},\alpha_{i_{2n}}]_\mp}{1 \mp e^{\lambda_{i_1}\beta\hbar e_{i_1}}}\mathrm{Tr}(\rho_G^0\alpha_{i_2}\alpha_{i_3}\cdots).
\end{aligned} \tag{4.2.44}$$

Now from Eq. (4.2.33) the contraction for any time order of the operators is defined as

$$\alpha_i^{\cdot}\alpha_j^{\cdot} = \langle T_\tau[\alpha_i\alpha_j]\rangle_0. \tag{4.2.45}$$

In our case $\tau_1 > \tau_2 > \tau_3 > \cdots > \tau_{2n}$, we can show

$$\alpha_i^{\cdot}\alpha_j^{\cdot} = \frac{[\alpha_i,\alpha_j]_\mp}{1 \mp e^{\lambda_i\beta e_i}}. \tag{4.2.46}$$

This is derived as follows: commutators with $i \neq j$ vanish and only nonzero commutators are

$$\begin{aligned}
\frac{[a_i^\dagger, a_i]_\mp}{1 \mp e^{\beta\hbar e_i}} &= \frac{\mp 1}{1 \mp e^{\beta\hbar e_i}} = \frac{1}{e^{\beta\hbar e_i} \mp 1} = f(e_i) \\
&= \langle a_i^\dagger a_i\rangle_0,
\end{aligned} \tag{4.2.47}$$

$$\frac{[a_i, a_i^\dagger]_\mp}{1 \mp e^{-\beta\hbar e_i}} = \frac{\mp 1}{1 \mp e^{-\beta\hbar e_i}} = 1 \pm f(e_i) = \langle a_i a_i^\dagger\rangle_0. \tag{4.2.48}$$

From Eq. (4.45), $\langle a_i^\dagger a_i\rangle_0$ and $\langle a_i a_i^\dagger\rangle_0$ show the contractions, $a_i^{\dagger\cdot} a_i^{\cdot}$ and $a_i^{\cdot} a_i^{\dagger\cdot}$, respectively. Thus we obtain

$$\begin{aligned}
&\mathrm{Tr}(\rho_{G_0}\alpha_{i_1}\alpha_{i_2}\alpha_{i_3}\cdots\alpha_{i_{2n}}) \\
&= \alpha_{i_1}^{\cdot}\alpha_{i_2}^{\cdot}\mathrm{Tr}(\rho_{G_0}\alpha_{i_3}\cdots\alpha_{i_{2n}}) \pm \alpha_{i_1}^{\cdot}\alpha_{i_3}^{\cdot}\mathrm{Tr}(\rho_{G_0}\alpha_{i_2}\cdots\alpha_{i_{2n}}) \\
&\quad + \cdots + \alpha_{i_1}^{\cdot}\alpha_{i_{2n}}^{\cdot}\mathrm{Tr}(\rho_{G_0}\alpha_{i_2}\alpha_{i_3}\cdots\alpha_{i_{2n-1}}) \\
&= \mathrm{Tr}(\rho_{G_0}\alpha_{i_1}^{\cdot}\alpha_{i_2}^{\cdot}\alpha_{i_3}\cdots\alpha_{i_{2n}}) + \mathrm{Tr}(\rho_{G_0}\alpha_{i_1}^{\cdot}\alpha_{i_2}\alpha_{i_3}^{\cdot}\cdots\alpha_{i_{2n}}) \\
&\quad + \cdots + \mathrm{Tr}(\rho_{G_0}\alpha_{i_1}^{\cdot}\alpha_{i_2}\alpha_{i_3}\cdots\alpha_{i_{2n}}^{\cdot}).
\end{aligned} \tag{4.2.49}$$

Inserting Eq. (4.2.49) into Eq. (4.2.38), we obtain

$$\begin{aligned}
&\langle\psi_1\psi_2\psi_3\cdots\psi_{2n}\rangle_0 \\
&= \sum_{i_1,i_2,i_3,\cdots,i_{2n}} \chi_{i_1}\chi_{i_2}\chi_{i_3}\cdots\chi_{i_{2n}}\{\mathrm{Tr}(\rho_{G_0}\alpha_{i_1}^{\cdot}\alpha_{i_2}^{\cdot}\alpha_{i_3}\cdots\alpha_{i_{2n}}) \\
&\quad + \mathrm{Tr}(\rho_{G_0}\alpha_{i_1}^{\cdot}\alpha_{i_2}\alpha_{i_3}^{\cdot}\cdots\alpha_{i_{2n}}) + \cdots + \mathrm{Tr}(\rho_{G_0}\alpha_{i_1}^{\cdot}\alpha_{i_2}\alpha_{i_3}\cdots\alpha_{i_{2n}}^{\cdot})\} \\
&= \langle\psi_1^{\cdot}\psi_2^{\cdot}\psi_3\cdots\psi_{2n}\rangle_0 + \langle\psi_1^{\cdot}\psi_2\psi_3^{\cdot}\cdots\psi_{2n}\rangle_0 \\
&\quad + \cdots + \langle\psi_1^{\cdot}\psi_2\psi_3\cdots\psi_{2n}^{\cdot}\rangle_0.
\end{aligned} \tag{4.2.50}$$

Since the contraction is a c number, we can contract the remaining operator in a similar manner so that we can obtain

$$\langle\psi_1\psi_2\psi_3\cdots\psi_{2n}\rangle_0 = \langle\dot{\psi}_1\dot{\psi}_2\ddot{\psi}_3\cdots\ddot{\psi}_{2n}\rangle_0 + \langle\dot{\psi}_1\ddot{\psi}_2\dot{\psi}_3\cdots\ddot{\psi}_{2n}\rangle_0 + \cdots, \quad (4.2.51)$$

where $\tau_1 > \tau_2 > \tau_3 > \cdots > \tau_{2n}$. We have proved Eq. (4.2.51) for any time ordering so that we could prove the Wick's theorem (4.2.35).

4.3 Feynman Diagrams

The Wick's theorem shows that all the terms in the perturbation expansions are calculated by summing over all possible fully contracted terms. These contractions are just expressed by the free particle Green's functions $\mathcal{G}^0$ so that the temperature Green's function $\mathcal{G}$ can be represented by the expansions of free particle Green's functions, $\mathcal{G}^0$. To represent these expansion processes, the Feynman diagram method is very convenient and practicable. Employing the Feynman diagram method, we present how to calculate the temperature Green's function $\mathcal{G}$.

4.3.1 *Expansions of the Numerator of the Temperature Green's Function, $\widetilde{\mathcal{G}}_{\alpha\beta}(x, x')$*

We investigate the numerator of Eq. (4.2.19) denoted as follows:

$$\begin{aligned}\widetilde{\mathcal{G}}_{\alpha\beta}(x, x') &= \langle T_\tau[\psi_\alpha(x)\psi^\dagger_\beta(x')\mathcal{U}(\beta\hbar)]\rangle_0 \\ &= -\sum_{n=0}^{\infty}\frac{1}{n!}\left(-\frac{1}{\hbar}\right)^n \int_0^{\beta\hbar} d\tau_1\cdots \\ &\quad\times\int_0^{\beta\hbar} d\tau_n\langle T_\tau[\mathcal{H}_I(\tau_1)\cdots\mathcal{H}_I(\tau_n)\psi_\alpha(x)\psi^\dagger_\beta(x')]\rangle_0. \end{aligned} \quad (4.3.1)$$

First we calculate the first order contribution of Eq. (4.3.1). We sum over all possible contractions and we obtain only the nonvanishing contractions; the pairs of ψ_α and $\psi^\dagger_\beta$;

$$\begin{aligned}\widetilde{\mathcal{G}}^{(1)}_{\alpha\beta}(x, x') &= \frac{1}{\hbar}\int_0^{\beta\hbar} d\tau_1 \mathrm{Tr}\{\rho_{G_0}[\mathcal{H}_I(\tau_1)\psi_\alpha(x)\psi^\dagger_\beta(x')]\} \\ &= \frac{1}{2\hbar}\sum_{\mu\nu}\int_0^{\beta\hbar} d^4x_1 d^4x_2 \\ &\quad\times\langle T_\tau[\psi^\dagger_\mu(x_1)\psi^\dagger_\nu(x_2)\mathcal{V}(x_1 - x_2)\psi_\nu(x_2)\psi_\mu(x_1)\psi_\alpha(x)\psi^\dagger_\beta(x')]\rangle_0 \\ &= \frac{1}{2\hbar}\sum_{\mu\nu}\int_0^{\beta\hbar} d^4x_1 d^4x_2 \mathcal{V}(x_1 - x_2) \\ &\quad\times\langle T_\tau[\psi^\dagger_\mu(x_1)\psi^\dagger_\nu(x_2)\psi_\nu(x_2)\psi_\mu(x_1)\psi_\alpha(x)\psi^\dagger_\beta(x')]\rangle_0, \end{aligned} \quad (4.3.2)$$

where in rewriting the interaction Hamiltonian

$$H_I = \frac{1}{2}\int d^3x_1 d^3x_2 \psi_\alpha^\dagger(\boldsymbol{x}_1)\psi_\beta^\dagger(\boldsymbol{x}_2)V(\boldsymbol{x}_1 - \boldsymbol{x}_2)\psi_\beta(\boldsymbol{x}_2)\psi_\alpha(\boldsymbol{x}_1), \tag{4.3.3}$$

by that in the interaction representation at finite temperature, we put

$$\begin{aligned}\mathcal{H}_I(\tau_1) &= \frac{1}{2}\int d^3x_1 d^3x_2 \\ &\quad\times \int_0^{\beta\hbar} d\tau_2 \psi_\alpha^\dagger(x_1)\psi^\dagger(x_2)V(\boldsymbol{x}_1 - \boldsymbol{x}_2)\delta(\tau_1 - \tau_2)\psi_\beta(x_2)\psi_\alpha(x_1) \\ &= \frac{1}{2}\int d^3x_1 d^3x_2 \\ &\quad\times \int_0^{\beta\hbar} d\tau_2 \psi_\alpha^\dagger(x_1)\psi_\beta^\dagger(x_2)\mathcal{V}(x_1 - x_2)\psi_\beta(x_2)\psi_\alpha(x_1),\end{aligned} \tag{4.3.4}$$

where

$$\mathcal{V}(x_1 - x_2) = V(\boldsymbol{x}_1 - \boldsymbol{x}_2)\delta(\tau_1 - \tau_2) \tag{4.3.5}$$

and $x_i = (\boldsymbol{x}_i, \tau_i)$. It should be noted that all field operators represent those in the interaction representation; we have omitted the subscript I indicating the interaction representation.

According to the Wick's theorem, the average $\langle\cdots\rangle_0$ can be written as:

$$\begin{aligned}&\langle T_\tau[\psi_\mu^\dagger(x_1)\psi_\nu^\dagger(x_2)\psi_\nu(x_2)\psi_\mu(x_1)\psi_\alpha(x)\psi_\beta^\dagger(x')]\rangle_0 \\ &= \langle T_\tau[\psi_\alpha(x)\psi_\mu^\dagger(x_1)]\rangle_0\langle T_\tau\psi_\nu^\dagger(x_2)\psi_\nu(x_2)\rangle_0\langle T_\tau[\psi_\mu(x_1)\psi_\beta^\dagger(x')\rangle_0 \quad (a)\\ &\pm\langle T_\tau[\psi_\alpha(x)\psi_\mu^\dagger(x_1)]\rangle_0\langle T_\tau\psi_\nu^\dagger(x_2)\psi_\mu(x_1)\rangle_0\langle T_\tau[\psi_\nu(x_2)\psi_\beta^\dagger(x')\rangle_0 \quad (b)\\ &+\langle T_\tau[\psi_\alpha(x)\psi_\nu^\dagger(x_2)]\rangle_0\langle T_\tau\psi_\mu^\dagger(x_1)\psi_\mu(x_1)\rangle_0\langle T_\tau[\psi_\nu(x_2)\psi_\beta^\dagger(x')\rangle_0 \quad (a')\\ &\pm\langle T_\tau[\psi_\alpha(x)\psi_\nu^\dagger(x_2)]\rangle_0\langle T_\tau\psi_\mu^\dagger(x_1)\psi_\nu(x_2)\rangle_0\langle T_\tau[\psi_\mu(x_1)\psi_\beta^\dagger(x')\rangle_0 \quad (b')\\ &+\langle T_\tau[\psi_\alpha(x)\psi_\beta^\dagger(x')\rangle_0\langle T_\tau\psi_\mu^\dagger(x_1)\psi_\mu(x_1)\rangle_0\langle T_\tau\psi_\nu^\dagger(x_2)\psi_\nu(x_2)\rangle_0 \quad (c)\\ &\pm\langle T_\tau[\psi_\alpha(x)\psi_\beta^\dagger(x')\rangle_0\langle T_\tau\psi_\mu^\dagger(x_1)\psi_\nu(x_2)\rangle_0\langle T_\tau\psi_\nu^\dagger(x_2)\psi_\mu(x_1)\rangle_0. \quad (d)\end{aligned} \tag{4.3.6}$$

Replacing the averages $\langle\cdots\rangle_0$ by suitable Green's functions $\mathcal{G}^0$, we obtain

$$\begin{aligned}&= \mp\mathcal{G}_{\alpha\mu}^0(x, x_1)\mathcal{G}_{\nu\nu}^0(x_2, x_2)\mathcal{G}_{\mu\beta}^0(x_1, x') \quad (a)\\ &-\mathcal{G}_{\alpha\mu}^0(x, x_1)\mathcal{G}_{\mu\nu}^0(x_1, x_2)\mathcal{G}_{\nu\beta}^0(x_2, x') \quad (b)\\ &\mp\mathcal{G}_{\alpha\nu}^0(x, x_2)\mathcal{G}_{\mu\mu}^0(x_1, x_1)\mathcal{G}_{\nu\beta}^0(x_2, x') \quad (a')\\ &-\mathcal{G}_{\alpha\nu}^0(x, x_2)\mathcal{G}_{\nu\mu}^0(x_2, x_1)\mathcal{G}_{\mu\beta}^0(x_1, x') \quad (b')\\ &-\mathcal{G}_{\alpha\beta}^0(x, x')\mathcal{G}_{\mu\mu}^0(x_1, x_1)\mathcal{G}_{\nu\nu}^0(x_2, x_2) \quad (c)\\ &\mp\mathcal{G}_{\alpha\beta}^0(x, x')\mathcal{G}_{\mu\nu}^0(x_1, x_2)\mathcal{G}_{\nu\mu}^0(x_2, x_1). \quad (d)\end{aligned} \tag{4.3.7}$$

It should be noted that $\mathcal{G}^0(\boldsymbol{x}_1 - \boldsymbol{x}_2, 0)$ is always calculated as $\lim_{\delta\to+0}\mathcal{G}^0(\boldsymbol{x}_1 - \boldsymbol{x}_2, -\delta)$.

4.3.2 *The Feynman Rules in Coordinate Space*

We first show an example of the Feynman diagrams for the first order expansions of the Green's functions $\widetilde{\mathcal{G}}^{(1)}$ in Eq. (4.3.2) as the following steps:

1. We start with bracket, $[\psi_\mu^\dagger(x_1)\psi_\nu^\dagger(x_2)\mathcal{V}(x_1-x_2)\psi_\nu(x_2)\psi_\mu(x_1)\psi_\alpha(x)\psi_\beta^\dagger(x')]$, in the second equation of Eq. (4.3.2). First we put the four points, x, x', x_1, x_2 illustrated as dots in Figure 4.1.
2. As illustrated in Figure (4.1): $\psi_\alpha(x)$ represents a particle with spin α leaving x, $\psi_\beta^\dagger(x')$ with β, entering x'. The interaction $\mathcal{V}(x_1 - x_2)$ represents a dotted line between x_1 and x_2, where $\psi_\mu^\dagger(x_1)$ and $\psi_\mu(x_1)$ represent a particle with μ simultaneously entering and leaving x_1, $\psi_\nu^\dagger(x_2)$ and $\psi_\nu(x_2)$ represent another particle with ν simultaneously entering and leaving x_2.
3. As illustrated in Figure 4.2, we connect the four points x, x', x_1, x_2 by three solid lines with the respective leaving and entering points in all the possible ways.
4. Each diagram has two external points x, x' leaving x and entering x', and two internal points x_1, x_2, each of which has both leaving and entering.

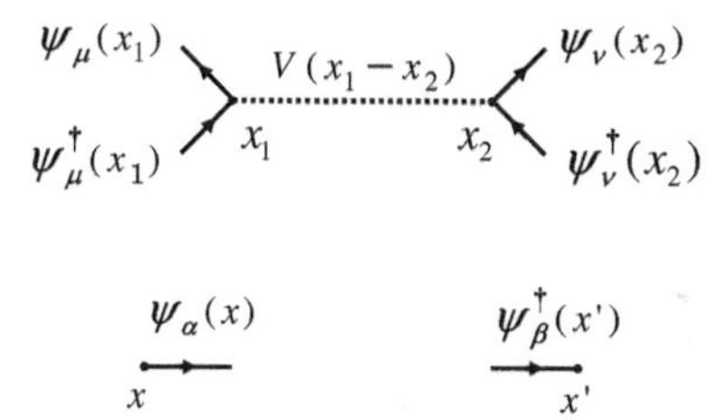

Figure 4.1 The bracket, $[\psi_\mu^\dagger(x_1)\psi_\nu^\dagger(x_2)\mathcal{V}(x_1-x_2)\psi_\nu(x_2)\psi_\mu(x_1)\psi_\alpha(x)\psi_\beta^\dagger(x')]$ in the second equation in Eq. (4.3.2), is schematically illustrated.

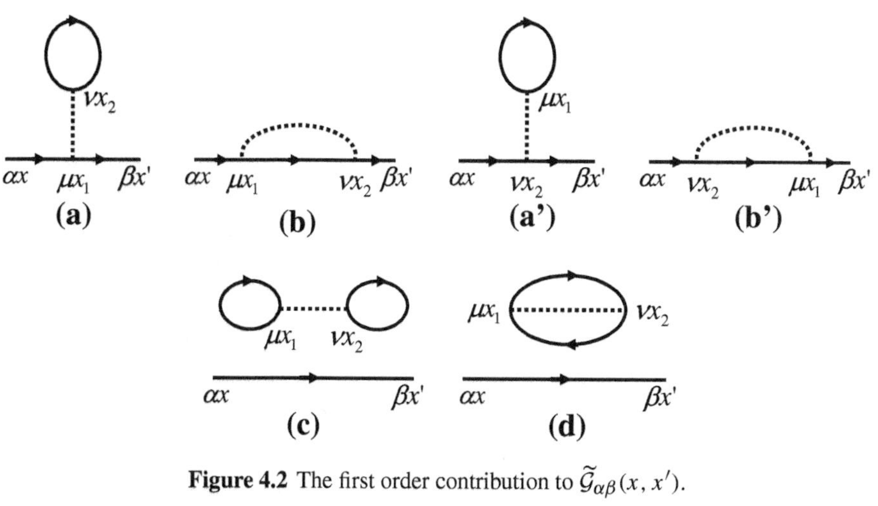

Figure 4.2 The first order contribution to $\widetilde{\mathcal{G}}_{\alpha\beta}(x, x')$.

5. We assign a solid line leaving x_l with γ and entering x_e with δ to $\mathcal{G}^0_{\gamma\delta}(x_l - x_e)$. Then each Feynman diagram, (a), (b), (a′), (b′), (c), and (d) in Figure (4.2), corresponds to the respective expansion, (a), (b), (a′), (b′), (c), and (d), in Eq. (4.3.7).
6. Finally, we integrate over the coordinates of the internal points and sum over the spin variables.

Following the same rules in the first order Green's function $\widetilde{\mathcal{G}}^{(1)}$, we can calculate the higher order expansions of the numerator of the Green's function $\widetilde{\mathcal{G}}$ in Eq. (4.3.4) illustrating the Feynman diagrams with the free field Green's function. In a similar manner to the numerator of the Green's function, $\widetilde{\mathcal{G}}$, we can calculate the denominator of the Green's function in Eq. (4.2.19).

4.3.3 Some Important Formulae

In calculating the Green's function in Eq. (4.2.19) by the Feynman diagrams, there are some important formulae, which play an important role in the calculations.

1. In Figure 4.2 for the first order of the diagrams, there are topologically the same diagrams: (a) and (a′), and (b) and (b′). The contributions of (a) and (a′), and (b) and (b′) are the same values under integrations over the coordinates x_1 and x_2, and summation over the corresponding spin variables in Eq. (4.3.2). This means that we can omit the factor $\frac{1}{2}$ in the interaction Hamiltonian in Eq. (4.3.4), only drawing topologically non-equivalent diagrams (a) and (b). This is true for any higher order of diagrams; we can neglect the factor $\frac{1}{2}$ in the interaction Hamiltonian, only drawing topologically non-equivalent Feynman diagrams. We show the Feynman diagrams for the first order Green's function in Figure 4.3.
2. All the Feynman diagrams for $\widetilde{\mathcal{G}}$ in Eq. (4.3.4) can be divided into two groups, connected diagrams and disconnected diagrams. In a connected diagram all the points connect with the external points x and $x\prime$ via lines. For example, in Figure 4.2, the diagrams (a), (b), (a′) and (b′), while (c) and (d) are disconnected. The significant formula asserts that the numerator of the Green's function $\widetilde{\mathcal{G}}$ is the product of the sum of all possible connected diagrams denoted as $\widetilde{\mathcal{G}}^{\text{con.}}$ and the sum of all possible disconnected diagrams identical with $\langle \mathcal{U}(\beta\hbar)\rangle_0$, the denominator of the Green's function $\mathcal{G}$ in Eq. (4.2.19):

$$\widetilde{\mathcal{G}}_{\alpha\beta}(x, x') = \widetilde{\mathcal{G}}_{\alpha\beta}(x, x')^{\text{con.}} \times \mathcal{U}(\beta\hbar), \tag{4.3.8}$$

as illustrated in Figure 4.4.
Thus the second factor $\mathcal{U}(\beta\hbar)$ cancels with the denominator of the Green's function so that the Green's function $\mathcal{G}$ is exactly the sum of all the possible connected

Figure 4.3 The Feynman diagrams of the first order Green's function $\widetilde{\mathcal{G}}^{(1)}_{\alpha\beta}(x, x')$ are illustrated.

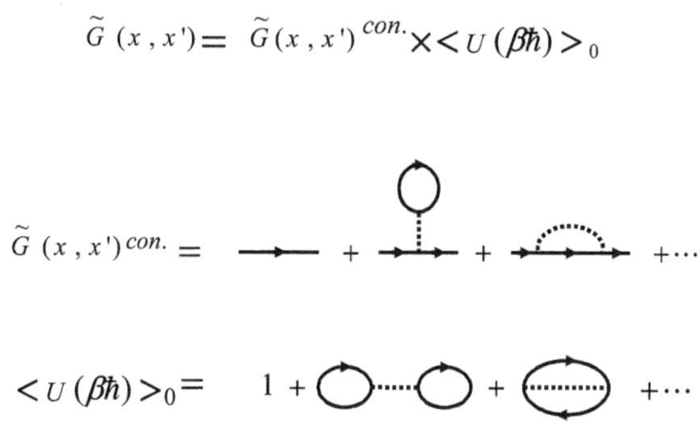

Figure 4.4 Factorization of the Green's function to $\widetilde{\mathcal{G}}(x, x')$. $\widetilde{\mathcal{G}}(x, x')^{\text{con.}}$ and $\langle\mathcal{U}(\beta\hbar)\rangle_0$ identical with the denominator of $\mathcal{G}(x, x')$. $\widetilde{\mathcal{G}}(x, x')^{\text{con.}}$ is the sum of all possible connected diagrams and $\langle\mathcal{U}(\beta\hbar)\rangle_0$ the sum of all possible disconnected diagrams. Here we show up to the first order diagrams.

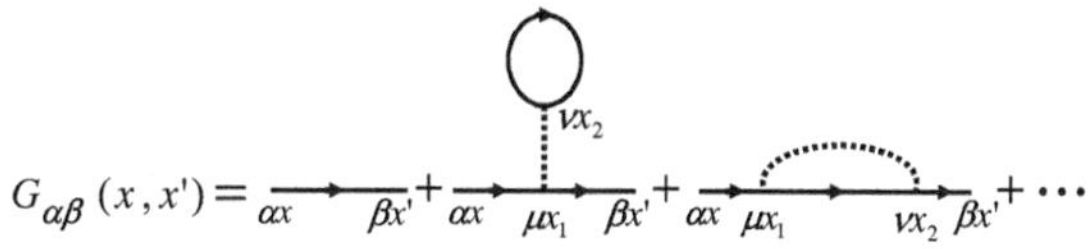

Figure 4.5 The Feynman diagrams of the Green's function $\mathcal{G}_{\alpha\beta}(x, x')$ are illustrated up to the first order diagrams.

diagrams in the perturbation series.

$$\mathcal{G}_{\alpha\beta}(x, x') = \widetilde{\mathcal{G}}_{\alpha\beta}(x, x')^{\text{con.}} = -\langle T_\tau[\psi_\alpha(x)\psi_\beta(x')]\mathcal{U}(\beta\hbar)\rangle_{\text{con.}}. \qquad (4.3.9)$$

We show the expansions of the Feynman diagrams of the Green's function $\mathcal{G}_{\alpha\beta}(x, x')$ in Figure 4.5.

Now we prove the significant formula (4.3.8). In order to prove Eq. (4.3.8) to all the orders, we investigate the nth order of the numerator of the Green's function, $\widetilde{\mathcal{G}}_{\alpha\beta}(x, x')$ in Eq. (4.3.1):

$$-\frac{1}{n!}(-\frac{1}{\hbar})^n \int_0^{\beta\hbar} d\tau_1 \cdots \int_0^{\beta\hbar} d\tau_n \langle T_\tau[\mathcal{H}_I(\tau_1)\cdots\mathcal{H}_I(\tau_n)\psi_\alpha(x)\psi_\beta^\dagger(x')]\rangle_0.$$

When we divide this term into mth connected and $(n-m)$th disconnected diagrams:

$$\begin{aligned}
&-\frac{1}{n!}(-\frac{1}{\hbar})^n \int_0^{\beta\hbar} d\tau_1 \cdots \\
&\quad\times \int_0^{\beta\hbar} d\tau_m \langle T_\tau[\mathcal{H}_I(\tau_1)\cdots\mathcal{H}_I(\tau_m)\psi_{I\alpha}(x)\psi_{I\beta}^\dagger(x')]\rangle_{con.} \\
&\quad\times \int_0^{\beta\hbar} d\tau_{m+1} \cdots \int_0^{\beta\hbar} d\tau_n \langle T_\tau[\mathcal{H}_I(\tau_{m+1})\cdots\mathcal{H}_I(\tau_n)]\rangle_0.
\end{aligned}$$

The Wick's theorem requires all the possible full contractions. The topologically same diagram can be obtained by exchanging $\mathcal{H}_I(\tau_i)$ by $\mathcal{H}_I(\tau_j)$ among the $\mathcal{H}_I(\tau_1)\cdots\mathcal{H}_I(\tau_n)$ so that the following factor is required:

$$\frac{n!}{m!(n-m)!}.$$

Thus we obtain

$$\begin{aligned}
&-\frac{1}{n!}(-\frac{1}{\hbar})^n\int_0^{\beta\hbar}d\tau_1\cdots\int_0^{\beta\hbar}d\tau_n\langle T_\tau[\mathcal{H}_I(\tau_1)\cdots\mathcal{H}_I(\tau_n)\psi_\alpha(x)\psi^\dagger_\beta(x')]\rangle_0\\
&=-\sum_{m=0}^{n}\frac{1}{m!}(-\frac{1}{\hbar})^m\int_0^{\beta\hbar}d\tau_1\cdots\\
&\times\int_0^{\beta\hbar}d\tau_m\langle T_\tau[\mathcal{H}_I(\tau_1)\cdots\mathcal{H}_I(\tau_m)\psi_\alpha(x)\psi^\dagger_\beta(x')]\rangle_{con.}\\
&\times\frac{1}{(n-m)!}(-\frac{1}{\hbar})^{n-m}\int_0^{\beta\hbar}d\tau_{m+1}\cdots\\
&\times\int_0^{\beta\hbar}d\tau_n\langle T_\tau[\mathcal{H}_I(\tau_{m+1})\cdots\mathcal{H}_I(\tau_n)]\rangle_0.
\end{aligned}$$

Then we obtain

$$\begin{aligned}
&\widetilde{\mathcal{G}}_{\alpha\beta}(x,x')\\
&=\sum_{n=0}^{\infty}\sum_{m=0}^{n}-\frac{1}{m!}(-\frac{1}{\hbar})^m\int_0^{\beta\hbar}d\tau_1\cdots\\
&\times\int_0^{\beta\hbar}d\tau_m\langle T_\tau[\mathcal{H}_I(\tau_1)\cdots\mathcal{H}_I(\tau_m)\psi_\alpha(x)\psi^\dagger_\beta(x')]\rangle_{con.}\\
&\times\frac{1}{(n-m)!}(-\frac{1}{\hbar})^{n-m}\int_0^{\beta\hbar}d\tau_{m+1}\cdots\\
&\times\int_0^{\beta\hbar}d\tau_n\langle T_\tau[\mathcal{H}_I(\tau_{m+1})\cdots\mathcal{H}_I(\tau_n)]\rangle_0
\end{aligned}\tag{4.3.10}$$

If we change the order of the summation $\sum_{n=0}^{\infty}\sum_{m=0}^{n}$ to $\sum_{m=0}^{\infty}\sum_{n=m}^{\infty}$, then change the variable n to $k=n-m$, we obtain

$$\begin{aligned}
&\widetilde{\mathcal{G}}_{\alpha\beta}(x,x')\\
&=\sum_{m=0}^{\infty}-\frac{1}{m!}(-\frac{1}{\hbar})^m\int_0^{\beta\hbar}d\tau_1\cdots\\
&\times\int_0^{\beta\hbar}d\tau_m\langle T_\tau[\mathcal{H}_I(\tau_1)\cdots\mathcal{H}_I(\tau_m)\psi_\alpha(x)\psi^\dagger_\beta(x')]\rangle_{con.}\\
&\times\sum_{k=0}^{\infty}\frac{1}{(k)!}\left(-\frac{1}{\hbar}\right)^k\int_0^{\beta\hbar}d\tau_{m+1}\cdots
\end{aligned}$$

$$\times \int_0^{\beta\hbar} d\tau_{m+k} \langle T_\tau [\mathcal{H}_I(\tau_{m+1}) \cdots \mathcal{H}_I(\tau_{m+k})] \rangle_0$$
$$= -\langle T_\tau [\psi_\alpha(x)\psi_\beta(x')\mathcal{U}(\beta\hbar)\rangle_{con.} < \mathcal{U}(\beta\hbar)\rangle_0. \tag{4.3.11}$$

Thus, we could prove Eq. (4.3.8).

3. Next we investigate a diagram of mth order of the connected Green's function $\widetilde{\mathcal{G}}$:

$$-\frac{1}{m!}\left(-\frac{1}{\hbar}\right)^m \int_0^{\beta\hbar} d\tau_1 \cdots \int_0^{\beta\hbar} d\tau_m \langle T_\tau [\mathcal{H}_I(\tau_1) \cdots \mathcal{H}_I(\tau_m)\psi_\alpha(x)\psi_\beta^\dagger(x')]\rangle_{con.} \tag{4.3.12}$$

The topologically same diagrams can be obtained by permutations among $\mathcal{H}_I(\tau_1) \cdots \mathcal{H}_I(\tau_m)$. Therefore when we draw the mth order Feynman diagrams we can omit the factor $\frac{1}{m!}$.

4. In order to derive the Green's function $\mathcal{G}_{\alpha\beta}(x, x')$ from the Feynman diagram in Figure 4.5, we must take into account the following rules: In fermion systems, the sign of a diagrams is associated with the permutation of the fermion operators ψ_α. A change of sign is connected with the formation of a closed loop in the diagram so that the sign of a diagram is determined by the factor $(-1)^F$, where F is the number of closed loops. In Figure 4.5, the number of loops is: $F = 0$ for (b); $F = 1$ for (a). Thus we obtain $\mathcal{G}_{\alpha\beta}(x, x')$ in Figure (4.5).

$$\begin{aligned}\mathcal{G}_{\alpha\beta}(x, x') &= \mathcal{G}^0_{\alpha\beta}(x, x') \\ &+\frac{1}{\hbar}\int_0^{\beta\hbar} d^4x_1 d^4x_2 \{\mp \mathcal{G}^0_{\alpha\mu}(x, x_1)\mathcal{G}^0_{\nu\nu}(x_2, x_2)\mathcal{G}^0_{\mu\beta}(x_1, x') \\ &-\mathcal{G}^0_{\alpha\mu}(x, x_1)\mathcal{G}^0_{\mu\nu}(x_1, x_2)\mathcal{G}^0_{\nu\beta}(x_2, x')\}\mathcal{V}(x_1 - x_2) \\ &+\cdots\end{aligned} \tag{4.3.13}$$

5. We investigate the case that the time arguments are the same, $t_x = t_{x'}$, in $G(x, x')$. This situation happens only when two operators from a single interaction Hamiltonian H_I are pairs. Since the order of the operators is specified such that $\psi^\dagger(x)$ appears to the $\psi(x)$ in the interaction Hamiltonian H_I, we must calculate as

$$\lim_{\delta\to+0} G(\boldsymbol{x}t, \boldsymbol{x}'t + \delta) = -\langle \psi^\dagger(\boldsymbol{x})\psi(\boldsymbol{x}')\rangle. \tag{4.3.14}$$

We now state the general rules to calculate the Green's functions from the nth order diagrams:

1. Draw all connected, topologically non-equivalent diagrams with n interaction lines and $2n + 1$ directed particle lines.
2. Assign a Green's function $\mathcal{G}^0_{\alpha\beta}(x_i - x_j)$ to each solid line starting from x_i, α to x_j, β.
3. Assign a potential function $\mathcal{V}(x_i - x_j)$ to each dotted line.
4. Integrate over all the vertex coordinates x_i, $\int_0^{\beta\hbar} d^4x_i$ and sum over all the spin variables α.

5. Multiply each factor $(-\frac{1}{\hbar})^n(\pm)^F$ for a diagram, where F is the number of closed loops.
6. Substitute $\lim_{\delta\to+0}\mathcal{G}^0(\boldsymbol{x}_i-\boldsymbol{x}_j,-\delta)$ for the Green's function $\mathcal{G}^0(x_i-x_j)$ with the same time arguments.

4.3.4 Feynman Rules in Momentum Space

Next we transform the temperature Green's function in the coordinate space $\mathcal{G}(x_1,x_2)=\mathcal{G}(\boldsymbol{x}_1,\tau_1,\boldsymbol{x}_2,\tau_2)$ to that in the momentum space. First, we investigate the transformation of imaginary time τ. The domain of the imaginary time τ is $0\leqslant\tau\leqslant\beta\hbar$. To periodically extend a function of the imaginary time τ, $\mathcal{G}(\tau)$, in the domain $[0,\beta\hbar]$ to that in the whole domain $[-\infty,\infty]$, we assume that the function $\mathcal{G}(\tau)$ is a periodic function in the domain $[-\beta\hbar,\beta\hbar]$ holding the same function as the original function in original domain $[0,\beta\hbar]$. Then we can expand the $\mathcal{G}(\tau)$ in a Fourier series:

$$\mathcal{G}(\tau)=\frac{1}{\beta\hbar}\sum_{\mathrm{i}\omega_n}e^{-\mathrm{i}\omega_n\tau}\mathcal{G}(\mathrm{i}\omega_n), \tag{4.3.15}$$

where

$$\omega_n=\frac{n\pi}{\beta\hbar}. \tag{4.3.16}$$

The associated Fourier coefficient is given by

$$\mathcal{G}(\mathrm{i}\omega_n)=\frac{1}{2}\int_{-\beta\hbar}^{\beta\hbar}d\tau e^{\mathrm{i}\omega_n\tau}\mathcal{G}(\tau). \tag{4.3.17}$$

Next we investigate the Fourier transformation of the position $\boldsymbol{x}$, which is defined as:

$$\mathcal{G}(\boldsymbol{x})=\frac{1}{V}\sum_{\boldsymbol{p}}e^{\mathrm{i}\boldsymbol{p}\cdot\boldsymbol{x}}\mathcal{G}(\boldsymbol{p}).$$

$$\mathcal{G}(\boldsymbol{p})=\int_V d^3x e^{-\mathrm{i}\boldsymbol{p}\cdot\boldsymbol{x}}\mathcal{G}(\boldsymbol{x}). \tag{4.3.18}$$

Now we investigate the temperature Green's function $\mathcal{G}(x_1.x_2)=\mathcal{G}(\boldsymbol{x}_1,\tau_1,\boldsymbol{x}_2,\tau_2)$. The translational invariance of $\mathcal{G}(\boldsymbol{x}_1,\tau_1,\boldsymbol{x}_2,\tau_2)$ with respect to x_1 and x_2, and τ_1 and τ_2 allows to replace $\mathcal{G}(\boldsymbol{x}_1,\tau_1,\boldsymbol{x}_2,\tau_2)$ by $\mathcal{G}(\boldsymbol{x}_1+\boldsymbol{x},\tau_1+\tau,\boldsymbol{x}_2+\boldsymbol{x},\tau_2+\tau)$ for any $\boldsymbol{x},\tau$. This fact leads to

$$\begin{aligned}
\mathcal{G}(x_1,x_2)&=\frac{1}{(\beta\hbar V)^2}\sum_{p_1}\sum_{p_2}\mathcal{G}(p_1,p_2)e^{\mathrm{i}(p_1x_1-p_2x_2)}\\
&=\frac{1}{(\beta\hbar V)^2}\sum_{p_1}\sum_{p_2}\mathcal{G}(\boldsymbol{x}_1,\tau_1,\boldsymbol{x}_2,\tau_2)e^{\mathrm{i}(\boldsymbol{p}_1\cdot\boldsymbol{x}_1-\boldsymbol{p}_2\cdot\boldsymbol{x}_2,-\omega_{n_1}\tau_1+\omega_{n_2}\tau_2)}\\
&=\frac{1}{(\beta\hbar V)^2}\sum_{p_1}\sum_{p_2}\frac{1}{2\beta\hbar}\int_{-\beta\hbar}^{\beta\hbar}d\tau\frac{1}{V}\int_V dx\\
&\quad\times\mathcal{G}(\boldsymbol{x}_1,\tau_1,\boldsymbol{x}_2,\tau_2)e^{\mathrm{i}[\boldsymbol{p}_1\cdot(x_1+\boldsymbol{x})-\boldsymbol{p}_2\cdot(x_2+\boldsymbol{x}),-\omega_{n_1}(\tau_1+\tau)+\omega_{n_2}(\tau_2+\tau)]}
\end{aligned}$$

$$= \frac{1}{(\beta\hbar V)^2} \sum_{p_1} \sum_{p_2} \mathcal{G}(\boldsymbol{x}_1, \tau_1, \boldsymbol{x}_2, \tau_2) \times e^{\mathrm{i}(\boldsymbol{p}_1 \cdot \boldsymbol{x}_1 - \boldsymbol{p}_2 \cdot \boldsymbol{x}_2, -\omega_{n_1}\tau_1 + \omega_{n_2}\tau_2)} \delta_{\boldsymbol{p}_1, \boldsymbol{p}_2} \delta_{\omega_{n_1}, \omega_{n_2}}$$
$$= \frac{1}{\beta\hbar V} \sum_p \mathcal{G}(\boldsymbol{x}_1, \tau_1, \boldsymbol{x}_2, \tau_2) e^{\mathrm{i}p(x_1 - x_2)}$$
$$\equiv \mathcal{G}(x_1 - x_2) = \mathcal{G}(\boldsymbol{x}_1 - \boldsymbol{x}_2, \tau_1 - \tau_2). \tag{4.3.19}$$

The translational invariant requires $\mathcal{G}(x_1, x_2) = \mathcal{G}(x_1 - x_2)$. This fact on the imaginary time τ yields the domain of $\tau_1 - \tau_2$ is $[-\beta\hbar, \beta\hbar]$ so that the Fourier transformation (4.3.15) and the inverse transformation (4.3.17) hold true in the actual domain $[-\beta\hbar, \beta\hbar]$. Thus we obtain

$$\mathcal{G}(\boldsymbol{x}, \mathrm{i}\omega_n) = \frac{1}{2} \int_{-\beta\hbar}^{\beta\hbar} d\tau e^{\mathrm{i}\omega_n \tau} \mathcal{G}(\boldsymbol{x}, \tau)$$
$$= \frac{1}{2} \int_{-\beta\hbar}^{0} d\tau e^{\mathrm{i}\omega_n \tau} \mathcal{G}(\boldsymbol{x}, \tau) + \frac{1}{2} \int_{0}^{\beta\hbar} d\tau e^{\mathrm{i}\omega_n \tau} \mathcal{G}(\boldsymbol{x}, \tau)$$
$$= \pm \frac{1}{2} \int_{-\beta\hbar}^{0} d\tau e^{\mathrm{i}\omega_n \tau} \mathcal{G}(\boldsymbol{x}, \tau + \beta\hbar) + \frac{1}{2} \int_{0}^{\beta\hbar} d\tau e^{\mathrm{i}\omega_n \tau} \mathcal{G}(\boldsymbol{x}, \tau)$$
$$= \frac{1}{2}(1 \pm e^{-\mathrm{i}\omega_n \beta\hbar}) \int_{0}^{\beta\hbar} d\tau e^{\mathrm{i}\omega_n \tau} \mathcal{G}(\boldsymbol{x}, \tau), \tag{4.3.20}$$

where the third equation is obtained by using the periodic nature for bosons and antiperiodic one for fermions with respect to τ in the domain $[-\beta\hbar, \beta\hbar]$, and the fourth equation by changing integral valuable. From Eq. (4.3.16) we obtain

$$\frac{1}{2}(1 \pm e^{-\mathrm{i}\omega_n \beta\hbar}) = \frac{1}{2}[1 \pm (-1)^n] = \left\{ \begin{array}{l} \left\{ \begin{array}{ll} 1 & n: \text{even}, \\ 0 & n: \text{odd}, \end{array} \right. \\ \left\{ \begin{array}{ll} 1 & n: \text{odd}, \\ 0 & n: \text{even}. \end{array} \right. \end{array} \right. \tag{4.3.21}$$

Thus we obtain

$$\mathcal{G}(\boldsymbol{x}, \mathrm{i}\omega_n) = \int_{0}^{\beta\hbar} d\tau e^{\mathrm{i}\omega_n \tau} \mathcal{G}(\boldsymbol{x}, \tau), \tag{4.3.22}$$

where

$$\omega_n = \left\{ \begin{array}{ll} \frac{2n\pi}{\beta\hbar} & \text{for bosons} \\ \frac{(2n+1)\pi}{\beta\hbar} & \text{for fermions.} \end{array} \right. \tag{4.3.23}$$

Next we investigate the Fourier transformation of the free particle Green's functions $\mathcal{G}^{(0)}(\boldsymbol{x}, \tau)$. Equation (4.1.42) for $\tau > 0$ leads to

$$\mathcal{G}^{0}_{\alpha\beta}(\boldsymbol{p}, \tau) = -\delta_{\alpha\beta}[1 \pm f(\epsilon^0_p)] e^{-\epsilon^0_p \tau}, \tag{4.3.24}$$

where

$$f(\epsilon_p^0) = \frac{1}{e^{\beta\hbar\epsilon_p^0} \mp 1}.$$

Substituting Eq. (4.3.24) to Eq. (4.3.20), we obtain

$$\begin{aligned}
\mathcal{G}_{\alpha\beta}^0(\boldsymbol{p}, \mathrm{i}\omega_n) &= \delta_{\alpha\beta}\mathcal{G}^0(\boldsymbol{p}, \mathrm{i}\omega_n) \\
&= -\delta_{\alpha\beta}[1 \pm f(\epsilon_k^0)] \int_0^{\beta\hbar} d\tau e^{(\mathrm{i}\omega_n - \epsilon_p^0)\tau} \\
&= -\frac{\delta_{\alpha\beta}}{(\mathrm{i}\omega_n - \epsilon_p^0)}[1 \pm f(\epsilon_p^0)][e^{(\mathrm{i}\omega_n - \epsilon_p^0)\beta\hbar} - 1] \\
&= \delta_{\alpha\beta}\frac{1}{\mathrm{i}\omega_n - \epsilon_p^0},
\end{aligned} \tag{4.3.25}$$

where we have considered

$$[1 \pm f(\epsilon_p^0)][e^{(\mathrm{i}\omega_n - \epsilon_p^0)\beta\hbar} - 1] = [1 \pm \frac{1}{e^{\beta\hbar\epsilon_p^0} \mp 1}][\pm e^{-\beta\hbar\epsilon_p^0} - 1] = -1. \tag{4.3.26}$$

To construct Feynman rules in momentum space, we summarize the Fourier representation of the Green's functions and the potential:

$$\mathcal{G}_{\alpha\beta}(\boldsymbol{x}, \tau) = \frac{1}{\beta\hbar V}\sum_p e^{\mathrm{i}(\boldsymbol{p}\cdot\boldsymbol{x} - \omega_n\tau)}\mathcal{G}_{\alpha\beta}(\boldsymbol{p}, \mathrm{i}\omega_n), \tag{4.3.27}$$

$$\mathcal{G}_{\alpha\beta}^0(\boldsymbol{x}, \tau) = \frac{1}{\beta\hbar V}\sum_p e^{\mathrm{i}(\boldsymbol{p}\cdot\boldsymbol{x} - \omega_n\tau)}\mathcal{G}_{\alpha\beta}^0(\boldsymbol{p}, \mathrm{i}\omega_n), \tag{4.3.28}$$

$$\mathcal{V}(\boldsymbol{x}, \tau) = \frac{1}{\beta\hbar V}\sum_q e^{\mathrm{i}(\boldsymbol{q}\cdot\boldsymbol{x} - \nu_n\tau)}\mathcal{V}(\boldsymbol{q}, \mathrm{i}\nu_n), \tag{4.3.29}$$

where $\sum_p$ means $\sum_{\boldsymbol{p}, \mathrm{i}\omega_n}$. Since

$$\delta(\tau) = \frac{1}{\beta\hbar}\sum_n e^{\mathrm{i}\frac{2n\pi}{\beta\hbar}\tau}, \tag{4.3.30}$$

we have

$$\mathcal{V}(\boldsymbol{q}, \mathrm{i}\nu_n) = V(\boldsymbol{q}), \qquad \nu_n = \frac{2n\pi}{\beta\hbar}: \quad \text{even.} \tag{4.3.31}$$

Thus the first order term of the Green's function $\mathcal{G}(x-x')$ in Eq. (4.3.13) leads to

$$
\begin{aligned}
\mathcal{G}^{(1)}_{\alpha\beta}(x-x') &= \frac{1}{\beta\hbar V}\sum_{p}\mathcal{G}^{(1)}_{\alpha\beta}(p)e^{\mathrm{i}p(x-x')} \\
&= \frac{1}{\hbar}\int_0^{\beta\hbar} d^4x_1 d^4x_2 \frac{1}{(\beta\hbar V)^4}\sum_{p,p',p_1,q} \\
&\quad\times\Big\{\mp\mathcal{G}^0_{\alpha\mu}(p)\mathcal{G}^0_{\nu\nu}(p_1)\mathcal{G}^0_{\mu\beta}(p')\mathcal{V}(q) \\
&\quad\times e^{\mathrm{i}[p(x-x_1)+p_1(x_2-x_2)^-+p'(x_1-x')+q(x_1-x_2)]} \\
&\quad-\mathcal{G}^0_{\alpha\mu}(p)\mathcal{G}^0_{\mu\nu}(p_1)\mathcal{G}^0_{\nu\beta}(p')\mathcal{V}(q) \\
&\quad\times e^{\mathrm{i}[p(x-x_1)+p_1(x_1-x_2)^-+p'(x_2-x')+q(x_1-x_2)]}\Big\} \\
&= \frac{1}{\hbar}\frac{1}{(\beta\hbar V)^2}\sum_{p,p',p_1,q}\Big\{\mp\mathcal{G}^0_{\alpha\mu}(p)\mathcal{G}^0_{\nu\nu}(p_1)e^{\mathrm{i}\omega_{n_1}\delta}\mathcal{G}^0_{\mu\beta}(p')\mathcal{V}(q)\delta_{p,p'}\delta_{q,0} \\
&\quad-\mathcal{G}^0_{\alpha\mu}(p)\mathcal{G}^0_{\mu\nu}(p_1)e^{\mathrm{i}\omega_{n_1}\delta}\mathcal{G}^0_{\nu\beta}(p')\mathcal{V}(q)\delta_{p,p_1+q}\delta_{p_1,p'-q}\Big\}e^{\mathrm{i}(px-p'x')} \\
&= \frac{1}{\hbar}\frac{1}{(\beta\hbar V)^2}\sum_{p,p_1}\Big\{\mp\mathcal{G}^0_{\alpha\mu}(p)\mathcal{G}^0_{\nu\nu}(p_1)e^{\mathrm{i}\omega_{n_1}\delta}\mathcal{G}^0_{\mu\beta}(p)\mathcal{V}(0)e^{\mathrm{i}\omega_{n_1}\delta} \\
&\quad-\mathcal{G}^0_{\alpha\mu}(p)\mathcal{G}^0_{\mu\nu}(p_1)e^{\mathrm{i}\omega_{n_1}\delta}\mathcal{G}^0_{\nu\beta}(p)\mathcal{V}(p-p_1)\Big\}e^{\mathrm{i}p(x-x')} \\
&= \frac{1}{\hbar(\beta\hbar V)^2}\sum_{p}\delta_{\alpha\beta}[\mathcal{G}^0(p)]^2 \\
&\quad\times\sum_{p_1}[\mp(2s+1)V(0)-V(\boldsymbol{p}-\boldsymbol{p}_1)]\mathcal{G}^0(p_1)e^{\mathrm{i}\omega_{n_1}\delta} \\
&\quad\times e^{\mathrm{i}p(x-x')},
\end{aligned}
\tag{4.3.32}
$$

where we have considered as follows:

$$
\begin{aligned}
\mathcal{G}^0(x_1-x_2)|_{\tau_1=\tau_2} &= \lim_{\delta\to 0}\mathcal{G}^0(\boldsymbol{x}_1-\boldsymbol{x}_2,-\delta) \\
&= \frac{1}{\beta\hbar V}\sum_{p_1}\mathcal{G}^0(p_1)e^{\mathrm{i}p_1(x_1-x_2)^-} \\
&= \frac{1}{\beta\hbar V}\sum_{p_1}\mathcal{G}^0(p_1)e^{\mathrm{i}\boldsymbol{p}_1\cdot(\boldsymbol{x}_1-\boldsymbol{x}_2)+\mathrm{i}\omega_{n_1}\delta}
\end{aligned}
\tag{4.3.33}
$$

and s is the spin of the particle, $s=0$ for bosons and $s=\frac{1}{2}$ for fermions. Thus we obtain

$$
\begin{aligned}
\mathcal{G}(p) &= \mathcal{G}^0(p)+\frac{1}{\hbar(\beta\hbar V)}[\mathcal{G}^0(p)]^2 \\
&\quad\times\sum_{p_1}[\mp(2s+1)V(0)-V(\boldsymbol{p}-\boldsymbol{p}_1)]\mathcal{G}^0(p_1)e^{\mathrm{i}\omega_{n_1}\delta}+\cdots,
\end{aligned}
\tag{4.3.34}
$$

where

$$\mathcal{G}_{\alpha\beta}(p) = \delta_{\alpha\beta}\mathcal{G}(p). \tag{4.3.35}$$

The integration with respect to $x_i = (\boldsymbol{x}_i, \tau_i)$ at a vertex at which an interaction with $q = (\boldsymbol{q}, \mathrm{i}\nu_m)$ ends, and a particle with $p = (\boldsymbol{p}, \mathrm{i}\omega_n)$ enters and one with $p' = (\boldsymbol{p}', \mathrm{i}\omega'_n)$ leaves leads to

$$\frac{1}{\beta\hbar V}\int_0^{\beta\hbar} d^4x_i e^{\mathrm{i}(p-p'\pm q)x_i} = \delta_{p-p'\pm q,0}, \tag{4.3.36}$$

where the sign $\pm$ means that an interaction entering or leaving at the vertex. Eq. (4.3.36) means that at each vertex the sum of momenta and imaginary frequencies conserves as illustrated in Figure 4.6.

We now state the general rules to construct the nth order diagrams in the Green's function $\mathcal{G}(p)$:

1. Draw all connected, topologically non-equivalent diagrams with n interaction lines and $2n + 1$ directed particle lines.
2. Assign a direction to each interaction line and assign an entering and a leaving Green's function lines at each vertex. Assign the wave vectors and frequencies to each line such that they conserve at each vertex.
3. Assign a Green's function $\mathcal{G}^0(p)$ to each external solid line.
4. Assign a Green's function $\mathcal{G}^0(p_i)$ to each internal solid line and a potential $\mathcal{V}(q) = V(\boldsymbol{q})$ to each dotted line such that at each vertex the sum of wave vectors and imaginary frequencies conserves.
5. Sum over all independent internal wave vectors and frequencies, and sum over all the spin variables.
6. Multiply each factor $\left(-\frac{1}{\hbar(\beta\hbar V)}\right)^n (2s+1)^F (\pm)^F$ for a diagram, where F is the number of closed loops.

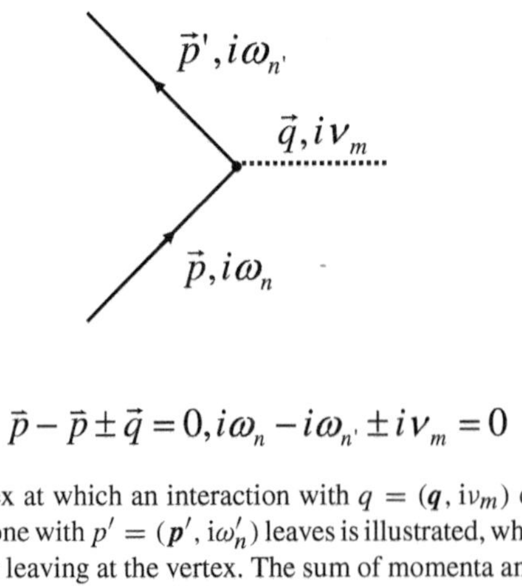

Figure 4.6 A basic vertex at which an interaction with $q = (\boldsymbol{q}, \mathrm{i}\nu_m)$ ends, and a particle with $p = (\boldsymbol{p}, \mathrm{i}\omega_n)$ enters and one with $p' = (\boldsymbol{p}', \mathrm{i}\omega'_n)$ leaves is illustrated, where the sign $\pm$ means that an interaction entering or leaving at the vertex. The sum of momenta and imaginary frequencies is conserved.

7. When both ends of a Green's function $\mathcal{G}^0(p_j)$ connect the same interaction line, attach a convergence factor $e^{i\omega_{n_j}\delta}$ to the Green's function $\mathcal{G}^0(p_j)$.
8. Sum over the independent momenta frequencies.

This procedure leads to the expansions of Feynman diagram of $\mathcal{G}_(p)$ as illustrated in Figure 4.7. The first order of the Feynman diagrams yields the corresponding Green's function expansions, Eq. (4.3.34).

All the second order, topologically non-equivalent connected diagrams are shown in Figure 4.8. For example, we represent the Green's function expansions for (g). In this case, $n = 2$ and $F = 1$. Thus we obtain

$$(g): \quad \pm\frac{2s+1}{h^2(\beta\hbar V)^2}\sum_{p_1,p_2}[\mathcal{G}^0(p)]^2[V(\boldsymbol{p}-\boldsymbol{p}_1)]^2\mathcal{G}^0(p_2)\mathcal{G}^0(p-p_1+p_2). \quad (4.3.37)$$

According to the general rule, we can represent the Green's function expansions for the other Feynman diagrams.

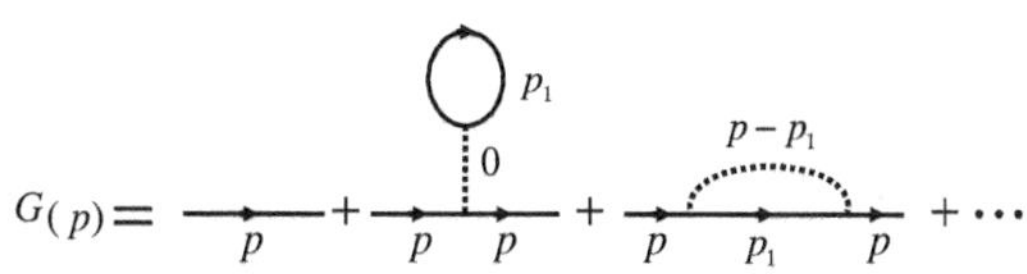

Figure 4.7 The Feynman diagrams of the Green's function $\mathcal{G}_{\alpha\beta}(p)$ are illustrated up to the first order diagrams.

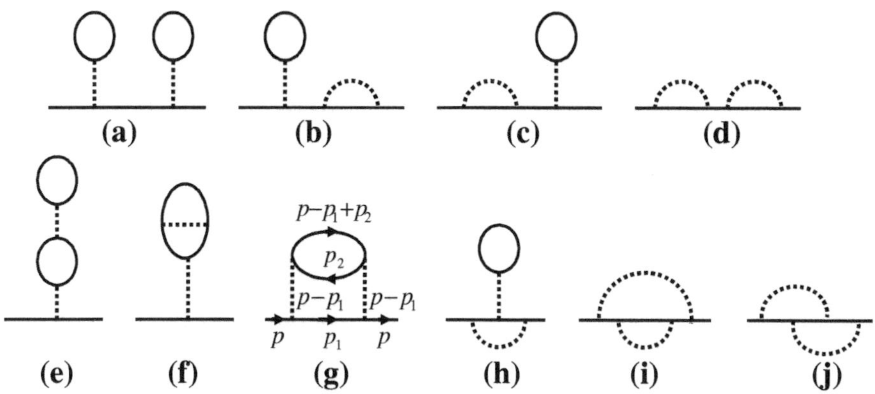

Figure 4.8 All the second order, topologically non-equivalent connected diagrams are illustrated.

4.3.5 The Calculation of Frequency Sum in the Free Particle Green's Function

Here we show the calculation of frequency sum in the free particle Green's function:

$$\sum_{i\omega_n} \mathcal{G}^0(p_1)e^{i\omega_{n_1}\delta} = \sum_{i\omega_n} \frac{e^{i\omega_{n_1}\delta}}{i\omega_n - \epsilon_p^0}. \tag{4.3.38}$$

If there is not the convergence factor $e^{i\omega_{n_1}\delta}$, the term $\sum_{i\omega_n} \frac{1}{i\omega_n - \epsilon_p^0}$ diverges logarithmically.

In order to calculate Eq. (4.3.38), we employ the contour integration, instead of the summing up of the frequencies $i\omega_n$: we multiply a function $g(z)$ with the poles at the frequencies $\{i\omega_n\}$, to $\frac{e^{z\delta}}{z-\epsilon_p^0}$, and then we integrate along a contour C involving all the poles illustrated in Figure 4.9, which yields $\sum_{i\omega_n} \frac{1}{i\omega_n - \epsilon_p^0}$.

To do so, we choose

$$g(z) = \pm\beta\hbar\frac{1}{e^{\beta\hbar z} \mp 1} \quad \text{for } \omega_n = \begin{cases} \text{even} \\ \text{odd} \end{cases} \tag{4.3.39}$$

as a function with the poles at $\{i\omega_n\}$. To see this, we consider the behavior of $g(z)$ around $i\omega_n$; $|z - i\omega_n| < \frac{2\pi}{\beta\hbar}$:

$$\begin{aligned} g(z) &= g(i\omega_n + z - i\omega_n) = \pm\beta\hbar\frac{1}{e^{\beta\hbar[i\omega_n+(z-i\omega_n)]} \mp 1} = \beta\hbar\frac{1}{e^{\beta\hbar(z-i\omega_n)} - 1} \\ &= \frac{1}{z - i\omega_n} + \cdots, \end{aligned} \tag{4.3.40}$$

where $\cdots$ is a regular function. $g(z)$ has all the poles at $i\omega_n$ with unit residue. Thus if C is a contour encircling the imaginary axis anticlockwise in Figure 4.9, the contour

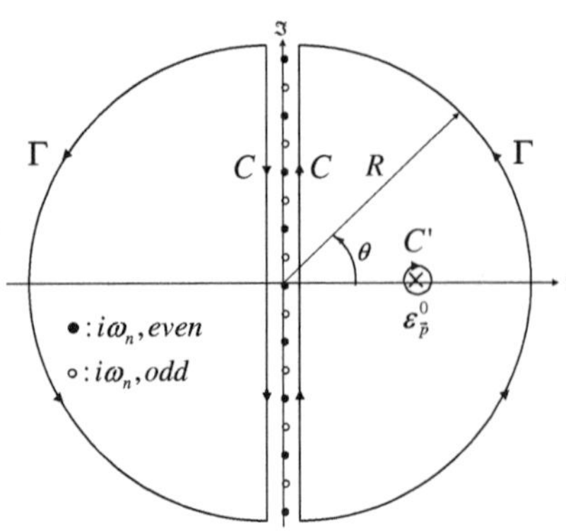

Figure 4.9 Contour integration for the sum of frequencies $i\omega_n$.

integral yields

$$\pm\frac{\beta\hbar}{2\pi \mathrm{i}}\int_C \frac{dz}{e^{\beta\hbar z}\mp 1}\frac{e^{z\delta}}{z-\epsilon_p^0}=\sum_{\mathrm{i}\omega_n}\frac{e^{\mathrm{i}\omega_n\delta}}{\mathrm{i}\omega_n-\epsilon_p^0}. \tag{4.3.41}$$

Now we analytically continue the contour C to C' and Γ shown in Figure 4.9. This process leads to

$$\int_C \frac{dz}{e^{\beta\hbar z}\mp 1}\frac{e^{z\delta}}{z-\epsilon_k^0}=\int_\Gamma \frac{dz}{e^{\beta\hbar z}\mp 1}\frac{e^{z\delta}}{z-\epsilon_p^0}+\int_{C'} \frac{dz}{e^{\beta\hbar z}\mp 1}\frac{e^{z\delta}}{z-\epsilon_p^0}.$$

On Γ, $z=Re^{\mathrm{i}\theta}=R(\cos\theta+\mathrm{i}\sin\theta)$. For $-\frac{\pi}{2}<\theta<\frac{\pi}{2}$; $\cos\theta>0$

$$\lim_{R\to\infty}\frac{1}{e^{\beta\hbar z}\mp 1}\frac{e^{z\delta}}{z-\epsilon_p^0}=\lim_{R\to\infty} e^{-(\beta\hbar-\delta)R\cos\theta}/R=0.$$

For $\frac{\pi}{2}<\theta<\frac{3\pi}{2}$; $\cos\theta<0$,

$$\lim_{R\to\infty}\frac{1}{e^{\beta\hbar z}\mp 1}\frac{e^{z\delta}}{z-\epsilon_p^0}=\lim_{R\to\infty} e^{\delta R\cos\theta}/R=0.$$

Since the integration on Γ vanishes, there remains the integration on C':

$$\int_{C'} \frac{dz}{e^{\beta\hbar z}\mp 1}\frac{e^{z\delta}}{z-\epsilon_p^0}=-2\pi\mathrm{i}\frac{1}{e^{\beta\hbar\epsilon_p^0}\mp 1}=-2\pi\mathrm{i}f(\epsilon_p^0),$$

where the contour C' encircles the pole ϵ_p^0 clockwise. Thus we obtain

$$\sum_{\mathrm{i}\omega_n}\frac{e^{\mathrm{i}\omega_n\delta}}{\mathrm{i}\omega_n-\epsilon_p^0}=\mp\beta\hbar f(\epsilon_p^0). \tag{4.3.42}$$

Combining Eqs. (4.3.34) and (4.3.42) leads to

$$\mathcal{G}^{(1)}_{\alpha\beta}(p)=\frac{1}{\hbar V}\delta_{\alpha\beta}[\mathcal{G}^0(p)]^2\sum_{\boldsymbol{p}_1}[(2s+1)V(0)\pm V(\boldsymbol{p}-\boldsymbol{p}_1)]f(\epsilon_{p_1}^0). \tag{4.3.43}$$

It should be noted that we have defined $\epsilon_p^0=(\hbar^2p^2/2M-\mu)/\hbar$ as a frequency.

4.4 Dyson's Equation

In the Feynman diagram method, the exact Green's functions are obtained by summing up all the order, topologically non-equivalent connected diagrams. In a perturbative sense, we may calculate a few orders of the Feynman diagrams for the Green's functions. Instead of calculating higher orders of the Feynman diagrams, we have a more accurate approximation procedure to systematically take into account the most important diagrams over all the orders. This is the Dyson's method. By introducing the self-energy part of the Green's function, the Dyson's equation shows that the exact Green's function is just the unperturbed Green's function replaced by the free particle frequency by (the free particle frequency + the proper self-energy part). Thus choosing the most important proper self-energy parts, we can carry out the more accurate approximation procedures systematically. We derive the Dyson's equation and show how to carry out the approximation procedures.

4.4.1 *Self-Energy Parts and the Dyson's Equation in the Coordinate Space*

It is very convenient to introduce the self-energy parts in the two different ways: the self-energy part and the proper self-energy part. This procedure yields the Dyson's equation.

4.4.1.1 The Self-energy part

First we define self-energy parts. A connected nth order Green's function $\mathcal{G}^{(n)}_{\alpha\beta}(x, x')$ is written as

$$\mathcal{G}^{(n)}_{\alpha\beta}(x, x') = \sum_{\mu\nu} \int dx_1^4 \int dx_2^4 \mathcal{G}^0_{\alpha\mu}(x - x_1) \tilde{\Sigma}^{(n)}_{\mu\nu}(x_1, x_2) \mathcal{G}^0_{\beta\nu}(x_2 - x').$$

We call $\tilde{\Sigma}^{(n)}_{\mu\nu}(x_1, x_2)$ the self-energy part of the Green's function $\mathcal{G}^{(n)}_{\alpha\beta}(x, x')$. The self-energy parts of the first and second order Green's functions $\mathcal{G}^{(1)}_{\alpha\beta}(x, x')$ and $\mathcal{G}^{(2)}_{\alpha\beta}(x, x')$ denoted as $\tilde{\Sigma}^{(1)}_{\mu\nu}(x_1, x_2)$ and $\tilde{\Sigma}^{(2)}_{\mu\nu}(x_1, x_2)$ are illustrated in Figure 4.10. Diagrammatically, the self-energy part of the Green's function $\mathcal{G}^{(n)}_{\alpha\beta}(x, x')$ is the Green's function subtracting the two unperturbed Green's functions leaving x and ending x'.

The exact self-energy part $\tilde{\Sigma}_{\mu\nu}(x_1, x_2)$ of the exact Green's function $\mathcal{G}_{\alpha\beta}(x, x')$ is given by the sum of all the order self-energy parts $\sum_n \tilde{\Sigma}^{(n)}_{\mu\nu}(x_1, x_2)$. Thus the exact Green's function $\mathcal{G}_{\alpha\beta}(x, x')$ consists of an unperturbed Green's function $\mathcal{G}^0_{\alpha\beta}(x, x')$ and all the orders, topologically non-equivalent connected diagrams with two unperturbed Green's functions ending x and x':

$$\mathcal{G}_{\alpha\beta}(x-x') = \mathcal{G}^0_{\alpha\beta}(x-x') + \int_0^{\beta\hbar} d^4x_1 d^4x_2 \mathcal{G}^0_{\alpha\mu}(x-x_1)\tilde{\Sigma}_{\mu\nu}(x_1-x_2)\mathcal{G}^0_{\nu\beta}(x_2-x'). \tag{4.4.1}$$

The diagrammatic expression of the exact self-energy part is illustrated in Figure 4.11.

4.4.1.2 The proper self-energy part

A proper self-energy part is a self-energy part which cannot be separated into two parts by cutting any unperturbed Green's function. For examples, in $\tilde{\Sigma}^{(2)}$ in Figure

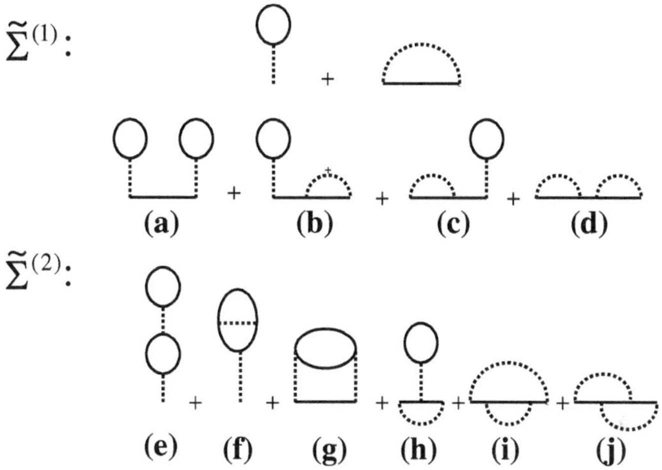

Figure 4.10 The self-energy parts of the first and second order Green's functions $\mathcal{G}^{(1)}$ and $\mathcal{G}^{(2)}$ are denoted as $\tilde{\Sigma}^{(1)}$ and $\tilde{\Sigma}^{(2)}$, respectively. All the free particle Green's functions are illustrated as thin lines.

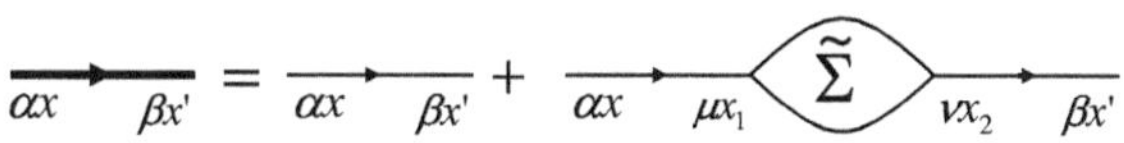

Figure 4.11 The exact self-energy part $\tilde{\Sigma}_{\mu\nu}(x_1 - x_2)$ of the exact Green's function $\mathcal{G}_{\alpha\beta}(x, x')$. The thick line denotes the exact Green's function $\mathcal{G}$ and the thin lines denote the free particle Green's functions $\mathcal{G}^0$.

4.10, the diagrams (e), (f), (g), (h), (i), and (j) are the proper self-energy parts, while the remaining diagrams (a), (b), (c), and (d) are the improper self-energy parts. The proper self-energy parts of the first and second order Green's functions $\mathcal{G}^{(1)}_{\alpha\beta}(x, x')$ and $\mathcal{G}^{(2)}_{\alpha\beta}(x, x')$ denoted as $\Sigma^{(1)}_{\mu\nu}(x_1, x_2)$ and $\Sigma^{(2)}_{\mu\nu}(x_1, x_2)$ are illustrated in Figure 4.12.

The exact self-energy part $\Sigma_{\mu\nu}(x_1, x_2)$ of the exact Green's function $\mathcal{G}_{\alpha\beta}(x, x')$ is given by the sum of all the order proper self-energy parts $\sum_n \tilde{\Sigma}^{(n)}_{\mu\nu}(x_1, x_2)$. By the definition, the exact self-energy part $\tilde{\Sigma}_{\mu\nu}(x_1 - x_2)$ is the sum of all possible repetitions of the proper self-energy part, which is expressed as:

$$\begin{aligned}\tilde{\Sigma}(x - x') = \Sigma(x - x') &+ \int_0^{\beta\hbar} d^4x_1 d^4x_2 \Sigma(x - x_1)\mathcal{G}^0(x_1 - x_2)\Sigma(x_2 - x') \\ &+ \int_0^{\beta\hbar} d^4x_1 d^4x_2 d^4x_3 d^4x_4 \Sigma(x - x_1)\mathcal{G}^0(x_1 - x_2) \\ &\times \Sigma(x_2 - x_3)\mathcal{G}^0(x_3 - x_4)\Sigma(x_4 - x') + \cdots,\end{aligned} \quad (4.4.2)$$

where we have expressed the equation in the matrix form in the spin indices. The diagrammatic expression is illustrated in Figure 4.13.

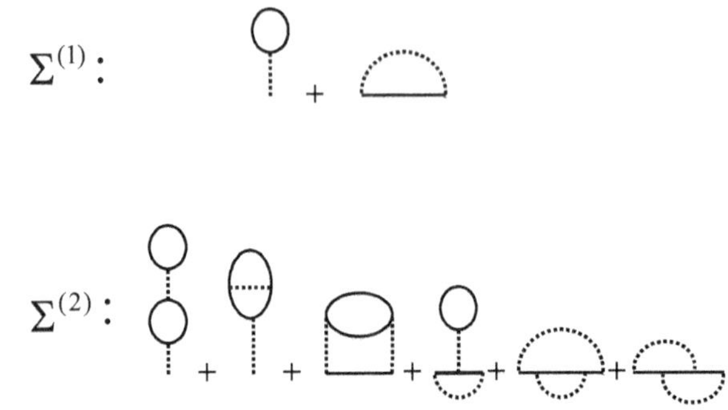

Figure 4.12 The first order proper self-energy part $\Sigma^{(1)}(p)$ and the second order proper self-energy part $\Sigma^{(2)}(p)$ are illustrated. All the Green's functions are free particle Green's functions illustrated as thin lines.

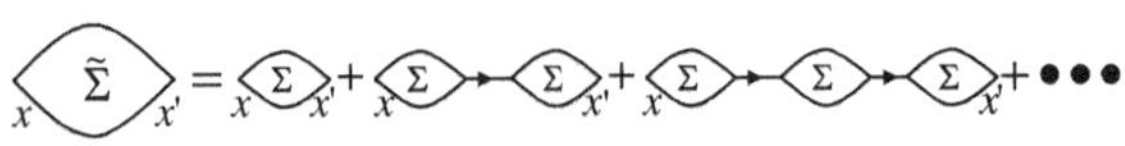

Figure 4.13 The relation between the exact self-energy part $\tilde{\Sigma}(x-x')$ and the proper self-energy part $\Sigma(x_x')$ is illustrated. All the Green's functions are free particle Green's functions illustrated as thin lines.

Using the proper self-energy part, we can write the exact Green's function as:

$$
\begin{aligned}
&\mathcal{G}(x-x') \\
&= \mathcal{G}^0(x-x') + \int_0^{\beta\hbar} d^4x_1 d^4x_2 \mathcal{G}^0(x-x_1)\Sigma(x_1-x_2)\mathcal{G}^0(x_2-x') \\
&\quad + \int_0^{\beta\hbar} d^4x_1 d^4x_2 d^4x_3 d^4x_4 \mathcal{G}^0(x-x_1) \\
&\quad \times \Sigma(x_1-x_2)\mathcal{G}^0(x_2-x_3)\Sigma(x_3-x_4)\mathcal{G}^0(x_4-x') \\
&\quad + \cdots \\
&= \mathcal{G}^0(x-x') + \int_0^{\beta\hbar} d_1^x d^4x_2 \mathcal{G}^0(x-x_1)\Sigma(x_1-x_2)\Bigg[\mathcal{G}^0(x_2-x') \\
&\quad + \int_0^{\beta\hbar} d^4x_3 d^4x_4 \mathcal{G}^0(x_2-x_3)\Sigma(x_3-x_4)\mathcal{G}^0(x_4-x') + \cdots\Bigg] \\
&= \mathcal{G}^0(x-x') + \int_0^{\beta\hbar} d_1^x d^4x_2 \mathcal{G}^0(x-x_1)\Sigma(x_1-x_2)\mathcal{G}(x_2-x'),
\end{aligned}
\tag{4.4.3}
$$

which yields

The Dyson's equation for the exact Green's function $\mathcal{G}_{\alpha\beta}(x-x')$:

$$
\begin{aligned}
\mathcal{G}_{\alpha\beta}(x-x') &= \mathcal{G}^0_{\alpha\beta}(x-x') \\
&\quad + \int_0^{\beta\hbar} d^4x_1 d^4x_2 \mathcal{G}^0_{\alpha\mu}(x-x_1)\Sigma_{\mu\nu}(x_1-x_2)\mathcal{G}_{\nu\beta}(x_2-x').
\end{aligned}
\tag{4.4.4}
$$

The diagrammatic expression for the Dyson's equation is shown in Figure 4.14.

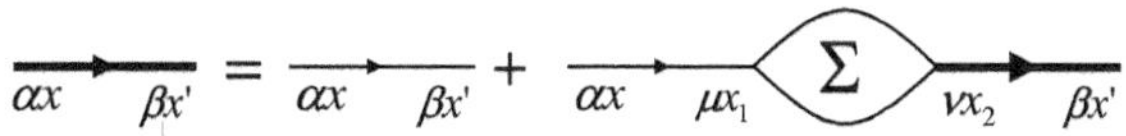

Figure 4.14 The Dyson's equation for $\mathcal{G}_{\alpha\beta}(x-x')$ is illustrated. The free particle Green's functions are illustrated as thin lines, while the exact Green's function are illustrated as the thick line.

4.4.2 The Self-Energy Parts and the Dyson's Equation in the Momentum Space

Now we transform the Dyson's Eq. (4.4.4) in the coordinate space to that in the momentum space. Using Eqs. (4.3.22) and (4.3.27), and the Fourier transformation of the proper self-energy part:

$$\Sigma_{\mu\nu}(\boldsymbol{x},\tau)=\frac{1}{\beta\hbar V}\sum_{\boldsymbol{p}\mathrm{i}\omega_n}e^{\mathrm{i}(\boldsymbol{p}\cdot\boldsymbol{x}-\omega_n\tau)}\Sigma_{\mu\nu}(\boldsymbol{p},\mathrm{i}\omega_n),\tag{4.4.5}$$

we obtain the Dyson's equation in the matrix form

$$\mathcal{G}(p)=\mathcal{G}^0(p)+\mathcal{G}^0(p)\Sigma(p)\mathcal{G}(p).\tag{4.4.6}$$

Then we have

$$\mathcal{G}(p)^{-1}=\mathcal{G}^0(p)^{-1}-\Sigma(p).\tag{4.4.7}$$

In the case of

$$\mathcal{G}^0_{\alpha\beta}(p)=\frac{\delta_{\alpha\beta}}{\mathrm{i}\omega_n-\epsilon^0_p},$$

we obtain

$$\mathcal{G}_{\alpha\beta}(p)=\frac{\delta_{\alpha\beta}}{\mathrm{i}\omega_n-\epsilon^0_p-\Sigma(p)},\tag{4.4.8}$$

where we have used the notation: $\Sigma_{\mu\nu}(p)=\delta_{\mu\nu}\Sigma(p)$. The Dyson's Eq. (4.4.8) says that the particle frequency is the sum of the free particle frequency ϵ^0_p and the proper self-energy $\Sigma(p)$.

Here we show the first and second order proper self-energy parts $\Sigma^{(1)}(p)$ and $\Sigma^{(2)}(p)$ in Figure 4.12. If we make the first order approximation $\Sigma(p)=\Sigma^{(1)}(p)$

$$\begin{aligned}\Sigma^{(1)}(p)&=\frac{1}{\hbar^2\beta V}\sum_{\boldsymbol{p}_1}[(2s+1)V(0)\pm V(\boldsymbol{p}-\boldsymbol{p}_1)]\mathcal{G}^0(p_1)\\&=\frac{1}{\hbar V}\sum_{\boldsymbol{p}_1}[(2s+1)V(0)\pm V(\boldsymbol{p}-\boldsymbol{p}_1)]f(\epsilon^0_{p_1})\\&=\frac{1}{\hbar}(2s+1)V(0)\rho\mp\frac{1}{\hbar V}\sum_{\boldsymbol{p}_1}V(\boldsymbol{p}-\boldsymbol{p}_1)]f(\epsilon^0_{p_1}),\end{aligned}\tag{4.4.9}$$

the frequency of a particle with a wave vector $\boldsymbol{p}$ is given by

$$\begin{aligned}\epsilon_p^{(1)} &= \epsilon_p^0 + \Sigma^{(1)}(p)\\ &= \epsilon_p^0 + \frac{1}{\hbar}(2s+1)V(0)\rho \mp \frac{1}{\hbar V}\sum_{\boldsymbol{p}_1} V(\boldsymbol{p}-\boldsymbol{p}_1)f(\epsilon_{p_1}^0),\end{aligned} \tag{4.4.10}$$

where ρ is the number density $\rho = \frac{N}{V}$. The first term is the shift of the frequency and the second term modifies the dispersion of the frequency ϵ_p^0. In this approximation, the Green's function $\mathcal{G}(p)$ has an infinite repetition of $\Sigma(p) = \Sigma^{(1)}(p)$ via the free particle Green's function $\mathcal{G}^0(p)$ illustrated in Figure 4.15.

The Dyson's equation allows us to calculate the Green's function systematically in the approximation procedure. The Hartree-Fock approximation is obtained by replacing the free particle Green's function $\mathcal{G}^0(p_1)$ in the first order self-energy $\Sigma^{(1)}(p)$ in Eq. (4.4.9) by the Green's function $\mathcal{G}^{\rm HF}(p_1)$:

$$\Sigma^{\rm HF}(p) = \frac{1}{\hbar(\beta\hbar V)}\sum_{\boldsymbol{p}_1}[(2s+1)V(0) \pm V(\boldsymbol{p}-\boldsymbol{p}_1)]\mathcal{G}^{\rm HF}(p_1), \tag{4.4.11}$$

the Feynman diagrams of which are illustrated in Figure 4.16. Thus the Dyson's equation in the Hartree-Fock approximation is a self-consistent equation with respect to the Hartree-Fock Green's function $\mathcal{G}^{\rm HF}(p)$ with the hartree-Fock proper self-energy $\Sigma^{\rm HF}(p)$ in Eq. (4.4.11):

$$\mathcal{G}^{\rm HF}(p) = \frac{1}{{\rm i}\omega_n - \epsilon_p^0 - \Sigma^{\rm HF}(p)}. \tag{4.4.12}$$

Thus the Dyson's equation plays a significant role in the approximation procedures.

In developing more accurate approximation procedures, there appear vertex corrections among particle Green's functions and the interaction potentials. In such situations, we must make approximations such that the self-energy parts and the vertex parts satisfy the Ward-Takahashi relations. We will show such examples in the later chapters in developing the unified theory of the liquid-glass transition.

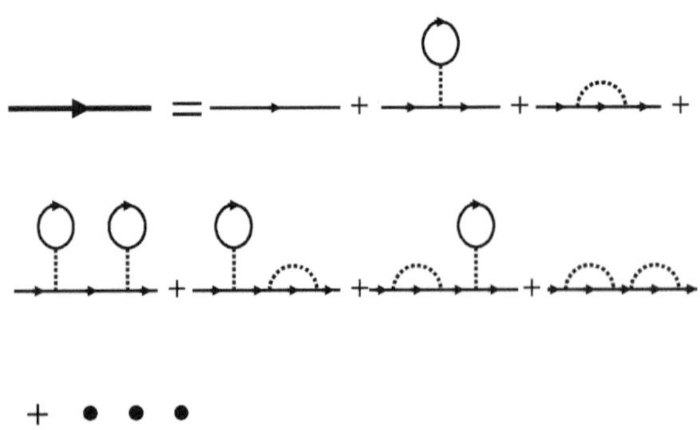

Figure 4.15 The Green's function $\mathcal{G}(p)$ for first order proper self-energy part $\Sigma(p) = \Sigma^{(1)}(p)$ is illustrated. All Green functions on the right-hand side are free particle Green's functions illustrated as thin lines.

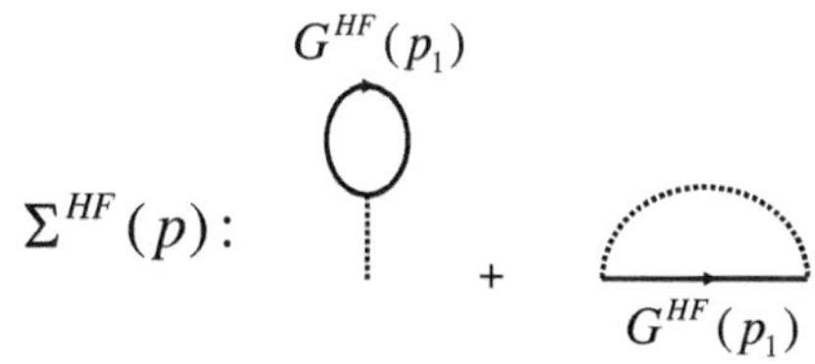

Figure 4.16 The proper self-energy in the Hartree-Fock approximation, $\Sigma^{\mathrm{HF}}(p)$, is illustrated. $\Sigma^{\mathrm{HF}}(p)$ is obtained by replacing free particle Green's functions by $\mathcal{G}^0(p_1)$ in the first order self-energy $\Sigma^{(1)}(p)$ in Figure 4.12 by the Hartree-Fock Green's function $\mathcal{G}^{\mathrm{HF}}(p_1)$ illustrated as thick lines.

References

[1] T. Matsubara, Prog. Theor. Phys. 14 (1955) 351.
[2] A.A. Abricosov, L.P. Gorkov, I.E. Dzyaloshinskii, Sov. Phys.-JETP 9 (1959) 636.
[3] E.S. Fradkin, Sov. Phys.-JETP 9 (1959) 912.
[4] P.C. Martin, J. Schwinger, Phys. Rev. 115 (1959) 1342.
[5] A.A. Abricosov, L.P. Gorkov, I.E. Dzyaloshinskii, Methods of Quantum Field Theory in Statistical Physics, Prentice-Hall, Englewood Cliffs, NJ, 1963.
[6] A.L. Fetter, J.D. Walecka, Quantum Theory of Many Particle System, MacGraw-Hill, New York,1971.
[7] R. Zwanzig, in: W.E. Britton, B.W. Downs, J. Downs (Eds.), Lectures in Theoretical Physics, vol. 3, Wiley Interscience, New York, 1961, p. 135.
[8] H. Mori, Prog. Theor. Phys. 33 (1965) 423.
[9] H. Mori, Prog. Theor. Phys. 34 (1965) 399.
[10] M. Ichiyanagi, J. Phys. Soc. Japan 32 (1972) 604.
[11] R. Kubo, Statistical Mechanics, North-Holland, 1965.

5 Real Time Green's Functions and Temperature Green's Functions

The thermodynamical states in an equilibrium system can be represented as the functions of the thermodynamical variables. In the equilibrium system, the temperature $\beta = \frac{1}{k_B T}$ plays the role of the indicator in holding the entropy maximum for the thermodynamical valuables; the statistical operators are controlled by the temperature β. This is why the temperature Green's functions play an essential role in calculating the thermodynamical system.

The nonequilibrium systems have no thermodynamical valuables; the nonequilibrium systems do not have such a parameter as the temperature indicating the systems. However, a nonequilibrium system has the entropy, which is fluctuating around the average entropy. The system finally reaches an equilibrium state when the entropy approaches a maximum value. There is no general theory of the nonequilibrium systems, but the nonequilibrium system deviated a little from an equilibrium state can be solved by the perturbative method. There are some approaches to such a nonequilibrium state: the Kubo's linear response theory [1], the Mori's projection operator method [2,3], the retarded Green's function method by Bogoliubov and Tyablikov [4], and so on. Here we present the Kubo's linear response theory presented by the retarded Green's functions. The equivalence between the Mori's and the retarded Green's function methods has been shown by Zwanzig [6] and Ichiyanagi [7].

We show a retarded Green's function is identical to the temperature Green's function when the imaginary temperature frequency in the temperature Green's function is analytically continued from above to the frequencies in the retarded Green's function. Thus the temperature Green's function method covers the perturbative method in weak nonequilibrium systems.

5.1 Various Kinds of Green's Functions

In this section, we investigate the characters of the various kinds of Green's functions appearing in the thermodynamical statistical mechanics. To do so, we calculate the Lehmann spectral representation of each Green's function and compare it with the temperature Green's function. We show that the retarded Green's function is identical with the temperature Green's function with the temperature frequencies replaced by the analytically continued frequencies. This fact is significant: the dynamical equations

Liquid Glass Transition. DOI: http://dx.doi.org/10.1016/B978-0-12-407177-3.00005-2

in the real time are obtained by analytical continuation of the temperature frequencies in those of the temperature Green's function method.

5.1.1 Real Time Green's Function

The real time Green's function is defined as

$$G_{\alpha\beta}(\boldsymbol{x}t, \boldsymbol{x}'t') \equiv -\mathrm{i}\,\mathrm{Tr}\{\rho_G T[\psi_\alpha(\boldsymbol{x}t)\psi^\dagger_\beta(\boldsymbol{x}'t')]\}, \tag{5.1.1}$$

where $\rho_G = e^{\beta(\Omega-\mathcal{H})}$ is the statistical operator in the grand canonical ensemble and $\psi_\alpha(\boldsymbol{x}t)$ is a true Heisenberg operator defined by

$$\psi_\alpha(\boldsymbol{x}t) \equiv e^{\mathrm{i}\mathcal{H}t/\hbar}\psi_\alpha(\boldsymbol{x})e^{-\mathrm{i}\mathcal{H}t/\hbar}. \tag{5.1.2}$$

T is the ordinary time ordering operator and $\mathcal{H} = H - \mu N$. The Green's function $G_{\alpha\beta}(\boldsymbol{x}t, \boldsymbol{x}'t')$ explicitly depends on the temperature β and the chemical potential μ.

For the Hamiltonian H independent of time and position (homogeneous), we can write the Green's function as

$$G_{\alpha\beta}(\boldsymbol{x}t, \boldsymbol{x}'t') = G_{\alpha\beta}(\boldsymbol{x} - \boldsymbol{x}', t - t'). \tag{5.1.3}$$

Here we discard the magnetic substances and magnetic field so that we can put

$$G_{\alpha\beta}(\boldsymbol{x}, t) = \delta_{\alpha\beta}G(\boldsymbol{x}, t). \tag{5.1.4}$$

In the homogeneous systems, the Heisenberg operator can be rewritten as

$$\psi_\alpha(\boldsymbol{x}t) = e^{-\mathrm{i}\boldsymbol{P}\cdot\boldsymbol{x}/\hbar}e^{\mathrm{i}\mathcal{H}t/\hbar}\psi_\alpha(0)e^{-\mathrm{i}\mathcal{H}t}e^{\mathrm{i}\boldsymbol{P}\cdot\boldsymbol{x}/\hbar}. \tag{5.1.5}$$

Here we choose the exact eigenstate of H, $\boldsymbol{P}$, and N as

$$\mathcal{H}|m\rangle = \mathcal{E}_n|m\rangle = (E_m - \mu N_m)|m\rangle \equiv \hbar e_m|m\rangle, \tag{5.1.6}$$

$$\boldsymbol{P}|m\rangle = \boldsymbol{P}_m|m\rangle \equiv \hbar\boldsymbol{p}_m|m\rangle, \tag{5.1.7}$$

where $e_m = (E_m - \mu N_m)/\hbar$ is a frequency. Then we obtain

$$\begin{aligned}\langle m|\psi_\alpha(\boldsymbol{x}t)|n\rangle &= e^{-\mathrm{i}\boldsymbol{p}_m\cdot\boldsymbol{x}}e^{\mathrm{i}e_m t}\langle m|\psi_\alpha(0)|n\rangle e^{-\mathrm{i}e_n t}e^{\mathrm{i}\boldsymbol{p}_n\cdot\boldsymbol{x}}\\ &= e^{-\mathrm{i}(\boldsymbol{p}_m-\boldsymbol{p}_n)\cdot\boldsymbol{x}}e^{\mathrm{i}(e_m-e_n)t}\langle m|\psi_\alpha(0)|n\rangle,\end{aligned} \tag{5.1.8}$$

where $N_m = N_n - 1$. Similarly we obtain

$$\langle m|\psi^\dagger_\alpha(\boldsymbol{x}t)|n\rangle = e^{-\mathrm{i}(\boldsymbol{p}_m-\boldsymbol{p}_n)\cdot\boldsymbol{x}}e^{\mathrm{i}(e_m-e_n)t}\langle m|\psi^\dagger_\alpha(0)|n\rangle, \tag{5.1.9}$$

where $N_m = N_n + 1$. Using Eqs. (5.1.8) and (5.1.9), we obtain

$$\begin{aligned} G^{>}_{\alpha\beta}(\boldsymbol{x}, t) &\equiv G_{\alpha\beta}(\boldsymbol{x}, t > 0) \\ &= -\mathrm{i}\theta(t)\mathrm{Tr}\{\rho_G[\psi_\alpha(\boldsymbol{x}t)\psi^\dagger_\beta(0)]\}, \\ &= -\mathrm{i}\theta(t)\sum_{mn} e^{\beta(\Omega-\mathcal{E}_m)}\, e^{\mathrm{i}(e_m-e_n)t-\mathrm{i}(\boldsymbol{p}_m-\boldsymbol{p}_n)\cdot\boldsymbol{x}} \\ &\quad \times\langle m|\psi_\alpha(0)|n\rangle\langle n|\psi^\dagger_\beta(0)|m\rangle, \end{aligned} \tag{5.1.10}$$

$$\begin{aligned} G^{<}_{\alpha\beta}(\boldsymbol{x}, t) &\equiv G_{\alpha\beta}(\boldsymbol{x}, t < 0) \\ &= \mp\mathrm{i}\theta(-t)\mathrm{Tr}\{\rho_G[\psi^\dagger_\beta(0)\psi_\alpha(\boldsymbol{x}, t)]\}, \\ &= \mp\mathrm{i}\theta(-t)\sum_{mn} e^{\beta(\Omega-\mathcal{E}_n)}\, e^{\mathrm{i}(e_m-e_n)t-\mathrm{i}(\boldsymbol{p}_m-\boldsymbol{p}_n)\cdot\boldsymbol{x}} \\ &\quad \langle n|\psi^\dagger_\beta(0)|m\rangle\langle m|\psi_\alpha(0)|n\rangle, \end{aligned} \tag{5.1.11}$$

where it should be noted that $N_n = N_m + 1$ in both Eqs. (5.1.10) and (5.1.11) and we define as $\mathcal{E}_n = \hbar e_n$.

Here we Fourier transform the Green's function to that in the momentum space:

$$G(\boldsymbol{p}, \omega) = \int_V d\boldsymbol{x}\, dt\, G(\boldsymbol{x}, t)e^{-\mathrm{i}\boldsymbol{p}\cdot\boldsymbol{x}+\mathrm{i}\omega t}. \tag{5.1.12}$$

In integrating Eqs. (5.1.10) and (5.1.11) in time, we substitute $\omega\pm\mathrm{i}\delta$ for ω, respectively, such that the integrations converge. Then we obtain

$$\begin{aligned} G^{>}_{\alpha\beta}(\boldsymbol{p}, \omega) &= V\sum_{mn} e^{\beta(\Omega-\mathcal{E}_m)}\langle m|\psi_\alpha(0)|n\rangle\langle n|\psi^\dagger_\beta(0)|m\rangle\delta_{\boldsymbol{p},\boldsymbol{p}_n-\boldsymbol{p}_m} \\ &\quad \times\frac{1}{\omega-(e_n-e_m)+\mathrm{i}\delta}, \end{aligned} \tag{5.1.13}$$

$$\begin{aligned} G^{<}_{\alpha\beta}(\boldsymbol{p}, \omega) &= \mp V\sum_{mn} e^{\beta(\Omega-\mathcal{E}_n)}\langle m|\psi_\alpha(0)|n\rangle\langle n|\psi^\dagger_\beta(0)|m\rangle\delta_{\boldsymbol{p},\boldsymbol{p}_n-\boldsymbol{p}_m} \\ &\quad \times\frac{1}{\omega-(e_n-e_m)-\mathrm{i}\delta}, \end{aligned} \tag{5.1.14}$$

so that we have

$$\begin{aligned} G_{\alpha\beta}(\boldsymbol{p}, \omega) &= V\sum_{mn} e^{\beta(\Omega-\mathcal{E}_m)}\langle m|\psi_\alpha(0)|n\rangle\langle n|\psi^\dagger_\beta(0)|m\rangle\delta_{\boldsymbol{p},\boldsymbol{p}_n-\boldsymbol{p}_m} \\ &\quad \times\left[\frac{1}{\omega-(e_n-e_m)+\mathrm{i}\delta}\mp\frac{e^{-\beta(\mathcal{E}_n-\mathcal{E}_m)}}{\omega-(e_n-e_m)-\mathrm{i}\delta}\right]. \end{aligned} \tag{5.1.15}$$

Here we consider Eq. (5.1.4) and the relation

$$\frac{1}{x\pm\mathrm{i}\delta} = \mathrm{i}\pi\delta(x)\mp P\frac{1}{x}, \tag{5.1.16}$$

where P means a principal value. Then Eqs. (5.1.15) and (5.1.16) lead to

$$\begin{aligned} G(\boldsymbol{p},\omega) &= \frac{1}{2s+1}\sum_{\alpha} G_{\alpha\alpha}(\boldsymbol{p},\omega) \\ &= \frac{V}{2s+1}\sum_{mn\alpha} e^{\beta(\Omega-\mathcal{E}_m)}|\langle m|\psi_\alpha(0)|n\rangle|^2 \delta_{\boldsymbol{p},\boldsymbol{p}_n-\boldsymbol{p}_m} \\ &\times\Bigg[P\frac{1}{\omega-(e_n-e_m)}(1\mp e^{-\beta(\mathcal{E}_n-\mathcal{E}_m)}) \\ &\quad -\mathrm{i}\pi\delta(\omega-(e_n-e_m))\left(1\pm e^{-\beta(\mathcal{E}_n-\mathcal{E}_m)}\right)\Bigg], \end{aligned} \tag{5.1.17}$$

where we have considered Eq. (5.1.4). The imaginary part of Eq. (5.1.17) are given by

$$\begin{aligned} \Im G(\boldsymbol{p},\omega) &= -\frac{V}{2s+1}\sum_{mn\alpha} e^{\beta(\Omega-\mathcal{E}_m)}|\langle m|\psi_\alpha(0)|n\rangle|^2 \delta_{\boldsymbol{p},\boldsymbol{p}_n-\boldsymbol{p}_m} \\ &\quad \times\pi\delta(\omega-(e_n-e_m))\left(1\pm e^{-\beta(\mathcal{E}_n-\mathcal{E}_m)}\right). \end{aligned} \tag{5.1.18}$$

Using $\Im G$, we obtain the real part of G:

$$\begin{aligned} \Re G(\boldsymbol{p},\omega) &= \frac{V}{2s+1}\sum_{mn\alpha} e^{\beta(\Omega-\mathcal{E}_m)}|\langle m|\psi_\alpha(0)|n\rangle|^2 \delta_{\boldsymbol{p},\boldsymbol{p}_n-\boldsymbol{p}_m} \\ &\quad \times P\frac{1}{\omega-(e_n-e_m)}\left(1\mp e^{-\beta(\mathcal{E}_n-\mathcal{E}_m)}\right) \\ &= -P\int_{-\infty}^{\infty}\frac{d\omega'}{\pi}\frac{\Im G(\boldsymbol{p},\omega')}{\omega-\omega'}\frac{1\mp e^{-\beta\hbar\omega'}}{1\pm e^{-\beta\hbar\omega'}} \\ &= -P\int_{-\infty}^{\infty}\frac{d\omega'}{\pi}\frac{\Im G(\boldsymbol{p},\omega')}{\omega-\omega'}\left[\tanh\left(\frac{1}{2}\beta\hbar\omega'\right)\right]^{\pm}. \end{aligned} \tag{5.1.19}$$

The real part of the Green's function is obtained by its imaginary part.

5.1.2 Retarded and Advanced Green's Functions

Now we introduce the retarded and advanced Green's function as

$$G^R_{\alpha\beta}(x,x') \equiv -\mathrm{i}\theta(t-t')\mathrm{Tr}\left\{\rho_G\left[\psi_\alpha(x),\psi^\dagger_\beta(x')\right]_\mp\right\}, \tag{5.1.20}$$

$$G^A_{\alpha\beta}(x,x') \equiv \mathrm{i}\theta(t'-t)\mathrm{Tr}\left\{\rho_G\left[\psi_\alpha(x),\psi^\dagger_\beta(x')\right]_\mp\right\}. \tag{5.1.21}$$

In a similar manner to the derivation of Eqs. (5.1.13) and (5.1.14), we obtain the Fourier transformed Green's functions to the momentum space as:

$$G^R(\boldsymbol{p}, \omega) = \frac{V}{2s+1} \sum_{mn\alpha} e^{\beta(\Omega - \mathcal{E}_m)} |\langle m|\psi_\alpha(0)|n\rangle|^2 \delta_{\boldsymbol{p}, \boldsymbol{p}_n - \boldsymbol{p}_m} \times \frac{1 \mp e^{-\beta(\mathcal{E}_n - \mathcal{E}_m)}}{\omega - (e_n - e_m) + \mathrm{i}\delta}, \tag{5.1.22}$$

$$G^A(\boldsymbol{p}, \omega) = \frac{V}{2s+1} \sum_{mn\alpha} e^{\beta(\Omega - \mathcal{E}_m)} |\langle m|\psi_\alpha(0)|n\rangle|^2 \delta_{\boldsymbol{p}, \boldsymbol{p}_n - \boldsymbol{p}_m} \times \frac{1 \mp e^{-\beta(\mathcal{E}_n - \mathcal{E}_m)}}{\omega - (e_n - e_m) - \mathrm{i}\delta}. \tag{5.1.23}$$

When we introduce the Lehmann spectral density [8]

$$\rho(\boldsymbol{p}, \omega) = \frac{V}{2s+1} \sum_{mn\alpha} e^{\beta(\Omega - \mathcal{E}_m)} |\langle m|\psi_\alpha(0)|n\rangle|^2 \delta_{\boldsymbol{p}, \boldsymbol{p}_n - \boldsymbol{p}_m} \delta(\omega - (e_n - e_m)) \times \left(1 \mp e^{-\beta(\mathcal{E}_n - \mathcal{E}_m)}\right), \tag{5.1.24}$$

we can rewrite Eqs. (5.1.22) and (5.1.23) as:

$$G^R(\boldsymbol{p}, \omega) = \int_{-\infty}^{\infty} d\omega' \frac{\rho(\boldsymbol{p}, \omega')}{\omega - \omega' + \mathrm{i}\delta}, \tag{5.1.25}$$

$$G^A(\boldsymbol{p}, \omega) = \int_{-\infty}^{\infty} d\omega' \frac{\rho(\boldsymbol{p}, \omega')}{\omega - \omega' - \mathrm{i}\delta}. \tag{5.1.26}$$

If we introduce a function of a complex variable z

$$\Gamma(\boldsymbol{p}, z) = \int_{-\infty}^{\infty} d\omega' \frac{\rho(\boldsymbol{p}, \omega')}{z - \omega'}. \tag{5.1.27}$$

$\Gamma(\boldsymbol{p}, z)$ has a branch cut on the real axis and, $G^R(\boldsymbol{p}, \omega)$ and $G^A(\boldsymbol{p}, \omega)$ are obtained by analytically continuing z in Γ to ω from above and below, respectively:

$$G^R(\boldsymbol{p}, \omega) = \Gamma(\boldsymbol{p}, \omega + \mathrm{i}\delta), \tag{5.1.28}$$

$$G^A(\boldsymbol{p}, \omega) = \Gamma(\boldsymbol{p}, \omega - \mathrm{i}\delta). \tag{5.1.29}$$

Next we derive a sum rule for ρ. To do so, we consider the following integral:

$$\int_{-\infty}^{\infty} d\omega\, G^R(\boldsymbol{p}, \omega) e^{-\mathrm{i}\omega\delta} = \int_{-\infty}^{\infty} \int_{-\infty}^{\infty} d\omega\, d\omega' \frac{\rho(\boldsymbol{p}, \omega') e^{-\mathrm{i}\omega\delta}}{\omega - \omega' + \mathrm{i}\delta}, \tag{5.1.30}$$

and at the last stage we put $\delta \to 0$. The right-hand side of this equation is given by

$$\int_{-\infty}^{\infty} \int_{-\infty}^{\infty} d\omega\, d\omega' \frac{\rho(\boldsymbol{p}, \omega') e^{-\mathrm{i}\omega\delta}}{\omega - \omega' + \mathrm{i}\delta} = -2\pi\mathrm{i} \int_{-\infty}^{\infty} d\omega'\, \rho(\boldsymbol{p}, \omega'), \tag{5.1.31}$$

where we have carried out the integral of ω along the clockwise contour in the lower half plane. According to the definition, the left-hand side of the equation leads to

$$\begin{aligned}\int_{-\infty}^{\infty} d\omega\, G^R(\boldsymbol{p},\omega)e^{-\mathrm{i}\omega\delta} &= \int_V d\boldsymbol{x}\, e^{-\mathrm{i}\boldsymbol{p}\cdot\boldsymbol{x}} G^R(\boldsymbol{x},\delta)\\ &= \frac{-2\pi\mathrm{i}}{2s+1}\int_V d\boldsymbol{x}\, e^{-\mathrm{i}\boldsymbol{p}\cdot\boldsymbol{x}} \sum_\alpha \mathrm{Tr}\left\{\rho_G\left[\psi_\alpha(\boldsymbol{x},0),\psi_\alpha^\dagger(0)\right]_\mp\right\}\\ &= -2\pi\mathrm{i}\int_V d\boldsymbol{x}\, e^{-\mathrm{i}\boldsymbol{p}\cdot\boldsymbol{x}}\delta(\boldsymbol{x})\mathrm{Tr}\rho_G\\ &= -2\pi\mathrm{i}. \end{aligned} \tag{5.1.32}$$

Equations (5.1.30)–(5.1.32) yield an important sum rule:

$$\int_{-\infty}^{\infty} d\omega\, \rho(\boldsymbol{p},\omega) = 1. \tag{5.1.33}$$

According to the sum rule, we can obtain the asymptotic behavior of the Green's function for large $|\omega|$:

$$G^R(\boldsymbol{p},\omega) = G^A(\boldsymbol{p},\omega) \sim \frac{1}{\omega}\int_{-\infty}^{\infty} d\omega'\, \rho(\boldsymbol{p},\omega') = \frac{1}{\omega} \quad \text{for } |\omega| \gg 1. \tag{5.1.34}$$

The retarded and advanced Green's functions, $G^R(\boldsymbol{p},\omega)$, $G^A(\boldsymbol{p},\omega)$ decays as $\frac{1}{\omega}$ for large frequencies.

5.1.3 The Temperature Green's Function and Analytic Continuation

Using the retarded Green's function method, Bogoliubov and Tyablikov, Zubarev [5] has approached to the nonequilibrium systems. On the other hand, using the projection operator method, Mori has approached to the nonequilibrium systems. The equivalence of both the approaches is clarified by Zubarev, Ichiyanagi. In this section, we prove the equivalence between the retarded Green's function method and the temperature Green's function method. By using the Lehmann spectral representation, we can prove that the analytical continuation of $z = \mathrm{i}\omega_n$ in the temperature Green's function $\mathcal{G}(\mathrm{i}\omega_n)$ to real frequency ω from the upper or lower plane in z leads to $G^R(\omega)$ or $G^A(\omega)$.

First we show the Lehmann spectral representation of the temperature Green's function $\mathcal{G}(x)$. According to the periodic nature of the temperature Green's function we only consider the imaginary time τ in the regions $[0, \beta\hbar]$. Therefore, we show

$$\begin{aligned}\mathcal{G}(\boldsymbol{x},\tau) &= -\frac{1}{2s+1}\sum_\alpha \mathrm{Tr}\left[\rho_G\psi_\alpha(\boldsymbol{x}\tau)\psi_\alpha^\dagger(0)\right]\\ &= -\frac{1}{2s+1}\sum_\alpha e^{\beta\Omega}\mathrm{Tr}\left[e^{-\beta\mathcal{H}}e^{-\mathrm{i}\boldsymbol{P}\cdot\boldsymbol{x}/\hbar}e^{\mathcal{H}\tau/\hbar}\psi_\alpha(0)e^{-\mathcal{H}\tau/\hbar}e^{\mathrm{i}\boldsymbol{P}\cdot\boldsymbol{x}/\hbar}\psi_\alpha^\dagger(0)\right]\\ &= -\frac{1}{2s+1}\sum_{mn\alpha}\left[e^{\beta(\Omega-\mathcal{E}_m)}e^{\mathrm{i}(\boldsymbol{p}_n-\boldsymbol{p}_m)\cdot\boldsymbol{x}}e^{-(e_n-e_m)\tau}|\langle m|\psi_\alpha(0)|n\rangle|^2\right].\end{aligned} \tag{5.1.35}$$

Its Fourier transformation is given by

$$\begin{aligned}
\mathcal{G}(\boldsymbol{p}, \mathrm{i}\omega_l) &= \int_0^{\beta\hbar} d^4x\, e^{-\mathrm{i}\boldsymbol{p}\cdot\boldsymbol{x}}\, e^{\mathrm{i}\omega_l\tau}\mathcal{G}(\boldsymbol{x}, \tau) \\
&= \frac{V}{2s+1}\sum_{mn\alpha} e^{\beta(\Omega-\mathcal{E}_m)}|\langle m|\psi_\alpha(0)|n\rangle|^2\delta_{\boldsymbol{p},\boldsymbol{p}_n-\boldsymbol{p}_m}\frac{1-e^{\beta\hbar\{\mathrm{i}\omega_l-(\epsilon_n-\epsilon_m)\}}}{\mathrm{i}\omega_l-(e_n-e_m)} \\
&= \frac{V}{2s+1}\sum_{mn\alpha} e^{\beta(\Omega-\mathcal{E}_m)}|\langle m|\psi_\alpha(0)|n\rangle|^2\delta_{\boldsymbol{p},\boldsymbol{p}_n-\boldsymbol{p}_m}\frac{1\mp e^{-\beta(\mathcal{E}_n-\mathcal{E}_m)}}{\mathrm{i}\omega_l-(e_n-e_m)},
\end{aligned} \tag{5.1.36}$$

where we have used $e^{\beta\hbar\mathrm{i}\omega_l} = \pm 1$ for bosons and fermions, respectively, and $\mathcal{E}_n - \mathcal{E}_m = \hbar(e_n - e_m)$.

Using the Lehmann spectral density (5.1.24), we obtain

$$\mathcal{G}(\boldsymbol{p}, \mathrm{i}\omega_n) = \int_{-\infty}^{\infty} d\omega' \frac{\rho(\boldsymbol{p}, \omega')}{\mathrm{i}\omega_n - \omega'}. \tag{5.1.37}$$

In $\mathcal{G}(\boldsymbol{p}, \mathrm{i}\omega_n)$, if we put $z = \mathrm{i}\omega_n$, $\mathcal{G}(\boldsymbol{p}, z)$ has a branch cut on the real axis so that $\mathcal{G}(\boldsymbol{p}, z)$ is regular on the upper or the lower half plane. If we analytically continue z from the upper or the lower plane into another sheet, $\mathcal{G}(\boldsymbol{p}, z)$ has a pole. This is the same situation as $\Gamma(\boldsymbol{p}, z)$ in Eq. (4.1.27). According to the unicity theorem in the theory of functions, we have

$$\mathcal{G}(\boldsymbol{k}, \omega + \mathrm{i}\delta) = G^R(\omega), \tag{5.1.38}$$

$$\mathcal{G}(\boldsymbol{p}, \omega - \mathrm{i}\delta) = G^A(\omega). \tag{5.1.39}$$

Thus we could clarify the equivalence between the temperature Green's function and the real time Green's function.

The Lehmann spectral density is given by

$$\rho(\boldsymbol{p}, \omega) = \frac{1}{2\pi\mathrm{i}}[\mathcal{G}(\boldsymbol{p}, \omega - \mathrm{i}\delta) - \mathcal{G}(\boldsymbol{p}, \omega + \mathrm{i}\delta)]. \tag{5.1.40}$$

For a free particle temperature Green's function

$$\mathcal{G}^0(\boldsymbol{p}, \mathrm{i}\omega_n) = \frac{1}{\mathrm{i}\omega_n - \epsilon_p^0}, \tag{5.1.41}$$

where $\epsilon_p^0 = (\hbar^2 p^2/2M - \mu)/\hbar$, we obtain

$$\begin{aligned}
\rho^0(\boldsymbol{p}, \omega) &= \frac{1}{2\pi\mathrm{i}}\left[\frac{1}{\omega - \epsilon_p^0 - \mathrm{i}\delta} - \frac{1}{\omega - \epsilon_p^0 + \mathrm{i}\delta}\right] \\
&= \delta(\omega - \epsilon_p^0).
\end{aligned} \tag{5.1.42}$$

Substituting $\rho^0(\boldsymbol{p}, \omega)$ into Eqs. (5.1.25) and (5.1.26), we obtain

$$G^R(\boldsymbol{p}, \omega) = \frac{1}{\omega - \epsilon_p^0 + \mathrm{i}\delta}, \tag{5.1.43}$$

$$G^A(\boldsymbol{p}, \omega) = \frac{1}{\omega - \epsilon_p^0 - \mathrm{i}\delta}. \tag{5.1.44}$$

These results are very important. When we obtain the frequency dispersion curves of free particles such as constituent particles, collective modes, and bound states in the temperature Green's functions, analytically continuing the temperature frequencies $\mathrm{i}\omega_n$ to the real frequencies $\omega \pm \mathrm{i}\delta$ in the temperature Green's functions, the temperature Green's functions reduce to the real time Green's functions. This fact allows us to derive the dynamical equations in the real time and space by continuing $\mathrm{i}\omega_n$ to $\omega \pm \mathrm{i}\delta$ in the equations obtained by putting the denominator of the temperature Green's function as zero. We use this fact frequently in the later chapters.

5.2 Linear Response and Density Correlation Function

When we add an external perturbation to an equilibrium system the equilibrium system is perturbed by the external field. We cannot solve such a nonequilibrium system in general, but if the perturbation is weak, we can obtain the linear response to the external field. The perturbation of the external field to the equilibrium yields its linear response to the external field associated with the dissipation. The dissipation relates with the fluctuations of the response quantities, the correlation functions of the response quantities. This is called the fluctuation and dissipation theorem. Here we present a general linear response theory, then we give an example: the external field of the mass potential and the response of the number density.

5.2.1 A General Theory of Linear Response to an External Perturbation at Zero Temperature

We consider an interacting many particle system with a time-independent Hamiltonian H. The Schrödinger equation is given by

$$\mathrm{i}\hbar\frac{\partial}{\partial t}|\Psi_S(t)\rangle = H|\Psi_S(t)\rangle \tag{5.2.1}$$

with the state vector

$$|\Psi_S(t)\rangle = e^{-\mathrm{i}Ht/\hbar}|\Psi_S(0)\rangle. \tag{5.2.2}$$

The system is perturbed at $t = t_0$ by turning on an additional time-dependent Hamiltonian

$$H^{\mathrm{ex}}(t) = \int d^3\boldsymbol{x}\, O(\boldsymbol{x})\varphi^{\mathrm{ex}}(x), \tag{5.2.3}$$

where $O(\boldsymbol{x})$ is an operator and $\varphi^{\mathrm{ex}}(x)$ is a scalar function. The modified Schrödinger equation for $t > t_0$ is given by

$$\mathrm{i}\hbar\frac{\partial}{\partial t}|\widetilde{\Psi}_S(t)\rangle = [H + H^{\mathrm{ex}}]|\widetilde{\Psi}_S(t)\rangle. \tag{5.2.4}$$

Now we find out a solution in a form:

$$|\widetilde{\Psi}_S(t)\rangle = e^{-\mathrm{i}Ht/\hbar}A(t)|\Psi_S(0)\rangle, \tag{5.2.5}$$

where the operator $A(t)$ obeys the causal boundary condition

$$A(t) = 1 \quad \text{for } t < t_0. \tag{5.2.6}$$

Equations (5.2.4) and (5.2.5) yield

$$\mathrm{i}\frac{\partial}{\partial t}A(t) = e^{\mathrm{i}Ht/\hbar}H^{\mathrm{ex}}(t)e^{-\mathrm{i}Ht/\hbar}A(t) = H_H^{\mathrm{ex}}(t)A(t), \tag{5.2.7}$$

where $H_H^{\mathrm{ex}}(t) = e^{\mathrm{i}Ht/\hbar}H^{\mathrm{ex}}(t)e^{-\mathrm{i}Ht/\hbar}$ is in the Heisenberg representation.

Equation (5.2.7) can be iteratively solved for $t > t_0$ as

$$A(t) = 1 - \frac{\mathrm{i}}{\hbar}\int_{t_0}^{t} dt'\, H_H^{\mathrm{ex}}(t') + \cdots, \tag{5.2.8}$$

which satisfies the causal boundary condition (5.2.6). The corresponding state vector is given by

$$|\widetilde{\Psi}_S(t)\rangle = e^{-\mathrm{i}Ht/\hbar}|\Psi_S(0)\rangle - \frac{\mathrm{i}}{\hbar}e^{-\mathrm{i}Ht/\hbar}\int_{t_0}^{t} dt'\, H_H^{\mathrm{ex}}(t')|\Psi_S(0)\rangle + \cdots. \tag{5.2.9}$$

Any physical operator O under the Schrödinger equation (5.2.1) is modified under that in (5.2.4) for $t > t_0$ as follows:

$$\begin{aligned}
\langle\widetilde{O}(t)\rangle &\equiv \langle\widetilde{\Psi}_S(t)|O|\widetilde{\Psi}_S(t)\rangle \\
&= \langle\Psi_S(0)|\left[1 + \frac{\mathrm{i}}{\hbar}\int_{t_0}^{t} dt'\, H_H^{\mathrm{ex}}(t') + \cdots\right]e^{\mathrm{i}Ht/\hbar}O\,e^{-\mathrm{i}Ht/\hbar} \\
&\quad\times\left[1 - \frac{\mathrm{i}}{\hbar}\int_{t_0}^{t} dt'\, H_H^{\mathrm{ex}}(t') + \cdots\right]|\Psi_S(0)\rangle \\
&= \langle\Psi_H(0)|O_H(t)|\Psi_H(0)\rangle \\
&\quad + \frac{\mathrm{i}}{\hbar}\int_{t_0}^{t} dt'\langle\Psi_H(0)|[H_H^{\mathrm{ex}}(t'), O_H(t)]|\Psi_H(0)\rangle + \cdots,
\end{aligned} \tag{5.2.10}$$

where $\langle\Psi_H(0)|O_H(t)|\Psi_H(0)\rangle = \langle\Psi_S(0)|O_H(t)|\Psi_S(0)\rangle$ is the expectation value in the system with the time-independent Hamiltonian H. If we put

$$\langle O(t)\rangle \equiv \langle\Psi_H(0)|O_H(t)|\Psi_H(0)\rangle = \langle\Psi_S(0)|O_H(t)|\Psi_S(0)\rangle, \tag{5.2.11}$$

we obtain the linear response of the expectation value of the operator O

$$\begin{aligned}\delta\langle\widetilde{O}(t)\rangle &\equiv \langle\widetilde{O}(t)\rangle - \langle O(t)\rangle \\ &= \frac{\mathrm{i}}{\hbar}\int_{t_0}^{t} dt' \langle\Psi_H(0)|\left[H_H^{\mathrm{ex}}(t'), O_H(t)\right]|\Psi_H(0)\rangle. \end{aligned} \tag{5.2.12}$$

This is the linear response to an external perturbation $H^{\mathrm{ex}}(t) = \int d^3\boldsymbol{x}\, O(\boldsymbol{x})\varphi^{\mathrm{ex}}(x)$.

5.2.2 The Linear Response to the External Scalar Mass Potential and Density Correlation Function

Here we consider a one-component system with the mass M being turned on the external scalar mass potential $\varphi^{\mathrm{ex}}(\boldsymbol{x}, t)$ at $t = t_0$. The corresponding external Hamiltonian is given by

$$H^{\mathrm{ex}}(t) = \int d^3x\, n(\boldsymbol{x}) M\varphi^{\mathrm{ex}}(t), \tag{5.2.13}$$

where $n(\boldsymbol{x})$ is the particle number density operator. The linear response of the density due to the external perturbation is given by

$$\begin{aligned}\delta\langle\tilde{n}(\boldsymbol{x}, t)\rangle &= \frac{\mathrm{i}}{\hbar}\int_{t_0}^{t} d^4x' \langle\Psi_H(0)|[n_H(\boldsymbol{x}'t'), n_H(\boldsymbol{x}t)]|\Psi_H(0)\rangle M\varphi^{\mathrm{ex}}(\boldsymbol{x}'t') \\ &= \frac{\mathrm{i}}{\hbar}\int_{t_0}^{t} d^4x' \langle\Psi_H(0)|[\tilde{n}_H(\boldsymbol{x}'t'), \tilde{n}_H(\boldsymbol{x}t)]|\Psi_H(0)\rangle M\varphi^{\mathrm{ex}}(\boldsymbol{x}'t'), \end{aligned} \tag{5.2.14}$$

where $\tilde{n}_H(x) \equiv n_H(x) - \langle n_H(x)\rangle$ and $\langle n_H(x)\rangle$ is a c-number. If we introduce the retarded density correlation function as

$$D^R(x, x') = -\mathrm{i}\theta(t - t')\langle\Psi_H(0)|[\tilde{n}_H(\boldsymbol{x}'t'), \tilde{n}_H(\boldsymbol{x}t)]|\Psi_H(0)\rangle, \tag{5.2.15}$$

Eq. (5.2.14) leads to

$$\delta\langle\tilde{n}(x)\rangle = \frac{1}{\hbar}\int_{-\infty}^{\infty} d^4x'\, D^R(x, x') M\varphi^{\mathrm{ex}}(x'), \tag{5.2.16}$$

where we have considered $\langle\Psi_H(0)|\Psi_H(0)\rangle = 1$.

The perturbation of the external mass potential yields the linear density response to the external potential associated with the diffusive dissipation. This diffusive dissipation relates with the number density fluctuations, the correlation function of the number densities.

5.3 A Linear Response Theory at Finite Temperature

Now we extend the linear response theory to finite temperature. To do so, we must investigate time-dependent statistical operator ρ_G for the grand canonical ensemble.

When the system is indicated by the temperature β, the equation of motion of the statistical operator ρ_G for the grand canonical ensemble satisfies the Neumann's formula. Adding the external perturbative Hamiltonian to the time-independent Hamiltonian of the system in the Neumann's formula, we can obtain the linear perturbative statistical operator. Using this linear perturbative statistical operator, we calculate the linear response to the external perturbation associated with the retarded Green's function in the time-independent Hamiltonian of the system. The retarded Green's function in momentum and frequency space is identical to the analytically continued temperature Green's function.

5.3.1 Statistical Operator and the Neumann's Formula

In order to obtain the linear response to an external perturbation at finite temperature, we first derive the Neumann's formula for the equation of motion for the operator for the grand canonical ensemble:

$$\rho_G = e^{\beta(\Omega-\mathcal{H})}, \quad \mathcal{H} = H - \mu N.$$

Considering the following eigenvalues and eigenstates:

$$\mathcal{H}|m\rangle = \mathcal{E}_m|m\rangle = (E_m - \mu N_m)|m\rangle, \tag{5.3.1}$$

$$\sum_m |m\rangle\langle m| = 1, \tag{5.3.2}$$

we obtain

$$\rho_G = \rho_G \sum_m |m\rangle\langle m| = \sum_m |m\rangle e^{\beta(\Omega-\mathcal{E}_m)}\langle m|. \tag{5.3.3}$$

Defining the probability for the system to be at the state $|m\rangle$ as P_m, we have

$$P_m = e^{\beta(\Omega-\mathcal{E}_m)}. \tag{5.3.4}$$

From $e^{-\beta\Omega} = \sum_m e^{-\beta\mathcal{E}_m}$, we obtain

$$\sum_m P_m = 1. \tag{5.3.5}$$

Thus we obtain the properties of ρ_G as follows:

1. The property of probability:

$$\operatorname{Tr}\rho_G = \sum_n \langle n|\rho_G|n\rangle = 1. \tag{5.3.6}$$

2. The property of the statistical average $\langle O\rangle$ for an operator O:

$$\begin{aligned}\langle O\rangle &= \sum_m P_m\langle m|O|m\rangle = \sum_{mn} P_m\langle m|O|n\rangle\langle n|m\rangle \\ &= \sum \langle n|m\rangle P_m\langle m|O|n\rangle = \sum_n \langle n|\rho_G O|n\rangle \\ &= \operatorname{Tr}[\rho_G O].\end{aligned} \tag{5.3.7}$$

Next we investigate the time-dependence of the statistical operator ρ_G:

$$i\hbar\frac{\partial}{\partial t}\rho_G = \sum_m \left[\left\{i\hbar\frac{\partial}{\partial t}|m\rangle\right\} e^{\beta(\Omega-\mathcal{E}_m)}\langle m| + |m\rangle e^{\beta(\Omega-\mathcal{E}_m)}\left\{i\hbar\frac{\partial}{\partial t}\langle m|\right\}\right]. \quad (5.3.8)$$

Using the Schrödinger equation for $|m\rangle$ and $\langle m|$

$$i\hbar\frac{\partial}{\partial t}|m\rangle = H|m\rangle, \quad -i\hbar\frac{\partial}{\partial t}\langle m| = \langle m|H, \quad (5.3.9)$$

we obtain

$$\begin{aligned} i\hbar\frac{\partial}{\partial t}\rho_G &= H\rho_G - \rho_G H \\ &= (H-\mu N)\rho_G - \rho_G(H-\mu N) \\ &= \mathcal{H}\rho_G - \rho_G\mathcal{H}, \end{aligned} \quad (5.3.10)$$

where from the second to the third line we have considered the number conservation: $[H, N] = 0$. Thus we obtain the well-known Neumann Formula:

$$i\hbar\frac{\partial}{\partial t}\rho_G = [\mathcal{H}, \rho_G]. \quad (5.3.11)$$

This equation is the dynamical equation of ρ_G for the time-dependent ensemble.

5.3.2 Linear Response at Finite Temperature

The system is perturbed by turning on an additional time-dependent Hamiltonian $H^{ex}(t)$ at $t = t_0$ so that the total Hamiltonian is given by

$$\mathcal{H}^T = \mathcal{H} + H^{ex}(t) \quad \text{for } t_0 < t, \quad (5.3.12)$$

$$H^{ex}(t) = \int d^3x\, O(\boldsymbol{x})\varphi^{ex}(x). \quad (5.3.13)$$

The total grand canonical operator ρ_G^H is given by adding the correction term $\rho'(t)$ to ρ_G:

$$\rho_G^T = \rho_G + \rho'(t). \quad (5.3.14)$$

Up to the first order of H^{ex}, Eq. (5.3.11) is approximated as

$$i\hbar\frac{\partial\rho'}{\partial t} = [\mathcal{H}, \rho'] + [H^{ex}, \rho_G]. \quad (5.3.15)$$

This solution is given by

$$\rho'(t) = -\frac{i}{\hbar}\int_{t_0}^{t} dt'\, e^{-i\mathcal{H}(t-t')/\hbar}[H^{ex}(t'), \rho_G]e^{i\mathcal{H}(t-t')/\hbar}. \quad (5.3.16)$$

We can easily show Eq. (5.3.16) satisfies Eq. (5.3.15): Operation $\mathrm{i}\hbar\frac{\partial}{\partial t}$ on the right-hand side of Eq. (5.3.16) yields the derivatives of the upper limit of the integral and the integrand;

$$\begin{aligned}
&\mathrm{i}\hbar\frac{\partial}{\partial t}\left[-\frac{\mathrm{i}}{\hbar}\int_{t_0}^{t} dt'\, e^{-\mathrm{i}\mathcal{H}(t-t')/\hbar}[H^{\mathrm{ex}}(t'),\rho_G]e^{\mathrm{i}\mathcal{H}(t-t')/\hbar}\right] \\
&= [H^{\mathrm{ex}}(t),\rho_G] + \left[\mathcal{H}, -\frac{\mathrm{i}}{\hbar}\int_{t_0}^{t} dt'\, e^{-\mathrm{i}\mathcal{H}(t-t')/\hbar}[H^{\mathrm{ex}}(t'),\rho_G]e^{\mathrm{i}\mathcal{H}(t-t')/\hbar}\right] \\
&= [H^{\mathrm{ex}}(t),\rho_G] + [\mathcal{H},\rho'(t)],
\end{aligned}$$

which yields the right-hand side of Eq. (5.3.15). Thus we could verify Eq. (5.3.16).

Considering

$$\begin{aligned}
e^{-\mathrm{i}\mathcal{H}(t-t')/\hbar}[H^{\mathrm{ex}}(t'),\rho_G]e^{\mathrm{i}\mathcal{H}(t-t')/\hbar} &= e^{-\mathrm{i}\mathcal{H}t/\hbar}[e^{\mathrm{i}\mathcal{H}t'}H^{\mathrm{ex}}(t')e^{-\mathrm{i}\mathcal{H}t'},\rho_G]e^{\mathrm{i}\mathcal{H}t/\hbar} \\
&= e^{-\mathrm{i}\mathcal{H}t/\hbar}[H^{\mathrm{ex}}_{\mathcal{H}}(t'),\rho_G]e^{\mathrm{i}\mathcal{H}t/\hbar},
\end{aligned}$$

where

$$H^{\mathrm{ex}}_{\mathcal{H}}(t) = e^{\mathrm{i}\mathcal{H}t}H^{\mathrm{ex}}(t)e^{-\mathrm{i}\mathcal{H}t},$$

from Eq. (5.3.16) we have

$$\rho'(t) = -\frac{\mathrm{i}}{\hbar}\int_{t_0}^{t} dt'\, e^{-\mathrm{i}\mathcal{H}t/\hbar}[H^{\mathrm{ex}}_{\mathcal{H}}(t'),\rho_G]e^{\mathrm{i}\mathcal{H}t/\hbar}. \tag{5.3.17}$$

Using Eq. (5.3.17), we obtain the ensemble average of O:

$$\begin{aligned}
\langle\widetilde{O}(t)\rangle &= \mathrm{Tr}[\rho_G^T O] \\
&= \mathrm{Tr}[\rho_G O] - \frac{\mathrm{i}}{\hbar}\int_{t_0}^{t} dt'\,\mathrm{Tr}\left\{e^{-\mathrm{i}\mathcal{H}t/\hbar}[H^{\mathrm{ex}}_{\mathcal{H}}(t'),\rho_G]e^{\mathrm{i}\mathcal{H}t/\hbar}O\right\} \\
&= \langle O\rangle - \frac{\mathrm{i}}{\hbar}\int_{t_0}^{t} dt'\,\mathrm{Tr}\left\{[H^{\mathrm{ex}}_{\mathcal{H}}(t'),\rho_G]O_{\mathcal{H}}(t)\right\} \\
&= \langle O\rangle + \frac{\mathrm{i}}{\hbar}\int_{t_0}^{t} dt'\,\mathrm{Tr}\left\{\rho_G[H^{\mathrm{ex}}_{\mathcal{H}}(t'),O_{\mathcal{H}}(t)]\right\},
\end{aligned} \tag{5.3.18}$$

where

$$O_{\mathcal{H}}(\boldsymbol{x}t) = e^{\mathrm{i}\mathcal{H}t/\hbar}O(\boldsymbol{x})e^{-\mathrm{i}\mathcal{H}t/\hbar}. \tag{5.3.19}$$

Thus we obtain the linear response of the expectation value of the operator O

$$\begin{aligned}
\delta\langle\widetilde{O}(t)\rangle &\equiv \langle\widetilde{O}(t)\rangle - \langle O\rangle \\
&= \frac{\mathrm{i}}{\hbar}\int_{t_0}^{t} dt'\,\mathrm{Tr}\left\{\rho_G[H^{\mathrm{ex}}_{\mathcal{H}}(t'),O_{\mathcal{H}}(t)]\right\}.
\end{aligned} \tag{5.3.20}$$

If we consider $H^{\text{ex}}(t) = \int d^3x\, O(\boldsymbol{x})\varphi^{\text{ex}}(x)$, we obtain

$$\begin{aligned}\delta\langle\widetilde{O}(x)\rangle &= -\frac{\mathrm{i}}{\hbar}\int_{t_0}^{t} d^4x\, \mathrm{Tr}\left\{\rho_G[O_{\mathcal{H}}(x), O_{\mathcal{H}}(x')]\right\}\varphi^{\text{ex}}(x') \\ &= \frac{1}{\hbar}\int_{t_0}^{t} d^4x'\, D^R(x, x')\varphi^{\text{ex}}(x'),\end{aligned} \tag{5.3.21}$$

where D^R is a retarded correlation function

$$D^R(x, x') \equiv -\mathrm{i}\theta(t - t')\mathrm{Tr}\left\{\rho_G[O_{\mathcal{H}}(x), O_{\mathcal{H}}(x')]\right\}. \tag{5.3.22}$$

If we assume that the operator O commutes with the number operator N, $[O(\boldsymbol{x}), N]$, we have

$$O_H(x) = e^{\mathrm{i}\mathcal{H}t}O(\boldsymbol{x})e^{-\mathrm{i}\mathcal{H}t} \equiv O_{\mathcal{H}}(x), \tag{5.3.23}$$

where the number of particles conserves.

5.3.3 The Relation Between the Retarded Green's Function and the Temperature Green's Function for an Operator O

Since the Hamiltonian H is time-independent, D^R is of the form $D^R(\boldsymbol{x}, \boldsymbol{x}', t - t')$. In order to obtain the Lehmann spectral density, we investigate

$$\begin{aligned}D^R(\boldsymbol{x}, t) &= -\mathrm{i}\theta(t)\mathrm{Tr}\left\{\rho_G[O_{\mathcal{H}}(\boldsymbol{x}, t)O_{\mathcal{H}}(\boldsymbol{x}, 0) - O_{\mathcal{H}}(\boldsymbol{x}, 0)O_{\mathcal{H}}(\boldsymbol{x}, t)]\right\} \\ &= -\mathrm{i}\theta(t)\sum_{mn}\Big\{e^{-\beta(\Omega-\mathcal{E}_m)}\, e^{\mathrm{i}(e_m-e_n)t-\mathrm{i}(\boldsymbol{q}_m-\boldsymbol{q}_n)\cdot\boldsymbol{x}}\langle m|O|n\rangle\langle n|O|m\rangle \\ &\quad - e^{-\beta(\Omega-\mathcal{E}_n)}\, e^{\mathrm{i}(e_m-e_n)t-\mathrm{i}(\boldsymbol{q}_m-\boldsymbol{q}_n)\cdot\boldsymbol{x}}\langle n|O|m\rangle\langle m|O|n\rangle\Big\} \\ &= -\mathrm{i}\theta(t)\sum_{mn} e^{-\beta(\Omega-\mathcal{E}_m)}e^{\mathrm{i}(e_m-e_n)t-\mathrm{i}(\boldsymbol{q}_m-\boldsymbol{q}_n)\cdot\boldsymbol{x}} \\ &\quad \times\left\{1 - e^{-\beta(\mathcal{E}_n-\mathcal{E}_m)}\right\}|\langle m|O|n\rangle|^2,\end{aligned} \tag{5.3.24}$$

where we have considered the operator O as a boson operator, because O is an hermitian. Thus we obtain

$$\begin{aligned}D^R(\boldsymbol{q}, \omega) &= \int_V d\boldsymbol{x}\, dt\, D^R(\boldsymbol{x}, t)e^{-\mathrm{i}\boldsymbol{q}\cdot\boldsymbol{x}+\mathrm{i}(\omega+\mathrm{i}\delta)t} \\ &= V\sum_{mn} e^{-\beta(\Omega-\mathcal{E}_m)}|\langle m|O|n\rangle|^2\delta_{\boldsymbol{q},\boldsymbol{q}_n-\boldsymbol{q}_m}\frac{1 - e^{-\beta(\mathcal{E}_n-\mathcal{E}_m)}}{\omega - (e_n - e_m) + \mathrm{i}\delta}.\end{aligned} \tag{5.3.25}$$

Now we introduce the corresponding temperature Green's function:

$$\mathcal{D}(x, x') \equiv -\mathrm{Tr}\left\{T_\tau[\rho_G O_{\mathcal{H}}(\boldsymbol{x}\tau)O_{\mathcal{H}}(\boldsymbol{x}'\tau')]\right\}. \tag{5.3.26}$$

According to the periodic nature of the temperature Green's function we only consider the imaginary time τ in the regions $[0, \beta\hbar]$. Then we have

$$\begin{aligned}\mathcal{D}(\boldsymbol{x}, \tau) &= -\mathrm{Tr}\{T_\tau[\rho_G O_{\mathcal{H}}(\boldsymbol{x}\tau) O_{\mathcal{H}}(0)]\} \\ &= -\sum_{mn} e^{-\beta(\Omega-\mathcal{E}_m)} e^{(e_m-e_n)\tau - \mathrm{i}(\boldsymbol{q}_m-\boldsymbol{q}_n)\cdot\boldsymbol{x}} |\langle m|O|n\rangle|^2, \end{aligned} \tag{5.3.27}$$

$$\begin{aligned}\mathcal{D}(\boldsymbol{q}, \mathrm{i}\nu_n) &= \int_0^{\beta\hbar} d^4x\, e^{-\mathrm{i}\boldsymbol{q}\cdot\boldsymbol{x}+i\nu_n\tau} \\ &= V\sum_{mn} e^{\beta(\Omega-\mathcal{E}_m)} |\langle m|O(0)|n\rangle|^2 \delta_{\boldsymbol{q},\boldsymbol{q}_n-\boldsymbol{q}_m} \frac{1-e^{-\beta(\mathcal{E}_n-\mathcal{E}_m)}}{\mathrm{i}\nu_n-(e_n-e_m)}, \end{aligned} \tag{5.3.28}$$

where we $\mathrm{i}\nu_n$ is even, because we have considered O as a boson operator.

If we introduce the Lehmann spectral density

$$\Delta(\boldsymbol{q}, \omega) = V\sum_{mn} e^{\beta(\Omega-\mathcal{E}_m)} |\langle m|O(0)|n\rangle|^2 \delta_{\boldsymbol{q},\boldsymbol{p}_n-\boldsymbol{p}_m} \delta(\omega-(e_n-e_m))(1\mp e^{-\beta\hbar\omega}), \tag{5.3.29}$$

we obtain

$$D^R(\boldsymbol{q}, \omega) = \int_{-\infty}^{\infty} d\omega' \frac{\Delta(\boldsymbol{q}\cdot\omega')}{\omega-\omega'+\mathrm{i}\delta}, \tag{5.3.30}$$

$$\mathcal{D}(\boldsymbol{q}, \mathrm{i}\nu_n) = \int_{-\infty}^{\infty} d\omega' \frac{\Delta(\boldsymbol{q}\cdot\omega')}{\mathrm{i}\nu_n-\omega'}. \tag{5.3.31}$$

Thus if we analytically continue $\mathrm{i}\nu_n \to \omega+\mathrm{i}\delta$ in $\mathcal{D}(\boldsymbol{q}, \mathrm{i}\nu_n)$, we have an important relation:

$$\mathcal{D}(\boldsymbol{q}, \omega+\mathrm{i}\delta) = D^R(\boldsymbol{q}, \omega). \tag{5.3.32}$$

This relation manifests that the correlation function $D^R(\boldsymbol{q}, \omega)$ is obtained by the analytically continued temperature Green's function $\mathcal{D}(\boldsymbol{q}, \mathrm{i}\nu_n)$. Thus we have shown the equivalence between the retarded Green's function and the temperature Green's function.

5.3.4 *Density Correlation Function*

Here we consider the case of external mass potential in the linear response:

$$H^{\mathrm{ex}}(t) = \int d^3x\, n(\boldsymbol{x}) M\varphi^{\mathrm{ex}}(t). \tag{5.2.13}$$

The retarded density correlation functions D^R_{nn} and the temperature correlation function $\mathcal{D}_{nn}$ are given by

$$D^R_{nn}(\boldsymbol{x}t, \boldsymbol{x}'t') = -\mathrm{i}\theta(t-t')\mathrm{Tr}\{\rho_G[\tilde{n}(\boldsymbol{x}t), \tilde{n}(\boldsymbol{x}'t')]\}, \tag{5.3.33}$$

$$\mathcal{D}_{nn}(\boldsymbol{x}\tau, \boldsymbol{x}'\tau') = -\mathrm{Tr}\{\rho_G T_\tau[\tilde{n}(\boldsymbol{x}\tau)\tilde{n}(\boldsymbol{x}'\tau')]\}, \tag{5.3.34}$$

where we have considered the deviated number density operator

$$\tilde{n}(\boldsymbol{x}) = n(\boldsymbol{x}) - \langle n(\boldsymbol{x})\rangle, \tag{5.3.35}$$

where $\langle n(\boldsymbol{x})\rangle$ is the ensemble average of the number density operator and depends on the temperature β and the chemical potential μ of the system.

Then we have

$$D^R_{nn}(\boldsymbol{q}, \omega) = \int_{-\infty}^{\infty} d\omega' \frac{\Delta_{nn}(\boldsymbol{q}.\omega')}{\omega - \omega' + \mathrm{i}\delta}, \tag{5.3.36}$$

$$\mathcal{D}_{nn}(\boldsymbol{q}, \mathrm{i}\nu_n) = \int_{-\infty}^{\infty} d\omega' \frac{\Delta_{nn}(\boldsymbol{q}.\omega')}{\mathrm{i}\nu_n - \omega'}, \tag{5.3.37}$$

where the Lehmann spectral density is given by

$$\Delta_{nn}(\boldsymbol{q}, \omega) = V \sum_{mn} e^{\beta(\Omega - \mathcal{E}_m)} |\langle m|\tilde{n}(0)|n\rangle|^2 \delta_{\boldsymbol{q}.\boldsymbol{q}_n - \boldsymbol{q}_m} \delta(\omega - (e_n - e_m))(1 \mp e^{-\beta\hbar\omega}). \tag{5.3.38}$$

Thus we obtain the equivalence between the retarded and the temperature Green's functions

$$D^R_{nn}(\boldsymbol{q}, \omega) = \mathcal{D}_{nn}(\boldsymbol{q}, \omega + \mathrm{i}\delta). \tag{5.3.39}$$

This fact shows that the dynamical equations of the density fluctuation modes in the real space are obtained by analytically continuing the temperature frequencies of the dynamical equations of density fluctuation modes found in the temperature Green's function method and vice versa. We frequently use this fact in the later chapters. The intra-band density fluctuation mode is sound. The inter-band density fluctuation modes are phonons.

References

[1] R. Kubo, J. Phys. Soc. Jpn. 12 (1957) 570.
[2] H. Mori, Prog. Theor. Phys. 33 (1965) 423.
[3] H. Mori, Prog. Theor. Phys. 34 (1965) 399.
[4] N.N. Bogoliubov, S.V. Tyablikov, Retarded and advanced Green's functions in statistical physics, Dokl. Akad. Nauk SSSR 126 (1959) 53.
[5] D.N. Zubarev, Double time Green's functions in statistical physics, Uspekhi Fiz. Nauk 71 (1960) 71.
[6] R. Zwanzig, in: W.E. Britton, B.W. Downs, J. Downs (Eds.), Lectures in Theoretical Physics, vol. 3, Wiley Interscience, New York, 1961, p. 135.
[7] M. Ichiyanagi, J. Phys. Soc. Jpn. 32 (1972) 604.
[8] H. Lehmann, Nuovo Cim. 11 (1954) 342.

6 The Structure of Glasses Associated with Phonons

The Hamiltonians in condensed states, the crystalline, glassy, and liquid states are the same, but each condensed state has its own structure. A crystal has a long range order, the periodic structure: the pair distribution function of particles, $g(\boldsymbol{R})$, has the same periodic structure at each Bravias lattice point. A glass has a kind of long range order; the pair distribution function of particles, $g(R)$, has the same radial distribution at each constituent particle in a sense of the configuration average. Since the Hamiltonian is invariant under the spatial translation, the symmetries of the spatial translation spontaneously break down. Such spontaneous breakdown of the symmetries requires the corresponding Nambu-Goldstone (NG) bosons, phonons. Here first we show the Ward-Takahashi relations associated with the spontaneous breakdown of symmetries and then we derive the Nambu-Goldstone theorem, which shows the existence of the NG bosons, phonons.

A liquid seems to have the same structure as the glass in the sense that the pair distribution function, $g(R)$, has the same as that in the glass at an instant time, but the distribution of particles changes with time. The spontaneous breakdown of the symmetry does not occur. Therefore, the hopping of particles plays an essential role in the liquid; sound is essential in the liquid state. Sound is not the NG boson.

However, in the liquid state, a kind of phonons appears according to the instantaneously same distribution functions as in the glassy state. But, the structure of the phonons is vague, because the phonon dispersion curves merge into the continuum of the inter-band elementary excitations so that the phonons decay soon.

6.1 The WT Relations at Finite Temperature

The Ward-Takahashi (WT) relations associated with the spontaneous breakdown of a symmetry at finite temperature [1–10] play an important role in developing the theory in the condensed states:

1. In the crystal and glass, the WT relation associated with the spontaneously broken symmetry represented by a pair distribution function, $g(\boldsymbol{R})$, leads to the Nambu-Goldstone (NG) theorem. The NG theorem manifests the existence of the NG bosons, phonons [11–15,8].
2. The WT relation associated with the NG theorem yields the two band model. The WT relation associated with the NG theorem restricts the formation of wave functions

Liquid Glass Transition. DOI: http://dx.doi.org/10.1016/B978-0-12-407177-3.00006-4

in the excited state; the restriction determines the interaction of Hamiltonian, which yields the two band model.

3. The WT relations play an essential role in the approximation procedures. The WT relations among the Green's functions restrict the structure of these Green's functions. Therefore the approximate Green's functions should satisfy the WT relations.

Here we derive the WT relations at finite temperature.

6.1.1 The WT Relations at Finite Temperature

To derive the WT relations at finite temperature, we add the infinitesimal symmetry breaking term H_ϵ to the invariant Hamiltonian H:

$$H_\epsilon = -\epsilon \int d^3x g(x)\psi^\dagger(x)\psi(x), \tag{6.1.1}$$

$$g(x) = \langle \psi^\dagger(x)\psi(x)\rangle_\epsilon, \tag{6.1.2}$$

$g(x)$ is a pair distribution function. The pair distribution function behaves in the same fashion when we fix any particle as an original point and ϵ is an infinitesimal number. We define the Heisenberg operator at finite temperature for any operator O:

$$O(x) \equiv e^{\frac{(H-\mu N)\tau}{\hbar}} O(\boldsymbol{x}) e^{-\frac{(H-\mu N)\tau}{\hbar}}, \qquad x \equiv (\boldsymbol{x}, \tau). \tag{6.1.3}$$

The commutation and the anti-commutation relations of a field are given by

$$\left[\psi(x), \psi^\dagger(x')\right]_\mp\Big|_{\tau=\tau'} = \delta(\boldsymbol{x} - \boldsymbol{x}'). \tag{6.1.4}$$

Hereafter, when $\pm$ or $\mp$ signs appear in an equation, the upper sign refers to bosons and the lower to fermions. Here we define the ensemble average as

$$\langle \cdots \rangle_\epsilon = \langle \mathrm{Tr} \cdots e^{-\beta(H+H_\epsilon-\mu N)}\rangle_c / \langle \mathrm{Tr} e^{-\beta(H+H_\epsilon-\mu N)}\rangle_c, \tag{6.1.5}$$

where the angular bracket $\langle \cdots \rangle_c$ means the configuration average of the position of particles and $\beta = \frac{1}{k_B T}$. The pair distribution function $g(x)$ in Eqs. (6.1.1) and (6.1.2) is a function of $x = |\boldsymbol{x}|$ for the glassy state and is a periodic function of $\boldsymbol{x}$ for the crystalline state.

Now we introduce the generator of the spatial translation:

$$\boldsymbol{N}(\tau) = -i\hbar \int d^3x \psi^\dagger(x)\nabla\psi(x), \tag{6.1.6}$$

which satisfies

$$[\boldsymbol{N}(\tau), \psi(x)] = i\hbar\nabla\psi(x), \tag{6.1.7}$$

$$-\frac{\partial}{\partial\tau}\boldsymbol{N}(\tau) = -i\hbar\epsilon \int d^3x g(x)\nabla n(x), \tag{6.1.8}$$

where $n(x) = \langle\psi^\dagger(x)\psi(x)\rangle_\epsilon$. Eq. (6.1.7) leads to

$$e^{\frac{-i\boldsymbol{r}\cdot\boldsymbol{N}}{\hbar}}\psi(\boldsymbol{x},\tau)e^{\frac{i\boldsymbol{r}\cdot\boldsymbol{N}}{\hbar}} = \psi(\boldsymbol{x}+\boldsymbol{r},\tau),$$

which is checked by expanding the equation with the displacement $\boldsymbol{r}$ replaced by an infinitesimal displacement $\delta\boldsymbol{r}$, which leads to Eq. (6.1.7). Therefore, $\boldsymbol{N}(\tau)$is the generator of the spatial translation.

Next we consider the derivative of the product of the operators in τ:

$$\begin{aligned}&\frac{\partial}{\partial\tau}T_\tau\{\boldsymbol{N}(\tau)\psi_1(x_1)\cdots\psi_n(x_n)\}\\&=\sum_{i=1}^{n}\delta(\tau-\tau_i)T_\tau\{\psi_1(x_1)\cdots[\boldsymbol{N}(\tau),\psi_i(x_i)]\cdots\psi_n(x_n)\}\\&\quad+T_\tau\left\{\frac{\partial}{\partial\tau}\boldsymbol{N}(\tau)\psi_1(x_1)\cdots\psi_n(x_n)\right\},\end{aligned}\tag{6.1.9}$$

where $\psi_i(x_i)$ stands for $\psi(x_i)$ or $\psi^\dagger(x_i)$ and T_τ is the imaginary time-ordering operator. If we integrate both sides of Eq. (6.1.9) in τ from 0 to $\beta\hbar$, we obtain

$$\begin{aligned}&\boldsymbol{N}(\beta\hbar)T_\tau\{\psi_1(x_1)\cdots\psi_n(x_n)\}-T_\tau\{\psi_1(x_1)\cdots\psi_n(x_n)\}\boldsymbol{N}(0)\\&=\sum_{i=1}^{n}T_\tau\{\psi_1(x_1)\cdots[\boldsymbol{N}(\tau_i),\psi_i(x_i)]\cdots\psi_n(x_n)\}\\&\quad+\int_0^{\beta\hbar}d\tau T_\tau\{\frac{\partial}{\partial\tau}\boldsymbol{N}(\tau)\psi_1(x_1)\cdots\psi_n(x_n)\}.\end{aligned}\tag{6.1.10}$$

Taking the ensemble average of the equation, using Eqs. (6.1.7) and (6.1.8) and the property of Tr operator, we obtain a general WT relation:

$$\begin{aligned}&\sum_i\hbar\langle T_\tau\{\psi_1(x_1)\cdots\nabla\psi_i(x_i)\cdots\psi_n(x_n)\rangle_\epsilon\\&=\epsilon\int d^3x\langle T_\tau\psi_1(x_1)\cdots\psi_n(x_n)n(x)\rangle_\epsilon\nabla g(x).\end{aligned}\tag{6.1.11}$$

We show two simple WT relations:

$$\begin{aligned}&\langle T_\tau\nabla\psi(x_1)\psi^\dagger(x_2)\rangle_\epsilon+\langle T_\tau\psi(x_1)\nabla\psi^\dagger(x_2)\rangle_\epsilon\\&=\frac{\epsilon}{\hbar}\int d^3x\langle T_\tau\psi(x_1)\psi^\dagger(x_2)n(x)\rangle_\epsilon\nabla g(x),\end{aligned}\tag{6.1.12}$$

$$\hbar\nabla\langle n(x)\rangle_\epsilon=\epsilon\int d^4y\langle T_\tau n(x)n(y)\rangle_\epsilon\nabla g(y).\tag{6.1.13}$$

These equations will be used to obtain the relations between particle and phonon Green's functions and determine the gap equation. Eq. (6.1.13) plays an important role in deriving the NG theorem and constructs the two band model.

6.2 The Two Band Model and Green's Functions

In the condensed state such as liquids, glasses, and crystals, A particle is located in a localized potential made up by the surrounding particles so that the energy of the particles are quantized. Each particle makes up and down transitions between the two levels in its own potential well. However, in liquids and glasses, a particle can hop a nearest neighbor vacancy, which is originated from the random distribution of particles. A particle makes up and down transitions for a while, and then hops to a surrounding vacancy. This picture reduces to the two band model; the hopping of particles yields the configurationally averaged two energy levels bands. The up and down transitions are essential for phonons in glasses, while the hopping is essential for sound in liquids.

6.2.1 The Two Band Model

Hereafter, we consider the spontaneously symmetry breaking state, but the subscript of the ensemble average ϵ in $\langle \cdots \rangle_\epsilon$ is replaced by that of the configurationally averaged state, c in $\langle \cdots \rangle_c$ for simplicity. We start with the following Hamiltonian [8]:

$$H = \int d^3x \psi^\dagger(x) \hbar\epsilon^0(-\mathrm{i}\nabla)\psi(x) + \frac{1}{2}\int d^3x d^3y n(x) V(\boldsymbol{x} - \boldsymbol{y}) n(y), \quad (6.2.1)$$

where ψ is an annihilation field operator of a particle, $\hbar\epsilon^0(-\mathrm{i}\nabla)$ is an energy operator of a free particle, $n(x)$ is a density operator of the particle, $n(x) = \psi^\dagger(x)\psi(x)$ and V are an interaction potential between particles. If we introduce an annihilation operator of the particle in a potential at the site $\boldsymbol{R}_m$, $b_{m\mu}$, we can write

$$\psi(x) = \sum_{m\mu} \tilde{w}_{m\mu}(\boldsymbol{x} - \boldsymbol{R}_m) b_{m\mu}, \quad \mu = 0,\ 1,\ 2,\ 3. \quad (6.2.2)$$

Here $\tilde{w}_{m\mu}$ is the wave function of the μth state at a site $\boldsymbol{R}_m$.

The Heisenberg equation for $b_{m\mu}$ is given by

$$-\hbar\frac{\partial}{\partial\tau} b_{m\mu} = \sum_{n\nu}\int d^3x \tilde{w}_{m\mu}(\boldsymbol{x} - \boldsymbol{R}_m) [\hbar\epsilon^0(-\mathrm{i}\nabla) + \Phi_{\boldsymbol{R}_n}(\boldsymbol{x})]\tilde{w}_{n\nu}(\boldsymbol{x} - \boldsymbol{R}_n) b_{n\nu}, \quad (6.2.3)$$

$$\Phi_{\boldsymbol{R}_n}(\boldsymbol{x}) = \int_{\{\boldsymbol{R}_n\}} d^3y V(\boldsymbol{x} - \boldsymbol{y}) n(y), \quad (6.2.4)$$

where $\{\boldsymbol{R}_n\}$ means summing the surrounding sites of the particle at $\boldsymbol{R}_n$ in $n(y)$. The term $\Phi_{\boldsymbol{R}_n}$ means a potential of the particle at a site $\boldsymbol{R}_n$ made up by the surrounding particles. Thus we obtain the Schrödinger equation:

$$\left\{\hbar\epsilon^0(-\mathrm{i}\nabla) + \Phi_{\boldsymbol{R}_n}(\boldsymbol{x})\right\} \tilde{w}_{n\mu}(\boldsymbol{x} - \boldsymbol{R}_n) = \hbar\tilde{\omega}_{n\mu}\tilde{w}_{n\mu}(\boldsymbol{x} - \boldsymbol{R}_n), \quad (6.2.5)$$

where $\tilde{\omega}_{n\mu}$ and $\tilde{w}_{n\mu}$ are the eigenfrequency and the eigenfunction of the μth state at the site $\boldsymbol{R}_n$; for the ground state, $\tilde{\omega}_{n0}$:

$$\tilde{w}_{n0}(\boldsymbol{x}) = w_{n0}(x_1)w_{n0}(x_2)w_{n0}(x_3), \quad \text{for} \quad \mu = 0,$$

and for the first excited states, $\tilde{\omega}_{ni}$:

$$\tilde{w}_{ni}(\boldsymbol{x}) = w_{n1}(x_i)w_{n0}(x_j)w_{n0}(x_k), \quad \text{for} \quad \mu \equiv i = 1,\ 2,\ 3,$$

where x_i, x_j and, x_k are three different Cartesian coordinates. Hereafter, $\mu,\ \nu$ mean $0, 1, 2, 3$ and $i,\ j$ mean $1, 2, 3$. Since the ground state wave function has a mount and the first excited state wave functions for the ith Cartesian coordinate has a node in the ith direction, we can expect:

$$w_{n1}(x_i) \equiv w_{ni}(x_i) = -2\zeta_n \nabla_i w_{n0}(x_i). \tag{6.2.6}$$

This equation is rigorously verified by the WT relation (6.1.13) associated with the Nambu-Goldstone theorem in the next section. Therefore these situations allow us to make the harmonic potential approximation. Thus, we consider a harmonic potential:

$$\Phi_{\boldsymbol{R}_n}(\boldsymbol{x}) \cong \frac{1}{2}M\omega_n^2(\boldsymbol{x} - \boldsymbol{R}_n)^2, \quad \tilde{\omega}_{n0} = \frac{3}{2}\omega_n, \quad \tilde{\omega}_{ni} = \frac{5}{2}\omega_n, \quad \zeta_n = \sqrt{\frac{\hbar}{2M\omega_n}}, \tag{6.2.7}$$

where ω_n is a harmonic frequency at the site $\boldsymbol{R}_n$, M the mass of a particle, and ζ_n the mean width of the zero point motion. For the harmonic potential the wave functions of the second excited states are written as $w_{n2ii} = \{w_{n0}(x_i) - 2\zeta_n \nabla_i w_{n1}(x_i)\}$. Since we are not concerned with the higher states than the first excited ones, we make the following approximation:

$$w_{n0}(x_i) = 2\zeta_n \nabla_i w_{n1}(x_i). \tag{6.2.8}$$

Eqs. (6.2.6) and (6.2.8) show that the operator ∇_i plays a role of changing the states like the spin rotational operator in magnetism. The magnitude of potentials $\Phi_{\boldsymbol{R}_n}$ is distributed randomly, so is that of the eigenfrequencies.

At this stage, we can take the following model Hamiltonian in terms of localized operators:

$$H = \sum_{m\mu} \hbar\tilde{\omega}_{m\mu} b_{m\mu}^{\dagger} b_{m\mu} + \sum_{m\mu n\nu} \hbar J_{m\mu n\nu} b_{m\mu}^{\dagger} b_{n\nu}, \tag{6.2.9}$$

where the hopping matrix is given by

$$\hbar J_{m\mu n\nu} = \int d^3x\, \tilde{w}_{m\mu}(\boldsymbol{x} - \boldsymbol{R}_m)\Phi_{\boldsymbol{R}_n}(\boldsymbol{x})\tilde{w}_{n\nu}(\boldsymbol{x} - \boldsymbol{R}_n). \tag{6.2.10}$$

Note that the interaction Hamiltonian is nonlinear through the term $n(y)$ in $\Phi_{\boldsymbol{R}_n}$ and the potential $\Phi_{\boldsymbol{R}_n}$ permits a particle to hop only to a vacancy.

Now we first establish the unperturbed Hamiltonian by taking the configurational average of the model Hamiltonian Eq. (6.2.9):

$$H_0 = \langle H \rangle_c = \sum_{\mu m} \hbar \tilde{\omega}_\mu b^\dagger_{m\mu} b_{m\mu} + \sum_{\mu mn} \hbar J_\mu(\boldsymbol{R}_m - \boldsymbol{R}_n) b^\dagger_{m\mu} b_{n\mu}, \tag{6.2.11}$$

$$\omega = \langle \omega_m \rangle_c, \quad \tilde{\omega}_\mu = \langle \tilde{\omega}_{m\mu} \rangle_c, \quad J_\mu(\boldsymbol{R}_m - \boldsymbol{R}_n) = \langle J_{m\mu n\mu} \rangle_c, \quad \zeta = \sqrt{\frac{\hbar}{2M\omega}} \tag{6.2.12}$$

where $\langle \cdots \rangle = \mathrm{Tr} e^{-\beta(H-\mu N)} \ldots / \mathrm{Tr} e^{-\beta(H-\mu N)}$ and the subscript c means the configuration average. The hopping between different states, $\langle J_{m\mu n\mu} \rangle_c$ with $\mu \neq \nu$ is neglected for simplicity.

Next we establish an alternative model Hamiltonian in terms of extended operators. If we introduce the following extended operators:

$$b_{m\mu} = \frac{1}{\sqrt{N}} \sum_{\boldsymbol{p}} e^{\mathrm{i}\boldsymbol{p}\cdot\boldsymbol{R}_m} a_{\mu\boldsymbol{p}}, \quad \mu = 0,\ 1,\ 2,\ 3, \tag{6.2.13}$$

we obtain

$$H_0 = \sum_{\mu\boldsymbol{p}} \hbar \epsilon_{\mu\boldsymbol{p}} a^\dagger_{\mu\boldsymbol{p}} a_{\mu\boldsymbol{p}}, \tag{6.2.14}$$

$$\epsilon_{\mu\boldsymbol{p}} = \tilde{\omega}_\mu + J_\mu(\boldsymbol{p}), \tag{6.2.15}$$

$$J_\mu(\boldsymbol{p}) \equiv \frac{1}{N} \int d^3R g(R) e^{\mathrm{i}\boldsymbol{p}\cdot\boldsymbol{R}} J_\mu(\boldsymbol{R}), \tag{6.2.16}$$

where $g(R)$ is a pair distribution function. When a μ-state particle at $\boldsymbol{R}_m$ hops to a surrounding vacancy at $\boldsymbol{R}_n$, the vacancy has raised the potential of the μ-state particle and has made a saddle point upwards convex in the direction $\boldsymbol{R} = \boldsymbol{R}_n - \boldsymbol{R}_m$. Since the magnitudes of the potential of the μth state particle is determined by the reduction of the potential made up by the surrounding particles, $J_\mu(\boldsymbol{p})$ should involve the term $g(R)$.

The field operator is written down as

$$\psi(x) = \frac{1}{\sqrt{N}} \sum_{\mu\boldsymbol{p}m} e^{\mathrm{i}\boldsymbol{p}\cdot\boldsymbol{R}_m} \tilde{w}_{m\mu}(\boldsymbol{x} - \boldsymbol{R}_m) a_{\mu\boldsymbol{p}}. \tag{6.2.17}$$

From Eq. (6.2.17), we obtain

$$\begin{aligned} n(x) &= \sum_m \left\{ \sum_\mu \tilde{w}^2_{m\mu}(\boldsymbol{x} - \boldsymbol{R}_m) b^\dagger_{m\mu} b_{m\mu} - \zeta \boldsymbol{B}_m \cdot \nabla \tilde{w}^2_{m0}(\boldsymbol{x} - \boldsymbol{R}_m) + \cdots \right\} \\ &= \frac{1}{N} \sum_{m\boldsymbol{q}} e^{-\mathrm{i}\boldsymbol{q}\cdot\boldsymbol{R}_m} \left[\sum_\mu \tilde{w}^2_{m\mu}(\boldsymbol{x} - \boldsymbol{R}_m) \rho^\dagger_{d\mu\boldsymbol{q}} \right. \\ &\qquad \left. - \sum_i \zeta \nabla_i \tilde{w}^2_{m0}(\boldsymbol{x} - \boldsymbol{R}_m) \rho^\dagger_{i\boldsymbol{q}} + \cdots \right] \end{aligned} \tag{6.2.18}$$

where

$$B_{mi} \equiv b_{mi}^{\dagger} b_{m0} + b_{m0}^{\dagger} b_{mi}, \tag{6.2.19}$$

$$\rho_{d\mu\boldsymbol{q}}^{\dagger} = \sum_{\boldsymbol{p}} a_{\mu\boldsymbol{p}}^{\dagger} a_{\mu\boldsymbol{p}-\boldsymbol{q}}, \tag{6.2.20}$$

$$\rho_{i\boldsymbol{q}}^{\dagger} = \sum_{\boldsymbol{p}} (a_{i\boldsymbol{p}}^{\dagger} a_{0\boldsymbol{p}-\boldsymbol{q}} + a_{0\boldsymbol{p}}^{\dagger} a_{i\boldsymbol{p}-\boldsymbol{q}}). \tag{6.2.21}$$

$\rho_{d\mu\boldsymbol{q}}^{\dagger}$ and $\rho_{i\boldsymbol{q}}^{\dagger}$ are intra-band and inter-band density fluctuation operators, respectively. Note that $\rho_{d\mu\boldsymbol{q}} = \sum_{\boldsymbol{p}} a_{\mu\boldsymbol{p}}^{\dagger} a_{\mu\boldsymbol{p}+\boldsymbol{q}}$.

Substitution of Eq. (6.2.18) into Eq. (6.2.1) yields

$$H = \sum_{\mu\boldsymbol{p}} \hbar\epsilon_{\mu\boldsymbol{p}} a_{\mu\boldsymbol{p}}^{\dagger} a_{\mu\boldsymbol{p}} + \frac{1}{2N} \sum_{\mu\nu\boldsymbol{q}}{}' V_{\mu\nu}^{d}(\boldsymbol{q}) \rho_{d\mu\boldsymbol{q}}^{\dagger} \rho_{d\nu\boldsymbol{q}} + \frac{1}{2N} \sum_{ij\boldsymbol{q}} V_{ij}^{od}(\boldsymbol{q}) \rho_{i\boldsymbol{q}}^{\dagger} \rho_{j\boldsymbol{q}}, \tag{6.2.22}$$

$$V_{\mu\nu}^{d} \equiv N V_{\mu\mu,\nu\nu}, \qquad V_{ij}^{od}(\boldsymbol{q}) \equiv N V_{i0,j0}, \tag{6.2.23}$$

$$V_{\mu\mu',\nu\nu'}(\boldsymbol{q}) = \frac{1}{N^2} \sum_{m \neq n} e^{-\mathrm{i}\boldsymbol{q}\cdot(\boldsymbol{R}_m - \boldsymbol{R}_n)} \times \int d^3x d^3y \tilde{w}_{m\mu}(\boldsymbol{x} - \boldsymbol{R}_m) \tilde{w}_{m\mu'}(\boldsymbol{x} - \boldsymbol{R}_m) V(\boldsymbol{x} - \boldsymbol{y}) \times \tilde{w}_{n\nu}(\boldsymbol{y} - \boldsymbol{R}_n) \tilde{w}_{n\nu'}(\boldsymbol{y} - \boldsymbol{R}_n), \tag{6.2.24}$$

where no cross term of the intra-band and inter-band density operators appears, because of the symmetry of the wave functions. Here it should be noted that since the unperturbed Hamiltonian H_0 in Eq. (6.2.11) involves the term $\int d^3xd^3yn(x)V(\boldsymbol{x} - \boldsymbol{y})\langle n(y)\rangle_c$, the interaction Hamiltonian H_I must exclude the term. Thus the prime on $\sum$ in Eq. (6.2.22) means that H_I excludes one loop diagrams.

6.2.2 Green's Functions for Particles, Intra-Band, and Inter-Band Density Fluctuations

Near the liquid-glass transition the majority of particles stay at the ground state. Therefore, for the intra-band density fluctuations we only consider the lowest band $\rho_{d0\boldsymbol{q}}^{\dagger} = \sum_{\boldsymbol{p}} a_{0\boldsymbol{p}}^{\dagger} a_{0\boldsymbol{p}-\boldsymbol{q}}$, so we abbreviate $\rho_{d0\boldsymbol{q}}^{\dagger}$ as

$$\rho_{0\boldsymbol{q}}^{\dagger} = \sum_{\boldsymbol{p}} a_{0\boldsymbol{p}}^{\dagger} a_{0\boldsymbol{p}-\boldsymbol{q}}. \tag{6.2.25}$$

Note that the inter-band density fluctuations $\rho^\dagger_{i\boldsymbol{q}} = \sum_{\boldsymbol{p}}(a^\dagger_{i\boldsymbol{p}}a_{0\boldsymbol{p}-\boldsymbol{q}} + a^\dagger_{0\boldsymbol{p}}a_{i\,\boldsymbol{p}-\boldsymbol{q}})$ holds. Hereafter, we consider the number density $n(x)$ and the Hamiltonian H as

$$n(x) = \frac{1}{N}\sum_{m\boldsymbol{q}} e^{-\mathrm{i}\boldsymbol{q}\cdot\boldsymbol{R}_m}[\tilde{w}^2_{m0}(\boldsymbol{x}-\boldsymbol{R}_m)\rho^\dagger_{0\boldsymbol{q}} - \sum_i \zeta\nabla_i\tilde{w}^2_{m0}(\boldsymbol{x}-\boldsymbol{R}_m)\rho^\dagger_{i\boldsymbol{q}} + \cdots], \tag{6.2.26}$$

$$H = \sum_{\mu\boldsymbol{p}} \hbar\epsilon_{\mu\boldsymbol{p}}a^\dagger_{\mu\boldsymbol{p}}a_{\mu\boldsymbol{p}} + \frac{1}{2N}{\sum_{\boldsymbol{q}}}' V^d_{00}(\boldsymbol{q})\rho^\dagger_{0\boldsymbol{q}}\rho_{0\boldsymbol{q}} + \frac{1}{2N}\sum_{ij\boldsymbol{q}} V^{od}_{ij}(\boldsymbol{q})\rho^\dagger_{i\boldsymbol{q}}\rho_{j\boldsymbol{q}}. \tag{6.2.27}$$

Now, we introduce particle Green's functions $G_\mu(p)$, Green's function $F(q)$ of intra-band density fluctuations, Green's functions $D_{ij}(q)$ of inter-band density fluctuations:

$$G_{\mu\boldsymbol{p}}(\tau_1-\tau_2) \equiv -\langle T_\tau a_{\mu\boldsymbol{p}}a^\dagger_{\mu\boldsymbol{p}}\rangle_c \equiv \frac{1}{\beta\hbar}\sum_{\mathrm{i}\omega_n} e^{-\mathrm{i}\omega_n(\tau_1-\tau_2)}G_\mu(p), \tag{6.2.28}$$

$$F_{\boldsymbol{q}}(\tau_1-\tau_2) \equiv -\frac{1}{\hbar N}\langle T_\tau \rho_{d0\boldsymbol{q}}(\tau_1)\rho^\dagger_{d0\boldsymbol{q}}(\tau_2)\rangle_c \equiv \frac{1}{\beta\hbar}\sum_{\mathrm{i}\nu_n} e^{-\mathrm{i}\nu_n(\tau_1-\tau_2)}F(q), \tag{6.2.29}$$

$$D_{ij\boldsymbol{q}}(\tau_1-\tau_2) \equiv -\frac{1}{\hbar N}\langle T_\tau \rho_{i\boldsymbol{q}}(\tau_1)\rho^\dagger_{j\boldsymbol{q}}(\tau_2)\rangle_c \equiv \frac{1}{\beta\hbar}\sum_{\mathrm{i}\nu_n} e^{-\mathrm{i}\nu_n(\tau_1-\tau_2)}D_{ij}(q). \tag{6.2.30}$$

We should not confuse the temperature frequency $\mathrm{i}\omega_n$ with the harmonic frequency ω_n. When we use the temperature frequencies $\mathrm{i}\omega_n$, the frequency is always accompanied with the imaginary number i.

In this book, we confine ourselves to the random phase approximation. In the approximation procedure, from Eq. (6.2.14) the particle Green's functions are given by

$$G_\mu(p) = \frac{1}{\mathrm{i}\omega_n - \epsilon_{\mu\boldsymbol{p}}}, \tag{6.2.31}$$

The Green's function of intra-band density fluctuations is given by

$$F(q) = P(q) + P(q)\langle V^d_{00}(q)\rangle_c F(q), \tag{6.2.32}$$

$$P(q) = \mp\frac{1}{\beta\hbar^2 N}\sum_p G_0(p)G_0(p-q), \tag{6.2.33}$$

where the interaction potential $\langle V^d_{00}(q)\rangle_c$ is spherical.

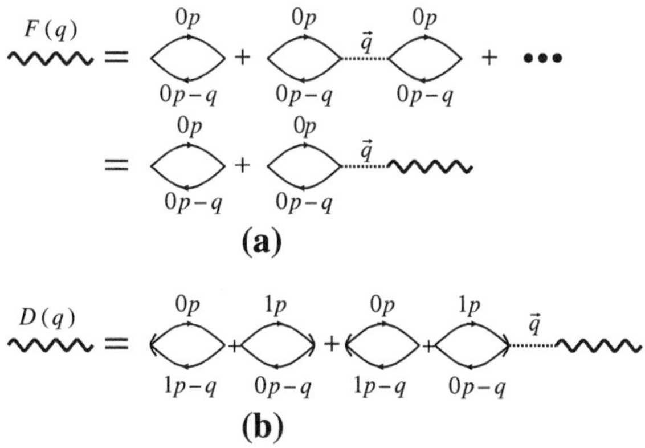

Figure 6.1 The Feynman diagrams of $F(q)$ and $D(q)$ in the random phase approximation are illustrated as wavy lines in (a) and (b), respectively. (a): The bubble diagrams of intra-band particle-hole pairs denote $P(q)$, and dotted lines denotes $\langle V^d_{00}(q)\rangle_c$. (b): All terms are illustrated in the matrix form. The bubble diagrams of inter-band particle-hole pairs denote $Q(q)$, and dotted line denotes $\langle V^{od}(\boldsymbol{q})\rangle_c$.

The Green's functions of inter-band density fluctuations are

$$D_{ij}(q) = Q(q)\delta_{ij} + Q(q)\sum_l \langle V^{od}_{il}(\boldsymbol{q})\rangle_c D_{lj}(q), \tag{6.2.34}$$

where

$$Q(q) = \mp\frac{1}{\beta\hbar^2 N}\sum_p \{G_0(p)G_i(p-q) + G_i(p)G_0(p-q)\}, \tag{6.2.35}$$

$Q(q)$ does not depend on the index i, because we are concerned with the isotropic system. $F(q)$ and $D(q)$ in the matrix form are illustrated in Figure 6.1.

6.3 The Nambu-Goldstone Theorem and Phonons

The Green's functions of inter-band density fluctuations yield phonons. Phonons are the Nambu-Goldstone (NG) bosons associated with the spontaneous breakdown of the translational invariance in the crystalline and glassy states. The NG theorem states that phonons have the following properties:

1. The phonon dispersion curves are gapless at zero wave vector.
2. Phonons are a longitudinal and two transverse modes.
3. Phonons are elastic waves.

Although sound is gapless at zero wave vector, sound is not a NG boson; sound is a longitudinal one, and no elastic wave. Sound originates from hopping of particles in liquids. Now we investigate the NG theorem.

6.3.1 The Nambu-Goldstone Theorem

To derive the NG theorem, we substitute

$$n(x) = \frac{1}{N}\sum_{mq} e^{-\mathrm{i}\boldsymbol{q}\cdot\boldsymbol{R}_m}[\tilde{w}^2_{m0}(\boldsymbol{x}-\boldsymbol{R}_m)\rho^\dagger_{0q} - \sum_i \zeta\nabla_i\tilde{w}^2_{m0}(\boldsymbol{x}-\boldsymbol{R}_m)\rho^\dagger_{iq} + \cdots] \tag{6.2.26}$$

into the WT relation

$$\hbar\nabla\langle n(x)\rangle_\epsilon = \epsilon\int d^4y\langle T_\tau n(x)n(y)\rangle_\epsilon\nabla g(y). \tag{6.1.13}$$

Since the majority of particles stay in the lower band, the left-hand side of equations leads to $\langle\sum_m\nabla_i\tilde{w}^2_{m0}(\boldsymbol{x}-\boldsymbol{R}_m\rangle_c$ so that the term $n(x)$ on the right-hand side should have $\langle\sum_m\nabla_i\tilde{w}^2_{m0}(\boldsymbol{x}-\boldsymbol{R}_m)\rangle_c$, that is, the term $n(x)$ on the right-hand side must have the second term of Eq. (6.2.26). On the other hand, the integration of y on the right hand side of Eq. (6.1.13) also requires that the term $n(x)$ must have the second term of Eq. (6.2.26).

To see this situation in the general WT relation (6.1.11), we investigate the right hand side of Eq. (6.1.11). There is a correlation among the particles with the positions of $0, \boldsymbol{x}_1, \cdots, \boldsymbol{x}_n$, and $\boldsymbol{x}$, when the positions are within a few particle distances; there is no correlation among the particles, when the positions are outside a few particle distances. The dominant contribution of the integration of $\boldsymbol{x}$ comes from the outside regions. Thus the integration of $\boldsymbol{x}$ over the regions truncates the correlation from the remaining particles with the positions $0, \boldsymbol{x}_1, \cdots, \boldsymbol{x}_n$ and reduces the rank of the many particle distribution functions. This procedure corresponds to the Kirkwood approximation. Therefore we obtain

$$\begin{aligned}
&\langle\cdots\int d^3x n(x)\rangle_c\nabla_i g(x) \\
&\cong\langle\cdots\frac{\zeta}{2N}\sum_{nmq}\int_{\{\boldsymbol{R}_n\}} d^3x e^{-\mathrm{i}\boldsymbol{q}\cdot\boldsymbol{R}_m}\nabla_i\tilde{w}^2_{m0}(\boldsymbol{x}-\boldsymbol{R}_m)\rho^\dagger_{iq}\rangle_c\nabla_i g(x) \\
&\cong\sum_q\langle\cdots\rho^\dagger_{iq}\rangle\delta_{q0}\frac{\zeta C}{2},
\end{aligned} \tag{6.3.1}$$

$$\int_{\{\boldsymbol{R}_n\}} d^3x\nabla_i g(x)\nabla_j g(x) = \delta_{ij}C, \tag{6.3.2}$$

$$g(x) = \langle\sum_m\tilde{w}^2_0(\boldsymbol{x}-\boldsymbol{R}_m)\rangle_c. \tag{6.3.3}$$

In Eq. (6.3.1), the subscript $\{\boldsymbol{R}_n\}$ means the summation of m is taken over the positions around n-particle. Since we can consider that the configurationally averaged distribution of particles $\langle \sum_m \nabla_i \tilde{w}_0^2(\boldsymbol{x} - \boldsymbol{R}_m)\rangle_c$ around $\boldsymbol{R}_n$ is of the same form, the summation of n leads to $\delta_{0,\boldsymbol{q}}$. The number 2 in the factor $\frac{\zeta}{2N}$ comes from the double count of variables n and m. Note that in Eq. (6.3.1) there does not appear the first term of $n(x)$, the intra-band density fluctuations, in Eq. (6.2.26). In Eq. (6.3.2) the subscript $\{\boldsymbol{R}_n\}$ of the integral means we have regard $g(x)$ as the actual pair distribution function, where the origin is taken at $\{\boldsymbol{R}_n\}$, and the pair distribution function is spherically symmetric.

Substituting Eq. (6.2.26) into Eq. (6.1.13), we obtain

$$\begin{aligned}
\nabla_i g(x) &= \frac{\epsilon}{\hbar}\sum_{ij}\nabla_j g(x)\int_0^{\beta\hbar} d\tau_y \sum_{\boldsymbol{q}}\langle T_\tau \rho_{j\boldsymbol{q}}(\tau_x)\rho_{i\boldsymbol{q}}^\dagger(\tau_y)\rangle_c \delta_{\boldsymbol{q}0}\frac{\zeta^2 C}{2N} \\
&= \frac{\epsilon}{\hbar}\sum_{ij}\nabla_j g(x)\hbar N\sum_{\mathrm{i}\nu_n \boldsymbol{q}}\frac{1}{\beta\hbar}\int_0^{\beta\hbar} d\tau_y e^{\mathrm{i}\nu_n(\tau_x-\tau_y)}D_{ji}(q)\delta_{\boldsymbol{q}0}\frac{\zeta^2 C}{2N} \\
&= \epsilon\sum_{ij}\nabla_j g(x)\sum_{\mathrm{i}\nu_n \boldsymbol{q}} D_{ji}(q)\delta_{\mathrm{i}\nu_n 0}\delta_{\boldsymbol{q}0}\frac{\zeta^2 C}{2}, \\
&= \epsilon\sum_{ij}\nabla_j g(x) D_{ji}(0)\frac{\zeta^2 C}{2},
\end{aligned} \tag{6.3.4}$$

where from the first line to the second we have used Eq. (6.2.30). Thus we have

$$\delta_{ij} = \epsilon D_{ij}(0)\frac{\zeta^2 C}{2}. \tag{6.3.5}$$

Here, we diagonalize $D_{ij}(q)$ by use of a unitary matrix γ:

$$D_{ij}(q) = \sum_\lambda \gamma_i^\lambda(\boldsymbol{q}) D_\lambda(q)\gamma_j^{\lambda *}(\boldsymbol{q}), \tag{6.3.6}$$

where $D_\lambda(q)$ has the form:

$$D_\lambda(q) = \frac{1}{(\mathrm{i}\nu_n)^2 - \omega_\lambda^2(\boldsymbol{q}) - \epsilon A_\lambda(\boldsymbol{q})}. \tag{6.3.7}$$

Here the term $\epsilon A_\lambda(\boldsymbol{q})$ is the shift of the phonon frequency $\omega_\lambda^2(\boldsymbol{q})$ due to H_ϵ. Substituting Eqs. (6.3.6) and (6.3.7) into Eq. (6.3.5), we obtain

$$\delta_{ij} = \epsilon\sum_\lambda \frac{\gamma_i^\lambda \gamma_j^{\lambda *}}{\omega_\lambda(0) + \epsilon A_\lambda(0)}\frac{\zeta^2 C}{2}. \tag{6.3.8}$$

In the limit $\epsilon \longrightarrow 0$, Eq. (6.3.8) leads to

$$\omega_\lambda(0) = 0, \tag{6.3.9}$$

$$\delta_{ij} = \sum_\lambda \Gamma_i^\lambda \Gamma_j^{\lambda *}, \qquad \Gamma_i^\lambda = \sqrt{\frac{\zeta^2 C}{2A_\lambda(0)}}\gamma_i^\lambda(0). \tag{6.3.10}$$

Eq. (6.3.9) means that NG bosons are gapless at zero wavenumber. Eq. (6.3.10) is the completeness relation, which shows that there are three kinds of the NG bosons.

If we substitute Eq. (6.2.34) into Eq. (6.3.5), we obtain

$$\delta_{ij} = Q(0)\langle V_{ij}^{od}(0)\rangle_c. \tag{6.3.11}$$

The term $Q(q)$ expresses a bubble of inter-band particle-hole pairs; there is a gap between the lower and upper bands. As the temperature increases, the gap becomes smaller satisfying Eq. (6.3.11): Eq. (6.3.11) determines the magnitude of the gap. Thus we call Eq. (6.3.11) the gap equation. The gap guarantees the rigidity.

Thus we obtain the Nambu-Goldstone theorem for the glassy state:

1. There appears the NG bosons; $\lim_{\boldsymbol{q}\longrightarrow 0}\omega_{\lambda\boldsymbol{q}} = 0$.
2. There are three kinds of the NG bosons; a longitudinal and two transverse phonons.
3. There is the gap Eq. (6.3.11), which guarantees the rigidity.

It should be noted that the Nambu-Goldstone theorem originates from the spontaneous breakdown of the translational invariance in the glassy state and there necessarily appear the three gapless phonons. The symmetry broken system is guaranteed by the gap equation; the gap between the lower and higher bands ensures rigidity. Gaps or discrete levels in quantum mechanics always guarantee the stable states in the systems.

The WT relation (6.1.13) leads to the Nambu-Goldstone theorem. The WT relation (6.1.13) requires that the correlation of the density fluctuations consists of the inter-band ones and the first excited state wave function is rigorously the spatial derivative of the ground state wave function. These requirements justify the harmonic potential approximation. On the other hand, the ground state wave function has a mount and the first excited state wave function has a node in a localized potential. This situation approximately justifies the harmonic potential approximation.

The inter-band density fluctuations yield phonons. The first excited state wave function spreads out from the ground state wave function. This spread of the first excited wave function corresponds to the displacement of the particle from the equilibrium position in the direction in the classical theory. Thus the inter-band density fluctuations exactly correspond to phonon operators [8,16]. The two level model of the particle-hole pairs in the harmonic potential approximation was first presented by W. Brenig [17], but independently we established the two band model by using the WT relations. The WT relations play an important role in deriving the Hamiltonian and the Nambu-Goldstone theorem.

The gap Eq. (6.3.11) guarantees the stability of the symmetry breaking state and the existence of the NG bosons associated with the symmetry breaking. At low temperatures, the band gap $\sim \omega$, where ω is the configurationally averaged harmonic frequency. The elastic constant $\sim M\omega^2$ means that the band gap, ω, represents the elasticity of the system. Since the gap Eq. (6.3.11) is a kernel equation of the dynamical equation of phonons (6.2.34), the gap equation plays an essential role in deriving the gapless phonon frequencies.

Sound originates from the intra-band density fluctuations due to the hopping. Although the sound has a gapless frequency at zero wavenumber, sound is not a NG boson. The liquid state has the same pair distribution function as the glassy state, but the pair distribution function changes with time because of the hopping; the pair distribution function guarantees no spontaneous breaking of the translational symmetry. Thus the sound plays an essential role in liquids. The sound disappears below the sound disappearing temperature T_0. The glass transition occurs at the glass transition temperature T_g close to T_0, where the phonon fluctuations are a dominant role.

6.4 The Structure of Phonons

Phonons are the NG bosons under the spontaneously broken symmetry of spatial translation associated with a long range order. In a crystal, the long range order is a periodicity; the pair distribution function has the same periodic structure at any of its own Bravais lattice point, while in a glass, there is a kind of long range order; the pair distribution function has the same radial distribution function at any of its own constituent particle. Phonon dispersion curves in the glass have the following features [18,19,8]:

1. In the long wavelength regime, phonon dispersion curves yield linear dispersions with respect to wavenumbers with a longitudinal and two transverse velocities.
2. Near a reciprocal particle distance, the longitudinal phonon dispersions have a roton type minimum, which is a vestige of the periodicity in the crystal.
3. In the short wavelength regime, phonon dispersions tend to the harmonic frequency, which corresponds to boson peak.

In a liquid, the hopping is essential. The hopping yields energy bands of particles so that the hopping causes the continuum of the excitations of inter-band particle-hole pairs (the inter-band elementary excitations). When phonon dispersion curves merge into the continuum of the excitations of inter-band particle-hole pairs, phonons have the width of the dispersion curves; the inverse of the width is proportional to the lifetime of the mode [20,21,8]. As the hopping increases, the width increases so that phonons are smeared out.

6.4.1 *The Dynamical Equation of Phonons*

In order to solve the dynamical equation of phonons:

$$D_{ij}(q) = Q(q)\delta_{ij} + Q(q)\sum_{l} \langle V_{il}^{od}(\mathbf{q})\rangle_c D_{lj}(q), \tag{6.2.34}$$

we must calculate the bubbles of inter-band particle-hole pairs:

$$Q(q) = \mp\frac{1}{\beta\hbar^2 N}\sum_{p}\{G_0(p)G_i(p-q) + G_i(p)G_0(p-q)\}. \tag{6.2.35}$$

with the Green's functions of particles:

$$G_\mu(p) = \frac{1}{\mathrm{i}\omega_n - \epsilon_{\mu p}}. \tag{6.2.31}$$

Therefore we first investigate the particle energy dispersion curves taking into account the hopping magnitude J_μ. After we obtain the particle energy dispersions, we calculate the bubble of the inter-band particle-hole pairs $Q(q)$. In solving the dynamical equation Eq. (6.2.34), we must use the gap equations:

$$\delta_{ij} = Q(0)\langle V_{ij}^{od}(0)\rangle_c, \tag{6.311}$$

The dynamical equation of phonons with the gap equation leads to the structure of phonons.

Since the potential functions $\langle V_{ij}^{od}(\boldsymbol{q})\rangle_c$ can be diagonalized, we can diagonalize the dynamical equation of phonons and the gap equation. In order to obtain the phonon dispersion curves, we solve the kernel of the diagonalized dynamical equation Eq. (6.2.34):

$$1 - Q(q)\langle V_\lambda^{od}(\boldsymbol{q})\rangle_c = 0, \tag{6.4.1}$$

together with the diagonalized gap equation, where λ indicates a longitudinal, l, and two transverse modes, ts.

6.4.2 The Bubbles of Inter-Band Particle-Hole Pairs

First we investigate the properties of $Q(q)$. Using the particle Green's function (6.2.31), we obtain

$$\begin{aligned}
Q(\boldsymbol{q}, \mathrm{i}\omega_n) &= \mp\frac{1}{\beta\hbar^2 N}\sum_{\boldsymbol{p},\mathrm{i}\omega_n}\left\{\frac{1}{\mathrm{i}\omega_n - \epsilon_{0\boldsymbol{p}}}\frac{1}{\mathrm{i}\omega_n - \mathrm{i}\nu_n - \epsilon_{1\boldsymbol{p}-\boldsymbol{q}}}\right.\\
&\quad\left.+\frac{1}{\mathrm{i}\omega_n - \epsilon_{1\boldsymbol{p}}}\frac{1}{\mathrm{i}\omega_n - \mathrm{i}\nu_n - \epsilon_{0\boldsymbol{p}-\boldsymbol{q}}}\right\}\\
&= \pm\frac{1}{\beta\hbar^2 N}\sum_{\boldsymbol{p},\mathrm{i}\omega_n}\left\{\frac{1}{\mathrm{i}\nu_n + (\epsilon_{1\boldsymbol{p}-\boldsymbol{q}} - \epsilon_{\boldsymbol{p}})}\left[\frac{1}{\mathrm{i}\omega_n - \epsilon_{\boldsymbol{p}}} - \frac{1}{\mathrm{i}\omega_n - \mathrm{i}\nu_n - \epsilon_{1\boldsymbol{p}-\boldsymbol{q}}}\right]\right.\\
&\quad\left.+\frac{1}{\mathrm{i}\nu_n - (\epsilon_{1\boldsymbol{p}-\boldsymbol{q}} - \epsilon_{\boldsymbol{p}})}\left[\frac{1}{\mathrm{i}\omega_n - \epsilon_{1\boldsymbol{p}-\boldsymbol{q}}} - \frac{1}{\mathrm{i}\omega_n - \mathrm{i}\nu_n - \epsilon_{\boldsymbol{p}}}\right]\right\}\\
&= \frac{1}{N\hbar}\sum_{\boldsymbol{p}}\{f(\epsilon_{0\boldsymbol{p}}) - f(\epsilon_{1\boldsymbol{p}-\boldsymbol{q}})\}\left\{\frac{1}{\mathrm{i}\nu_n - (\epsilon_{1\boldsymbol{p}-\boldsymbol{q}} - \epsilon_{0\boldsymbol{p}})}\right.\\
&\quad\left.-\frac{1}{\mathrm{i}\nu_n + (\epsilon_1\boldsymbol{p} - \boldsymbol{q} - \epsilon_{0\boldsymbol{p}})}\right\},
\end{aligned} \tag{6.4.2}$$

where we have considered the spherical symmetry of $\epsilon_{\mu\boldsymbol{p}}$ and $\epsilon_{i\boldsymbol{p}} = \epsilon_{1\boldsymbol{p}}$. From the first to the second line in the second term: first we transform variables $\boldsymbol{p} \to \boldsymbol{p}+\boldsymbol{q}$, next use $\epsilon_{\boldsymbol{p}+\boldsymbol{q}} = \epsilon_{-\boldsymbol{p}-\boldsymbol{q}}$, and then transform valiables $\boldsymbol{p} \to -\boldsymbol{p}$. From the second to the third line we have considered

$$\mp\frac{1}{\beta\hbar}\sum_{\mathrm{i}\omega_n}\frac{1}{\mathrm{i}\omega_n - \epsilon_{\mu\boldsymbol{p}}} = f(\epsilon_{\mu\boldsymbol{p}}),$$

where f is the distribution function for bosons or fermions

Analytical continuation $\mathrm{i}\nu_n \to q_0 + \mathrm{i}\delta$ leads to

$$Q(\boldsymbol{q}, q_0) \equiv \frac{1}{N\hbar}\sum_{\boldsymbol{p}} \{f(\epsilon_{0\boldsymbol{p}}) - f(\epsilon_{1\boldsymbol{p}-\boldsymbol{q}})\} \times \left\{\frac{1}{q_0 + \mathrm{i}\delta - (\epsilon_{1\boldsymbol{p}-\boldsymbol{q}} - \epsilon_{0\boldsymbol{p}})} - \frac{1}{q_0 + \mathrm{i}\delta + (\epsilon_{1\boldsymbol{p}-\boldsymbol{q}} - \epsilon_{0\boldsymbol{p}})}\right\}. \quad (6.4.3)$$

We separate $Q(\boldsymbol{q}, q_0)$ into the real and imaginary parts:

$$Q(\boldsymbol{q}, q_0) = Q'(\boldsymbol{q}, q_0) + \mathrm{i}Q''(\boldsymbol{q}, q_0). \quad (6.4.4)$$

Phonon frequencies are determined by putting

$$q_0 \equiv \omega_{\lambda\boldsymbol{q}} - \mathrm{i}\gamma_{\lambda\boldsymbol{q}} \quad (6.4.5)$$

in Eq. (6.4.1) with Eq. (6.4.4), where $\omega_{\lambda\boldsymbol{q}}$ is the phonon frequency and $\gamma_{\lambda\boldsymbol{q}}$ is the width of the phonon which is proportional to the inverse lifetime $\tau_{\lambda\boldsymbol{q}}$; $\gamma_{\lambda\boldsymbol{q}} \sim \hbar/\tau_{\lambda\boldsymbol{q}}$. It should be noted that phonon picture is valid under the condition $\omega_{\lambda\boldsymbol{q}} > \gamma_{\lambda\boldsymbol{q}}$; this condition guarantees the existence of the phonons during the life time. Expanding $Q(\boldsymbol{q}, \omega_{\lambda\boldsymbol{q}} - i\gamma_{\lambda\boldsymbol{q}})$ with respect to $\frac{\gamma_{\lambda\boldsymbol{q}}}{\omega_{\lambda\boldsymbol{q}}}$, we obtain up to the order of $\frac{\gamma_{\lambda\boldsymbol{q}}}{\omega_{\lambda\boldsymbol{q}}}$:

$$1 - Q'(\boldsymbol{q}, \omega_{\lambda\boldsymbol{q}})\langle V_\lambda^{od}(\boldsymbol{q})\rangle_c = 0, \quad (6.4.6)$$

$$\gamma_{\lambda\boldsymbol{q}} = \left.\frac{Q''(\boldsymbol{q}, q_0)}{\frac{\partial Q'(\boldsymbol{q}, q_0)}{\partial q_0}}\right|_{q_0=\omega_{\lambda\boldsymbol{q}}}. \quad (6.4.7)$$

Eqs. (6.4.6) and (6.4.7), determine phonon dispersion curves and their width, respectively.

6.4.3 Energy Dispersion Curves of Particles

In order to calculate Eqs. (6.4.6) and (6.4.7), first we investigate energy dispersions of particles:

$$\epsilon_{\mu\boldsymbol{p}} = \tilde{\omega}_\mu + J_\mu(\boldsymbol{p}), \quad (6.2.15)$$

$$J_\mu(\boldsymbol{p}) \equiv \frac{1}{N}\int d^3R g(R) e^{\mathrm{i}\boldsymbol{p}\cdot\boldsymbol{R}} J_\mu(\boldsymbol{R}), \quad (6.2.16)$$

If we make an approximation; $g(R) \cong \rho\delta(R - a)$ and denote the polar coordinates

$$\boldsymbol{p} = p(\sin\theta\cos\varphi, \sin\theta\sin\varphi, \cos\theta),$$
$$\boldsymbol{R} = R(\sin\theta_1\cos\varphi_1, \sin\theta_1\sin\varphi_1, \cos\theta_1),$$

and the angle between $\boldsymbol{p}$ and $\boldsymbol{R}$ as Θ, Eq. (6.2.16) leads to

$$\begin{aligned} J_\mu(\boldsymbol{p}) &= \frac{1}{N}\int R^2 dR \sin\theta_1 d\theta_1 d\varphi_1 e^{\mathrm{i}pR\cos\Theta} g(R) J_\mu(\boldsymbol{R}) \\ &= \frac{4\pi a^2\rho}{N}\int \sin\theta_1 d\theta_1 d\varphi_1 e^{\mathrm{i}\alpha\cos\Theta} J_\mu(\boldsymbol{R}) \\ &= \frac{4\pi a^2\rho}{N}\int \sin\theta_1 d\theta_1 d\varphi_e^{\mathrm{i}(x\sin\theta_1\cos\varphi_1 + y\sin\theta_1\sin\varphi_1 + z\cos\theta_1)} J_\mu(\boldsymbol{R}), \end{aligned} \tag{6.4.8}$$

where

$$\alpha = pa, \qquad (x, y, z) = \alpha(\sin\theta\cos\varphi, \sin\theta\sin\varphi, \cos\theta). \tag{6.4.9}$$

In the hopping matrix:

$$\hbar J_{m\mu n\nu} = \int d^3x \tilde{w}_{m\mu}(\boldsymbol{x} - \boldsymbol{R}_m)\Phi_{\boldsymbol{R}_n}(\boldsymbol{x})\tilde{w}_{n\nu}(\boldsymbol{x} - \boldsymbol{R}_n), \tag{6.4.10}$$

if we take the first excited state in the z-direction, and denote

$$J_0(R) = \frac{1}{\hbar}\int d^3x \tilde{w}_0(\boldsymbol{x})\Phi_{\boldsymbol{R}}(\boldsymbol{x})\tilde{w}_0(\boldsymbol{x} - \boldsymbol{R})|_{R=a} \equiv J_0(a), \tag{6.4.11}$$

$$\begin{aligned} J_z(\boldsymbol{R}) &= \frac{1}{\hbar}\int d^3x \tilde{w}_z(\boldsymbol{x})\Phi_{\boldsymbol{R}}(\boldsymbol{x})\tilde{w}_z(\boldsymbol{x} - \boldsymbol{R})|_{R=a} \\ &\equiv J_\sigma(a)\cos^2\theta_1 + J_\pi(a)\sin^2\theta_1, \end{aligned} \tag{6.4.12}$$

where we have defined

$$J_\sigma(a) = \frac{1}{\hbar}\int d^3x \tilde{w}_z(\boldsymbol{x})\Phi_{\boldsymbol{R}}(\boldsymbol{x})\tilde{w}_z(\boldsymbol{x} - \boldsymbol{R})|_{\boldsymbol{R}=a\boldsymbol{e}_z}, \tag{6.4.13}$$

$$J_\pi(a) = \frac{1}{\hbar}\int d^3x \tilde{w}_x(\boldsymbol{x})\Phi_{\boldsymbol{R}}(\boldsymbol{x})\tilde{w}_x(\boldsymbol{x} - \boldsymbol{R})|_{\boldsymbol{R}=a\boldsymbol{e}_z}, \tag{6.4.14}$$

and $\boldsymbol{e}_z$ is a unit vector in the z-direction as illustrated in Figure 6.2.

Using the properties of the spherical Bessel function:

$$\begin{aligned} j_0\left(\sqrt{x^2+y^2+z^2}\right) &= \frac{1}{4\pi}\int_0^{2\pi} d\varphi_1 \\ &\quad \times \int_0^{\pi} d\theta_1 \sin\theta_1 e^{\mathrm{i}\alpha(x\sin\theta_1\cos\varphi_1 + y\sin\theta_1\sin\varphi_1 + z\cos\theta_1)} \\ &= j_0(\alpha) = \frac{\sin\alpha}{\alpha}, \end{aligned} \tag{6.4.15}$$

$$\begin{aligned} \frac{d^n}{d(\mathrm{i}z)^n} j_0(\alpha) &= \frac{1}{4\pi}\int_0^{2\pi} d\varphi_1 \int_0^{\pi} d\theta_1 \sin\theta_1 \cos^n\theta_1 \\ &\quad \times e^{\mathrm{i}\alpha(x\sin\theta_1\cos\varphi_1 + y\sin\theta_1\sin\varphi_1 + z\cos\theta_1)}, \end{aligned} \tag{6.4.16}$$

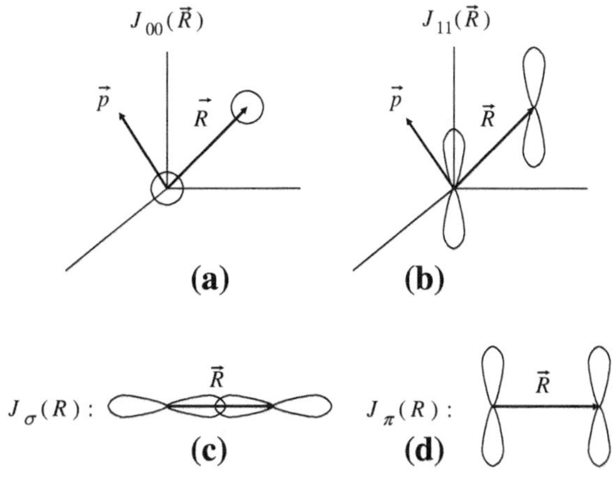

Figure 6.2 The configuration of the vectors $\boldsymbol{p}$ and $\boldsymbol{R}$ and the hopping amplitudes $J_{00}(\boldsymbol{p})$, $J_{11}(\boldsymbol{p})$ are schematically shown in (a) and (b), respectively. The σ, π-couplings $J_\sigma(R)$, $J_\pi(R)$ are also schematically shown in (c) and (d), respectively.

$$\frac{d}{dz} j_0(\alpha) = j_0'(\alpha)\frac{d\sqrt{x^2+y^2+z^2}}{dz}$$
$$= j_0'(\alpha)\frac{z}{\sqrt{x^2+y^2+z^2)}} = j_0'(\alpha)\cos\theta$$
$$\frac{d^2}{dz^2} j_0(\alpha) = j_0{}''\cos^2\theta + j_0{}'(\alpha)\sin^2\theta, \tag{6.4.17}$$

we obtain

$$J_0(\boldsymbol{p}) \cong \bar{J}_0 j_0(\alpha), \qquad \bar{J}_0 \cong \frac{4\pi a^2}{N}\rho J_0(a), \tag{6.4.18}$$
$$J_z(\boldsymbol{p}) \cong -(\bar{J}_\sigma - \bar{J}_\pi)\{j_0{}'(\alpha)\cos^2\theta + j_0{}'(\alpha)\sin^2\theta\} + \bar{J}_\pi j_0(\alpha),$$
$$\bar{J}_{\sigma,\pi} = \frac{(4\pi a)^2\rho}{N} J_{\sigma,\pi}(a), \tag{6.4.19}$$

where a is the mean particle distance and the primes on the 0th Bessel function j_0 means the derivative. $J_\sigma = J_{zz}$ and $J_\pi = J_{xx}$. $\bar{J}_0 < 0$ and $\bar{J}_{\sigma,\pi} > 0$. $\epsilon_{0\boldsymbol{p}}$ is isotropic, but $\epsilon_{z\boldsymbol{p}}$ depends on the angle θ between the z-axis and $\boldsymbol{p}$. The directional dependence plays a crucial role in constructing the structure of the crystal. The term $\frac{\sin\alpha}{\alpha}$ behaves as a decreasingly oscillatory function. It has the maximum at $\alpha = 0$ and the minimum near $\alpha = \frac{3\pi}{2a}$. In the periodic system, $\epsilon_{\mu\boldsymbol{p}}$ behaves as a periodic function with a reciprocal periodic distance $\frac{2\pi}{a}$. The extension of the minimum point to $\frac{3\pi}{2a}$ from $\frac{\pi}{a}$ are due to the superposition of all directional waves. We show particle energy dispersion curves $\epsilon_{\mu\boldsymbol{p}}$ in Figure 6.3, where the upper band, $\epsilon_{z\boldsymbol{p}}$, is averaged over the angle θ, only σ-coupling is considered, and we take $\bar{J}_0 = -\frac{1}{5}$ and $\bar{J}_\sigma = \frac{2}{5}$. There is a band gap between ϵ_i and ϵ_0 at $\alpha \cong \frac{3\pi}{2a}$. We denote the minimum point as $\widetilde{K}$.

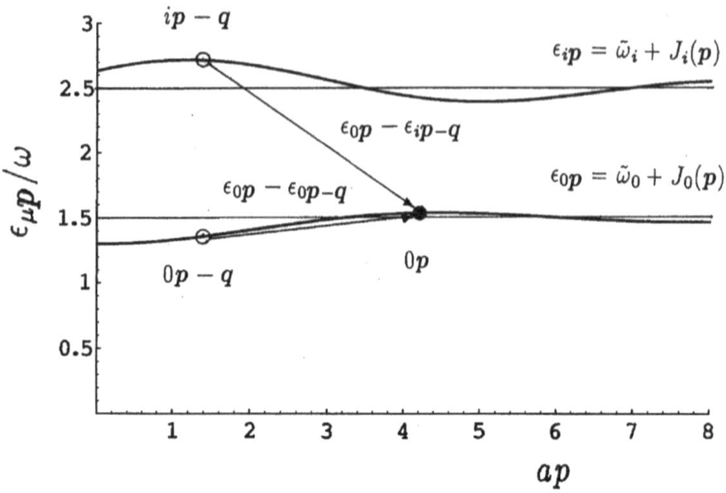

Figure 6.3 Energy dispersion curves normalized by the configurationally averaged frequency ω as a function of ap. Here we put $g(R) = \rho\delta(R - a)$. $J_0(\boldsymbol{p})$ is isotropic, while $J_i(\boldsymbol{p})$ depends on the direction, so we have shown the averaged value over the directions. The minimum gap is near a reciprocal inter-particle distance. An intra-band elementary excitation, $\epsilon_{0\boldsymbol{p}} - \epsilon_{0\boldsymbol{p}-\boldsymbol{q}}$, and an inter-band elementary excitation, $\epsilon_{0\boldsymbol{p}} - \epsilon_{i\boldsymbol{p}-\boldsymbol{q}}$ are also shown.

Using the energy dispersion curves of particles, we can calculate the term $Q(\boldsymbol{q}, q_0)$ in Eq. (6.4.3) so that we can calculate phonon dipersion curves from Eq. (6.4.6) and the phonon widths from Eq. (6.4.7). The phonon dispersion curves are derived from the real part $Q'(\boldsymbol{q}, q_0)$, while the phonon width from the imaginary part $Q''(\boldsymbol{q}, q_0)$, which originates from the phonon frequencies $q_0 = \omega_{\lambda\boldsymbol{q}}$ merging into the inter-band particle-hole pair excitations $\epsilon_{1\boldsymbol{p}-\boldsymbol{q}} - \epsilon_{0\boldsymbol{p}}$;

$$\Im\left\{\frac{1}{q_0 + i\delta - (\epsilon_{1\boldsymbol{p}-\boldsymbol{q}} - \epsilon_{0\boldsymbol{p}})} - \frac{1}{q_0 + i\delta + (\epsilon_{1\boldsymbol{p}-\boldsymbol{q}} - \epsilon_{0\boldsymbol{p}})}\right\}$$
$$= -\pi\left\{\delta(q_0 - (\epsilon_{1\boldsymbol{p}-\boldsymbol{q}} - \epsilon_{0\boldsymbol{p}})) + \delta(q_0 + (\epsilon_{1\boldsymbol{p}-\boldsymbol{q}} - \epsilon_{0\boldsymbol{p}}))\right\}.$$

The bandwidths of the particle energy yield the phonon widths, which smear the phonon dispersion curves.

Now we are ready to solve Eqs. (6.4.6) and (6.4.7). In solving the equations, we investigate two things:

1. We solve Eq. (6.4.6) for no width, $\gamma_{\lambda\boldsymbol{q}} = 0$ to obtain the pure phonon dispersion curves.
2. We solve Eq. (6.4.7) to obtain the width $\gamma_{\lambda\boldsymbol{q}}$.

6.4.4 Phonon Dispersion Curves

First we investigate phonon dispersion curves of Eq. (6.4.6). Since the smearing of the phonon dispersion curves comes from the bandwidth of particle energy, we put

$J_\mu(\boldsymbol{p}) = 0$:

$$\epsilon_{1\boldsymbol{p}-\boldsymbol{q}} - \epsilon_{0\boldsymbol{p}} = \omega. \tag{6.4.20}$$

Under the low temperature regions $\beta\hbar\omega < 1$, we can put

$$f(\epsilon_{0p}) \cong 1, \qquad f(\epsilon_{1\boldsymbol{p}}) \cong 0. \tag{6.4.21}$$

Thus we obtain

$$Q'(q) = Q(q) = \frac{2\omega/\hbar}{q_0^2 - \omega^2}, \qquad Q(0) = -\frac{2}{\hbar\omega}. \tag{6.4.22}$$

Combining Eq. (6.4.1) with the gap equation

$$1 = Q(0)\langle V_\lambda^{od}(0)\rangle_c, \tag{6.4.23}$$

and using a relation

$$Q(q) - Q(0) = -\frac{q_0^2/\hbar}{2\omega} Q(q)Q(0), \tag{6.4.24}$$

we obtain

$$\begin{aligned} 1 - Q(q)\langle V_\lambda^{od}(\boldsymbol{q})\rangle_c &= Q(0)\langle V_\lambda^{od}(0)\rangle_c - Q(q)\langle V_\lambda^{od}(\boldsymbol{q})\rangle_c \\ &= \{Q(0) - Q(q)\}\langle V_\lambda^{od}(0)\rangle_c + Q(q)\langle V_\lambda^{od}(0) - V_\lambda^{od}(\boldsymbol{q})\rangle_c \\ &= \zeta^2 Q(q)\left\{Mq_0^2 + \zeta^{-2}\langle V_\lambda^{od}(0) - V_\lambda^{od}(\boldsymbol{q})\rangle_c\right\}, \end{aligned} \tag{6.4.25}$$

where $\zeta = \sqrt{\frac{\hbar}{2M\omega}}$.

To calculate the potential function $\langle V_{ij}^{od}(\boldsymbol{q})\rangle_c$, we introduce the σ and π couplings like the chemical bondings:

$$V_\sigma(R) = -\int d^3x d^3y \nabla_{x_3}\tilde{w}_0^2(\boldsymbol{x})V(\boldsymbol{x}-\boldsymbol{y})\nabla_{y_3}\tilde{w}_0^2(\boldsymbol{y} - R\boldsymbol{e}_3), \tag{6.4.26}$$

$$V_\pi(R) = -\int d^3x d^3y \nabla_{x_1}\tilde{w}_0^2(\boldsymbol{x})V(\boldsymbol{x}-\boldsymbol{y})\nabla_{y_1}\tilde{w}_0^2(\boldsymbol{y} - R\boldsymbol{e}_3). \tag{6.4.27}$$

The potential function is essentially written as

$$V_{ij}(\boldsymbol{q}) \equiv \langle V_{ij}^{od}(\boldsymbol{q})\rangle_c = \zeta^2 \int d^3R g(R) e^{-i\boldsymbol{q}\cdot\boldsymbol{R}} V_{ij}(\boldsymbol{R}), \tag{6.4.28}$$

$$V_{ij}(\boldsymbol{R}) \equiv \int d^3x d^3y \nabla_i \tilde{w}_0^2(\boldsymbol{x})V(\boldsymbol{x}-\boldsymbol{y})\nabla_j\tilde{w}_0^2(\boldsymbol{y}-\boldsymbol{R}). \tag{6.4.29}$$

It should be noted that $V_{ij}(\boldsymbol{q})$ is not the Fourier component of $V_{ij}(\boldsymbol{R})$. Hereafter we use $V_{ij}(\boldsymbol{q})$ instead of $\langle V_{ij}^{od}(\boldsymbol{q})\rangle_c$. If we take the direction of $\boldsymbol{q}$ as the z-axis and the unit vector of $\boldsymbol{R}$ as $\boldsymbol{e}_R$;

$$\boldsymbol{q} = q\boldsymbol{e}_z, \qquad \boldsymbol{R} = R\boldsymbol{e}_R = R(\sin\theta\cos\varphi, \sin\theta\sin\varphi, \cos\varphi), \tag{6.4.30}$$

$V_{ij}(\boldsymbol{q})$ is diagonalized as follows: first, we introduce a unit vector $\boldsymbol{p}_i$, on the plane composed of the unit vectors $\boldsymbol{e}_R$ and $\boldsymbol{e}_i$, and perpendicular to the vector $\boldsymbol{R}$.

$$\boldsymbol{p}_z = (-\cos\theta\cos\varphi, -\cos\theta\sin\varphi, \sin\theta), \tag{6.4.31a}$$

$$\boldsymbol{p}_x = \frac{1}{\sqrt{1-\sin^2\theta\cos^2\varphi}}\Big(1-\sin^2\theta\cos^2\varphi, -\sin^2\theta\cos\varphi\sin\varphi, -\sin\theta\cos\theta\cos\varphi\Big), \tag{6.4.31b}$$

$$\boldsymbol{p}_y = \frac{1}{\sqrt{1-\sin^2\theta\sin^2\varphi}}\Big(-\sin^2\theta\cos\varphi\sin\varphi, 1-\sin^2\theta\sin^2\varphi, -\sin\theta\cos\theta\sin\varphi\Big), \tag{6.4.31c}$$

then we obtain

$$\begin{aligned} V_{zz}(\boldsymbol{q}) &= \zeta^2 \int R^2 dR g(R) \int_0^{\pi} d\theta \sin\theta \\ &\quad \times \int_0^{2\pi} d\varphi e^{\mathrm{i}\alpha\cos\theta}\Big[V_\sigma(R)(\boldsymbol{e}_R\cdot\boldsymbol{e}_z)^2 + V_\pi(R)(\boldsymbol{p}_z\cdot\boldsymbol{e}_z)^2\Big] \\ &= 2\pi\zeta^2 \int R^2 dR g(R) \int_0^{\pi} d\theta \sin\theta e^{\mathrm{i}\alpha\cos\theta}\Big[V_\sigma(R)\cos^2\theta + V_\pi(R)\sin^2\theta\Big] \\ &= 4\pi\zeta^2 \int R^2 dR g(R) \Big\{V_\sigma(R)\Big(\frac{\sin\alpha}{\alpha} + \frac{2\cos\alpha}{\alpha^2} - \frac{2\sin\alpha}{\alpha^3}\Big) \\ &\quad + 2V_\pi(R)(-\frac{\cos\alpha}{\alpha^2} + \frac{\sin\alpha}{\alpha^3})\Big\}, \end{aligned} \tag{6.4.32}$$

$$\begin{aligned} V_{xx}(\boldsymbol{q}) &= V_{yy}(\boldsymbol{q}) \\ &= \zeta^2 \int R^2 dR g(R) \int_0^{\pi} d\theta \sin\theta \\ &\quad \times \int_0^{2\pi} d\varphi e^{\mathrm{i}\alpha\cos\theta}\Big[V_\sigma(R)(\boldsymbol{e}_R\cdot\boldsymbol{e}_x)^2 + V_\pi(R)(\boldsymbol{p}_x\cdot\boldsymbol{e}_x)^2\Big] \\ &= \zeta^2 \int R^2 dR g(R) \int_0^{\pi} d\theta \sin\theta \\ &\quad \times \int_0^{2\pi} d\varphi e^{\mathrm{i}\alpha\cos\theta}\Big[V_\sigma(R)\sin^2\theta\cos^2\varphi + V_\pi(R)\Big(1-\sin^2\theta\sin^2\varphi\Big)\Big] \\ &= \pi\zeta^2 \int R^2 dR g(R) \\ &\quad \times \int_0^{\pi} d\theta \sin\theta e^{\mathrm{i}\alpha\cos\theta}\Big[V_\sigma(R)\sin^2\theta + V_\pi(R)\Big(2-\sin^2\theta\Big)\Big] \\ &= 4\pi\zeta^2 \int R^2 dR g(R) \Big\{V_\sigma\Big(-\frac{\cos\alpha}{\alpha^2} + \frac{\sin\alpha}{\alpha^3}\Big) \\ &\quad + V_\pi\Big(\frac{\sin\alpha}{\alpha} + \frac{\cos\alpha}{\alpha^2} - \frac{\sin\alpha}{\alpha^3}\Big)\Big\}, \end{aligned} \tag{6.4.33}$$

$$
\begin{aligned}
V_{xy}(\boldsymbol{q}) &= \zeta^2 \int R^2 dR g(R) \int_0^{\pi} d\theta \sin\theta \\
&\quad \times \int_0^{2\pi} d\varphi e^{\mathrm{i}\alpha\cos\theta} \left[V_\sigma(R)(\boldsymbol{e}_R \cdot \boldsymbol{e}_x)(\boldsymbol{e}_R \cdot \boldsymbol{e}_y) \right. \\
&\qquad\qquad \left. + V_\pi(R)(\boldsymbol{p}_x \cdot \boldsymbol{e}_x)(\boldsymbol{p}_y \cdot \boldsymbol{e}_y)(\boldsymbol{p}_x \cdot \boldsymbol{p}_y) \right] \\
&= \zeta^2 \int R^2 dR g(R) \int_0^{\pi} d\theta \sin\theta \\
&\quad \times \int_0^{2\pi} d\varphi e^{\mathrm{i}\alpha\cos\theta} \left[V_\sigma(R) - V_\pi(R) \right] \sin^2\theta \cos\varphi \sin\varphi \\
&= 0,
\end{aligned} \tag{6.4.34}
$$

$$
\begin{aligned}
V_{yz}(\boldsymbol{q}) &= V_{zx}(\boldsymbol{q}) \\
&= \zeta^2 \int R^2 dR g(R) \int_0^{\pi} d\theta \sin\theta \\
&\quad \times \int_0^{2\pi} d\varphi e^{\mathrm{i}\alpha\cos\theta} \left[V_\sigma(R)(\boldsymbol{e}_R \cdot \boldsymbol{e}_y)(\boldsymbol{e}_R \cdot \boldsymbol{e}_z) \right. \\
&\qquad\qquad \left. + V_\pi(R)(\boldsymbol{p}_y \cdot \boldsymbol{e}_y)(\boldsymbol{p}_z \cdot \boldsymbol{e}_z)(\boldsymbol{p}_y \cdot \boldsymbol{p}_z) \right] \\
&= \zeta^2 \int R^2 dR g(R) \int_0^{\pi} d\theta \sin\theta \\
&\quad \times \int_0^{2\pi} d\varphi e^{\mathrm{i}\alpha\cos\theta} \left[V_\sigma(R) - V_\pi(R) \right] \sin\theta \cos\theta \sin\varphi \\
&= 0,
\end{aligned} \tag{6.4.35}
$$

where we have put $\alpha = qR$.

Thus we denote

$$
V_{zz}(\boldsymbol{q}) = V_l(\boldsymbol{q}), \qquad V_{xx}(\boldsymbol{q}) = V_{yy}(\boldsymbol{q}) = V_t(\boldsymbol{q}),
$$

which yield a longitudinal $\omega_{l\boldsymbol{q}}$ and two transverse phonons $\omega_{t\boldsymbol{q}}$, respectively.

6.4.4.1 The phonon Green's functions and the phonon dispersion curves

Eq. (6.4.25) leads to

$$
\begin{aligned}
1 - Q(q)V_\lambda(\boldsymbol{q}) &= \zeta^2 Q(q)\{Mq_0^2 + \zeta^{-2}(V_\lambda(0) - V_\lambda(\boldsymbol{q}))\} \\
&= M\zeta^2 Q(q)\{q_0^2 + \frac{\zeta^{-2}}{M}(V_\lambda(0) - V_\lambda(\boldsymbol{q}))\}.
\end{aligned}
$$

Then substituting the above equation into $\{1 - Q(q)V_\lambda(\boldsymbol{q})\}D_\lambda(q) = Q(q)$ yields phonon Green's function and phonon frequencies:

$$
D_\lambda(q) = \frac{\zeta^{-2}}{M} \frac{1}{q_0^2 - \omega_{\lambda q}^2}, \tag{6.4.36}
$$

$$\omega_{\lambda q}^2 = -\frac{\zeta^{-2}}{M}[V_\lambda(0) - V_\lambda(\boldsymbol{q})],$$
$$= -\frac{1}{M}\int d^3Rg(R)\{1 - e^{-i\boldsymbol{q}\cdot\boldsymbol{R}}\}V_\lambda(\boldsymbol{R}), \tag{6.4.37}$$

In order to obtain explicit phonon dispersion curves, we first show $V_{l,t\boldsymbol{q}}$ in the long wavelength regime in Eqs. (6.4.32) and (6.4.33):

$$V_l(\boldsymbol{q}) \cong 4\pi\zeta^2 \int R^2 dRg(R)\left\{V_\sigma\left(\frac{1}{3} - \frac{\alpha^2}{10} + \cdots\right) + V_\pi\left(\frac{2}{3} - \frac{\alpha^2}{15} + \cdots\right)\right\}, \tag{6.4.38}$$

$$V_t(\boldsymbol{q}) \cong 4\pi\zeta^2 \int R^2 dRg(R)\left\{V_\sigma\left(\frac{1}{3} - \frac{\alpha^2}{30} + \cdots\right) + V_\pi\left(\frac{2}{3} - \frac{2\alpha^2}{15} + \cdots\right)\right\}. \tag{6.4.39}$$

Thus we obtain explicit phonon frequencies:

$$\omega_{l\boldsymbol{q}}^2 = -\frac{4\pi}{M}\int R^2 dRg(R)\left\{V_\sigma(R)\left[\frac{1}{3} - \left(\frac{\sin\alpha}{\alpha} + \frac{2\cos\alpha}{\alpha^2} - \frac{2\sin\alpha}{\alpha^3}\right)\right]\right.$$
$$\left. + V_\pi(R)\left[\frac{2}{3} - 2\left(-\frac{\cos\alpha}{\alpha^2} + \frac{\sin\alpha}{\alpha^3}\right)\right]\right\}, \tag{6.4.40}$$

$$\omega_{t\boldsymbol{q}}^2 = -\frac{4\pi}{M}\int R^2 dRg(R)\left\{V_\sigma\left[\frac{1}{3} - \left(-\frac{\cos\alpha}{\alpha^2} + \frac{\sin\alpha}{\alpha^3}\right)\right]\right.$$
$$\left. + V_\pi\left[\frac{2}{3} - \left(\frac{\sin\alpha}{\alpha} + \frac{\cos\alpha}{\alpha^2} - \frac{\sin\alpha}{\alpha^3}\right)\right]\right\}. \tag{6.4.41}$$

From the gap Eq. (6.4.23) with $Q(0) = -\frac{2}{\hbar\omega}$, we obtain the harmonic frequency:

$$\omega^2 = -\frac{\zeta^{-2}}{M}\int d^3Rg(R)V_\lambda(\boldsymbol{R})$$
$$= \frac{4\pi}{M}\int R^2 dRg(R)\left\{\frac{V_\sigma(R)}{3} + \frac{2V_\pi(R)}{3}\right\}. \tag{6.4.42}$$

$M\omega^2$ corresponds to the spring constant in the classical theory, so ω expresses the elasticity of the system.

Now we illustrate the phonon dispersion curves. To do so, we make the following approximations:

1. Since the strength of the coupling V_σ is larger than that of V_π : $|V_\sigma| \gg |V_\pi|$, we put $V_\pi = 0$.
2. We put $g(R) = \rho\delta(R - a)$.

Thus we obtain

$$\omega_{l\boldsymbol{q}} = \frac{4\pi a^2\rho}{M}V_\sigma(a)\left[\frac{1}{3} - \left(\frac{\sin\alpha}{\alpha} + \frac{2\cos\alpha}{\alpha^2} - \frac{2\sin\alpha}{\alpha^3}\right)\right]$$
$$= \omega^2\left[1 - 3\left(\frac{\sin\alpha}{\alpha} + \frac{2\cos\alpha}{\alpha^2} - \frac{2\sin\alpha}{\alpha^3}\right)\right], \tag{6.4.43}$$

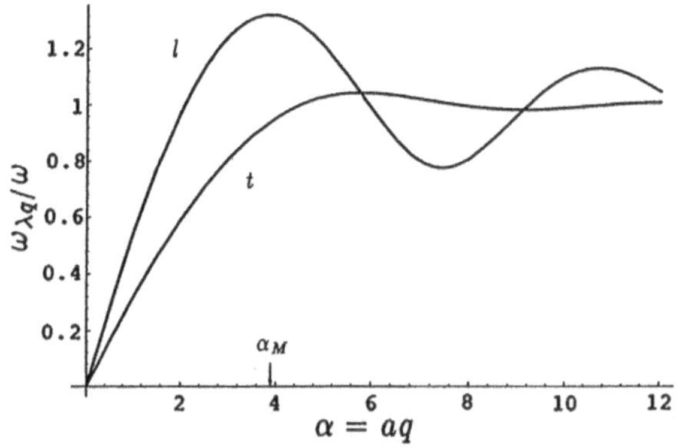

Figure 6.4 A longitudinal and two transverse phonon dispersion curves $\omega_{\lambda q}$ normalized by the configuratonal averaged harmonic frequency ω are illustrated as functions of aq. α_M is the Debye cut off. In the short wavelength limit, $\lim_{q\longrightarrow\infty}\omega_{\lambda q} = \omega$: this frequency corresponds to boson peak.

$$\begin{aligned}\omega_{tq} &= \frac{4\pi a^2\rho}{M}V_\sigma(a)\left[\frac{1}{3} - (-\frac{\cos\alpha}{\alpha^2} + \frac{\sin\alpha}{\alpha^3})\right]\\ &= \omega^2\left[1 - 3(-\frac{\cos\alpha}{\alpha^2} + \frac{\sin\alpha}{\alpha^3})\right],\end{aligned} \tag{6.4.44}$$

where

$$\omega^2 = \frac{4\pi a^2\rho}{M}\frac{1}{3}V_\sigma(a). \tag{6.4.45}$$

In Figure 6.4, we show the phonon dispersion curves reduced by ω.

6.4.4.2 *The properties of phonon dispersion curves*

Next we investigate properties of the phonon dispersion curves.

1. $aq \ll 1$: in the long wavelength regime.
   From Eqs. (6.4.38) and (6.4.39), we obtain

$$\omega_{lq}^2 = \frac{4\pi}{M}\int R^2 dR g(R)(\frac{V_\sigma}{10} + \frac{V_\pi}{15})\alpha^2, \tag{6.4.46}$$

$$\omega_{tq}^2 = \frac{4\pi}{M}\int R^2 dR g(R)(\frac{V_\sigma}{30} + \frac{2V_\pi}{15})\alpha^2. \tag{6.4.47}$$

   Phonon velocities $c_{l,t}$ are given by

$$\omega_{l,tq} = c_{l,t}q. \tag{6.4.48}$$

   If we put $g(R) = \rho\delta(R-a)$, we obtain the phonon velocities and the harmonic frequency:

$$c_l^2 = \frac{4\pi a^4\rho}{M}(\frac{V_\sigma}{10} + \frac{V_\pi}{15}), \tag{6.4.49}$$

$$c_t^2 = \frac{4\pi a^4}{M}(\frac{V_\sigma}{30} + \frac{2V_\pi}{15}), \tag{6.4.50}$$

$$\omega^2 = \frac{4\pi a^2}{M}(\frac{V_\sigma}{3} + \frac{2V_\pi}{3}). \tag{6.4.51}$$

2. $\widetilde{K} \cong \frac{3\pi}{2a}$: at a reciprocal particle distance. There is a roton type minimum for the longitudinal mode, which is a vestige of a periodic nature of the crystal. This roton type minimum was first derived by Takeno et al. in the self-consistent phonon theory [22,23,18,8].
3. $aq \gg 1$: in the short wavelength regime. We obtain

$$\lim_{q\to\infty} \omega_{\lambda q} = \omega. \tag{6.4.52}$$

The harmonic frequency ω corresponds to the boson peak [24,25,18,8]. These flat phonon dispersion curves mean the localization of phonons and makes a peak of the density states. It should be noted that the density of states is proportional to the inverse of $\nabla\omega_{\lambda \boldsymbol{q}}$.

6.4.5 *The Width of Phonons*

In calculating the width of phonons (6.4.7), energy dispersion curves $\epsilon_{\mu \boldsymbol{p}}$ play an essential role. It is hard task to calculate $Q(q)$ in Eq. (6.4.3) with complex dispersion curves of $\epsilon_{\mu \boldsymbol{p}}$. Therefore, we make the following approximation:

1. For simplicity, we assume that the energy dispersion curves $\epsilon_{\mu \boldsymbol{p}}$ for the lower and the upper bands behave in the same fashion but the hopping amplitudes J_μ:

$$\epsilon_{\mu \boldsymbol{p}} = \tilde{\omega}_\mu + \bar{J}_\mu j_0(\alpha), \qquad j_0(\alpha) = \frac{\sin\alpha}{\alpha}, \qquad \alpha = ap. \tag{6.4.53}$$

2. The main contribution of the integrand to $Q(q)$ comes from the minimum region of the particle-hole pair excitation; $\epsilon_{i\,\boldsymbol{p}-\boldsymbol{q}} - \epsilon_{0\boldsymbol{p}}$, where the density of states is also very large compared to that near the original point. So we replace $\epsilon_{\mu \boldsymbol{p}}$ by a parabolic spectrum with the maximum point at $\widetilde{K}$. The larger the momentum p becomes, the larger the density of states becomes, but the parabolic spectrum cancel out the larger part of the density. Here we take the origin of the p-space at $\widetilde{K} \cong \frac{3\pi}{2a}$;

$$\epsilon_{\mu k} \equiv \tilde{\omega}_\mu - d_\mu + \alpha_\mu k^2, \qquad p = \widetilde{K} + k. \tag{6.4.54}$$

Thus for $\widetilde{K} = \frac{3\pi}{2a}$, we obtain

$$\epsilon_{\mu\widetilde{K}} = \tilde{\omega}_\mu - d_\mu = \tilde{\omega}_\mu - \frac{2}{3\pi}\bar{J}_\mu; \qquad d_\mu = \frac{2}{3\pi}\bar{J}_\mu. \tag{6.4.55}$$

Since $j_0(\alpha = \pi) = 0$, we determine the coefficient of k^2 such that $\epsilon_{\mu p=\frac{\pi}{a}} = \omega_\mu = \epsilon_{\mu k=\frac{\pi}{2a}}$;

$$\epsilon_{\mu k=\frac{\pi}{2a}} = \tilde{\omega}_\mu - d_\mu + \alpha_\mu(\frac{\pi}{2a})^2 = \tilde{\omega}_\mu; \qquad d_\mu = \alpha_\mu(\frac{\pi}{2a})^2. \tag{6.4.56}$$

From Eqs. (6.4.55) and (6.4.56), we obtain

$$\epsilon_{\mu k} \cong \tilde{\omega}_\mu - \frac{2}{3\pi}\bar{J}_\mu + \frac{8a^2}{3\pi^3}\bar{J}_\mu k^2; \qquad \alpha_\mu = \frac{8a^2}{3\pi^3}\bar{J}_\mu, \tag{6.4.57}$$

In calculating the bubbles of the inter-band particle-hole pairs $Q''(q)$, we use these particle dispersion curves.

3. Now we investigate the term $\epsilon_{i\boldsymbol{p}-\boldsymbol{q}} - \epsilon_{0\boldsymbol{p}}$ in $Q(q)$. In Eq. (6.4.54), we have transformed $\boldsymbol{p} = \tilde{\boldsymbol{K}} + \boldsymbol{k}$, where we have considered $\boldsymbol{p}, \boldsymbol{k}, \tilde{\boldsymbol{K}}$ are parallel. Therefore, when we consider the behavior of $\boldsymbol{q}$ in the term $\epsilon_{i\boldsymbol{k}-\boldsymbol{q}} - \epsilon_{0\boldsymbol{k}}$, we must consider $\epsilon_{i\tilde{\boldsymbol{K}}+\boldsymbol{k}-\boldsymbol{q}} - \epsilon_{0\tilde{\boldsymbol{K}}+\boldsymbol{k}}$. Thus we put

$$\epsilon_{\mu\boldsymbol{k}-\boldsymbol{q}} \cong \epsilon_{k-q\eta}, \qquad \eta = \cos\theta, \tag{6.4.58}$$

where θ is the angle between $\boldsymbol{k}$ and $\boldsymbol{q}$. Here, we assume

$$\bar{J}_i = -\bar{J}_0 \equiv \bar{J}. \tag{6.4.59}$$

Putting $\boldsymbol{k} \to \boldsymbol{k} + \frac{\boldsymbol{q}}{2}$, we obtain the particle-hole excitation:

$$\epsilon_{i\boldsymbol{k}-\frac{\boldsymbol{q}}{2}} - \epsilon_{0\boldsymbol{k}+\frac{\boldsymbol{q}}{2}} = \omega_G + \tilde{\alpha}\left(k^2 + \frac{q^2}{4}\eta^2\right), \tag{6.4.60}$$

$$\omega_G \equiv \omega - \frac{4}{3\pi}\bar{J}, \qquad \tilde{\alpha} \equiv \frac{16a^2}{3\pi^3}\bar{J}, \tag{6.4.61}$$

where ω_G is the band gap of the energy spectrum of particles.

4. We consider the low temperature regions $\beta^{-1} < \hbar\omega_G$ and the narrow bands. so that we put

$$f(\epsilon_{0k}) \cong 1, \qquad f(\epsilon_{ik}) \cong 0. \tag{6.4.62}$$

5. The most dominant contribution of the integrand comes from $p \sim \tilde{K}$. So we replace the density of states that at $\boldsymbol{k} = 0$. The integral region of k is taken to be from 0 to $\tilde{K}$. Thus we obtain

$$\begin{aligned} Q(q) &\cong \frac{1}{N}\sum_{\boldsymbol{k}}\left\{\frac{1}{q_0+i\delta-(\epsilon_{1\boldsymbol{k}-\frac{\boldsymbol{q}}{2}}-\epsilon_{0\boldsymbol{k}+\frac{\boldsymbol{q}}{4}})} - \frac{1}{q_0+i\delta+(\epsilon_{1\boldsymbol{k}-\frac{\boldsymbol{q}}{2}}-\epsilon_{0\boldsymbol{k}+\frac{\boldsymbol{q}}{4}})}\right\} \\ &\cong \frac{1}{N}\frac{V}{(2\pi)^3}2\pi\tilde{K}^2\int_0^{\tilde{K}}dk\int_{-1}^{1}d\eta\times\left\{\frac{1}{q_0+i\delta-\omega_G-\tilde{\alpha}(k^2+\frac{q^2}{4}\eta^2)}\right. \\ &\qquad \left. - \frac{1}{q_0+i\delta+\omega_G+\tilde{\alpha}(k^2+\frac{q^2}{4}\eta^2)}\right\}. \end{aligned} \tag{6.4.63}$$

First, we change variables $\tilde{k} = \sqrt{\tilde{\alpha}}k$, $\tilde{\eta} = \sqrt{\tilde{\alpha}}\frac{q}{2}\eta$. Secondly, we change the variables $(\tilde{k}, \tilde{\eta})$ to the polar axis (r, θ); $\tilde{k} = r\cos\theta$, $\tilde{\eta} = r\sin\theta$ and since the dominant contribution of the integrand comes from $r \sim 0$, we make the following approximation:

$$\begin{aligned}
\int_0^{\sqrt{\widetilde{\alpha}}\widetilde{K}} d\widetilde{k} \int_{\frac{-\sqrt{\widetilde{\alpha}}q}{2}}^{\frac{\sqrt{\widetilde{\alpha}}q}{2}} d\widetilde{\eta}\{\cdots\} &= \int_0^{\sqrt{\langle\rangle}} r\,dr \int_{-\sin^{-1}(\frac{\sqrt{\widetilde{\alpha}}q}{2}/r)}^{\sin^{-1}(\frac{\sqrt{\widetilde{\alpha}}q}{2}/r)} d\theta\{\cdots\} \\
&\cong \int_0^{\sqrt{\langle\rangle}} r\,dr \int_{-\frac{\pi}{2}}^{\frac{\pi}{2}} d\theta\{\cdots\} \\
&\cong \pi \int_0^{\sqrt{\langle\rangle}} r\,dr\{\cdots\}, \qquad \langle\rangle = \widetilde{\alpha}\left(\widetilde{K}^2 + \frac{q^2}{4}\right).
\end{aligned} \tag{6.4.64}$$

Finally putting the variable $t = r^2$, we obtain

$$Q(q) = \frac{1}{N\hbar}\frac{V\widetilde{K}^2}{4\pi\widetilde{\alpha}q}\int_0^{\langle\rangle} dt \left\{\frac{1}{q_0 + \mathrm{i}\delta - \omega_G - t} - \frac{1}{q_0 + \mathrm{i}\delta + \omega_G + t}\right\}. \tag{6.4.65}$$

The real part of $Q(q)$ is given by

$$\begin{aligned}
Q'(q) = -\frac{1}{N\hbar}\frac{V\widetilde{K}^2}{4\pi\widetilde{\alpha}q}\{&\ln|q_0 - \omega_G - \langle\rangle| - \ln|q_0 - \omega_G| \\
&+ \ln|q_0 + \omega_G + \langle\rangle| - \ln|q_0 + \omega_G|\}.
\end{aligned} \tag{6.4.66}$$

Then we obtain

$$\frac{dQ'(q)}{dq_0} = -\frac{1}{N\hbar}\frac{V\widetilde{K}^2}{2\pi\widetilde{\alpha}q}\left[\frac{q_0}{q_0^2 - (\omega_G + \langle\rangle)^2} - \frac{q_0}{q_0^2 - \omega_G^2}\right]. \tag{6.4.67}$$

The imaginary part of $Q(q)$ is given by

$$\begin{aligned}
Q''(q) &= \frac{1}{N\hbar}\frac{V\widetilde{K}^2}{4\pi\widetilde{\alpha}q}\int_0^{\langle\rangle} dt\pi\delta(q_0 - \omega_G - t) \\
&= \frac{1}{N\hbar}\frac{V\widetilde{K}^2}{2\pi\widetilde{\alpha}q} I(q), \\
I(q) &= \begin{cases} 0 & \text{for } 0 < q_0 < \omega_G, \\ \frac{\pi}{2} & \text{for } \omega_G < q_0 < \omega_G + \langle\rangle, \end{cases}
\end{aligned} \tag{6.4.68}$$

where we should note the sign of the imaginary part of $q_0 = \omega_{\lambda \boldsymbol{q}} - \mathrm{i}\delta$ due to $q_0 = \omega_{\lambda \boldsymbol{q}} - \mathrm{i}\gamma_{\lambda \boldsymbol{q}}$.

Thus we obtain the width of phonon dispersion curves [20,21,8]:

$$\gamma_{\lambda \boldsymbol{q}} \cong I(q)\left(\frac{\omega_{\lambda \boldsymbol{q}}}{(\omega_G + \langle\rangle)^2 - \omega_{\lambda \boldsymbol{q}}^2} + \frac{\omega_{\lambda \boldsymbol{q}}}{\omega_{\lambda \boldsymbol{q}}^2 - \omega_G^2}\right)^{-1}. \tag{6.4.69}$$

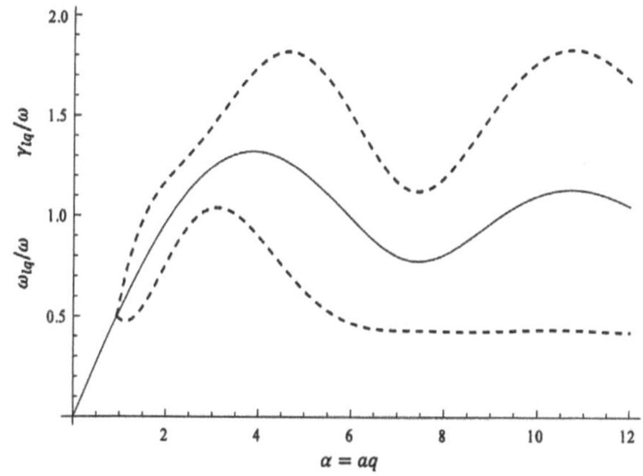

Figure 6.5 A longitudinal phonon dispersion curves ω_{lq} and the width γ_{lq} reduced by the harmonoic frequency ω are schematically illustrated. The thick line shows ω_{lq}/ω. The upper and lower dotted lines show $(\omega_{lq} \pm \frac{1}{2}\gamma_{lq})/\omega$, respectively. So the difference between the upper and lower lines shows γ_{lq}/ω.

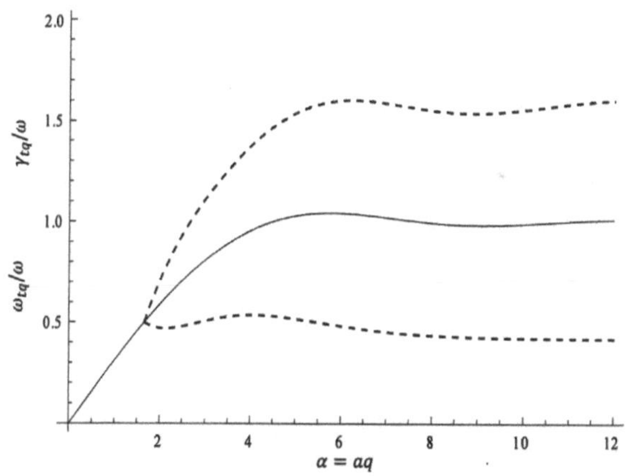

Figure 6.6 Two transverse phonon dispersion curves ω_{tq} and the width γ_{tq} reduced by the harmonoic frequency ω are schematically illustrated. The thick line shows ω_{tq}/ω. The upper and lower dotted lines show $(\omega_{tq} \pm \frac{1}{2}\gamma_{lq})/\omega$, respectively. So the difference between the upper and lower lines shows γ_{tq}/ω.

The width of phonons $\gamma_{\lambda q}$ comes from the delta function in Eq. (6.4.68), which means that a phonon frequency merges into the continuum of the inter-band particle-hole excitations, $\epsilon_{ik-\frac{q}{2}} - \epsilon_{0k+\frac{q}{2}} = \omega_G + \widetilde{\alpha}(k^2 + \frac{q^2}{4}\eta^2)$. The continuum of the inter-band particle-hole pairs originates from the two factors: $\widetilde{\alpha}$ in Eq. (6.4.61) and η in Eq. (6.4.58). The factor $\widetilde{\alpha}$ comes from the hopping magnitudes of particles J_μ and the factor η from angle of $\boldsymbol{k}$ and $\boldsymbol{q}$. J_μ yields the dispersion curves of excitations of inter-band particle-hole pairs, $\widetilde{\alpha}k^2$, while combining J_μ and η yields the width of the continuum of the excitation of inter-band particle-hole pairs, $\widetilde{\alpha}\frac{q^2}{4}\eta$. Therefore, both the hopping of particles and the vector property of $\boldsymbol{q}$ are essential for the width of the phonon dispersion curves, $\gamma_{\lambda \boldsymbol{q}}$. If there is no factor η but J_μ, $\gamma_{\lambda \boldsymbol{q}}$ disappears. Since the lifetime of phonons is proportional to $\gamma_{\lambda \boldsymbol{q}}^{-1}$, the lifetime is intrinsic in the liquid state contrary to the glassy state.

In Figures 6.5 and 6.6, we schematically illustrate the width of a longitudinal and two transverse phonon frequencies reduced by the harmonic frequencies, $\gamma_{l,tq}/\omega$. Here we take $\omega_G = \frac{1}{2}\omega$ so that $\bar{J} = \frac{3\pi}{8}\omega$ and $\widetilde{\alpha} = \frac{2}{\pi^2}\omega a^2$. The figures show the following features [20,21,8]:

1. In the long wavelength regime, $\gamma_{l,tq} = 0$ so that phonon dispersion curves are clear.
2. In the intermediate regime, the width is finite so that phonon picture scarcely holds.
3. In the short wavelength regime, the width become to the phonon frequencies, $\gamma_{l,tq} \sim \omega_{l,tq}$, so that phonon picture becomes destroyed, but as the wave vectors q becomes large, phonon frequencies $\omega_{l,tq}$ tend to the harmonic frequency ω, $\lim_{\alpha=aq\to\infty} \omega_{l,tq} = \omega$, so that phonon width tends to a constant value. Thus we can expect the boson peak.

References

[1] E. Noether, Goett. Nachr. (1918) 235.
[2] J.C. Ward, Phys. Rev. 78 (1950) 182.
[3] Y. Takahashi, Nuovo Chim. 6 (1957) 370.
[4] Y. Takahashi, Phys. Rev. D 15 (1977) 1589.
[5] H. Umezawa, H. Matsumoto, M. Tachiki, Thermo Field Dynamics and Condensed States, North-Holland, Amsterdam, 1982.
[6] T. Matsubara, Prog. Theor. Phys. 14 (1955) 351.
[7] T. Toyoda, Ann. Phys. 173 (1987) 226.
[8] T. Kitamura, Phys. Rep. 383 (2003) 1.
[9] A.A. Abrikosov, L.P. Gorkov, I.E. Dzyloshinski, Methods of Quantum Field Theory in Statistical Physics, Prentice-Hall, NJ, 1963.
[10] A.L. Fetter, J.D. Walecka, Quantum Theory of Many-Particle Systems, McGraw-Hill, New York, 1971).
[11] Y. Nambu, G. Jona-Lasinio, Phys. Rev. 122 (1961) 345.
[12] Y. Nambu, G. Jona-Lasinio, Phys. Rev. 124 (1961) 246.
[13] J. Goldstone, Nuovo Cim. 19 (1961) 154.
[14] J. Goldstone, A. Salam, S. Weinberg, Phys. Rev. 127 (1962) 965.
[15] P.W. Anderson, Basic Notions of Condensed Matter Physics, Benjamin, 1984.
[16] J.P. Hansen, I.R. MacDonald, Theory of Simple Liquids, second ed., Academic Press, London, 1990.
[17] W. Brenig, Zeitschrift für Physik 171 (1963) 60.
[18] T. Kitamura, Physica A 167 (1990) 509.
[19] T. Kitamura, Physica A 167 (1990) 527.
[20] T. Kitamura, Phys. Lett. A 152 (1991) 218.
[21] T. Kitamura, Physica A 223 (1996) 227.
[22] S. Takeno, Prog. Theor. Phys. Suppl. 45 (1970) 137.
[23] S. Takeno, M. Goda, Prog. Theor. Phys. 45 (1971) 331.

[24] B. Frick and D. Richter, Science 267 (1995) 1939.
[25] E. Rössler, V.N. Novikov, A.R. Sokolov, Phase Transition 63 (1997) 201.

7 The Liquid-Glass Transition

The liquid state is a condensed state, but not an ordered state. A liquid seems to have a kind of the long range order, the pair distribution function, as the glass, but the pair distribution function changes with time; the hopping is essential. The hopping causes the density fluctuations, which yields sound. The sound plays an essential role in the liquid-glass transition. The sound is formed by the sum of an infinite series of interacting bubbles of intra-band particle-hole pairs. At higher temperature, the sound picture holds only in the long wavelength regime, but near the liquid-glass transition temperature, the bands of particles are very narrow so that the sound picture holds even in the short wavelength regime. As the temperature decreases, the sound disappears at a disappearing temperature T_0 and a reciprocal particle distance K.

Under supercooling process, enormous bubbles of intra-band particle-hole pairs not constituting the sound, which make the fluctuation entropy, are created. The mixing between the fluctuation entropy and the sound entropy yields the Kauzmann entropy, which is of the Curie law with the critical temperature T_0 and a negative sign. The hopping amplitude J is given by the probability corresponding to the Kauzmann entropy per particle, from the Einstein relation. The hopping amplitude J represents the VTF law. The energy bands are the hopping magnitudes proportional to the hopping amplitude J; the energy bands and the hopping amplitudes are self-consistently determined.

Liquids and glasses are random systems. In treating the random systems, we consider the two stages: At the first stage, we treat the dynamics of sound and phonons in the configurationally averaged systems as developed in the previous chapter. At the second stage, we take into account the randomness of the systems as the scattering processes of particles from the random potentials due to the random harmonic frequencies and the random hopping magnitudes.

At the first stage, we have calculated phonons in the previous chapter. Now we have to calculate sound. In case of phonons, the gap of the inter-band elementary excitations makes the calculation of the bubbles of inter-band particle-hole pairs very easy, while the gapless of the intra-band elementary excitations makes the calculation of the bubbles of intra-band particle-hole pairs very hard. But if we consider the relaxation time in particle frequencies, the relaxation time in the intra-band elementary excitations plays the same role as the gap in the inter-band ones. So we first investigate the random scattering processes.

Liquid Glass Transition. DOI: http://dx.doi.org/10.1016/B978-0-12-407177-3.00007-6

7.1 Random Scattering Processes and the Bethe–Salpeter Equation

In order to treat the scattering processes, we introduce the random scattering potentials due to the random harmonic frequencies and the hopping magnitudes [1–6]. In the scattering processes, we only consider the elementary scattering processes taking into account the simultaneously scattering processes of two particles by the same potentials. The vertex corrections in the bubble diagrams of intra-band and inter-band particle-hole pairs are considered by the Bethe–Salpeter equation. We prove the WT relations for the elementary scattering processes, which the vertex parts and the self-energy parts of Green's functions must satisfy. We illustrate how the vertex parts and the self-energy parts of the Green's functions satisfy the WT relations.

7.1.1 The Elementary Scattering Potentials

At the second stage, to include the randomness of the system, we take into account the random scattering processes due to the random harmonic frequencies and the random hopping magnitudes [1–6]. Thus we introduce the random scattering potentials by subtracting the configurationally averaged Hamiltonian $H_0 =< H >_c$ in Eq. (6.2.11) from the model Hamiltonian H in Eq. (6.2.9):

$$\begin{aligned} H_I &= H - H_0 \\ &= \sum_{m\mu} \hbar(\tilde{\omega}_{m\mu} - \tilde{\omega}_\mu) b^\dagger_{m\mu} b_{m\mu} \\ &\quad + \sum_{mn\mu} \hbar\{J_{m\mu n\mu} - J_\mu(\boldsymbol{R}_m - \boldsymbol{R}_n)\} b^\dagger_{m\mu} b_{n\mu}. \end{aligned} \tag{7.1.1}$$

Note that since the term $J_{m\mu n\mu}$ involves operators $n(y)$, H_I is essentially nonlinear. But at low temperatures particles stay longer at the same sites. Thus we make the approximation that the term $J_{m\mu n\mu}$ is a random c-number. Using the extended operator

$$b_{m\mu} = \frac{1}{N} \sum_{\boldsymbol{p}} e^{\mathrm{i}\boldsymbol{p}\cdot\boldsymbol{R}_{m\mu}} a_{m\mu\boldsymbol{p}},$$

we rewrite Eq. (7.1.1) as

$$\begin{aligned} H_I = \frac{1}{N} \sum_{\mu\boldsymbol{p}\boldsymbol{p}'} \Bigg[& \sum_m \hbar(\tilde{\omega}_{\mu m} - \tilde{\omega}_\mu) e^{-\mathrm{i}(\boldsymbol{p}-\boldsymbol{p}')\cdot\boldsymbol{R}_m} \\ & + \sum_{mn} \hbar\{J_{m\mu n\mu} - J_\mu(\boldsymbol{R}_m - \boldsymbol{R}_n)\} e^{-\mathrm{i}\boldsymbol{p}\cdot\boldsymbol{R}_m + \mathrm{i}\boldsymbol{p}'\cdot\boldsymbol{R}_n} \Bigg] a^\dagger_{\mu\boldsymbol{p}} a_{\mu\boldsymbol{p}'}. \end{aligned} \tag{7.1.2}$$

In including the random scattering processes, we only take into account the dominant scattering processes due to simultaneously scattering processes of two particles by the same random potentials. We call these simultaneously scattering processes of two particles by the same random potentials, the elementary scattering processes. The elementary scattering processes come from the scattering processes due to random harmonic frequencies and random hopping magnitudes, Eq. (7.1.1) or Eq. (7.1.2). An elementary scattering process: the particle-hole pairs $\mu \boldsymbol{p}, \nu \boldsymbol{p}-\boldsymbol{q}$ scattered to $\mu \boldsymbol{p}', \nu \boldsymbol{p}'-\boldsymbol{q}$ by the same random potentials due to random harmonic frequencies and random hopping magnitudes are calculated by summing the configurationally averaged component of $a^{\dagger}_{\mu \boldsymbol{p}} a_{\mu \boldsymbol{p}'} a_{\nu \boldsymbol{p}-\boldsymbol{q}} a^{\dagger}_{\nu \boldsymbol{p}'-\boldsymbol{q}}$ of the second order perturbation with respect to H_I over all random particle positions. The elementary scattering process due to random harmonic frequencies and random hopping amplitudes are given by

$$\begin{aligned}
&U^{\mu\nu}_{\omega} a^{\dagger}_{\mu \boldsymbol{p}} a_{\mu \boldsymbol{p}'} a_{\nu \boldsymbol{p}-\boldsymbol{q}} a^{\dagger}_{\nu \boldsymbol{p}'-\boldsymbol{q}} \\
&\quad = \frac{1}{\hbar^2 N^2} \sum_{m} < \hbar^2 (\tilde{\omega}_{m\mu} - \tilde{\omega}_{\mu})(\tilde{\omega}_{m\nu} - \tilde{\omega}_{\nu}) >_c \\
&\quad \times e^{\mathrm{i}(\boldsymbol{p}-\boldsymbol{p}')\cdot \boldsymbol{R}_m - \mathrm{i}(\boldsymbol{p}-\boldsymbol{p}')\cdot \boldsymbol{R}_m} a^{\dagger}_{\mu \boldsymbol{p}} a_{\mu \boldsymbol{p}'} a_{\nu \boldsymbol{p}-\boldsymbol{q}} a^{\dagger}_{\nu \boldsymbol{p}'-\boldsymbol{q}} \\
&\quad = \frac{1}{N^2} \sum_{m} < \hat{\omega}_{m\mu} \hat{\omega}_{m\nu} >_c a^{\dagger}_{\mu \boldsymbol{p}} a_{\mu \boldsymbol{p}'} a_{\nu \boldsymbol{p}-\boldsymbol{q}} a^{\dagger}_{\nu \boldsymbol{p}'-\boldsymbol{q}}, \qquad \hat{\omega}_{m\mu} \equiv \tilde{\omega}_{m\mu} - \tilde{\omega}_{\mu}.
\end{aligned}$$

and

$$\begin{aligned}
&U^{\mu\nu}_{J}(\boldsymbol{q}) a^{\dagger}_{\mu \boldsymbol{p}} a_{\mu \boldsymbol{p}'} a_{\nu \boldsymbol{p}-\boldsymbol{q}} a^{\dagger}_{\nu \boldsymbol{p}'-\boldsymbol{q}} \\
&\quad = \frac{1}{\hbar N^2} \sum_{mn} < \hbar^2 (J_{m\mu n\mu} - J_{\mu}(\boldsymbol{R}_m - \boldsymbol{R}_n))(J_{m\nu n\nu} - J_{\nu}(\boldsymbol{R}_m - \boldsymbol{R}_n)) >_c \\
&\quad \times e^{\mathrm{i}(\boldsymbol{q}\cdot \boldsymbol{R}_m - \mathrm{i}\boldsymbol{q}\cdot \boldsymbol{R}_n)} a^{\dagger}_{\mu \boldsymbol{p}} a_{\mu \boldsymbol{p}'} a_{\nu \boldsymbol{p}-\boldsymbol{q}} a^{\dagger}_{\nu \boldsymbol{p}'-\boldsymbol{q}} \\
&\quad = \frac{1}{N} \int d^3 R g(R) e^{\mathrm{i}\boldsymbol{q}\cdot \boldsymbol{R}} < \hat{J}_{m\mu n\mu} \hat{J}_{m\nu n\nu} >_c \Big|_{\boldsymbol{R}=\boldsymbol{R}_m - \boldsymbol{R}_n} \\
&\quad \times a^{\dagger}_{\mu \boldsymbol{p}} a_{\mu \boldsymbol{p}'} a_{\nu \boldsymbol{p}-\boldsymbol{q}} a^{\dagger}_{\nu \boldsymbol{p}'-\boldsymbol{q}}, \qquad \hat{J}_{m\mu n\nu} \equiv J_{m\mu n\nu} - J_{\mu}(\boldsymbol{R}_m - \boldsymbol{R}_n).
\end{aligned}$$

which are shown in Figure 7.1. Here the crosses on the dotted line means summing the random scattering potentials over random positions. Thus we obtain the elementary scattering amplitudes:

$$U^{\mu\nu}_{\omega} = \frac{1}{N^2} \sum_{m} < \hat{\omega}_{m\mu} \hat{\omega}_{m\nu} >_c, \tag{7.1.3}$$

$$U^{\mu\nu}_{J}(\boldsymbol{q}) = \frac{1}{N} \int d^3 R g(R) e^{\mathrm{i}\boldsymbol{q}\cdot \boldsymbol{R}} < \hat{J}_{m\mu n\mu} \hat{J}_{m\nu n\nu} >_c \Big|_{\boldsymbol{R}=\boldsymbol{R}_m - \boldsymbol{R}_n}. \tag{7.1.4}$$

Hereafter we take into account only the two particle simultaneously scattering processes as the elementary scattering amplitudes in the random scattering processes.

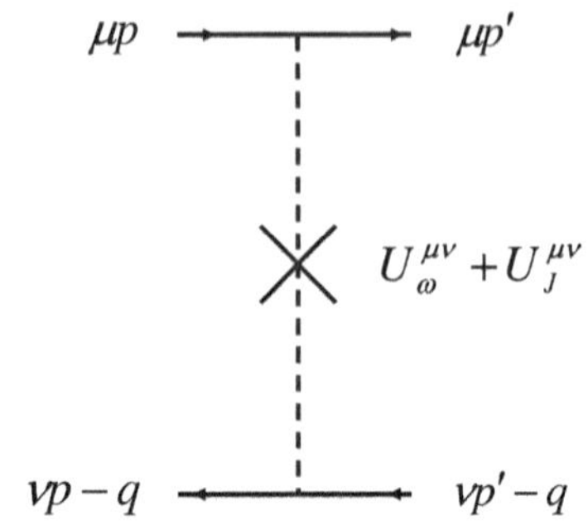

Figure 7.1 The elementary scattering amplitudes, $U_\omega^{\mu\nu}$ due to the random harmonic frequencies and $U_J^{\mu\nu}$ due to random hopping magnitudes. The cross on the dotted line means summing up the scattering process over random positions.

Here we investigate the vertex corrections for bubble diagrams for intra-band and inter-band particle-hole pairs. In order to do so, we derive the Bethe–Salpeter equation [7,8]. First, we introduce particle Green's functions:

$$\begin{aligned} G^{\mu}_{\boldsymbol{pp'}}(\tau - \tau') &\equiv - < T_\tau a_{\mu \boldsymbol{p}}(\tau) a^{\dagger}_{\mu \boldsymbol{p'}}(\tau') > \\ &= \frac{1}{\beta\hbar} \sum_{\mathrm{i}\omega_n} e^{-\mathrm{i}\omega_n(\tau-\tau')} G^{\mu}_{\boldsymbol{pp'}}(\mathrm{i}\omega_n). \end{aligned} \tag{7.1.5}$$

Note that Eq. (7.1.5) is not configurationally averaged.

7.1.2 The Bethe–Salpeter Equation

Now we derive the Bethe–Salpeter equation for configurationally averaged two Green's functions, a bubble diagram of μth particle-νth hole pairs:

$$\phi^{\mu\nu}_{\boldsymbol{pp'}}(\mathrm{i}\omega_n, q) \equiv \left\langle G^{\mu}_{\boldsymbol{p},p'}(\mathrm{i}\omega_n) G^{\nu}_{\boldsymbol{p-q},p'-\boldsymbol{q}}(\mathrm{i}\omega_n - \mathrm{i}\nu_n) \right\rangle_c . \tag{7.1.6}$$

Using the vertex function, we can write Eq. (7.1.6) as

$$\begin{aligned} \phi^{\mu\nu}_{\boldsymbol{pp'}}(\mathrm{i}\omega_n, q) \equiv\ & G^{\mu}(p) G^{\nu}(p' - q) \delta_{\boldsymbol{pp'}} \\ & + G^{\mu}(p) G^{\nu}(p - q) \Gamma^{\mu\nu}(p, p - q; p', p' - q) \\ & \times G^{\mu}(p') G^{\nu}(p' - q), \end{aligned} \tag{7.1.7}$$

where $G^{\mu}(p)$ is the configuration average of the Green's function (7.1.5):

$$< G^{\mu}_{\boldsymbol{pp'}}(\mathrm{i}\omega_n) >_c = G^{\mu}_{\boldsymbol{p}}(\mathrm{i}\omega_n) \delta_{\boldsymbol{pp'}} = G^{\mu}(p) \delta_{\boldsymbol{pp'}}, \tag{7.1.8}$$

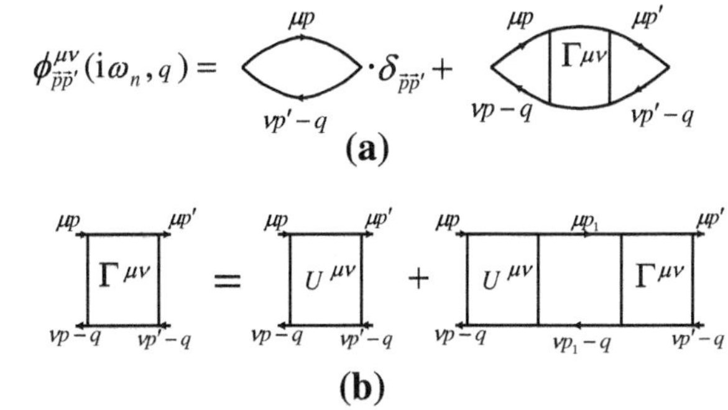

Figure 7.2 The configurationally averaged two Green's functions, $\phi^{\mu\nu}_{\boldsymbol{pp}'}(\mathrm{i}\omega_n, q)$ and the vertex part, $\Gamma^{\mu\nu}(p, p-q; p', p'-q)$ are diagrammatically illustrated in (a) and (b), respectively. (a) $\phi^{\mu\nu}_{\boldsymbol{pp}'}(\mathrm{i}\omega_n, q)$ consists of an uncorrected bubble diagram and corrected bubble diagrams by the vertex part. (b) The vertex function $\Gamma^{\mu\nu}$ are corrected by the elementary scattering amplitudes due to $U^{\mu\nu}_{\omega}$ and $U^{\mu\nu}_{J}$. The vertex part $U^{\mu\nu}$ is the irreducible vertex part.

and the abbreviations $p = (\boldsymbol{p}, \mathrm{i}\omega_n), q = (\boldsymbol{q}, \mathrm{i}\nu_n)$ have been used. Note that $p' = (\boldsymbol{p}', \mathrm{i}\omega_n)$. The vertex function is written down as:

$$\begin{aligned}\Gamma^{\mu\nu}&(p, p-q; p', p'-q)\\&= U^{\mu\nu}(p, p-q; p', p'-q)\\&\quad+\sum_{\boldsymbol{p}_1} U^{\mu\nu}(p, p-q; p_1, p_1-q)G^{\mu}(p_1)G^{\nu}(p_1-q)\\&\quad\times\Gamma^{\mu\nu}(p_1, p_1-q; p', p'-q),\end{aligned} \tag{7.1.9}$$

where the term $U^{\mu\nu}$ is the irreducible vertex part, which is constructed by summing up the diagrams not separated by cutting any two internal particle Green's functions. The configurationally averaged two Green's functions, $\phi^{\mu\nu}_{\boldsymbol{pp}'}(\mathrm{i}\omega_n, q)$ and the vertex part $\Gamma^{\mu\nu}(p, p-q; p', p'-q)$, are diagrammatically illustrated in Figure 7.2. In the vertex part and the irreducible vertex part, since $p = (\boldsymbol{p}, \mathrm{i}\omega_n)$ and $p' = (\boldsymbol{p}', \mathrm{i}\omega_n)$ do not depend on $\mathrm{i}\omega_n$, we can abbreviate:

$$\Gamma^{\mu\nu}_{\boldsymbol{pp}'}(q) \equiv \Gamma^{\mu\nu}(p, p-q; p', p'-q), \tag{7.1.10}$$

$$U^{\mu\nu}_{\boldsymbol{pp}'}(q) \equiv U^{\mu\nu}(p, p-q; p', p'-q). \tag{7.1.11}$$

Therefore we can rewrite Eqs. (7.1.7) and (7.1.9) as

$$\begin{aligned}\phi^{\mu\nu}_{\boldsymbol{pp}'}(\mathrm{i}\omega_n, q) \equiv{}& G^{\mu}(p)G^{\nu}(p'-q)\delta_{\boldsymbol{pp}'}\\&+G^{\mu}(p)G^{\nu}(p-q)\Gamma^{\mu\nu}_{\boldsymbol{pp}'}(q)G^{\mu}(p')G^{\nu}(p'-q),\end{aligned} \tag{7.1.12}$$

$$\Gamma^{\mu\nu}_{\boldsymbol{pp}'}(q) = U^{\mu\nu}_{\boldsymbol{pp}'}(q) + \sum_{\boldsymbol{p}_1} U^{\mu\nu}_{\boldsymbol{pp}_1}(q)G^{\mu}(p_1)G^{\nu}(p_1-q)\Gamma^{\mu\nu}_{\boldsymbol{p}_1\boldsymbol{p}'}(q). \tag{7.1.13}$$

Substituting Eq. (7.1.13) into Eq. (7.1.12), we obtain

$$\begin{aligned}
\phi^{\mu\nu}_{\boldsymbol{pp}'}(\mathrm{i}\omega_n, q) &= G^{\mu}(p)G^{\nu}(p-q)\left[\delta_{\boldsymbol{pp}'} + \Gamma^{\mu\nu}_{\boldsymbol{pp}'}(q)G^{\mu}(p')G^{\nu}(p'-q)\right] \\
&= G^{\mu}(p)G^{\nu}(p-q) \\
&\quad \times\left[\delta_{\boldsymbol{pp}'} + \left\{U^{\mu\nu}_{\boldsymbol{pp}'}(q) + \sum_{\boldsymbol{p}_1} U^{\mu\nu}_{\boldsymbol{pp}_1}(q)G^{\mu}(p_1)\right.\right. \\
&\quad \left.\left. G^{\nu}(p_1-q)\Gamma^{\mu\nu}_{\boldsymbol{p}_1 p'}(q)\right\}G^{\mu}(p')G^{\nu}(p'-q)\right] \\
&= G^{\mu}(p)G^{\nu}(p-q) \\
&\quad \times\left[\delta_{\boldsymbol{pp}'} + \sum_{\boldsymbol{p}_1} U^{\mu\nu}_{\boldsymbol{pp}_1}(q)\left\{G^{\mu}(p_1)G^{\nu}(p_1-q)\delta_{\boldsymbol{p}_1 p'}\right.\right. \\
&\quad \left.\left. +G^{\mu}(p_1)G^{\nu}(p_1-q)\Gamma^{\mu\nu}_{\boldsymbol{p}_1 p'}(q)G^{\mu}(p')G^{\nu}(p'-q)\right\}\right].
\end{aligned}$$

Thus we obtain the Bethe–Salpeter equation:

$$\phi^{\mu\nu}_{\boldsymbol{pp}'}(\mathrm{i}\omega_n, q) = G^{\mu}(p)G^{\nu}(p-q)\left[\delta_{\boldsymbol{pp}'} + \sum_{\boldsymbol{p}_1} U^{\mu\nu}_{\boldsymbol{pp}_1}(q)\phi^{\mu\nu}_{\boldsymbol{p}_1\boldsymbol{p}'}(\mathrm{i}\omega_n, q)\right]. \quad (7.1.14)$$

In this section, we assume

$$U^{\mu\mu}_{\omega,J}(\boldsymbol{q}) \sim U^{\nu\nu}_{\omega,J}(\boldsymbol{q}) \sim U^{\mu\nu}_{\omega,J}(\boldsymbol{q}),$$

so we abbreviate

$$U^{\mu\nu}_{\omega} + U^{\mu\nu}_{J})(\boldsymbol{q}) = U(\boldsymbol{q}).$$

The irreducible vertex parts $U^{\mu\nu}_{\boldsymbol{p},\boldsymbol{p}'}(q)$ up to the second order are as follows:

$$\begin{aligned}
U^{\mu\nu}_{\boldsymbol{pp}'}(q) &= U^{\mu\nu}_{\boldsymbol{pp}'(1)}(q) + U^{\mu\nu}_{\boldsymbol{pp}'(2)}(q) + \cdots, && && (7.1.15) \\
U^{\mu\nu}_{\boldsymbol{pp}'(1)}(q) &= U(\boldsymbol{p}-\boldsymbol{p}'), && \longrightarrow (a) && (7.1.16) \\
U^{\mu\nu}_{\boldsymbol{pp}'(2)}(q) &= \sum_{\boldsymbol{p}_1}[U(\boldsymbol{p}-\boldsymbol{p}_1)U(\boldsymbol{p}-\boldsymbol{p}') \\
&\qquad G^{\mu}(p_1)G^{\mu}(p'-p+p_1) && \longrightarrow (b) \\
&\quad +U(\boldsymbol{p}_1-\boldsymbol{p}')U(\boldsymbol{p}-\boldsymbol{p}_1) \\
&\qquad G^{\mu}(p_1)G^{\nu}(p'-p_1+p-q) && \longrightarrow (c) \\
&\quad +U(\boldsymbol{p}-\boldsymbol{p}')U(\boldsymbol{p}-\boldsymbol{p}_1)G^{\nu}(p_1-q) \\
&\qquad G^{\nu}(p'-p+p_1-q)], && \longrightarrow (d) && (7.1.17)
\end{aligned}$$

which are illustrated in Figure 7.3. The vertex part $\Gamma^{\mu\nu(1)}_{\boldsymbol{pp}'}(q)$ for the first order irreducible vertex part $U^{\mu\nu(1)}_{\boldsymbol{pp}'} = U(\boldsymbol{p}-\boldsymbol{p}')$ is shown in Figure 7.4. This approximation is called the ladder approximation.

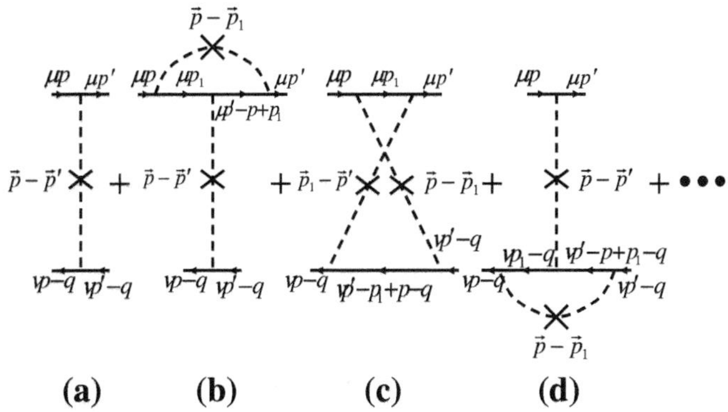

Figure 7.3 The irreducible vertex parts $U^{\mu\nu}$ up to the second order are illustrated. (a) is the first order irreducible vertex part. (b), (c), and (d) are the second order ones.

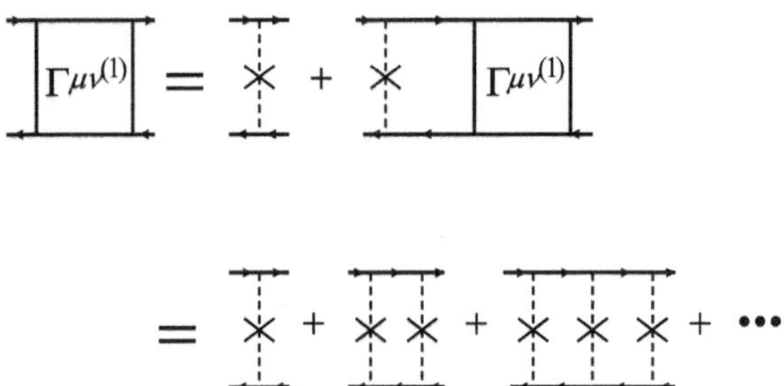

Figure 7.4 The vertex part in the ladder approximation $\Gamma^{\mu\nu(1)}$ is shown. The irreducible vertex part is taken as $U^{\mu\nu(1)} = U^{\mu\nu}_{\omega} + U^{\mu\nu}_{J}$ in the first order scattering potential.

If we introduce the self-energy parts of the Green's function as

$$\{G^{\mu}(p)\}^{-1} \equiv \mathrm{i}\omega_n - \epsilon_{\mu\mathbf{p}} - \Sigma^{\mu}(p), \tag{7.1.18}$$

from Eq. (7.1.14) we obtain

$$\begin{aligned} &\left\{-\mathrm{i}\nu_n + \epsilon_{\mu\mathbf{p}} - \epsilon_{\nu\mathbf{p}-\mathbf{q}} + \Delta\Sigma^{\mu\nu}(p,q)\right\}\phi^{\mu\nu}(p,q) \\ &\quad = \Delta G^{\mu\nu}(p,q)\left[1 + \sum_{\mathbf{p}_1} U^{\mu\nu}_{\mathbf{p}\mathbf{p}_1}(q)\phi^{\mu\nu}(p_1,q)\right], \end{aligned} \tag{7.1.19}$$

where we have put

$$\Delta G^{\mu\nu}(p,q) = G^{\mu}(p) - G^{\nu}(p-q), \tag{7.1.20}$$

$$\Delta\Sigma^{\mu\nu}(p,q) = \Sigma^{\mu}(p) - \Sigma^{\nu}(p-q), \tag{7.1.21}$$

$$\phi^{\mu\nu}(p,q) \equiv \sum_{\mathbf{p}'} \phi^{\mu\nu}_{\mathbf{p}\mathbf{p}'}(\mathrm{i}\omega_n, q). \tag{7.1.22}$$

The WT relation for the self-energy parts and the vertex parts in the elementary scattering processes provides us with a simple equation instead of Eq. (7.1.19).

7.1.3 The WT Relation for the Self-Energy Parts and the Vertex Parts

Here under the following assumption:

$$U^{\mu\nu} \sim U^{\mu\mu} \sim U^{\nu\nu},$$

we prove the WT relation for the self-energy parts and the vertex parts:

$$\Delta\Sigma^{\mu\nu}(p,q)\}\phi^{\mu\nu}(p,q) = \sum_{p'} U^{\mu\nu}_{pp'}(q)\Delta G^{\mu\nu}(p',q). \tag{7.1.23}$$

The proof of the WT relation as follows:

The proper self-energy parts are defined by the sum of self-energy parts which cannot be separated into two parts by cutting any internal line. Therefore we denote the nth order proper self-energy part as follows:

$$\Sigma^{\mu}(p) \equiv \sum_{np_1\cdots p_n} U(1,\ldots,n)G^{\mu}(p_1)\cdots G^{\mu}(p_i)\cdots G^{\mu}(p_n), \tag{7.1.24}$$

where $U(1,\ldots,n)$ is an ensemble of n elementary scattering potentials, and the wave vectors conserve at each point entering or leaving of an elementary scattering potential. Then we obtain:

$$\begin{aligned}
\Delta\Sigma^{\mu\nu}(p,q) &= \sum_{np_1\cdots p_n} U(1,\ldots,n) \\
&\quad \times [G^{\mu}(p_1)\cdots G^{\mu}(p_i)\cdots G^{\mu}(p_n) - G^{\nu}(p_1-q) \\
&\quad + \cdots + G^{\nu}(p_i-q)\cdots G^{\nu}(p_n-q)] \\
&= \sum_{np_1\cdots p_n} U(1,\ldots,n) \\
&\quad \times [G^{\mu}(p_1)\cdots G^{\mu}(p_i)\cdots G^{\mu}(p_{n-2})G^{\mu}(p_{n-1}) \\
&\quad \times \{G^{\mu}(p_n) - G^{\nu}(p_n-q)\} + G^{\mu}(p_1)\cdots G^{\mu}(p_i)\cdots G^{\mu}(p_{n-2}) \\
&\quad \times \{G^{\mu}(p_{n-1}) - G^{\nu}(p_{n-1}-q)\}G^{\nu}(p_n-q) \\
&\quad + \cdots + G^{\mu}(p_1)\cdots G^{\mu}(p_{i-1}) \\
&\quad \times \{G^{\mu}(p_i) - G^{\nu}(p_i-q)\}G^{\nu}(p_{i+1}-q)\cdots G^{\nu}(p_n-q) \\
&\quad + \cdots + \{G^{\mu}(p_1) - G^{\nu}(p_1)\}G^{\nu}(p_2-q) \\
&\quad + \cdots + G^{\nu}(p_i-q)\cdots G^{\nu}(p_n-q)] \\
&= \sum_{ni\,p_1,\cdots,p_n} U_i(1,\cdots,n) \times G^{\mu}(p_1)\cdots G^{\mu}(p_{i-1})\cdot G^{\nu}(p_{i+1}-q) \\
&\quad + \cdots + \cdots G^{\nu}(p_n-q)\Delta G^{\mu\nu}(p_n,q),
\end{aligned} \tag{7.1.25}$$

where from the first equation to the second one, starting with $G^{\mu}(p_1)\cdots G^{\mu}(p_i)\cdots G^{\mu}(p_n)$ in Eq. (7.1.24), we have carried out successively subtraction and addition processes: in the ith line in the second equation

$$\begin{aligned}
&G^{\mu}(p_1)\cdots G^{\mu}(p_{i-1})\{G^{\mu}(p_i)-G^{\nu}(p_i-q)\}G^{\mu}(p_i-q)\cdots G^{\nu}(p_n-q)\\
&\quad = G^{\mu}(p_1)\cdots G^{\mu}(p_{i-1})G^{\mu}(p_i)G^{\mu}(p_i-q)\cdots G^{\nu}(p_n-q)\\
&\quad G^{\mu}(p_1)\cdots G^{\mu}(p_{i-1})G^{\mu}(p_i-q)G^{\nu}(p_{i+1}-q)\cdots G^{\nu}(p_n-q).
\end{aligned}$$

the first and the second terms on the right-hand side equation are the addition and the substraction terms, respectively. The first term cancels with the subtraction term in the $(i-1)$th line and the subtraction term cancels with the addition term in $(i+1)$th line. Next we replace each term $\Delta G^{\mu\nu}(p_i, q)$ in the last equation of Eq. (7.1.25) by $\Delta G^{\mu\nu}(p', q)$ and arranging the $U(1, \ldots, n)$ to conserve momentum at each vertex connecting two Green's functions and an elementary scattering potential denoted as $U_i(1, \ldots, n)$. In these procedures (bootstrapping), we can summarize as follows:

1. We start with the self-energy part in Eq. (7.1.24).
2. We divide $G^{\mu}(p_i)$ into $\Delta G^{\mu\nu}(p_i, q) = G^{\mu}(p_i) - G^{\nu}(p_i - q)$ and replace all Green's functions $G^{\mu}(p_k)$; $i < k \leqslant N$ by $G^{\nu}(p_k - q)$.
3. We flip internal lines $G^{\nu}(p_{i+1})\cdots G^{\nu}(p_N - q)$ in the Feynman diagrams.
4. We replace $\Delta G^{\mu\nu}(p_i, q)$ by $\Delta G^{\mu\nu}(p', q)$ together with arranging the internal lines to conserve momenta at every vertex.

Since the proper self-energy part cannot be separated into two parts, the vertex part in front of the term $\Delta G^{\mu\nu}(p_i, q)$ are irreducible so that we have proved the WT relation (7.1.23).

Here we show the WT relations up to the second order. The proper self-energy parts up to the second order are

$$\Sigma^{\mu}(p) = \Sigma^{\mu(1)}(p) + \Sigma^{\mu(2)}(p) + \cdots, \tag{7.1.26}$$

$$\Sigma^{\mu(1)}(p) = \sum_{p_1} U(p - p_1)G^{\mu}(p_1), \tag{7.1.27}$$

$$\Sigma^{\mu(2)}(p) = \sum_{p_1 p_2, p_3 = p - p_1 + p_2} U(p - p_1)U(p_1 - p_2) G^{\mu}(p_1)G^{\mu}(p_2)G^{\mu}(p_3). \tag{7.1.28}$$

The proper self-energy parts up to the second order are illustrated in Figure 7.5. The WT relation (7.1.23) plays an important role in the approximation procedures; in each approximation procedure, we must choose the self-energy part and the vertex part such that they satisfy the WT relation. In the first approximation we must choose the set of the self-energy $\Sigma^{\mu,\nu(1)}(p)$ in Eq. (7.1.24) as shown in Figure 7.5(a) and the vertex part $U^{\mu\nu}_{pp'(1)}$ in Eq. (7.1.16) as shown in Figure 7.3(a). In the second approximation we must choose the set of the self-energy $\Sigma^{\mu,\nu(2)}(p)$ in Eq. (7.1.28) as shown in Figure 7.5(b), (c), (d) and the vertex part $U^{\mu\nu(2)}_{pp'}$ in Eq. (7.1.17) as shown in Figure 7.3(b), (c), (d). The WT relations are given by:

$$\begin{aligned}
\Delta\Sigma^{\mu\nu(1)}(p, q) &= \sum_{p_1} U(p - p_1)(G^{\mu}(p_1) - G^{\nu}(p_1 - q))\\
&= \sum_{p'} U(p - p')\Delta G^{\mu\nu(1)}(p', q), \quad \longrightarrow (a)
\end{aligned} \tag{7.1.29}$$

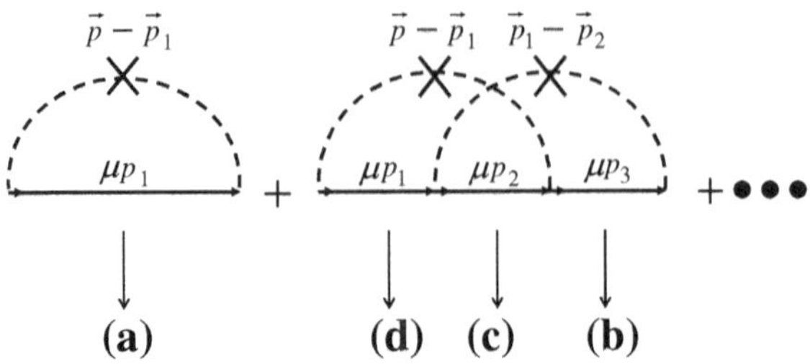

Figure 7.5 The self-energy parts $\Sigma^{\mu\mu(1)}(p)$ and $\Sigma^{\mu\mu}(2)(p)$ are illustrated. In each bootstrapping diagram, the procedures (1)–(4) mentioned above in proving the WT-relation (7.1.23) yield the respective vertex part in Figure 7.3.

$$\begin{aligned}
\Delta\Sigma^{\mu\nu(2)}(p) = & \sum_{p_1 p_2, p_3 = p - p_1 + p_2} U(\boldsymbol{p}-\boldsymbol{p}_1)U(\boldsymbol{p}_1-\boldsymbol{p}_2) \\
& \times [G^{\mu\mu}(p_1)G^{\mu\mu}(p_2)\{G^{\mu\mu}(p_3) - G^{\nu\nu}(p_3-q)\} \\
& + G^{\mu\mu}(p_1)\{G^{\mu\mu}(p_2) - G^{\nu\nu}(p_2-q)\}G^{\nu\nu}(p_3-q) \\
& + \{G^{\mu\mu}(p_1) - G^{\nu\nu}(p_1-q)\}G^{\nu\nu}(p_2-q)G^{\nu\nu}(p_3-q)] \\
= & \sum_{p_1, p'} [U(\boldsymbol{p}-\boldsymbol{p}_1)U(\boldsymbol{p}-\boldsymbol{p}')G^{\mu\mu}(p_1)G^{\nu\nu}(p'-p+p_1) \quad \rightarrow (b) \\
& + U(\boldsymbol{p}_1-\boldsymbol{p}')U(\boldsymbol{p}-\boldsymbol{p}_1)G^{\mu\mu}(p_1)G^{\nu\nu}(p'-p_1+p-q) \rightarrow (c) \\
& + U(\boldsymbol{p}-\boldsymbol{p}')U(\boldsymbol{p}-\boldsymbol{p}_1)G^{\nu\nu}(p_1-q) \\
& \quad G^{\nu\nu}(p'-p+p_1-q)]\Delta G^{\mu\nu(2)}(p', q). \quad \rightarrow (d)
\end{aligned} \tag{7.1.30}$$

In the case satisfying the WT relation (7.1.23), since

$$\sum_{pp'} U^{\mu\nu}_{pp'}(q)\Delta G^{\mu\nu}(p', q) = \sum_{pp'} U^{\mu\nu}_{p'p}(q)\Delta G^{\mu\nu}(p', q),$$

we obtain a compact relation

$$\sum_p \{-i\nu_n + \epsilon_{\mu p} - \epsilon_{\nu p-q}\}\phi^{\mu\nu}(p, q) = \sum_p \Delta G^{\mu\nu}(p, q). \tag{7.1.31}$$

This relation shows that the complicated vertex corrections are taken into account in the self-energy part of the Green's function. This is due to the WT relation (7.1.23). We use this relation to obtain sound and phonons in dissipative processes.

7.2 Intra-Band Density Fluctuations: Sound and Diffusion

We have established the model Hamiltonian (6.2.27) for the intra-band and the inter-band density fluctuations. The collective excitations for the intra-band and the inter-band density fluctuations are sound and phonons, respectively. In phonons, the gap

(∼the harmonic frequency ω) of the inter-band elementary excitations makes the calculation of the bubbles of inter-band particle-hole pairs very easy. While in sound, the gapless of the intra-band elementary excitations makes the calculation of the bubbles of intra-band particle-hole pairs very hard. Here taking into account the vertex corrected bubble diagrams of intra-band particle-hole pairs, we calculate the sound in the random phase approximation. This procedure makes not only the calculation of sound easier, but also clarifies the relaxation and dissipative processes.

At high temperature, the sound picture holds only in the long wavelength regime and decays into multi-particles in the short wavelength regime. On the other hand, near the liquid-glass transition temperature, the intra-band elementary excitations are very narrow so that the sound mode holds even in the short wavelength regime. The sound mode disappears at a sound disappearing temperature T_0 and at a reciprocal particle distance K.

7.2.1 Sound in the Relaxation and Dissipative Processes

In order to calculate the sound waves, we investigate the correlation functions for intra-band density fluctuations. We are concerned with the temperature regions near the glass transition; we only consider the lower temperatures so that we restrict ourselves to the lower band density fluctuations in the intra-band density fluctuations. Therefore we put:

$$\rho^{\dagger}_{0\boldsymbol{q}} \equiv \rho^{\dagger}_{d0\boldsymbol{q}} = \sum_{\boldsymbol{p}} a^{\dagger}_{0\boldsymbol{p}} a_{0\boldsymbol{p}-\boldsymbol{q}}, \tag{7.2.1}$$

Hereafter the index $\mu = 0$ means the intra-band density fluctuations and $\mu = i$, $(i = 1, 2, 3)$ means the inter-band density fluctuations. The correlation function for intra-band density fluctuations is given by

$$\begin{aligned} F_{\boldsymbol{q}}(\tau_1 - \tau_2) &\equiv -\frac{1}{\hbar N} < T_\tau \rho_{0\boldsymbol{q}}(\tau_1)\rho^{\dagger}_{0\boldsymbol{q}}(\tau_2) >_c \\ &\equiv \frac{1}{\beta\hbar}\sum_{\mathrm{i}\nu_n} e^{-\mathrm{i}\nu_n(\tau_1-\tau_2)}F(q). \end{aligned} \tag{7.2.2}$$

In the case of non-dissipative process, employing the model Hamiltonian obtained by the configuration average, we can obtain the dynamical equation of $F(q)$ in the random phase approximation. In the case of the relaxation and dissipative processes, we must take into account the elementary scattering potentials $U^{00}_{\omega} + U^{00}_{J}$, where U^{00}_{ω} due to the random frequencies in Eq. (7.1.3) and U^{00}_{J} due to the potentials in Eq. (7.1.4).

The scattering processes consist of two parts: the vertex corrections of intra-bubble diagrams and the inter-bubble diagrams. The vertex correction of intra-bubble diagrams has been investigated in the previous section. But since the vertex corrections of the inter-bubble diagrams do not contribute to construct the sound, we can neglect such corrections. The vertex corrections of intra-bubble diagrams are taken into account by using the WT relation (7.1.23). Thus in the random phase approximation we obtain

$$F(q) = P(q) + P(q)V_{00}(\boldsymbol{q})F(q), \tag{7.2.3}$$

where $P(q)$ is the vertex corrected bubble diagrams in Eq. (7.1.22) for intra-band density fluctuations.

$$P(q) \equiv \mp \frac{1}{\beta \hbar^2 N} \sum_p \phi^{00}(p, q), \tag{7.2.4}$$

The configurationally averaged diagonal potential function is given by

$$V_{00}(q) \equiv < V_{00}^d(\boldsymbol{q}) >_c = \int d^3 R g(R) e^{-\mathrm{i}\boldsymbol{q}\cdot\boldsymbol{R}} V_{00}(\boldsymbol{R}), \tag{7.2.5}$$

$$V_{00}(\boldsymbol{R}) \equiv \int d^3 x d^3 y \tilde{w}_0^2(\boldsymbol{x}) V(\boldsymbol{x} - \boldsymbol{y}) \tilde{w}_0^2(\boldsymbol{y} - \boldsymbol{R}), \tag{7.2.6}$$

where $\tilde{w}_0(\mathbf{x})$ is configurationally averaged and $V_{00}(q)$ isotropic. Then we obtain

$$F(q) = \frac{P(q)}{1 - V_{00}(q) P(q)}. \tag{7.2.7}$$

Here we investigate the vertex corrections of the bubble diagrams. From Eq. (7.1.19) with $\mu = \nu = 0$, we obtain

$$\begin{aligned} &\left\{ -\mathrm{i}\nu_n + \epsilon_{0\boldsymbol{p}} - \epsilon_{0\boldsymbol{p}-\boldsymbol{q}} + \Delta\Sigma^{00}(p, q) \right\} \phi^{00}(p, q) \\ &\quad = \Delta G^{00}(p, q) \left\{ 1 + \sum_{p_1} U_{\boldsymbol{p}\boldsymbol{p}_1}^{00}(q) \phi^{00}(p_1, q) \right\}. \end{aligned} \tag{7.2.8}$$

The WT relation (7.1.23) leads to

$$\Delta\Sigma^{00}(p, q) = \sum_{p'} U_{\boldsymbol{p}\boldsymbol{p}'}^{00}(q) \Delta G^{00}(p', q). \tag{7.2.9}$$

In deriving the WT relation (7.1.23), we have assumed the same magnitude of elementary scattering magnitudes between the intra-bands and between the inter-band: $U_{\omega,J}^{\mu\mu}(\boldsymbol{q}) \sim U_{\omega,J}^{\nu\nu}(\boldsymbol{q}) \sim U_{\omega,J}^{\mu\nu}(\boldsymbol{q})$ so that we have put $U_\omega^{\mu\nu} + U_J^{\mu\nu}(\boldsymbol{q}) = U(q)$. In the case of $F(q)$ since the elementary scattering magnitude $U_\omega^{00} + U_J^{00}(\boldsymbol{q})$ holds the same, Eq. (7.2.9) is the exact WT relation. Summing both sides of Eq. (7.7) over $\boldsymbol{p}$, we obtain another exact equation:

$$\sum_{\boldsymbol{p}} \{-\mathrm{i}\nu_n + \epsilon_{0\boldsymbol{p}} - \epsilon_{0\boldsymbol{p}-\boldsymbol{q}}\} \phi^{00}(p, q) = \sum_{\boldsymbol{p}} \Delta G^{00}(p, q). \tag{7.2.10}$$

In calculating the vertex corrected bubble diagrams, $P(q)$, we use Eqs. (7.2.8) and (7.2.10).

7.2.2 The Approximation Procedure

In order to obtain the vertex corrected bubble diagram of intra-band particle-hole pairs, $P(q)$, we must solve Eq. (7.2.8). It is very hard to solve Eq. (7.2.8) rigorously. Since we are concerned with the low temperature regime: $\beta\hbar\omega \gg 1$, $\beta|\tilde{J}| \ll 1$ and $\beta\hbar/\tau_0 \ll 1$. Thus we make the following approximations:

1. In the self-energy part Σ^0, we take account of only the relaxation processes, not the modification of frequencies; we consider the imaginary part, not the real part. Therefore, when we analytically continue $\mathrm{i}\omega_n \to p_0 \pm \mathrm{i}\delta$, we put

$$\Sigma^0(p_0 \pm \mathrm{i}\delta) \cong \mp \frac{\mathrm{i}}{2\tau_0(p_0)}. \tag{7.2.11}$$

2. In summation of the particle Green's functions with respect to the temperature frequencies,

$$\sum_{\mathrm{i}\omega_n} G^0(p-q) = \sum_{\mathrm{i}\omega_n} \frac{1}{\mathrm{i}\omega_n - \mathrm{i}\nu_n - \epsilon_{0\boldsymbol{p}-\boldsymbol{q}} - \Sigma^0(\mathrm{i}\omega_n - \mathrm{i}\nu_n)},$$

we carry out the contour integral method; instead of the summing over the temperature frequencies, we pick up all the poles of $f(z) = \frac{1}{e^{\beta\hbar z} \mp 1}$ enclosing $G^0(z - \mathrm{i}\nu_n, \boldsymbol{p}-\boldsymbol{q})f(z)$ by the contour C and then deform the contour C to enclose the pole of the Green's function $G^0(z - \mathrm{i}\nu_n, \boldsymbol{p}-\boldsymbol{q})$. Thus we obtain

$$\begin{aligned}
&\sum_{\mathrm{i}\omega_n} \frac{1}{\mathrm{i}\omega_n - \mathrm{i}\nu_n - \epsilon_{0\boldsymbol{p}-\boldsymbol{q}} - \Sigma^0(\mathrm{i}\omega_n - \mathrm{i}\nu_n)} \\
&\quad = \int_C G^0(z - \mathrm{i}\nu_n, \boldsymbol{p}-\boldsymbol{q}) f(z) \\
&\quad = \mp\beta\hbar f(\epsilon_{0\boldsymbol{p}-\boldsymbol{q}} + \Sigma^0(\epsilon_{0\boldsymbol{p}-\boldsymbol{q}} - \mathrm{i}\nu_n)).
\end{aligned}$$

Analytic continuation: $\mathrm{i}\nu_n \to q_0 + \mathrm{i}\delta$ leads to

$$\begin{aligned}
&\sum_{\mathrm{i}\omega_n} \frac{1}{\mathrm{i}\omega_n - \mathrm{i}\nu_n - \epsilon_{0\boldsymbol{p}-\boldsymbol{q}} - \Sigma^0(\mathrm{i}\omega_n - \mathrm{i}\nu_n)} \xrightarrow{\mathrm{i}\nu_n \to q_0 + \mathrm{i}\delta} \\
&\quad \mp\beta\hbar f\left(\epsilon_{0\boldsymbol{p}-\boldsymbol{q}} - \frac{\mathrm{i}}{2\tau_0(\epsilon_{0\boldsymbol{p}})}\right).
\end{aligned} \tag{7.2.12}$$

Considering the low temperature, $\beta\hbar\epsilon_{0\boldsymbol{p}} \ll 1$ and $\beta\hbar/\tau_0 \ll 1$ we expand the distribution function f around $\epsilon_{0\boldsymbol{p}}$

$$\begin{aligned}
&\sum_{\mathrm{i}\omega_n} \Delta G^0(p, q) \\
&\quad \cong \mp(\beta\hbar)\left\{ f\left(\epsilon_{0\boldsymbol{p}} - \frac{\mathrm{i}}{2\tau_0(\epsilon_{0\boldsymbol{p}})}\right) - f\left(\epsilon_{0\boldsymbol{p}-\boldsymbol{q}} - \frac{\mathrm{i}}{2\tau_0(\epsilon_{0\boldsymbol{p}-\boldsymbol{q}})}\right)\right\} \\
&\quad \cong \mp(\beta\hbar)^2 \left(\epsilon_{0\boldsymbol{p}} - \epsilon_{0\boldsymbol{p}-\boldsymbol{q}} - \frac{\mathrm{i}}{\tau_0}\right) f'(\epsilon_{0\boldsymbol{p}}) \\
&\quad \cong \mp(\beta\hbar)^2 \left(\boldsymbol{q}\cdot\nabla\epsilon_{0\boldsymbol{p}} - \frac{\mathrm{i}}{\tau_0}\right) f'(\epsilon_{0\boldsymbol{p}}),
\end{aligned} \tag{7.2.13}$$

where $f'(x) = \frac{df(x)}{d(\beta\hbar x)}$. Here the difference of intra-band elementary excitations, $\epsilon_{0\boldsymbol{p}} - \epsilon_{0\boldsymbol{p}-\boldsymbol{q}}$, has been considered up to the first order expansion of the wave vector $\boldsymbol{q}$. This is because in case of the narrow band $\beta\hbar\epsilon_{0\boldsymbol{p}} \ll 1$, the higher order expansions can be neglected compared with the first order. This fact allows us to extend the sound dispersion curve in the short wavelength regime.

3. Since the energy band is very narrow $\beta\hbar\epsilon_{0\boldsymbol{p}} \ll 1$, $f(\epsilon_{0\boldsymbol{p}})$ almost does not depends on $\boldsymbol{p}$; $f(\epsilon_{0\boldsymbol{p}}) \cong$ constant. Considering the fact $\sum_{\mathbf{p}} f(\epsilon_{0\mathbf{p}}) \cong N$ we obtain

$$f(\epsilon_{0\boldsymbol{p}}) \cong 1. \tag{7.2.14}$$

Since $f'(x) = \frac{df(x)}{d(\beta\hbar x)} = -f(x)\{1 \pm f(x)\}$, we obtain

$$f'(\epsilon_{0\mathbf{p}}) \cong \begin{cases} -2, & \text{for bosons} \\ \ \ 0, & \text{for fermions} \\ -1, & \text{for the Boltzmann distribution} \end{cases} . \tag{7.2.15}$$

Hereafter, we abbreviate $f'(\epsilon^0_{\boldsymbol{p}})$ as f'.

4. It is very difficult to solve Eq. (7.2.8) rigorously. But in case of the narrow bands $\beta\hbar\epsilon_{0\boldsymbol{p}} \ll 1$, the first order expansion of the wave vector $\boldsymbol{q}$ in Eq. (7.2.13) is rather good approximation. Since we need $\sum_p \phi^{00}(p,q)$ to obtain $P(q) = \mp\frac{1}{\beta\hbar^2 N} \sum_p \phi^{00}(p,q)$ in Eq. (7.2.4), in an accuracy of the first order of the wave vector $\boldsymbol{q}$, we investigate the following terms:

$$\sum_{\mathrm{i}\omega_n,\boldsymbol{p}} \phi^{00}(p,q) \equiv \phi, \quad \sum_{\mathrm{i}\omega_n,\boldsymbol{p}} \boldsymbol{q}\cdot\nabla\epsilon_{\boldsymbol{p}}\phi^{00}(p,q) = \boldsymbol{q}\cdot\boldsymbol{\Phi}. \tag{7.2.16}$$

In order to obtain the bubble diagram of the intra-band particle-hole pairs, $P(q)$, we solve Eqs. (7.2.8) and (7.2.10) using ϕ and $\boldsymbol{q}\cdot\boldsymbol{\Phi}$.

7.2.3 The Calculation of the Vertex Corrected Bubble Diagram of Intra-Band Particle-Hole Pairs

First we rewrite Eq. (7.2.10) up to the first order of the wave vector $\boldsymbol{q}$. Considering the above approximation and analytically continuing $\mathrm{i}\nu_n \to q_0 + \mathrm{i}\delta$, we obtain

$$-q_0\phi + \boldsymbol{q}\cdot\boldsymbol{\Phi} \cong \pm N(\beta\hbar)^2 f' \frac{\mathrm{i}}{\tau_0}, \tag{7.2.17}$$

To obtain ϕ, we need another equation composed of ϕ and $\boldsymbol{q}\cdot\boldsymbol{\Phi}$. To do so, we multiply the term $\boldsymbol{q}\cdot\nabla\epsilon_{0\boldsymbol{q}}$ on both sides of Eq. (7.2.8), sum up with respect to $\mathrm{i}\omega_n$, $\boldsymbol{p}$ and then analytically continue $\mathrm{i}\nu_n \to q_0 + \mathrm{i}\delta$, we obtain

$$\sum_{\mathrm{i}\omega_n \boldsymbol{p}} \left\{-q_0\boldsymbol{q}\cdot\nabla\epsilon_{0\boldsymbol{p}} + (\boldsymbol{q}\cdot\nabla\epsilon_{0\boldsymbol{p}})^2 + \boldsymbol{q}\cdot\nabla\epsilon_{0\boldsymbol{p}}\Delta\Sigma^{00}(p,q)\right\}\phi^{00}(p,q)$$

$$= \sum_{\mathrm{i}\omega_n \boldsymbol{p}} \boldsymbol{q}\cdot\nabla\epsilon_{0\boldsymbol{p}}\Delta G^{00}(p,q)\left[1 + \sum_{p_1} U^{00}_{\boldsymbol{p}\boldsymbol{p}_1}(q)\phi^{00}(p_1,q)\right]. \tag{7.2.18}$$

We denote each term on the left- and right-hand sides of Eq. (7.2.18) as L1, L2, L3, and R1, R2. Thus we obtain these terms as follows:

L1 is given by

$$\mathrm{L1} \cong -q_0 \boldsymbol{q} \cdot \boldsymbol{\Phi}.$$

L2 seems to be the second order term of $\boldsymbol{q}$. We must take into account the following term:

$$\sum_{p} (\boldsymbol{q} \cdot \nabla \epsilon_{0\boldsymbol{p}})^2,$$

multiplying to the constant term $\frac{1}{N}\phi$ in $\phi^{00}(p,q)$. It should be noted that $\phi^{00}(p,q)$ consists ϕ and $\boldsymbol{q} \cdot \nabla \epsilon_{0\boldsymbol{p}}$. So we introduce a mean intra-band elementary excitation ω_q as

$$\omega_q^2 \equiv \frac{1}{N} \sum_{p} (\epsilon_{0\boldsymbol{p}} - \epsilon_{0\boldsymbol{p}-\boldsymbol{q}})^2 \cong \frac{1}{N} \sum_{p} (\boldsymbol{q} \cdot \nabla \epsilon_{0\boldsymbol{p}})^2 = \frac{v_p^2}{3} q^2,$$

$$v_p = \frac{1}{N} \sum_{p} \frac{\partial \epsilon_{0\boldsymbol{p}}}{\partial p}, \tag{7.2.19}$$

v_p is a mean velocity of particles. Then L2 becomes the zeroth order term:

$$\mathrm{L2} = \omega_q^2 \phi.$$

Now we estimate L3 term. Since we have made the assumption (7.2.11): $\Sigma^0(q_0) \cong \frac{\mathrm{i}}{2\tau_0}$, we obtain

$$\mathrm{L3} \cong -\frac{\mathrm{i}}{\tau_0} \boldsymbol{q} \cdot \boldsymbol{\Phi}.$$

In obtaining R1, we consider the first term of the right-hand side of Eq. (7.2.13), since the second term cancels in integration of $\boldsymbol{q}$, we have

$$\mathrm{R1} \cong \mp (\beta\hbar)^2 N f' \omega_q^2.$$

Next we estimate R2 term. Since R2 term involves the factors $\boldsymbol{q} \cdot \nabla$ and $\phi^{00}(p,q)$, it becomes the higher order of $\boldsymbol{q}$ term and the corrections of $\boldsymbol{q} \cdot \nabla$. The former is out of our consideration and the latter is only higher order correction to L3, so we neglect these terms. Thus we obtain

$$\mathrm{R2} \cong 0.$$

Substituting above results into Eq. (7.2.18), we obtain

$$-q_0 \boldsymbol{q} \cdot \Phi + \omega_q^2 \phi - \frac{\mathrm{i}}{\tau_0} \boldsymbol{q} \cdot \boldsymbol{\Phi} = \mp (\beta\hbar)^2 N f' \omega_q^2. \tag{7.2.20}$$

From Eqs. (7.2.17) and (7.2.20), we obtain

$$\left\{\left(q_0+\frac{\mathrm{i}}{\tau_0}\right)q_0-\omega_q^2\right\}\phi=\pm N(\beta\hbar)^2 f'\left[\omega_q^2+\left(q_0+\frac{\mathrm{i}}{\tau_0}\right)\frac{\mathrm{i}}{\tau_0}\right].$$

Since the left-hand side of the above equation means the amplitude of the bubble of the intra-band particle-hole pairs, the imaginary part $\left(q_0+\frac{\mathrm{i}}{\tau_0}\right)\frac{\mathrm{i}}{\tau_0}$ should vanish. Thus we have

$$\phi=\sum_p \phi^{00}(p,q)=\pm N(\beta\hbar)^2 f'\frac{\omega_q^2}{\left(q_0+\frac{\mathrm{i}}{\tau_0}\right)q_0-\omega_q^2}. \tag{7.2.21}$$

Substituting Eq. (7.2.21) into Eq. (7.2.4), we obtain

$$P(q)=-\frac{\beta f'\omega_q^2}{\left(q_0+\frac{\mathrm{i}}{\tau_0}\right)q_0-\omega_q^2}. \tag{7.2.22}$$

Since we obtain the bubble of intra-band particle-hole pairs, we obtain the correlation function of intra-band density fluctuation, $F(q)$. Before we discuss on $F(q)$, we investigate the self-energy part of the particle Green's function corresponding to $\phi^{00}(p,q)$.

7.2.4 The Self-Energy Part of Particle Green's Function: $\Sigma^0(p_0+\mathrm{i}\delta)$

Now we calculate the self-energy part. Here we make the ladder approximation to the vertex part for the two Green's functions:

$$U_{\boldsymbol{pp}'}^{00}(q)\cong U_\omega^{00}+U_J^{00}(q), \tag{7.2.23}$$

which is illustrated in Figure 7.3(a). From the WT relation this corresponds to the following renormalized self-energy part:

$$\begin{aligned}\Sigma^0(p_0+\mathrm{i}\delta)&\cong(U_\omega^{00}+U_J^{00}(0))\sum_{\boldsymbol{p}}G^0(\boldsymbol{p},p_0+\mathrm{i}\delta)\\&=(U_\omega^{00}+U_J^{00}(0))\sum_{\boldsymbol{p}}\frac{1}{p_0+\mathrm{i}\delta-\epsilon_{0\boldsymbol{p}}-\Sigma^0(p_0+\mathrm{i}\delta)},\end{aligned} \tag{7.2.24}$$

which is illustrated in Figure 7.5(a). It should be noted that $G^0(p)$ is the true Green's function.

Here we consider only the imaginary part of the self-energy:

$$\Sigma^0(p_0\pm\mathrm{i}\delta)\cong\mp\frac{\mathrm{i}}{2\tau_0(p_0)}. \tag{7.2.25}$$

Considering the dispersion curve of a free particle, we can expand

$$\epsilon_{0\boldsymbol{p}}\cong|J_0|\left(1-\frac{\sin\alpha}{\alpha}\right),\quad \alpha=ap,$$

where a: the mean atomic distance. The dominant contribution of wave vectors to the density of states comes from the short wavelength regime, where $\epsilon_{0\boldsymbol{p}} \cong |J_0|$. Therefore we put:

$$\sum_{\boldsymbol{p}} G^0(\boldsymbol{p}, p_0 + \mathrm{i}\delta) \cong N/\{p_0 - |J_0| - \Sigma^0(p_0 + \mathrm{i}\delta)\}. \tag{7.2.26}$$

Defining

$$\frac{1}{2\tau_0} \equiv \sqrt{N(U_\omega^{00} + U_J^{00})}. \tag{7.2.27}$$

Substituting Eqs. (7.2.25)–(7.2.27), into Eq. (7.2.24), we obtain

$$-\frac{\mathrm{i}}{2\tau_0(p_0)} = \left(\frac{1}{2\tau_0}\right)^2 \frac{1}{p_0 - |J_0| + \frac{\mathrm{i}}{2\tau_0(p_0)}},$$

which leads to

$$\frac{\mathrm{i}}{2\tau_0(p_0)} = \frac{1}{2}\left\{p_0 - |J_0| + \mathrm{i}\sqrt{\left(\frac{1}{\tau_0}\right)^2 - (p_0 - |J_0|)^2}\right\}.$$

Thus neglecting the imaginary part, we put

$$\frac{1}{2\tau_0(p_0)} \cong \frac{1}{2}\sqrt{\left(\frac{1}{\tau_0}\right)^2 - (p_0 - |J_0|)^2} \cong \frac{1}{2}\sqrt{\left(\frac{1}{\tau_0}\right)^2 - p_0^2}, \tag{7.2.28}$$

Finally we make the following approximation:

$$\frac{1}{2\tau_0(p_0)} \cong \frac{1}{2\tau_0}. \tag{7.2.29}$$

Now we investigate the validity of Eq. (7.2.29). When we consider the frequency dependent relaxation time $1/\tau_0(p_0)$, it is always associated with p_0 such as $p_0 + \mathrm{i}/\tau_0(p_0)$ in a dynamical equation. Thus for low frequencies $p_0\tau_0 < 1$, we can put $p_0 + \mathrm{i}/\tau_0(p_0) \cong p_0 + \mathrm{i}/\tau_0 \cong \mathrm{i}/\tau_0$ and for high frequencies $p_0 + \mathrm{i}/\tau_0(p_0) \cong p_0 + \mathrm{i}/\tau_0 \cong p_0$. Thus we can justify the validity of the approximation (7.2.29).

7.2.5 The Correlation Functions of the Intra-Band Density Fluctuations; Isothermal Sound and Diffusivity

Substituting Eq. (7.2.22) into Eq. (7.2.7), we obtain the correlation function of the intra-band density fluctuations:

$$F(q) = -\frac{\beta f' \omega_q^2}{q_0(q_0 + \frac{\mathrm{i}}{\tau_0}) - \omega_{sq}^2}, \tag{7.2.30}$$

where ω_{sq} is the isothermal sound frequency given by

$$\omega_{sq}^2 = (1 - \beta f' V_{00}(q))\omega_q^2$$
$$= \frac{v_p^2}{3S(q)} q^2. \tag{7.2.31}$$

Here v_p is the velocity of particle and $S(q)$ is the structure factor, which is given by

$$S(q) = \frac{1}{1 - \beta f' V^{00}(q)}. \tag{7.2.32}$$

The correlation function $F(q)$ plays an important role in the dynamical processes of the intra-band density fluctuations.

The denominator of Eq. (7.2.30) yields the dynamical equation of the intra-band density fluctuations:

$$q_0 - \frac{\omega_{sq}^2}{q_0 + \frac{i}{\tau_0}} = 0. \tag{7.2.33}$$

Equation (7.2.33) corresponds to the dynamical equation in Mode Coupling Theory (MCT). In MCT, the term ω_{sq} corresponds to a local frequency in a cage, $\Omega_{\boldsymbol{q}}^2$, and the term i/τ_0 corresponds to $M(q_0, \boldsymbol{q})$, the memory function, where $M(q_0, \boldsymbol{q})$ is self-consistently determined as a nonlinear term. However, there are some important differences between our theory and MCT:

1. The term ω_{sq}^2 is self-consistently determined in Eq. (7.2.31), where $\omega_{sq} \propto \omega_q \propto J = e^{\frac{E}{T-T_0}}$, while in MCT $\Omega_{\boldsymbol{q}}^2$ is only replaced by the frequency $= k_B T/\hbar$.
2. As discussed later, the memory function $M(q_0, \boldsymbol{q})$ can show the characteristic behavior in the intermediate scattering function, but it cannot explain the drastic changes of the inverse relaxation time $\frac{1}{\tau_0} \propto J = e^{\frac{E}{T-T_0}}$ naturally derived from our theory. On the other hand, our theory can only explain the general situation of the intermediate scattering function. This is because our approximation (7.2.29) is valid in the low and high frequency regimes, but in the intermediate regime $\frac{1}{\tau_0}$ should be replaced by the frequency dependent relaxation time $\frac{1}{\tau_0(q_0)}$ in Eq. (7.2.28).

Now we investigate Eq. (7.2.33) in the two limiting cases:

I. $q_0\tau_0 \gg 1$: isothermal sound

Neglecting the term $\frac{i}{\tau_0}$ in the denominator of the second term in Eq. (7.2.33), we obtain the equation of isothermal sound:

$$q_0^2 - c_T^2 q^2 = 0, \tag{7.2.34}$$

where

$$\omega_{s\boldsymbol{q}}{}^2 \equiv c_T^2 q^2, \tag{7.2.35}$$

and c_T is the velocity of the isothermal sound given by

$$c_T \equiv \frac{v_p}{\sqrt{3S(0)}}. \tag{7.2.36}$$

If we consider the following relation:

$$q_0 \longleftrightarrow \mathrm{i}\frac{\partial}{\partial t}, \quad \boldsymbol{q} \longleftrightarrow -\mathrm{i}\nabla, \tag{7.2.37}$$

we obtain the dynamical equation of isothermal sound:

$$\left\{\frac{\partial^2}{\partial t^2} - c_T^2\nabla^2\right\}\phi(x,t) = 0. \tag{7.2.38}$$

This equation corresponds to the dynamical equation of isothermal sound. In the fluid, the classical dynamical equation of sound is given by

$$\left\{\frac{\partial^2}{\partial t^2} - c_S^2\nabla^2\right\}\phi(x,t) = 0, \tag{7.2.39}$$

where c_S is the adiabatic sound velocity given by

$$c_S = \gamma c_T, \quad \gamma = \frac{C_P}{C_V}. \tag{7.2.40}$$

Here C_p and C_T are specific heats at constant pressure and volume, respectively.

II. $q_0\tau_0 \ll 1$: diffusivity

Neglecting q_0 in the denominator of the second term in Eq. (7.2.33), using Eq. (7.2.35), we obtain the diffusion equation:

$$q_0 + \mathrm{i}\tau_0 c_T^2 q^2 = 0, \tag{7.2.41}$$

where the diffusion constant D is given by

$$D = \tau_0 c_T^2. \tag{7.2.42}$$

Using the relations in Eq. (7.2.37), we obtain the classical diffusion equation:

$$\left\{\frac{\partial}{\partial t} - D\nabla^2\right\}\phi(x,t) = 0. \tag{7.2.43}$$

The diffusion equation uniforms the gradient of the density.

7.2.6 The Disappearing of Sound

Next we investigate the disappearing of sound. For $g(R) = \rho\delta(R-a)$, a: the mean atomic distance, we obtain

$$V_{00}(q) = 4\pi a^2 \rho V_{00}(a)\frac{\sin\alpha}{\alpha}, \tag{7.2.44}$$

which has the minimum value at $K \cong \frac{3\pi}{2a}$, $\alpha = aq$. K is a reciprocal particle distance [9]. The sound disappears at the sound disappearing temperature $T_0 = \frac{1}{k_B \beta_0}$ and at the reciprocal particle distance K satisfying

$$1 - \beta_0 f' V_{00}(K) = 0, \tag{7.2.45}$$

where sound disappears;

$$\omega_{sK}^2 = \left\{1 - \beta_0 f' V_{00}(q)\right\} \omega_q^2 \Big|_{q=K} = 0. \tag{7.2.46}$$

This equation is used in deriving the Kauzmann entropy S_K.

7.3 Inter-Band Density Fluctuations: Phonons and Viscosity

In Chapter 6, we have investigated phonons as the Nambu-Goldstone bosons originated from the spontaneous break-down of the translational symmetry, a kind of the long range order in the glasses. There we have not considered the random scattering processes, but we could calculate the bubbles of inter-band particle-hole pairs easily, because the inter-band elementary excitations have the energy gap (~the harmonic frequency ω), which makes the calculation easier. On the other hand, we have investigated the dissipative process associated with phonon decaying into the inter-band elementary excitations, where the phonon frequencies merge into the inter-band elementary excitations. Here taking into account the random scattering processes and using the Bethe–Salpeter equation in the ladder approximation, we calculate the vertex corrected bubble diagrams of the inter-band particle-hole pairs and investigate another relaxation and dissipative processes associated with viscosity.

7.3.1 Phonons in the Relaxation and Dissipative Processes

The intra-band density fluctuations are given by

$$\rho_i^\dagger = \sum_p \{a_{ip}^\dagger a_{0p-q} + a_{0p}^\dagger a_{ip-q}\}. \tag{7.3.1}$$

The correlation functions for inter-band density fluctuations are given by

$$\begin{aligned} D_{ijq}(\tau_1 - \tau_2) &\equiv -\frac{1}{\hbar N} < T_\tau \rho_{iq}(\tau_1)\rho_{jq}^\dagger(\tau_2) >_c \\ &\equiv \sum_{i\nu_n} e^{-i\nu_n(\tau_1-\tau_2)} D_{ij}(q). \end{aligned} \tag{7.3.2}$$

We make the random phase approximation in a similar manner to sound

$$D_{ij}(q) = Q(q)\delta_{ij} + Q(q)\sum_l V_{il}(\boldsymbol{q}) D_{lj}(q), \tag{7.3.3}$$

where the vertex corrected bubble diagram of inter-band elementary excitations $Q_{ij}(q) = Q(q)$ is assumed to be isotropic with respect to i, j:

$$Q(q) \equiv \mp \frac{1}{\beta \hbar^2 N} \sum_p \{\phi^{10}(p,q) + \phi^{01}(p,q)\}, \tag{7.3.4}$$

where $\phi^{\mu\nu}(p,q)$ is defined as Eq. (7.1.22), which satisfy Eq. (7.1.19). In calculating the vertex corrected bubble diagram of intra-band elementary excitations, $\phi^{00}(p,q)$ of $P(q)$ in Eq. (7.2.4), we have solved Eq. (7.2.10) using the WT relation (7.2.9). When we have derived the WT relation, we have assumed

$$U^{\mu\mu}_{\omega,J} \sim U^{\mu\nu}_{\omega,J} \sim U^{\nu\nu}_{\omega,J}.$$

In the case of the vertex corrected bubble diagrams $\phi^{00}(p,q)$, the assumption is satisfied rigorously but in the case of the vertex corrected bubble diagrams $\phi^{10}(p,q)$ and $\phi^{01}(p,q)$, since the amplitudes of wave functions of excited particles are larger than those of ground state particles, we can expect

$$|U^{11}_{\omega.J}| > |U^{10}_{\omega.J}|, \quad |U^{01}_{\omega.J}| > |U^{00}_{\omega.J}|,$$

so that the WT relation (7.2.9) is not rigorously satisfied.

It is very hard to obtain the rigorous WT relations under the existence of the different magnitudes among $U^{\mu\mu}_{\omega,J}$, $U^{\mu\nu}_{\omega,J}$ and $U^{\nu\nu}_{\omega,J}$. Therefore we assume

$$\left(U^{\mu\mu}_{\omega} + U^{\mu\mu}_{J}\right) \sim \left(U^{\mu\nu}_{\omega} + U^{\mu\nu}_{J}\right) \sim \left(U^{\nu\nu}_{\omega} + U^{\nu\nu}_{J}\right), \tag{7.3.5}$$

and then we obtain the WT relation (7.2.9). Then in calculating phonons, we make the approximation such that the vertex and the self-energy parts satisfy the WT relation. For example, we make the ladder approximation for the vertex parts (7.1.16) and the self-energy part (7.1.27) as illustrated in Figure 7.3(a) and Figure 7.5(a), which satisfy the WT relation.

Under the ladder approximation, we solve the Bethe–Salpeter equation:

$$\begin{aligned} &\big\{ -\mathrm{i}\nu_n + \epsilon_{\mu \boldsymbol{p}} - \epsilon_{\nu \boldsymbol{p}-\boldsymbol{q}} + \Delta\Sigma^{\mu\nu}(p,q) \big\} \phi^{\mu\nu}(p,q) \\ &\quad = \Delta G^{\mu\nu}(p,q) \left[1 + \sum_{\boldsymbol{p}_1} U^{\mu\nu}_{\boldsymbol{p}\boldsymbol{p}_1}(q) \phi^{\mu\nu}(p_1,q) \right], \end{aligned} \tag{7.1.19}$$

for $\mu\nu = 01$ and 10.

In order to obtain the bubbles of the inter-band particle-hole pairs $Q(q)$ in Eq. (7.3.4), we carry out the summation of p. In summing Eq. (7.1.19a) over $\mathrm{i}\omega_n$, the main contribution of $\phi^{\mu\nu}$ comes from the very narrow band poles of $G_\mu(p)G_\nu(p-q)$ as seen in Eq. (7.1.14):

$$\phi^{\mu\nu}_{\boldsymbol{p}\boldsymbol{p}'}(\mathrm{i}\omega_n, q) = G^{\mu}(p) G^{\nu}(p-q) \left[\delta_{\boldsymbol{p}\boldsymbol{p}'} + \sum_{\boldsymbol{p}_1} U^{\mu\nu}_{\boldsymbol{p}\boldsymbol{p}_1}(q) \phi^{\mu\nu}_{\boldsymbol{p}_1\boldsymbol{p}'}(\mathrm{i}\omega_n, q) \right]. \tag{7.1.14}$$

Thus we can replace the terms such as $\Delta\Sigma^{\mu\nu}(p,q)$ and $\Delta G^{\mu\nu}(p,q)U^{\mu\nu}_{\boldsymbol{p}\boldsymbol{p}_1}(q)$ in front of $\phi^{\mu\nu}(p,q)$ by those at the narrow energy bands. So we make the following approximation:

$$\Delta\Sigma^{\mu\nu}(p,q) \cong (U^{\mu}_{\omega}+U^{\mu}_{J})\sum_{p'} G^{\mu}(p') - (U^{\nu}_{\omega}+U^{\nu}_{J})\sum_{p'} G^{\nu}(p'-q)$$
$$\cong -\frac{\mathrm{i}}{2\tau_0} - \frac{\mathrm{i}}{2\tau_1}, \tag{7.3.5}$$

$$\sum_{p} \Delta G^{\mu\nu}(p,q)U^{\mu\nu}_{\boldsymbol{p}\boldsymbol{p}_1} \cong \sum_{p} \Delta G^{\mu\nu}(p,q)(U^{01}_{\omega}+U^{01}_{J})$$
$$\cong -\frac{\mathrm{i}}{2\tau^{o}_0} - \frac{\mathrm{i}}{2\tau^{o}_1}, \tag{7.3.6}$$

where $\mu\nu = 01$, or 10. Equation (7.3.5) comes from the same approximation as Eq. (7.2.11). τ_0 and τ_1 are the relaxation times of the particles in the lower and upper bands, respectively. Equation (7.3.6) comes from the simultaneously scattering processes between the lower and upper bands. This term corresponds to the cosine-function term with the scattering angle of an electron in the electric conductivity.

Substituting Eqs. (7.3.5) and (7.3.6) into Eq. (7.1.19), summing over $\mathrm{i}\omega_n$ and $\boldsymbol{p}$, and analytically continuing $\mathrm{i}\nu_n \to q_0 + \mathrm{i}\delta$, we obtain

$$\sum_{p}\left\{-q_0 + \epsilon_{\mu\boldsymbol{p}} - \epsilon_{\nu\boldsymbol{p}-\boldsymbol{q}} - \frac{\mathrm{i}}{2\tau_M}\right\}\phi^{\mu\nu}(p,q)$$
$$= \mp\beta\hbar\sum_{p}\left\{f\left(\epsilon_{\mu\boldsymbol{p}} - \frac{\mathrm{i}}{2\tau_\mu}\right) - f\left(\epsilon_{\nu\boldsymbol{p}-\boldsymbol{q}} + \frac{\mathrm{i}}{2\tau_\nu}\right)\right\}, \tag{7.3.7}$$

$$\frac{1}{2\tau_M} \equiv \frac{1}{2\tau_0} + \frac{1}{2\tau_1} - \frac{1}{2\tau^{o}_0} - \frac{1}{2\tau^{o}_1}. \tag{7.3.8}$$

Under the condition of $|J_\mu|\beta\hbar \ll 1$ and $\beta\hbar/\tau_\mu \ll 1$, considering $f(\epsilon_0) \cong 1$ and $f(\epsilon_1) \cong 0$, we obtain

$$\left(-q_0 + \omega - \frac{\mathrm{i}}{2\tau_M}\right)\phi^{10} = \pm\beta\hbar N f(\epsilon_0) \cong \pm\beta\hbar N, \tag{7.3.9}$$

$$\left(-q_0 - \omega - \frac{\mathrm{i}}{2\tau_M}\right)\phi^{01} = \mp\beta\hbar N f(\epsilon_0) \cong \mp\beta\hbar N, \tag{7.3.10}$$

where

$$\phi^{\mu\nu} = \sum_{p}\phi^{\mu\nu}(p,q). \tag{7.3.11}$$

The calculation of $\phi^{\mu\nu}$ in Eqs. (7.3.9) and (7.3.10) are very easier than $\phi = \sum_p \phi^{00}(p,q)$ in Eq. (7.2.21), because the lower and upper bands are separated by the gap $\cong \omega$.

Substituting Eqs. (7.3.9) and (7.3.10) into Eq. (7.3.4), we obtain

$$Q(q) = \frac{2\omega/\hbar}{\left(q_0 + \frac{\mathrm{i}}{2\tau_M}\right)^2 - \omega^2}. \tag{7.3.12}$$

If we do not consider the random scattering processes $\frac{1}{\tau_M} = 0$, $Q(q)$ reduces to that in the pure phonon state.

The translational invariance requires the gap equation:

$$\delta_{ij} = Q(0)V_{ij}(\boldsymbol{q} = 0), \tag{6.3.11}$$

where

$$V_{ij}(\boldsymbol{q}) = \zeta^2 \int d^3Rg(R)e^{-\mathrm{i}\boldsymbol{q}\cdot\boldsymbol{R}}V_{ij}(R), \quad \zeta^2 = \frac{\hbar}{2M\omega}. \tag{6.4.28}$$

The gap equation leads to

$$\delta_{ij} = -\frac{\frac{2\omega}{\hbar}}{\omega^2 + \frac{1}{4\tau_M^2}}V_{ij}(0) = -\frac{1}{M\left(\omega^2 + \frac{4}{\tau_M^2}\right)}\int d^3Rg(R)V_{ij}(R). \tag{7.3.13}$$

If we take $\boldsymbol{q} = q\boldsymbol{e}_3$, we can diagonalize $V_{ij}(\boldsymbol{q})$ as $V_{ii}(\boldsymbol{q}) \equiv V_\lambda(\boldsymbol{q})$. Equation (7.3.3) leads to

$$\begin{aligned}
D_\lambda(q) &= \frac{Q(q)}{1 - Q(q)V_\lambda(q)} = \frac{\frac{2\omega}{\hbar}}{\left(q_0 + \frac{\mathrm{i}}{2\tau_M}\right)^2 - \omega^2 - \frac{2\omega}{\hbar}V_\lambda(q)} \\
&= \frac{\frac{2\omega}{\hbar}}{q_0\left(q_0 + \frac{\mathrm{i}}{\tau_M}\right) - \frac{1}{4\tau_M^2} - \omega^2 - \frac{2\omega}{\hbar}V_\lambda(q)} \\
&= \frac{\frac{2\omega}{\hbar}}{q_0\left(q_0 + \frac{\mathrm{i}}{\tau_M}\right) + \frac{2\omega}{\hbar}V_\lambda(0) - \frac{2\omega}{\hbar}V_\lambda(q)} \\
&= \frac{\frac{2\omega}{\hbar}}{q_0\left(q_0 + \frac{\mathrm{i}}{\tau_M}\right) - \omega_{\lambda\boldsymbol{q}}^2},
\end{aligned} \tag{7.3.14}$$

where from the second line to the third one, we have used the gap equation (7.3.13). The gap equation plays an important role in constructing phonons in the dissipative process.

The phonon dispersion curves are given by

$$\begin{aligned}
\omega_{\lambda\boldsymbol{q}}^2 &= \frac{2\omega}{\hbar}\{V_\lambda(0) - V_\lambda(\boldsymbol{q})\} \\
&= -\frac{1}{M}\int d^3Rg(R)(1 - e^{\mathrm{i}\boldsymbol{q}\cdot\boldsymbol{R}})V_\lambda(\boldsymbol{R}),
\end{aligned} \tag{7.3.15}$$

$\omega_{\lambda\mathbf{q}}$ are a longitudinal $\lambda = l$ for $i = z$ and two transverse modes $\lambda = t$ for $i = x, y$.

The denominator of the phonon Green's functions (7.3.14) yields the dynamical equations of phonons with random scattering processes:

$$q_0 - \frac{\omega^2_{\lambda \boldsymbol{q}}}{q_0 + \frac{\mathrm{i}}{\tau_M}} = 0. \tag{7.3.16}$$

Here τ_M is called the Maxwell relaxation time [10]. Equation (7.3.16) for phonons is just the dynamical equations in MCT. It should be noted the term $\omega^2_{\lambda \boldsymbol{q}}$ is a phonon frequency contrary to $\Omega^2_{\boldsymbol{q}}$, a local frequency of a particle in a cage in MCT. The term i/τ_M corresponds to $M(q_0, \boldsymbol{q})$, the memory function and should be more accurately replaced by the frequency-dependent term proportional to Eq. (7.2.28).

Now we investigate Eq. (7.3.16) in the two limiting cases:

I. $\tau_M q_0 \gg 1$: phonons

Neglecting $\frac{\mathrm{i}}{\tau_M}$, we obtain the equation of phonons:

$$q_0^2 = \omega^2_{\lambda \boldsymbol{q}}, \tag{7.3.17}$$

where $\omega_{\lambda \boldsymbol{q}}$ are phonons or elastic waves as shown in Figure 6.3. In the long wavelength regime, if we put $g(R) = \rho \delta(R - a)$, and consider only the V_σ-coupling in Eq. (6.4.26), we obtain

$$\omega^2_{\lambda \mathbf{q}} \cong c^2_\lambda q^2, \tag{7.3.18}$$

$$c_t^2 = \frac{\omega^2 a^2}{10}, \quad c_l^2 = \frac{3\omega^2 a^2}{10}, \quad \omega^2 = -\frac{4\pi a^2}{3M} V_\sigma. \tag{7.3.19}$$

where c_l, c_t, and ω are a longitudinal, two transverse phonon velocities and the harmonic frequency given in Eqs. (6.4.49–51) by putting $V_\pi = 0$, respectively. If we use the relations (7.2.37):

$$q_0 \longleftrightarrow \mathrm{i}\frac{\partial}{\partial t}, \quad \boldsymbol{q} \longleftrightarrow -\mathrm{i}\nabla,$$

we have

$$\left\{ \frac{\partial^2}{\partial t^2} - c^2_\lambda \nabla^2 \right\} \phi_\lambda(x, t), \tag{7.3.20}$$

which correspond to the classical dynamical equations of elastic waves.

II. $\tau_M q_0 \ll 1$: viscosity

In the long wavelength regime, we obtain the equation of kinetic viscosity:

$$q_0^2 = \mathrm{i}\tau_M c^2_\lambda q^2. \tag{7.3.21}$$

Thus we obtain the kinetic viscosity ν_λ:

$$\frac{\nu_\lambda}{M\rho} \equiv \eta_\lambda = \tau_M c^2_\lambda, \quad \nu_t = \eta, \quad \nu_l = \frac{4}{3}\eta + \zeta. \tag{7.3.22}$$

Using Eq. (7.3.19), we obtain kinetic shear and longitudinal viscosity:

$$\frac{\eta}{\rho M} = \tau_M c_t^2 = \frac{\tau_M \omega^2 a^2}{10}, \quad \frac{\frac{4}{3}\eta + \zeta}{M\rho} = \tau_M c_l^2 = \frac{3\tau_M \omega^2 a^2}{10}, \quad \omega^2 = -\frac{4\pi a^2}{3M} V_\sigma. \tag{7.3.23}$$

If we use the relations (7.2.37), we have the classical kinetic equations of viscosity:

$$\left\{ M\rho \frac{\partial}{\partial t} - \nu_\lambda \nabla^2 \right\} \phi_\lambda(x, t) = 0. \tag{7.3.24}$$

Since $\bar{J}_0 < 0$ and $\bar{J}_i > 0$, τ_μ are positive in Eq. (7.3.5) and τ_μ^o negative in Eq. (7.3.6), where μ takes 0 or 1. Thus we obtain $\tau_M > 0$ in Eq. (7.3.8). Since from Eqs. (7.3.5) and (7.3.6), we obtain

$$\frac{1}{\tau_\mu^o} = \frac{U_\omega^{01} + U_J^{01}}{U_\omega^\mu + U_J^\mu} \frac{1}{\tau_\mu}, \quad |\tau_\mu^o| \propto \tau_\mu \propto \tau_0.$$

Thus all inverse relaxation times obey the same hopping amplitude J, which plays a significant role in the liquid-glass transition as discussed later.

7.4 The Kauzmann Entropy Crisis and the VTF Law; Specific Heat, Relaxation Times, and Transport Coefficients

In the previous sections, introducing the random scattering potentials, we have constructed the dynamical and dissipative equations for intra-band density fluctuations in liquids and those for the inter-band density fluctuations in glasses. Sound and diffusion modes play an essential role in the liquid state, where sound velocity c_T, the inverse relaxation time τ_0^{-1}, and the diffusivity $D = \tau_0 c_T^2$ are proportional to the hopping amplitude J; $c_T, \tau_0^{-1}, D \propto J$, while phonons and viscosities play an essential role in the glassy state, where the phonon velocities c_λ is independent of temperature, but the inverse Maxwell relaxation time τ_M^{-1} and the inverse viscosities $\eta_\lambda^{-1} = (\tau_M c_\lambda^2)^{-1}$ are proportional to J; $\tau_M^{-1}, \eta_\lambda^{-1} \propto J$.

Experimental results show the following universal features in the liquid-glass transition:

1. *The Kauzmann entropy crisis:* $S_K = -\frac{N_0 k_B}{2}\frac{T_0}{T-T_0}$.
   The entropy of the system seems to have a negative value below the so-called Kauzmann temperature T_K. This phenomenon is called the Kauzmann paradox.
2. The gap of specific heat at the glass transition temperature T_g.
3. *The VTF law:* $v_p, c_T, \tau_0{}^{-1}, \tau_M^{-1}, D, \eta_\lambda^{-1} \propto J = e^{-\frac{E}{T-T_0}}$.
   The particle velocity v_p, the sound velocity c_T, the inverse relaxation times $\tau_0{}^{-1}$, $\tau_M{}^{-1}$, the diffusivity D, and the inverse viscosities η_λ^{-1} drastically drop to zero near the liquid-glass transition. The Stokes law is satisfied; $D\eta_\lambda =$constant.

We elucidate these universal features as follows:

1. Near the liquid-glass transition, the Free energies of sound and intra-band density fluctuations due to the extra bubbles of the intra-band particle-hole pairs mix. The mixing Free energy is proportional to $\sum_{q\sim K} \ln \frac{\omega_{sq}}{\omega_q} = \sum_{q\sim K} \ln[1-\beta_0 f' V_{00}(K)] \propto \ln(1-\frac{T_0}{T})$, where sound disappears at T_0; $1-\beta_0 f' V_{00}(K) = 0$. The T-derivative of the mixing Free energy leads to the Kauzmann entropy.
2. The T-derivative of the Kauzmann entropy yields the gap of the specific heat at the liquid-glass transition temperature.
3. The hopping probability is proportional to the configuration number, which is e^{S_K/k_B} from the Einstein relation. The hopping probability of a particle is given by the hopping amplitude: $J = e^{\frac{zS_K}{Nk_B}} = e^{-\frac{E}{T-T_0}}$, where $E = zN_0T_0/2N$ and z is of the order of the number of the surrounding particles. J manifests the VTF law.

Thus we can develop the unified theory of the liquid-glass transition [1–6].

7.4.1 The Kauzmann Entropy and the VTF Law

First, we investigate the entropy due to intra-band density fluctuations. Since sound depends on the temperature, we must carefully investigate the thermodynamical quantities. So we start with the interaction Hamiltonian. Here we consider the thermodynamical function. We introduce the Hamiltonian with a parameter λ [11,12]:

$$H(\lambda) = H_0 + \lambda H_I. \tag{7.4.1}$$

The thermodynamical function Ω_λ for the above Hamiltonian is given by

$$\Omega_\lambda = -\frac{1}{\beta} \ln\{\mathrm{Tr} e^{-\beta(H_0-\mu N+\lambda H_I)}\}. \tag{7.4.2}$$

Here we calculate the thermodynamical function due to intra-band density fluctuations denoted by $\Omega_{\mathrm{intra}\lambda}$. From Eq. (6.2.22), we obtain

$$\frac{\partial \Omega_{\mathrm{intra}\lambda}}{\partial \lambda} = \frac{1}{\lambda} < \lambda H_\lambda >_\lambda \cong \frac{1}{2} \sum_q V_{00}(q) < \rho_{0q}^\dagger \rho_{0q} >_\lambda, \tag{7.4.3}$$

where $< \cdots >_\lambda$ is given by the equation replaced H_I by λH_I in $< \cdots >_c$. Inserting Eq. (7.2.7) replaced $V_{00}(q)$ by $\lambda V_{00}(q)$ into Eq. (7.4.3), neglecting the λ-dependence of $P_0(q)$ because of the condition $\beta|J_\mu| \ll 1$, integrating Eq. (7.4.3) in λ and putting $\lambda \to 1$, we obtain the thermodynamical function due to the intra-band density fluctuations, Ω_{intra}:

$$\Omega_{\mathrm{intra}} = \frac{1}{2\beta} \sum_{\mathrm{i}\nu_n q} \ln\{1 - V_{00}(q)P_0(q)\} + \Omega_{\mathrm{intra}0}, \tag{7.4.4}$$

where the term $\Omega_{\mathrm{intra}\,0}$ is added such that Ω_{intra} is constructed only by sound mode. In order to calculate the entropy we neglect $1/\tau_0$ terms under the condition $\beta\hbar/\tau_0 \ll 1$. Since Eq. (7.2.22) leads to

$$P_0(q) = \beta f'(\omega_q^0)^2/(\mathrm{i}\nu_n^2 - \omega_q^2), \tag{7.4.5}$$

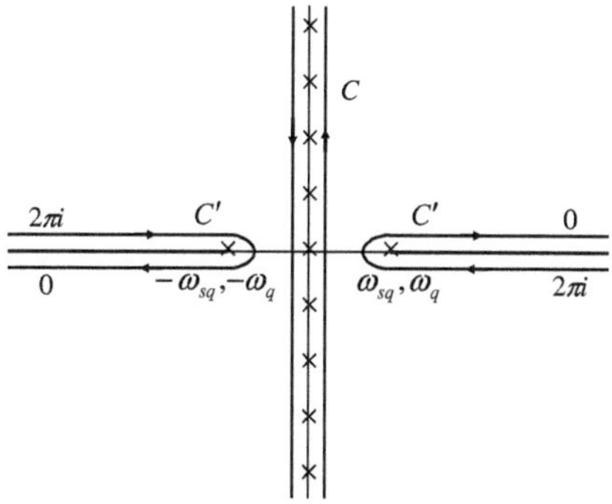

Figure 7.6 The contours C and C' in z-plane are shown. The contour C encloses the poles at $i\nu_n$. C' encloses the branch cuts of $\ln(z-\omega_{sg},\omega_q)$ ended at ω_{sq} or ω_q $\ln(z+\omega_{sq},\omega_q)$ ended at $-\omega_{sq}$ or ω_q. 0 and $2\pi i$ are the arguments of $\ln(z-\omega_{sq},\omega_q)$ and $\ln(z+\omega_{sq},\omega_q)$.

we obtain

$$\Omega_{\text{intra}} = \frac{1}{2\beta}\sum_q \ln\frac{i\nu_n^2-\omega_{sq}^2}{i\nu_n^2-\omega_q^2} + \Omega_{\text{intra0}}. \tag{7.4.6}$$

Analytically continuing the contour C to C' as shown in Figure 7.6, we calculate the first term:

$$\begin{aligned}
\sum_{i\nu_n}\ln\frac{i\nu_n^2-\omega_{sq}^2}{i\nu_n^2-(\omega_q^0)^2} &= \frac{\beta\hbar}{2\pi i}\int_C \frac{dz}{e^{\beta\hbar z}-1}\ln\frac{z^2-\omega_{sq}}{z^2-\omega_q^2}e^{\eta z} \\
&= \beta\hbar\left[\int_{-\infty}^{-\omega_{sq}}+\int_{\infty}^{\omega_{sq}}-\int_{-\infty}^{-\omega_q}-\int_{\infty}^{\omega_q}\right]\frac{dx}{e^{\beta\hbar x}-1} \\
&= -\beta\hbar\int_{\omega_q}^{\omega_{sq}} dx\left[\frac{1}{e^{-\beta\hbar x}-1}-\frac{1}{e^{\beta\hbar x}-1}\right] \\
&= \ln\frac{1-e^{\beta\hbar\omega_{sq}}}{1-e^{\beta\hbar\omega_q}}+\ln\frac{1-e^{-\beta\hbar\omega_{sq}}}{1-e^{-\beta\hbar\omega_q}} \\
&= \beta\hbar(\omega_{sq}-\omega_q)+2\ln\frac{1-e^{-\beta\hbar\omega_{sq}}}{1-e^{-\beta\hbar\omega_q}},
\end{aligned}$$

where from the third line to the fourth line we have used the relation

$$\frac{d}{dx}\ln(1-e^{\pm\beta\hbar x}) = \mp\beta\hbar\frac{1}{e^{\mp\beta\hbar x}-1)}.$$

Thus we obtain

$$\begin{aligned}
\Omega_{\text{intra}} &= \frac{1}{\beta}\sum_{\boldsymbol{q}}[\ln(1-e^{-\beta\hbar\omega_{sq}})-\ln(1-e^{-\beta\hbar\omega_q})]+\Omega_{\text{intra0}} \\
&\equiv \Omega_{\text{sound}}+\Omega_{\text{intra-fluct.}}+\Omega_{\text{intra0}},
\end{aligned} \tag{7.4.7}$$

$$\Omega_{\text{sound}} = \frac{1}{\beta}\sum_{q} \ln(1 - e^{-\beta\hbar\omega_{sq}}), \tag{7.4.8}$$

$$\Omega_{\text{intra-fluct.}} = -\frac{1}{\beta}\sum_{q} \ln(1 - e^{-\beta\hbar\omega_{q}}), \tag{7.4.9}$$

$$\Omega_{\text{intra0}} = -\Omega_{\text{intra-fluct.}} = \frac{1}{\beta}\sum_{q} \ln(1 - e^{-\beta\hbar\omega_{q}}), \tag{7.4.10}$$

where we have neglected the term $\frac{1}{2}\sum_{q}(\hbar\omega_{sq} - \hbar\omega_{q})$.

The first term, Ω_{sound}, the second term, $\Omega_{\text{intra-fluct.}}$, and the third term, Ω_{intra0} in Eq. (7.4.7) correspond to the thermodynamical function for the sound, the fluctuations due to extra bubbles of the intra-band particle-hole pairs, and the dissipation due to the diffusion, respectively. In an equilibrium, bubbles of the intra-band particle-hole pairs are always created and sum of an infinite series of interacting bubbles of the intra-band particle-hole pairs constitutes sound mode. Therefore, extra bubbles of the intra-band particle-hole pairs non-constructing sound generate the fluctuations of the system, $\Omega_{\text{intra-fluct.}}$. The dissipative processes Ω_{intra0} compensate the fluctuations, $\Omega_{\text{intra-fluct.}}$ so that the system becomes an local equilibrium.

However, in a supercooling process, enormous bubbles of intra-band particle-hole pairs non-constituting sound are created so that the dissipative processes cannot completely compensate the bubbles of the intra-band particle-hole pairs; the thermodynamical function of the remaining part of bubbles of intra-band particle-hole pairs mixes with the thermodynamical function of sound. We denote the thermodynamical function of the mixing as Ω_{mixing}. Near the liquid-glass transition, since $\beta\hbar|J_{\mu}| \ll 1$, so $\beta\hbar\omega_{sq}, \beta\hbar\omega_{q} \ll 1$, from Eq. (7.2.31) we obtain

$$\begin{aligned}\Omega_{mixing} &= \frac{1}{\beta}\sum_{q} \ln(1 - e^{-\beta\omega_{sq}}) - \frac{1}{\beta}\sum_{q} \ln(1 - e^{-\beta\omega_{q}}) \\ &\cong \frac{1}{\beta}\sum_{q} \ln\frac{\omega_{sq}}{\omega_{q}} \\ &= \frac{1}{\beta}\sum_{q} \ln\frac{\sqrt{\{1-\beta f' V_{00}(q)\}}\omega_{q}}{\omega_{q}} = \frac{1}{2\beta}\sum_{q} \ln\{1-\beta f' V_{00}(q)\}\end{aligned} \tag{7.4.11}$$

Since the sound instability occurs at the sound instability temperature $T_0 = 1/k_B\beta_0$ and at a reciprocal particle distance K:

$$1 - \beta_0 f' V_{00}(K) = 0. \tag{7.2.44}$$

The dominant contribution of the mixing (7.4.11) to the thermodynamical function comes from the regions $q \sim K$. Near the glass transition, If we denote the number of states of the regions $q \sim K$ as N_0, $\boldsymbol{q} \in N_0$, we obtain

$$\Omega_{\text{mixing}} \cong \sum_{q \in N_0} \frac{1}{2\beta} \ln\{1 - \beta f' V_{00}(q)\} = \frac{N_0}{2\beta} \ln\{1 - \beta f' V_{00}(K)\}$$
$$= \frac{N_0}{2\beta} \ln\left\{1 - \frac{T_0}{T}\right\}. \tag{7.4.12}$$

At the regions outside $\boldsymbol{q} \in N_0$, which we denote as $\boldsymbol{q} \notin N_0$, the thermodynamical function of sound remains and that of the extra bubbles of intra-band particle-hole pairs are compensated by the dissipative process. Thus we obtain

$$\Omega_{\text{intra}} \cong \Omega_{\text{mixing}} + \Omega'_{\text{sound}}, \tag{7.4.13}$$

$$\Omega'_{\text{sound}} \cong \frac{1}{\beta} \sum_{\boldsymbol{q} \notin N_0} \ln(1 - e^{-\beta \hbar \omega_{sq}}), \tag{7.4.14}$$

where the prime on Ω_{sound} means that $\boldsymbol{q}$ are restricted in $\boldsymbol{q} \notin N_0$; the number of $\boldsymbol{q}$-states, $\boldsymbol{q} \notin N_0$, is about $N - N_0$.

Thus we obtain the entropy due to the intra-band density fluctuations, S_{intra}:

$$S_{\text{intra}} = \frac{\partial \Omega_{\text{mixing}}}{\partial T} + \frac{\partial \Omega'_{\text{sound}}}{\partial T}$$
$$\equiv S_{\text{K}} + S'_{\text{sound}}, \tag{7.4.15}$$

$$S_{\text{K}} \cong -\frac{N_0 k_B}{2} \frac{T_0}{T - T_0}, \tag{7.4.16}$$

$$S'_{\text{sound}} \cong -k_B \sum_{\boldsymbol{q} \notin N_0} \ln(1 - e^{-\beta \hbar \omega_{sq}}) + \frac{1}{T} \sum_{\boldsymbol{q} \notin N_0} \frac{\hbar \omega_{sq}}{e^{\beta \hbar \omega_{sq}} - 1}, \tag{7.4.17}$$

where S_{K} comes from Ω_{mixing}. We call S_{K} the Kauzmann entropy, which manifests the Entropy crisis. S'_{sound} comes from Ω'_{sound}. The number of $\boldsymbol{q}$-states, $\boldsymbol{q} \notin N_0$, explains the specific heat anomaly at the low temperature as discussed later.

A state of N-particles distributed randomly in space corresponds to a minimum of N-particle potential in a configuration space. A hopping of a particle from a site to a vacancy corresponds to a jump from a deep valley to another deep valley in the multi-dimensional configuration space in ELM as the α-relaxation process. The successive hoppings constitutes a configuration space. A hopping also generates intra-band elementary excitations and the successive hoppings yield the fluctuation entropy S_K due to the intra-band density fluctuations. The hopping probability is proportional to the configuration number, which is e^{S_K/k_B} from the Einstein relation. The hopping probability of a particle is given by the hopping amplitude:

$$J = e^{\frac{zS_{\text{K}}}{Nk_B}} = e^{-\frac{E}{T - T_0}}, \tag{7.4.18}$$

where $E = zN_0T_0/2N$ and z is of the order of the number of the surrounding particles. This equation manifests the Vogel-Tamman-Fulcher (VTF) law. Since a μ-state particle temporarily stays at a site $\boldsymbol{R}_m$ and hops to a surrounding vacancy at $\boldsymbol{R}_n$, the fluctuation magnitude of the hopping becomes from 0 to $J_{m\mu n\mu}$, the configurationally averaged value of which is $J_\mu(\boldsymbol{R}_m - \boldsymbol{R}_n)$. Thus we can put

$$|J_\mu| \propto J, \tag{7.4.19}$$

$$\sqrt{U_J^{\mu\nu}} \propto J. \tag{7.4.20}$$

It should be noted that the hopping J generates the intra-band elementary excitations, the sound ω_{sq}, and the Kauzmann entropy S_K, and that S_{K} determines J in Eq. (7.4.18); $J \longrightarrow \omega_{sq} \longrightarrow S_{\mathrm{K}} \longrightarrow J$. Thus ω_{sq} is self-consistently determined.

7.4.2 The Entropy of Bubbles of Inter-Band Particle-Hole Pairs and the Magnitude of the Randomness of Harmonic Frequencies

Next we investigate the entropy due to inter-band density fluctuations. The thermodynamical function due to the inter-band density fluctuations is obtained in a similar manner to that due to the intra-band density fluctuations:

$$\begin{aligned}\Omega_{\text{inter}} &= \frac{1}{2\beta}\sum_{\lambda\boldsymbol{q}} \ln \frac{\mathrm{i}\nu_n^2 - \omega_{\lambda\boldsymbol{q}}^2}{\mathrm{i}\nu_n^2 - \omega^2} + \Omega_{\text{inter}\,0} \\ &= \frac{1}{\beta}\sum_{\lambda\boldsymbol{q}} \{\ln(1 - e^{-\beta\hbar\omega_{\lambda\boldsymbol{q}}}) - \ln(1 - e^{-\beta\hbar\omega})\} + \Omega_{\text{inter}\,0} \\ &= \Omega_{\text{phonon}} + \Omega_{\text{inter-fluct.}} + \Omega_{\text{inter0}} \end{aligned} \tag{7.4.21}$$

$$\Omega_{\text{phonon}} = \frac{1}{\beta}\sum_{\lambda\boldsymbol{q}} \ln(1 - e^{-\beta\hbar\omega_{\lambda\boldsymbol{q}}}), \tag{7.4.22}$$

$$\Omega_{\text{inter-fluct.}} = -\frac{1}{\beta}\sum_{\lambda\boldsymbol{q}} \ln(1 - e^{-\beta\hbar\omega}), \tag{7.4.23}$$

$$\Omega_{\text{inter}\,0} = -\Omega_{\text{inter-fluct.}} = \frac{1}{\beta}\sum_{\lambda\boldsymbol{q}} \ln(1 - e^{-\beta\hbar\omega}), \tag{7.4.24}$$

where we have neglected the term $\frac{1}{2}\sum_{\lambda\boldsymbol{q}}(\hbar\omega_{\lambda\boldsymbol{q}} - \hbar\omega)$. In this case, contrary to the intra-band density fluctuations, the mixing between Ω_{phonon} and $\Omega_{\text{inter-fluct.}}$ does not occur. The entropy due to the inter-band density fluctuations is given by

$$\begin{aligned} S_{\text{inter}} &= \frac{\partial\Omega_{\text{phonon}}}{\partial T} + \frac{\partial\Omega_{\text{inter-fluct.}}}{\partial T} + \frac{\partial\Omega_{\text{inter0}}}{\partial T} \\ &= S_{\text{phonon}} + S_\Omega + -S_{\text{inter0}}, \end{aligned} \tag{7.4.25}$$

$$S_{\text{phonon}} = -k_B\sum_{\lambda\boldsymbol{q}} \ln(1 - e^{-\beta\hbar\omega_{\lambda\boldsymbol{q}}}) + \frac{1}{T}\sum_{\lambda\boldsymbol{q}} \frac{\hbar\omega_{\lambda\boldsymbol{q}}}{e^{\beta\hbar\omega_{\lambda\boldsymbol{q}}} - 1}. \tag{7.4.26}$$

$$S_\Omega = k_B \sum_{\lambda q} \ln(1 - e^{-\beta\hbar\omega}) - \frac{1}{T}\sum_{\lambda q} \frac{\hbar\omega}{e^{\beta\hbar\omega} - 1}. \tag{7.4.27}$$

$$S_{\text{inter0}} = -S_\Omega = -k_B \sum_{\lambda q} \ln(1 - e^{-\beta\hbar\omega}) + \frac{1}{T}\sum_{\lambda q} \frac{\hbar\omega}{e^{\beta\hbar\omega} - 1}. \tag{7.4.28}$$

S_{phonon}, S_Ω, and S_{inter0} correspond to the entropies for the phonons, the fluctuations due to the extra bubbles of inter-band particle-hole pairs, and the dissipation due to the viscosity, respectively. The inter-band fluctuation entropy S_Ω lowers the equilibrium entropy. The dissipation entropy compensates the fluctuation entropy and the system becomes a local equilibrium.

A propagation of an up and down transition of a particle at a site to a surrounding site corresponds to a jump of a shallow valley to another shallow valley in the multi-dimensional configuration space in the ELM as the β-relaxation process. The successive propagations constitute another configuration space different from hoppings. The probability of the magnitude of the randomness of harmonic frequencies is proportional to the configuration number e^{S_Ω/k_B}. The probability of the magnitude of randomness of harmonic frequencies of a particle is proportional to

$$\Omega = e^{zS_\Omega/Nk_B} = \exp\left\{-\frac{3z\beta\hbar\omega}{e^{\beta\hbar\omega} - 1}\right\}. \tag{7.4.29}$$

Thus we obtain

$$\sqrt{U_\omega^{\mu\nu}} \propto \Omega. \tag{7.4.30}$$

Ω increases smoothly as the temperature decreases.

7.4.3 The Liquid-Glass Transition

In liquids, sound is essential, while phonons are not essential, but near the liquid-glass transition temperature, the phonon fluctuation entropy (the fluctuation entropies of the extra bubbles of inter-band particle-hole pairs) plays an essential role in causing the liquid-glass transitions. Near the liquid-glass transition, as the temperature decreases, the Kauzmann entropy S_K due to the fluctuations of bubbles of intra-band particle-hole pairs rapidly drops according to the Curie law with the critical temperature T_0 and a negative sign. The Kauzmann entropy S_K diverges negatively at T_0 so that the entropy of the system seems to occur the entropy crisis. This is called the Kauzmann paradox. However, the phonon fluctuation entropy S_Ω increases with the temperature decreasing and S_Ω dominates S_K below a crossover temperature above T_0 so that S_Ω prevents the entropy crisis as shown in Figure 7.7, where the Kauzmann entropy S_K and the inter-band fluctuation entropy S_Ω normalized by Nk_B are shown with respect to $\frac{k_BT}{\hbar\omega}$, where we take $\frac{N_0}{N} = \frac{1}{5}$. We show two cases: a fragile case, $\frac{k_BT_0}{\hbar\omega} = \frac{1}{4}$ and a less fragile case, $\frac{1}{10}$.

Thus we can identify the crossover temperature with the liquid-glass transition temperature T_g:

$$S_K = S_\Omega|_{T=T_g}. \tag{7.4.31}$$

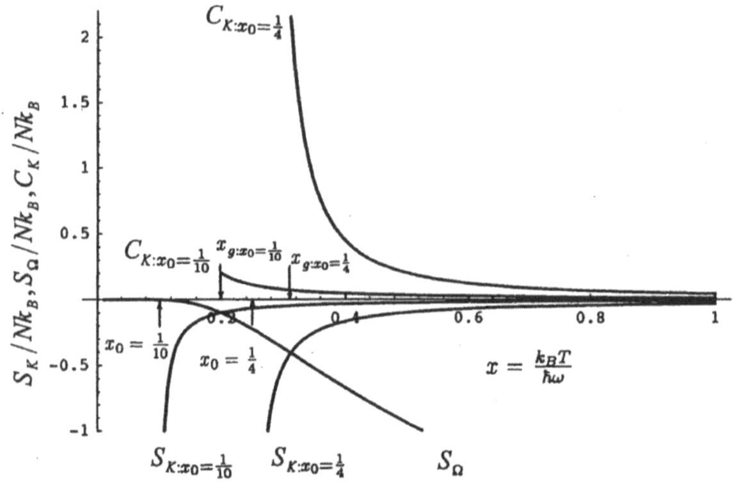

Figure 7.7 The Kauzmann entropy S_K, phonon fluctuation entropy S_Ω, and the specific heat C_K due to S_K normalized by Nk_B are illustrated in $x = k_\mathrm{B}T/\hbar\omega$ for $x_0 = 1/4$ and $1/10$.

Sound is density fluctuation waves essential in a liquid, while phonons are elastic waves essential in a solid. The glass transition is a sort of dynamical transition.

The ratio of the value of S_K at the crossover temperature T_g to that in the liquid state is of the order of 10, $\frac{S_\mathrm{K}|_{T_g}}{S_\mathrm{K}|_{T_\mathrm{liquid}}} \sim O(10)$, while the ratio of the value of S_Ω is of the order of 1, $\frac{S_\Omega|_{T_g}}{S_\Omega|_{T_\mathrm{liquid}}} \sim O(1)$, as shown in Figure 7.7. Thus the ratio of J^{-1} is exponential of the order of 10, while Ω^{-1} is exponential of the order of 1, so that we can reconfirm Eq. (7.4.31).

7.4.4 Specific Heat

Next we investigate the specific heat. The specific heat C of a system with entropy S is given by

$$C = T\frac{\partial S}{\partial T}. \tag{7.4.32}$$

We calculate the specific heat due to the intra-band and inter-band density fluctuations.

i. The specific heat due to intra-band density fluctuations

First, we calculate the specific heat due to the entropy of the intra-band density fluctuations, which consists of the Kauzmann entropy S_K and the entropy due to the sound, S'_sound. The specific heat due to S_K, C_K is given by

$$C_K = \begin{cases} \frac{N_0 k_B T}{2}\frac{T_0}{(T-T_0)^2} & \text{for } T_g < T \\ 0 & \text{for } T < T_g \end{cases}. \tag{7.4.33}$$

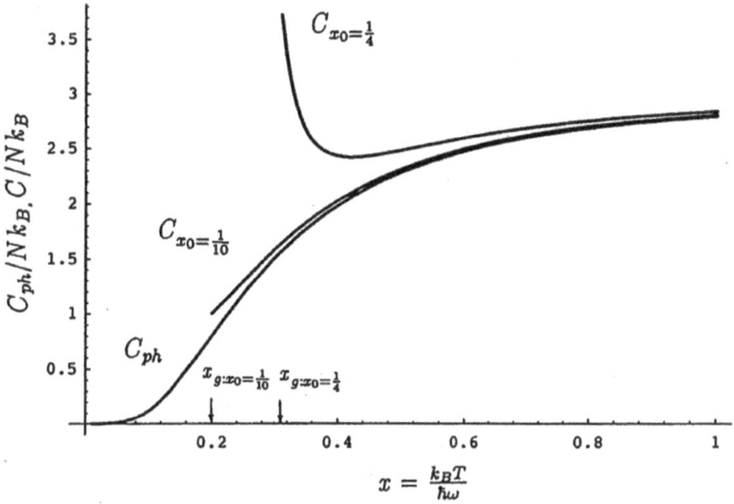

Figure 7.8 The specific heat of phonons, C_{phonon} and the summed specific heat $C = C_{\text{phonon}} + C_{\text{K}}$ normalized Nk_{B} for a fragile liquid $x_0 = k_{\text{B}}T/\hbar\omega = \frac{1}{4}$ and a strong liquid $x_0 = \frac{1}{10}$ are illustrated in $x = k_{\text{B}}T/\hbar\omega$.

The specific heat due to the sound, C_s is given by

$$C_s = \frac{k_B}{4}\sum_q{}' \frac{(\beta\hbar\omega_{sq})^2}{\sinh^2\frac{\beta\hbar\omega_{sq}}{2}} \cong k_B(N - N_0), \tag{7.4.34}$$

where we have used the relation $\beta\hbar\omega_{sq} \ll 1$. It should be noted that the state number N_0 near $q \sim K$ is included in C_{K}. The specific heats C_K for a fragile $x_0 = \frac{1}{4}$ and a strong $x_0 = \frac{1}{10}$ are shown in Figure 7.8. The contribution of the fragile liquid to the specific heat is much larger than that of the strong liquid. The specific heat C_{sound} is independent of the temperature, nearly equal to $N - N_0$ so that the behavior of C_s plays no important role in the glass transition.

ii. The specific heat due to the inter-band density fluctuations

Next we investigate the specific heat due to the inter-band density fluctuations. In this case, since the inter-band fluctuation entropy cancels with the dissipative entropy, there remains only the specific heat of phonons. This is because the structure of the phonons does change little above and below the glass transition, while the structure of the sound essentially depends on the temperature. We obtain the conventional specific heat of phonons, C_{phonon} as

$$C_{\text{phonon}} = \frac{k_B}{4}\sum_{\lambda q} \frac{(\beta\hbar\omega_{\lambda q})^2}{\sinh^2\frac{\beta\hbar\omega_{\lambda q}}{2}}. \tag{7.4.35}$$

Next we estimate the specific heat C_{phonon}. To obtain the upper limit of the wave vectors corresponding to the Debye cut off, we replace the summation in $\boldsymbol{q}$ by the integration in $\alpha = aq$, i.e.

$$\sum_{\mathbf{q}} \cdots = \frac{V}{(2\pi)^3} \int_0^{q_M} \cdots d^3q = \frac{V}{a^3(2\pi)^3} \int_0^{\alpha_M} 4\pi\alpha^2 \cdots d\alpha$$
$$= \frac{N}{2\pi^2} \int_0^{\alpha_M} \cdots \alpha^2 d\alpha,$$

and we put

$$\frac{N}{2\pi^2} \int_0^{\alpha_M} \alpha^2 d\alpha = N, \quad \alpha_M = (6\pi^2)^{\frac{1}{3}} \cong 3.9,$$

we obtain the numerical value of the specific heat, C_{phonon} and the summed specific heat, $C_{\text{K}} + C_{\text{phonon}}$ for a fragile liquid, $\frac{k_BT}{\hbar\omega} = \frac{1}{4}$ and a strong liquid, $\frac{k_BT}{\hbar\omega} = \frac{1}{10}$ as shown in Figure 7.8.

We show two limiting cases of the specific heat of phonons:

1. At low temperatures; $\frac{k_{\text{B}}T}{\hbar\omega} \ll 1$
   At low temperatures, since the main contribution of the integration in Eq. (7.4.35) comes from the long wavelength regions, we put $\omega_{\lambda q} \cong c_\lambda q = \frac{c_\lambda \alpha}{a}$. Thus we obtain

$$C_{\text{phonon}} = \frac{k_{\text{B}}}{4} \frac{N}{2\pi^2} \sum_\lambda \int_0^{\alpha_M} \alpha^2 d\alpha \frac{\left(\frac{\beta\hbar c_\lambda \alpha}{a}\right)^2}{\sinh^2 \frac{\beta\hbar c_\lambda \alpha}{2a}}$$
$$= \frac{k_{\text{B}}N}{8\pi^2} \sum_\lambda \left(\frac{a}{\beta\hbar c_\lambda}\right)^3 \int_0^\infty \xi^2 d\xi \frac{\xi^2}{\sinh^2 \frac{\xi}{2}}$$
$$\propto T^3, \tag{7.4.36}$$

   where we have put $\xi_M = \frac{\beta\hbar c_\lambda \alpha_M}{a} \to \infty$.

2. At high temperatures; $\frac{k_{\text{B}}}{\hbar\omega} \gg 1$
   Since $\sinh \beta\hbar\omega_{\lambda q} \cong \beta\hbar\omega_{\lambda q}$, from Eq. (7.4.35) we obtain

$$C_{\text{phonon}} \cong \frac{k_B}{4} \sum_{\lambda \boldsymbol{q}} \frac{(\beta\hbar\omega_{\lambda q})^2}{\left(\frac{\beta\hbar\omega_{\lambda q}}{2}\right)^2}$$
$$= 3Nk_{\text{B}}. \tag{7.4.37}$$

These results are well known as the specific heat of phonons and illustrated in Figure 7.8.

iii. Specific heat anomaly in the low temperature glasses

At low temperatures, Eq. (7.4.36) shows that the specific heat is due to phonons in glasses $C_{\text{phonon}} \propto T^3$. However, Zeller and Pohl's experimental results [13] showed that the specific heat in the low temperature glasses behaves in a linear and a cubic dependence. This linear dependence phenomenon in the specific heat is well known as the specific heat anomaly. Anderson et al. [14] proposed the two level system; the tunneling between two minimum potentials yields a linear term. Siever and Takeno [15] also showed that anharmonic resonant modes can yield a linear and a cubic term. Here we show that the linear dependence specific heat anomaly originates from the low temperature specific heat due to the intra-band elementary excitations [16].

The specific heat due to the intra-band density fluctuations consists of the specific heat due to the Kauzmann entropy C_K in Eq. (7.4.33) and due to the sound C_{sound} in Eq. (7.4.34) for $\beta\hbar\omega_{sq} \ll 1$. In deriving Eqs. (7.4.34) we have neglected the inverse relaxation times $\frac{1}{\tau_0}$ due to the random scattering processes compared with ω_{sq} under the condition $\beta\hbar/\tau_0 \ll 1$ and the sound mode disappears in the regions $T < T_0$. However, at low temperatures: $\frac{\beta\hbar}{\tau_0} \gg 1$, another situation appears. There remains the bandwidth of the lower band ϵ_{0p} from the uncertainty principles; the bandwidth $\sim \hbar/\tau_0$. So, there are intra-band elementary excitations ω_q. Assuming the uniform density of states D, we can estimate the thermodynamical function of the intra-band density fluctuations $\Omega_{\text{intra-fluct}}$ as:

$$\Omega_{\text{intra-fluct}} \sim \frac{D}{\beta}\int_0^{\hbar/\tau_0} d\omega \ln(1 - e^{-\beta\hbar\omega}). \tag{7.4.38}$$

Then we obtain the specific heat of the intra-band density fluctuations $C_{\text{intra-fluct.}}$:

$$C_{\text{intra-fluct.}} \sim \frac{k_B D}{4}\int_0^{\hbar/\tau_0} d\omega \frac{(\beta\hbar\omega)^2}{\sinh^2\frac{\beta\hbar\omega}{2}}. \tag{7.4.39}$$

Changing variable $x = \beta\hbar\omega$ and putting the upper bound of the integration $x_{upper} = \frac{\beta\hbar}{\tau_0} \to \infty$, we obtain

$$C_{\text{intra-fluct.}} \cong \frac{k_B D}{\beta\hbar}\int_0^{\infty} dx \frac{x^2}{\sinh^2\frac{x}{2}} \propto T. \tag{7.4.40}$$

Thus we show that the low temperature specific heat anomaly in glasses comes from the specific heat due to the remaining bandwidth of the lower band ϵ_{0p} from the uncertainty principles.

Zeller and Pohl's experimental results [13] also showed the low temperature thermal conductivity in glasses. We will explain this anomaly in later chapter.

7.4.5 The Estimation of the Glass Transition and the Melting Temperature

Phonon dispersion curves essentially depend on the harmonic frequency ω, which is the order of the glass transition or the melting temperature $T_{g,m}$. Since the glass transition temperature T_g is less than the melting temperature T_m; $T_m > T_g$, the harmonic frequency in the glass is larger than that in the crystal. The temperature dependence of C_{ph} is essentially similar to that of the crystal. Here we show that the order of the melting temperature and that of the harmonic frequency in glass or crystal is the same; $k_B T_{g,m} \sim \hbar\omega$ [17–24].

Phonons are the Nambu-Goldstone bosons originating from the spontaneous symmetry breaking of the system. The gap Eq. (6.3.11) guarantees the symmetry breaking of the system. In the crystal, the pair distribution function $g(R)$ in $V_{ij}(\boldsymbol{q})$ in Eq. (6.4.28) is replaced by the delta function type distribution function of the nearest neighbor particles depending on the crystalline structure, $g(\boldsymbol{R})$. Considering the stable state; $\frac{\mathrm{i}}{\tau_M} \cong 0$, we obtain

$$Q(0) = -\frac{2}{\hbar\omega}. \tag{7.4.41}$$

As the temperature increases, the band gap decreases and at least at the glass transition or the melting temperature the band gap vanishes, where two bands join to a free particle energy dispersion. The up and down transitions of particles between the inter-bands in $Q(0)$ reduce to the transition between the intra-bands in $P(0)$ in Eq. (7.2.22):

$$Q(0) \longrightarrow P(0) = \beta f'. \tag{7.4.42}$$

At the glass transition or the melting temperature $T_{g,m}$, since the interaction potential $V_{ij}(\boldsymbol{q})$ in the gap Eq. (6.3.11) holds constant, we obtain

$$\hbar\omega \sim \beta^{-1}|_{T=T_{g,m}}. \tag{7.4.43}$$

Thus we obtain the harmonic frequency $\hbar\omega$ and the glass transition or the melting temperature $k_B T_{g,m}$ is the same order. In actual crystals, since the melting is of the first order transition, before the band gap vanishes in the crystalline phase, the melting takes place. But we can estimate the order of the melting temperature. This criterion corresponds to the Lindemann' criterion [25].

7.4.6 The VTF Law in the Velocities of Modes, the Relaxation Times, and the Transport Coefficients

Now, we investigate the VTF law in the velocities of modes, the relaxation times, and the transport coefficients. The VTF law $J = e^{-\frac{E}{T-T_0}}$ in Eq. (7.4.18) reflects $|J_\mu| \propto J$ in Eq. (7.4.19) and $\sqrt{U_J^{\mu\nu}} \propto J$ in Eq. (7.4.20). Therefore, the VTF law governs the velocities of modes and the relaxation times in different ways:

1. The velocities of particles and sound depend on J, while the phonon velocities do not depend on J:

$$v_p, c_T \propto J \qquad \text{for } T_0 < T, \tag{7.4.44}$$

$$c_\lambda \cong \text{constant} \qquad \text{for } T < T_g. \tag{7.4.45}$$

Here it should be noted the phonon velocities c_λ is constant.

Table 7.1 The Vogel-Fulcher law of transport coefficients, relaxation times, and velocity of modes $J = \exp\left\{-\frac{E}{T-T_0}\right\}$, $\Omega = \exp\left\{-\frac{3z\beta\hbar\omega}{e^{\beta\hbar\omega}-1}\right\}$.

	Transport coefficient				Relaxation time				Velocity			
	D	ν_λ	κ_s	κ_λ	τ_0	τ_M	τ_{es}	$\tau_{e\lambda}$	c_T	c_λ	u_s	u_λ
$T_g < T$	J	J^{-1}	–	J^{-1}	J^{-1}	J^{-1}	J^{-2}	J^{-1}	J	–	J	–
$T < T_g$	J^2	Ω^{-1}	J	Ω^{-1}	Ω^{-1}	Ω^{-1}	J^{-1}	Ω^{-1}	J	–	J	–

2. Since the relaxation times of sound and phonons

$$\frac{1}{\tau_0}, \frac{1}{\tau_M} \propto \sqrt{U_J^{00} + U_\omega^{00}},$$

and the term $\sqrt{U_J^{\mu\nu}} \propto J$ dominates for $T_0 < T_g < T$; the α-relaxation, while the term $\sqrt{U_\omega^{\mu\nu}} \propto \Omega$ dominates for $T_0 < T < T_g$; the β-relaxation, we obtain

$$\frac{1}{\tau_0}, \frac{1}{\tau_M} \propto \begin{cases} J & \text{for } T_0 < T_g < T, \\ \Omega & \text{for } T < T_g. \end{cases} \tag{7.4.46}$$

Considering the diffusivity D in Eq. (7.2.42) and the viscosity ν_λ in Eq. (7.3.22):

$$D = \tau_0 c_T^2, \quad \frac{\nu_\lambda}{M\rho} = \tau_M c_\lambda^2,$$

we obtain

1. $T_0 < T_g < T$:

$$D^{-1}, \nu_\lambda \propto J. \tag{7.4.47}$$

It should be noted that since the isothermal sound velocity $c_T \propto J$ the diffusivity $D \propto J$, while since the phonon velocities $c_\lambda =$ constant, the viscosities $\nu_\lambda \propto J^{-1}$. These facts explain that the Stokes law $D\nu_\lambda =$ constant satisfies the Stokes law.

2. $T_0 < T < T_g$:

$$D \propto J^2, \quad \nu_\lambda \propto \Omega^{-1}. \tag{7.4.48}$$

These results are summarized in Table 7.1.

7.4.7 The Adam–Gibbs Formula on the VTF Law

The VTF law is constructed from the Einstein relation on entropy and probability so that the VTF law includes naturally the Kauzmann entropy in exponent; $1/\tau_0 \propto J = \exp\{zS_K/Nk_B\}$. The Kauzmann entropy explains the Kauzmann paradox, the jump of the specific heat, and the VTF law on the relaxation times and the transport coefficients so that the present theory elucidates the universal features self-consistently.

On the other hand, the well-known Adam-Gibbs formula for the VTF law [26] includes inversely the configuration entropy, S_c corresponding to the Kauzmann entropy, in exponent; $1/\tau_{\mathrm{AG}} \propto \exp\{-A/(TS_c)\}$, where τ_{AG} is the relaxation time in the Adam-Gibbs formula and A is a constant. The Adam-Gibbs formula has been usually used in arranging experimental data; the transport coefficients and the relaxation times.

However, the Adam-Gibbs formula cannot explain the configuration entropy, the Kauzmann paradox, and the jump of the specific heat well. The Adam-Gibbs formula faces the conflicting points between the configuration entropy and, the transport coefficients and the relaxation times. Furthermore, the Adam-Gibbs formula cannot explain the relations among the transport coefficients and the relaxation times as shown in Table 7.1. This is because the Adam-Gibbs formula is only a phenomenological theory.

7.5 The Intermediate Scattering Function

The intermediate scattering function can be observed by the experiments. Near the glass transition, the behavior of the intermediate scattering function drastically changes with the temperature decreasing toward the glass transition temperature. The intermediate scattering function is derived from the fluctuation-dissipation theorem for the dynamical structure factor [27,6]. The fluctuation-dissipation theorem implies that the density fluctuations relate with the dissipation of the density fluctuations, the dissipative part of the retarded Green's function, which is obtained from the temperature Green's function of the intra-band density fluctuations replaced the imaginary temperature frequencies by analytically continued frequencies.

7.5.1 The Fluctuation-Dissipation Theorem for the Dynamical Structure Factor

In order to show the fluctuation-dissipation theorem for the dynamical structure factor, we connect the dynamical structure factor with the retarded Green's function for the intra-band density fluctuations. The dynamical structure factor is given by

$$S(\boldsymbol{q}, q_0) = \frac{1}{2\pi}\int_{-\infty}^{\infty} dt e^{\mathrm{i}q_0 t}\frac{1}{N} < \rho_{0\boldsymbol{q}}(t)\rho_{0\boldsymbol{q}}^{\dagger}(0) >_c . \tag{7.5.1}$$

There is a relation:

$$\begin{aligned} < \rho_{0\boldsymbol{q}}(0)\rho_{0\boldsymbol{q}}^{\dagger}(t) >_c &= < \mathrm{Tr}e^{-\beta\hbar\mathcal{H}}\rho_{0\boldsymbol{q}}e^{\mathrm{i}\mathcal{H}t}\rho_{0\boldsymbol{q}}^{\dagger}e^{-\mathrm{i}\mathcal{H}t} >_c \\ &= < \mathrm{Tr}e^{\mathrm{i}\mathcal{H}t}\rho_{0\boldsymbol{q}}^{\dagger}e^{-\mathrm{i}\mathcal{H}t}e^{-\beta\hbar\mathcal{H}}\rho_{0\boldsymbol{q}} >_c \\ &= < \mathrm{Tr}e^{-\beta\hbar\mathcal{H}}e^{\mathrm{i}\mathcal{H}(t-\mathrm{i}\beta\hbar)}\rho_{0\boldsymbol{q}}^{\dagger}e^{-\mathrm{i}\mathcal{H}(t-\mathrm{i}\beta\hbar)}\rho_{0\boldsymbol{q}} >_c \\ &= < \rho_{0\boldsymbol{q}}^{\dagger}(t-\mathrm{i}\beta\hbar)\rho_{0\boldsymbol{q}}(0) >_c \\ &= < \rho_{0\boldsymbol{q}}(t-\mathrm{i}\beta\hbar)\rho_{0\boldsymbol{q}}^{\dagger}(0) >_c , \end{aligned} \tag{7.5.2}$$

where from the first to the second line we have used the property of Tr and, from the fourth to the last line we have used the space inversion symmetry:

$$\rho^{\dagger}_{0-\boldsymbol{q}} = \sum_{p} a^{\dagger}_{p} a_{p+q} = \sum_{p} a^{\dagger}_{p-q} a_{p} = \rho_{0\boldsymbol{q}}.$$

Using the relation (7.5.2), we have

$$\begin{aligned}
S(\boldsymbol{q}, -q_0) &= \frac{1}{2\pi}\int_{-\infty}^{\infty} dt e^{-\mathrm{i}q_0 t}\frac{1}{N} < \rho_{0\boldsymbol{q}}(t)\rho^{\dagger}_{0\boldsymbol{q}}(0) >_c \\
&= \frac{1}{2\pi}\int_{-\infty}^{\infty} dt e^{\mathrm{i}q_0 t}\frac{1}{N} < \rho_{0\boldsymbol{q}}(0)\rho^{\dagger}_{0\boldsymbol{q}}(t) >_c \\
&= \frac{1}{2\pi}\int_{-\infty}^{\infty} dt e^{\mathrm{i}q_0 t}\frac{1}{N} < \rho_{0\boldsymbol{q}}(t - \mathrm{i}\beta\hbar)\rho^{\dagger}_{0\boldsymbol{q}}(0) >_c \\
&= e^{-\beta\hbar q_0}\frac{1}{2\pi}\int_{-\infty}^{\infty} dt e^{\mathrm{i}q_0(t-\mathrm{i}\beta\hbar)}\frac{1}{N} < \rho_{0\boldsymbol{q}}(t - \mathrm{i}\beta\hbar)\rho^{\dagger}_{0\boldsymbol{q}}(0) >_c \\
&= e^{-\beta\hbar q_0}\frac{1}{2\pi}\int_{-\infty-\mathrm{i}\beta\hbar}^{\infty-\mathrm{i}\beta\hbar} d(t - \mathrm{i}\beta\hbar) e^{\mathrm{i}q_0(t-\mathrm{i}\beta\hbar)} \\
&\quad \times \frac{1}{N} < \rho_{0q}(t - \mathrm{i}\beta\hbar)\rho^{\dagger}_{0\boldsymbol{q}}(0) >_c .
\end{aligned}$$

Thus we obtain

$$S(\boldsymbol{q}, -q_0) = e^{-\beta\hbar q_0} S(\boldsymbol{q}, q_0). \tag{7.5.3}$$

Here we introduce the retarded Green's function, $F^R_{\boldsymbol{q}}$ and a function, $K(\boldsymbol{q}, t)$:

$$\begin{aligned}
F^R_{\boldsymbol{q}}(t) &= -\frac{\mathrm{i}}{\hbar N} < [\rho_{0\boldsymbol{q}}(t), \rho^{\dagger}_{0\boldsymbol{q}}(0)] >_c \theta(t) \\
&= \frac{1}{2\pi}\int_{-\infty}^{\infty} dt e^{-\mathrm{i}q_0 t} F^R(\boldsymbol{q}, q_0),
\end{aligned} \tag{7.5.4}$$

$$K(\boldsymbol{q}, t) = -\frac{\mathrm{i}}{N} < [\rho_{0\boldsymbol{q}}(t), \rho^{\dagger}_{0\boldsymbol{q}}(0)] >_c . \tag{7.5.5}$$

The function $K(\boldsymbol{q}, t)$ has the following properties [28]:

1. Real function: $K^*(\boldsymbol{q}, t) = K(\boldsymbol{q}, t)$
   This is derived as follows:

$$\begin{aligned}
K^*(\boldsymbol{q}, t) &= \frac{\mathrm{i}}{N} < \rho_{0\boldsymbol{q}}(t)\rho^{\dagger}_{0\boldsymbol{q}}(0) - \rho^{\dagger}_{0\boldsymbol{q}}(0)\rho_{0\boldsymbol{q}}(t) >^*_c \\
&= \frac{\mathrm{i}}{N} < \rho_{0\boldsymbol{q}}(0)\rho^{\dagger}_{0\boldsymbol{q}}(t) - \rho^{\dagger}_{0\boldsymbol{q}}(t)\rho_{0\boldsymbol{q}}(0) >_c \\
&= \frac{\mathrm{i}}{N} < \rho^{\dagger}_{0\boldsymbol{q}}(0)\rho_{0\boldsymbol{q}}(t) - \rho_{0\boldsymbol{q}}(t)\rho^{\dagger}_{0\boldsymbol{q}}(0) >_c \\
&= K(\boldsymbol{q}, t),
\end{aligned} \tag{7.5.6}$$

where from the first to the second line, we have used the relation: $< m|AB|n >^* = < n|B^\dagger A^\dagger|m >$ for any operator A and B, and from the second to the third line, we have used the space inversion symmetry: $\rho^\dagger_{0-\boldsymbol{q}} = \rho_{0\boldsymbol{q}}$.

2. Odd function in t : $K(\boldsymbol{q}, -t) = -K(\boldsymbol{q}, t)$,
This is derived as follows:

$$\begin{aligned} K(\boldsymbol{q}, -t) &= -\frac{\mathrm{i}}{N} < \rho_{0\boldsymbol{q}}(0)\rho^\dagger_{0\boldsymbol{q}}(t) - \rho^\dagger_{0\boldsymbol{q}}(t)\rho_{0\boldsymbol{q}}(0) >_c \\ &= \frac{\mathrm{i}}{N} < \rho_{0\boldsymbol{q}}(0)\rho^\dagger_{0\boldsymbol{q}}(t) - \rho^\dagger_{0\boldsymbol{q}}(t)\rho_{0\boldsymbol{q}}(0) >^*_c \\ &= \frac{\mathrm{i}}{N} < \rho_{0\boldsymbol{q}}(t)\rho^\dagger_{0\boldsymbol{q}}(0) - \rho^\dagger_{0\boldsymbol{q}}(0)\rho_{0\boldsymbol{q}}(t) >_c \\ &= -K(\boldsymbol{q}, t), \end{aligned} \tag{7.5.7}$$

where we have used the property (7.5.6) from the first to the second line.

Equations (7.5.1) and (7.5.3) lead to

$$\begin{aligned} \int_{-\infty}^{\infty} dt K(\boldsymbol{q}, t) e^{\mathrm{i}q_0 t} &= -\frac{\mathrm{i}}{N}\int_{-\infty}^{\infty} dt e^{\mathrm{i}q_0 t} < \rho_{0\boldsymbol{q}}(t)\rho^\dagger_{0\boldsymbol{q}}(0) - \rho^\dagger_{0\boldsymbol{q}}(0)\rho_{0\boldsymbol{q}}(t) >_c \\ &= -2\pi\mathrm{i}(1 - e^{-\beta\hbar q_0})S(\boldsymbol{q}, q_0), \end{aligned} \tag{7.5.8}$$

Next we relate the dynamical structure factor $S(\boldsymbol{q}, q_0)$ to the retarded Green's function of intra-band density fluctuations $F^R(\boldsymbol{q}, q_0)$:

$$\begin{aligned} \int_{-\infty}^{\infty} dt K(\boldsymbol{q}, t) e^{\mathrm{i}q_0 t} &= \int_0^{\infty} dt K(\boldsymbol{q}, t) e^{\mathrm{i}q_0 t} + \int_{-\infty}^{0} dt K(\boldsymbol{q}, t) e^{\mathrm{i}q_0 t} \\ &= \hbar F^R(\boldsymbol{q}, q_0) - \int_0^{\infty} dt e^{-\mathrm{i}q_0 t} K(\boldsymbol{q}, t) \\ &= \hbar F^R(\boldsymbol{q}, q_0) - \hbar F^R(\boldsymbol{q}, q_0)^* \\ &= 2\mathrm{i}\hbar\Im F^R(\boldsymbol{q}, q_0), \end{aligned} \tag{7.5.9}$$

where we have changed the variable; $t \longrightarrow -t$ and used Eq. (7.5.7) from the first to the second line at the second term, and we have used Eq. (7.5.6) from the second to the third line. Thus we obtain the fluctuation-dissipation theorem:

$$S(\boldsymbol{q}, q_0) = -\frac{1}{1 - e^{-\beta\hbar q_0}}\frac{\hbar}{\pi}\Im F^R(\boldsymbol{q}, q_0). \tag{7.5.10}$$

This theorem implies that the intra-band density fluctuations $S(\boldsymbol{q}, q_0)$ relate with the dissipation of the intra-band density fluctuations, $\Im F^R(\boldsymbol{q}, q_0)$.

7.5.2 The Intermediate Scattering Function

Using the relation between the retarded and the temperature Green's functions:

$$F^R(\boldsymbol{q}, q_0) = F(\boldsymbol{q}, \mathrm{i}\nu_n \rightarrow q_0 + \mathrm{i}\delta), \tag{7.5.11}$$

Equation (7.5.10) leads to

$$S(\boldsymbol{q}, q_0) = -\frac{1}{1 - e^{-\beta\hbar q_0}} \frac{\hbar}{\pi} \Im F(\boldsymbol{q}, q_0 + \mathrm{i}\delta). \tag{7.5.12}$$

Since we are concerned with the lower frequency regime, we obtain

$$\Im F(\boldsymbol{q}, q_0 + \mathrm{i}\delta) = -\frac{\beta f' \omega_q^2}{2\mathrm{i}} \left\{ \frac{1}{q_0^2 - \omega_{sq}^2 + \mathrm{i}q_0/\tau_0} - \frac{1}{q_0^2 - \omega_{sq}^2 - \mathrm{i}q_0/\tau_0} \right\}. \tag{7.5.13}$$

Under the frequency independent relaxation time approximation, substitution of Eq. (7.5.13) into Eq. (7.5.12) leads to

$$S(\boldsymbol{q}, q_0) \cong -\frac{\beta\hbar\omega_q^2}{2\pi\mathrm{i}} \frac{1}{1 - e^{-\beta\hbar q_0}} \left\{ \frac{1}{q_0^2 - \omega_{sq}^2 + \mathrm{i}q_0/\tau_0} - \frac{1}{q_0^2 - \omega_{sq}^2 - \mathrm{i}q_0/\tau_0} \right\}, \tag{7.5.14}$$

where we have put $f' = -1$. Hereafter in this chapter, we restrict ourselves to the classical system, we take the Boltzmann distribution function.

The intermediate wave function is defined by

$$S(\boldsymbol{q}, t) \equiv \int_{-\infty}^{\infty} dq_0 e^{-\mathrm{i}q_0 t} S(\boldsymbol{q}, q_0). \tag{7.5.15}$$

To carry out the contour integration of Eq. (7.5.15), we investigate the poles of $S(\boldsymbol{q}, q_0)$ in Eq. (7.5.14):

$$\text{The poles of } \frac{1}{1 - e^{-\beta\hbar q_0}}: \qquad q_0 = \frac{2n\pi\mathrm{i}}{\beta\hbar}, \quad n: \text{integer}.$$

$$\text{The poles of } \frac{1}{q_0^2 - \omega_{sq}^2 + \mathrm{i}q_0/\tau_0}: \qquad q_0 = -\frac{\mathrm{i}}{2\tau_0} \pm \Omega_q,$$

$$\text{The poles of } \frac{1}{q_0^2 - \omega_{sq}^2 - \mathrm{i}q_0/\tau_0}: \qquad q_0 = \frac{\mathrm{i}}{2\tau_0} \pm \Omega_q,$$

where

$$\Omega_q \equiv \sqrt{\omega_{sq}^2 - \frac{1}{4\tau_0^2}}. \tag{7.5.16}$$

Under the approximation $\beta\hbar/\tau_0 \ll 1$, $\beta\hbar\omega_{sq} \ll 1$, we can take only $n = 0$ as the pole of $\frac{1}{1-e^{-\beta\hbar q_0}}$, so we make the following approximation:

$$S(\boldsymbol{q}, q_0) \cong -\frac{\omega_q^2}{2\pi\mathrm{i}q_0} \left\{ \frac{1}{q_0^2 - \omega_{sq}^2 + \mathrm{i}q_0/\tau_0} - \frac{1}{q_0^2 - \omega_{sq}^2 - \mathrm{i}q_0/\tau_0} \right\}, \tag{7.5.17}$$

The contour integration yields the intermediate scattering function:

$$\begin{aligned} S(\boldsymbol{q},t) &= \int_{\Gamma} dq_0 e^{-\mathrm{i}q_0 t} S(\boldsymbol{q},q_0) \\ &= \frac{\omega_q^2}{\omega_{sq}^2} e^{-\frac{t}{2\tau_0}} \left\{ \cos \Omega_q t + \frac{1}{2\Omega_q \tau_0} \sin \Omega_q t \right\} \\ &= \frac{1}{1+\beta V(q)} e^{-\frac{t}{2\tau_0}} \left\{ \cos \Omega_q t + \frac{1}{2\Omega_q \tau_0} \sin \Omega_q t \right\} \\ &= S(q) e^{-\frac{t}{2\tau_0}} \left\{ \cos \Omega_q t + \frac{1}{2\Omega_q \tau_0} \sin \Omega_q t \right\}, \end{aligned} \tag{7.5.18}$$

where we have taken the contour Γ clockwisely along the real axis and the lower half circle with the radius $R \longrightarrow \infty$ for $t > 0$ putting $q_0 \longrightarrow q_0 + \mathrm{i}\delta$.

Now we investigate the intermediate scattering functions $S(\boldsymbol{q},t)$ in (7.5.18).

1. $T_g < T$:
From Eqs. (7.4.44) and (7.4.46), we obtain

$$\Omega_q \propto J, \quad \frac{1}{\tau_0} \propto J, \quad \frac{1}{2\tau_0 \Omega_q} = \text{const.} \tag{7.5.19}$$

We illustrate $S(\mathbf{q},t)$ for different temperatures $x = k_B T/(\hbar\omega)$ in Figure 7.9 with $\Omega_q = J$, $1/(2\tau_0 \Omega_q) = 1$ and a fragile liquid $x_0 = k_B T_0/(\hbar\omega) = 1/4$, $z = 30$ and $N/N_0 = 1/5$. This is just the α-relaxation.

2. $T < T_g$:

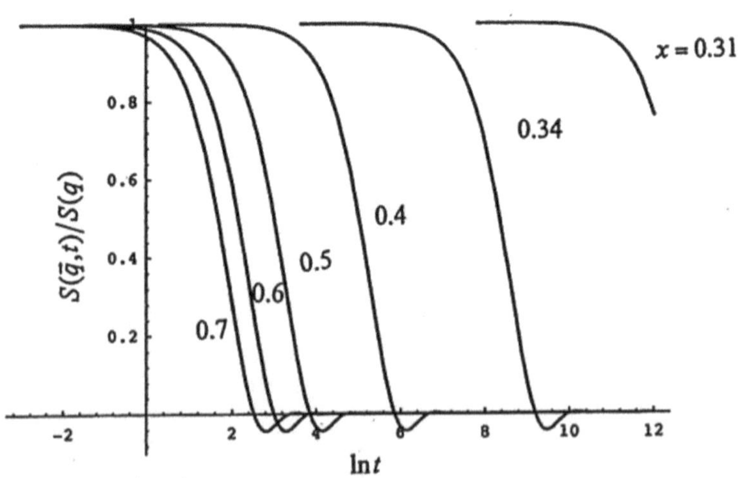

Figure 7.9 The intermediate scattering function $S(\boldsymbol{q},t)$ normalized by the static structure factor $S(\boldsymbol{q})$ is shown as a function of $\ln t$ for different temperatures $x = k_B T/(\hbar\omega)$. We take $\Omega_q = J$, $1/(2\tau_0 \Omega_q) = 1$, a fragile liquid $x_0 = k_B T_0/(\hbar\omega) = 1/4$, $z = 30$ and $N/N_0 = 1/5$.

From Eqs. (7.4.44) and (7.4.46) we obtain

$$\omega_{sq} \propto J^2, \quad \Omega_q \cong \mathrm{i}/(2\tau_0) \propto \mathrm{i}\Omega. \tag{7.5.20}$$

Since

$$\cos \Omega_q t + \frac{1}{2\tau_0 \Omega_g} \sin \Omega_q t \cong \cos \frac{\mathrm{i}t}{2\tau_0} - \mathrm{i} \sin \frac{\mathrm{i}t}{2\tau_0} = e^{\frac{t}{2\tau_0}},$$

the intermediate scattering function is as follows:

$$S(\boldsymbol{q}, t) \cong S(q). \tag{7.5.21}$$

This is the β-relaxation.

We have presented the intermediate scattering function near the liquid-glass transition considering the correlation function of intra-band density fluctuations under the frequency independent relaxation time approximation. The frequency independent relaxation time approximation misses the second plateau, but we have explained the extraordinary slowing down of the intermediate scattering function in the α-relaxation regime due to $1/\tau_0 \propto J$ and rather smooth change in the β-relaxation regime due to $1/\tau_0 \propto \Omega$. The former $S(q, t)$ has a ripple in the tail [29], while the latter does not. If we take into account the frequency dependent relaxation time Eq. (7.2.28), we can expect the second plateau in $S(q, t)$.

7.6 A Generalized Navier-Stokes Equation

The Navier-Stokes equation is derived by including the internal stress due to the viscosity in the Euler equation in the classical fluid mechanics [30]. In fluids, sound and its associated diffusivity are essential, while in elastic glasses, phonons and their associated viscosities are essential. The dynamical equation of sound corresponds to the Euler equation in an ideal fluid, but in an actual fluid the current of the fluid consists of the two parts: the normal current and the displacement current which originates from bound particles in the harmonic potentials made up by the surrounding particles. The normal current relates with the Euler equation, but the latter current relates with the current due to viscosity associated with the glassy state. Therefore the generalized Navier-Stokes equation represents both the aspects of fluid and glass; the generalized Navier-Stokes equation reduces to the Navier-Stokes equation in the fluid state, and to the dynamical equation of elastic waves in the glassy state [31,6].

7.6.1 The Continuity Equation

The density fluctuations consist of the intra-band and the inter-band ones so that the current also consists of the two parts due to the intra- and inter-band density fluctuations. In order to see these situations, we investigate the continuity equation. Now we investigate the number operator:

$$n(x) = \frac{1}{N} \sum_{m\boldsymbol{q}} e^{-\mathrm{i}\boldsymbol{q}\cdot\boldsymbol{R}_m} \left\{ \sum_{\mu} \tilde{w}_{m\mu}^2 (\boldsymbol{x} - R_m) \rho_{\mu\boldsymbol{q}}^{\dagger} - \sum_i \zeta_m \nabla_i \tilde{w}_{m0}^2 (\boldsymbol{x} - R_m) \rho_{i\boldsymbol{q}}^{\dagger} + \cdots \right\}. \tag{7.6.1}$$

Since we construct the continuity equation in the classical regime, we make the following approximation:

1. We confine ourselves to the lower band of the intra-band density fluctuations $\rho_{0\boldsymbol{q}}$ in Eq. (7.6.1) because near the liquid-glass transition the majority of particles stay in the lower state:

$$\sum_{\mu} \tilde{w}_{m\mu}^2 (\boldsymbol{x} - R_m) \rho_{\mu\boldsymbol{q}}^{\dagger} \longrightarrow \tilde{w}_{m0}^2 (\boldsymbol{x} - R_m) \rho_{0\boldsymbol{q}}^{\dagger}.$$

2. We replace the random wave functions $\tilde{w}_{m0}(\boldsymbol{x} - \boldsymbol{R_m})$ and ζ_m as follows by the averaged wave function:

$$\tilde{w}_{m0}(\boldsymbol{x} - \boldsymbol{R_m}) \longrightarrow \tilde{w}_0(\boldsymbol{x} - \boldsymbol{R_m}),$$

$$\zeta_m \longrightarrow \zeta_M = \sqrt{\frac{\hbar}{2M\omega}}.$$

3. We consider the wave function $\tilde{w}_0^2(\boldsymbol{x} - \boldsymbol{R}_m)$ in the classical limit:

$$\tilde{w}_0^2(\boldsymbol{x} - \boldsymbol{R}_m) \longrightarrow \delta(\boldsymbol{x} - \boldsymbol{R}_m).$$

Thus we obtain

$$\begin{aligned} n(x) &= \frac{1}{N} \sum_{m\boldsymbol{q}} e^{-\mathrm{i}\boldsymbol{q}\cdot\boldsymbol{R}_m} \left\{ \delta(\boldsymbol{x} - \boldsymbol{R}_m) \rho_{0\boldsymbol{q}}^{\dagger} - \sum_i \zeta_m \nabla_i \delta(\boldsymbol{x} - \boldsymbol{R}_m) \rho_{i\boldsymbol{q}}^{\dagger} + \cdots \right\} \\ &= \frac{1}{N} \sum_{m\boldsymbol{q}} e^{-\mathrm{i}\boldsymbol{q}\cdot\boldsymbol{R}_m} \frac{1}{(2\pi)^3} \int d\boldsymbol{q}' e^{\mathrm{i}\boldsymbol{q}'\cdot(\boldsymbol{x}-\boldsymbol{R}_m)} \left\{ \rho_{0\boldsymbol{q}}^{\dagger} - \sum_i \zeta_M \mathrm{i} q_i' \rho_{i\boldsymbol{q}}^{\dagger} + \cdots \right\} \\ &= \frac{1}{(2\pi)^3} \int d\boldsymbol{q} e^{\mathrm{i}\boldsymbol{q}\cdot\boldsymbol{x}} \left\{ \rho_{0\boldsymbol{q}} - \mathrm{i} \sum_i q_i \zeta_M \rho_{i\boldsymbol{q}}^{\dagger} + \cdots \right\}. \end{aligned} \tag{7.6.2}$$

Since the conventional density current comes from

$$\boldsymbol{v}_{0\boldsymbol{q}} = \boldsymbol{\nabla} \epsilon_{0\boldsymbol{q}}, \tag{7.6.3}$$

the continuity equation is given by

$$\frac{\partial \rho_{0\boldsymbol{q}}}{\partial t} + \mathrm{i}\boldsymbol{q} \cdot \boldsymbol{j}_{\boldsymbol{q}} = 0, \tag{7.6.4}$$

$$j_{i\boldsymbol{q}} \cong \rho v_{0i\boldsymbol{q}} - \zeta_M \frac{\partial \rho_{i\boldsymbol{q}}}{\partial t}, \tag{7.6.5}$$

where ρ is the density and the second term of the current $j_{i\boldsymbol{q}}$ in Eq. (7.6.5) comes from the time derivative of the second term of the density $n(x)$ in Eq. (7.6.2). The first term $\rho v_{0i\boldsymbol{q}}$ in the current corresponds to the conventional current and the second term $-\zeta_M \frac{\partial \rho_{i\boldsymbol{q}}}{\partial t}$ corresponds to a Fourier component of the displacement current $\rho \frac{\partial u_i}{\partial t}$:

$$\rho u_{i\boldsymbol{q}} = -\zeta_M \rho_{i\boldsymbol{q}}, \tag{7.6.6}$$

because $-\zeta_M \rho_{i\boldsymbol{q}}$ represents the spread of the wave functions of the ith component of the first excited states. Thus we can rewrite Eqs. (7.6.4) and (7.6.5) as

$$\frac{\partial \rho}{\partial t} + \nabla \cdot \boldsymbol{j} = 0, \tag{7.6.7}$$

$$\boldsymbol{j} \equiv \rho \boldsymbol{v}, \quad \boldsymbol{v} = \boldsymbol{v}_0 + \frac{\partial \boldsymbol{u}}{\partial t}, \tag{7.6.8}$$

where ρ is the density in the real space. If we consider a dielectric matter, $\rho \boldsymbol{v}_0$ corresponds to electric current and $\rho \frac{\partial \boldsymbol{u}}{\partial t}$ to a displacement current.

7.6.2 The Euler Equation

We derive the Euler equation from the dynamical equation of intra-band density fluctuations in the classical and the ideal fluid approximation. We first investigate the Green's function for the intra-band density fluctuations, Eq. (7.2.30):

$$F(q) = -\frac{\beta f' \omega_q^2}{q_0(q_0 + \frac{i}{\tau_0}) - \omega_{sq}^2}. \tag{7.2.30}$$

In the limiting case: $\frac{1}{\tau_0} \to 0$, the imaginary part of Eq. (7.2.30a) leads to

$$\begin{aligned} \Im F(\boldsymbol{q}, q_0 + i\delta) &\longrightarrow -\Im \frac{\beta f' \omega_q^2}{2\omega_{sq}} \left\{ \frac{1}{q_0 + i\delta - \omega_{sq}} - \frac{1}{q_0 + i\delta + \omega_{sq}} \right\} \\ &= \frac{\beta f' \pi \omega_{\boldsymbol{q}}^2}{2\omega_{s\boldsymbol{q}}} [\delta(q_0 - \omega_{s\boldsymbol{q}}) - \delta(q_0 + \omega_{s\boldsymbol{q}})]. \end{aligned} \tag{7.6.9}$$

Insertion of Eq. (7.6.9) into Eq. (7.5.12) leads to

$$S(\boldsymbol{q}, q_0) = -\frac{1}{1 - e^{-\beta \hbar q_0}} \frac{\beta \hbar f' \omega_q^2}{2\omega_{sq}} \{\delta(q_0 - \omega_{sq}) - \delta(q_0 + \omega_{sq})\}. \tag{7.6.10}$$

Thus we obtain

$$S(q) = \int_{-\infty}^{\infty} dq_0 S(\boldsymbol{q}, q_0) = \frac{\beta \hbar f' \omega_q^2}{2\omega_{s\boldsymbol{q}}} \coth \frac{\beta \hbar \omega_{s\boldsymbol{q}}}{2}. \tag{7.6.11}$$

In the case $\beta \hbar \omega_{s\boldsymbol{q}} \ll 1$ and in the classical limit $f' = -1$, we obtain

$$S(q) \cong \frac{1}{1 + \beta V(q)}. \tag{7.6.12}$$

In the classical limit, using the

$$\frac{v_p^2}{3M} \cong k_B T, \tag{7.6.13}$$

$$S(0) = \rho k_B T \chi_T, \tag{7.6.14}$$

we can identify the isothermal sound velocity c_T with

$$c_T^2 = \frac{v_p^2}{3S(0)} \cong \frac{1}{M\rho\chi_T} = \frac{1}{M}\left(\frac{\partial p}{\partial \rho}\right)_T, \tag{7.6.15}$$

where χ_T is isothermal compressibility and p is pressure.

The denominator of Eq. (7.2.30) leads to the dynamical equation of intra-band density fluctuations:

$$q_0\left(q_0 + \frac{\mathrm{i}}{\tau_0}\right) - \omega_{sq}^2 = 0 \tag{7.6.16}$$

Under the condition $q_0\tau_0 \gg 1$, Eqs. (7.6.15) and (7.6.16) yield

$$M\frac{\partial^2 \rho}{\partial t^2} - \left(\frac{\partial p}{\partial \rho}\right)_T \nabla^2 \rho = 0. \tag{7.6.17}$$

Since the continuity Eq. (7.6.7) with Eq. (7.6.8) yields

$$M\frac{\partial^2 \rho}{\partial t^2} = M\nabla \cdot \rho\frac{\partial \boldsymbol{v}}{\partial t}$$

and the relation $\nabla^2 = \nabla \cdot \nabla$ yields

$$\left(\frac{\partial p}{\partial \rho}\right)_T \nabla^2 \rho = \nabla \cdot \left(\frac{\partial p}{\partial \rho}\right)_T \nabla \rho,$$

Equation (7.6.17) leads to

$$M\rho\frac{\partial \boldsymbol{v}}{\partial t} = -\left(\frac{\partial p}{\partial \rho}\right)_T \nabla \rho. \tag{7.6.18}$$

If we consider the pressure gradient comes through the density gradient,

$$\left(\frac{\partial p}{\partial \rho}\right)_T \nabla \rho = \nabla p, \tag{7.6.19}$$

we obtain

$$M\rho\frac{\partial \boldsymbol{v}}{\partial t} = -\nabla p, \tag{7.6.20}$$

This equation is valid in the classical limit; $v_p^2/3M \cong k_B T$ and under no displacement current. Equation (7.6.20) is just the Euler equation. Thus we could derive the Euler equation from the dynamical equation of intra-band density fluctuations in the classical and the ideal fluid approximation.

7.6.3 A Generalized Navier-Stokes Equation

Next we investigate the displacement current. To do so, we investigate the Green's function for the inter-band density fluctuations (7.3.14),

$$D_\lambda(\boldsymbol{q}, q_0) = \frac{\frac{2\omega}{\hbar}}{q_0\left(q_0 + \frac{\mathrm{i}}{\tau_M}\right) - \omega_{\lambda\boldsymbol{q}}^2}, \tag{7.3.14}$$

where $\lambda = l, t$ represent a longitudinal and two transverse phonons or corresponding elastic wave velocities, respectively. The denominator of Eq. (7.3.14) leads to the dynamical equations for the velocity of the displacement of particles from assigned positions:

$$q_0\left(q_0 + \frac{\mathrm{i}}{\tau_M}\right) - \omega_{\lambda\boldsymbol{q}}^2 = 0. \tag{7.6.21}$$

If we consider the long wavelength regions and $\boldsymbol{q} = q\boldsymbol{e}_z$, we obtain

$$\omega_{\lambda\boldsymbol{q}}^2 = c_\lambda^2 q^2, \tag{7.6.22}$$

where for $\lambda = t$, $\lambda = x$, or y and for $\lambda = l$, $\lambda = z$. The condition $\boldsymbol{q} = q\boldsymbol{e}_z$ means that the displacement operator $\boldsymbol{u}$ depends only on z. Thus we obtain

$$\left(\frac{\partial^2}{\partial t^2} + \frac{1}{\tau_M}\frac{\partial}{\partial t} - c_\lambda^2\frac{\partial^2}{\partial z^2}\right)\frac{\partial u_\lambda}{\partial t} = 0. \tag{7.6.23}$$

Multiplying $M\rho\tau_M$ on Eq. (7.6.23), we obtain

$$M\rho\left(\frac{\partial}{\partial t} + \tau_M\frac{\partial^2}{\partial t^2}\right)\frac{\partial u_\lambda}{\partial t} = \nu_\lambda^2\frac{\partial^2}{\partial z^2}\frac{\partial u_\lambda}{\partial t}, \tag{7.6.24}$$

where

$$\frac{\nu_\lambda}{M\rho} = \tau_M c_\lambda^2, \tag{7.6.25}$$

$$\nu_t = \eta, \tag{7.6.26}$$

$$\nu_l = \zeta + \frac{4}{3}\eta, \tag{7.6.27}$$

η is a kinetic shear viscosity and $\zeta + \frac{4}{3}\eta$ is a kinetic bulk viscosity. Equation (7.6.24) can be rewritten in a matrix form as follows:

$$M\rho\left(\frac{\partial}{\partial t} + \tau_M \frac{\partial^2}{\partial t^2}\right)\begin{pmatrix} \frac{\partial u_x}{\partial t} \\ \frac{\partial u_y}{\partial t} \\ \frac{\partial u_z}{\partial t} \end{pmatrix}$$

$$= \begin{pmatrix} \eta \frac{\partial^2}{\partial z^2}\frac{\partial u_x}{\partial t} \\ \eta \frac{\partial^2}{\partial z^2}\frac{\partial u_y}{\partial t} \\ \left(\zeta + \frac{4}{3}\eta\right)\frac{\partial^2}{\partial z^2}\frac{\partial u_z}{\partial t} \end{pmatrix}$$

$$= \eta \frac{\partial^2}{\partial z^2}\begin{pmatrix} \frac{\partial u_x}{\partial t} \\ \frac{\partial u_y}{\partial t} \\ \frac{\partial u_z}{\partial t} \end{pmatrix} + \left(\zeta + \frac{1}{3}\eta\right)\frac{\partial^2}{\partial z^2}\begin{pmatrix} 0 \\ 0 \\ \frac{\partial u_z}{\partial t} \end{pmatrix}. \tag{7.6.28}$$

If we take $\boldsymbol{q}$ in an arbitrary direction, we have

$$M\rho\left(\frac{\partial}{\partial t} + \tau_M \frac{\partial^2}{\partial t^2}\right)\frac{\partial \boldsymbol{u}}{\partial t} = \eta\nabla^2\frac{\partial \boldsymbol{u}}{\partial t} + \left(\zeta + \frac{1}{3}\eta\right)\boldsymbol{\nabla\nabla}\cdot\frac{\partial \boldsymbol{u}}{\partial t}, \tag{7.6.29}$$

which yields phonons or elastic waves for $q_0\tau_M \gg 1$ and kinetic viscosity for $\tau_M q_0 \ll 1$.

At higher temperatures, the viscosity comes mainly from the exchange of the conventional velocity $\boldsymbol{v}_0$, while at lower temperatures, the viscosity comes mainly from the exchange of the displacement velocity $\frac{\partial \boldsymbol{u}}{\partial t}$. Near the liquid-glass transition, since the viscosity due to the conventional velocity $\boldsymbol{v}_0$ is of the order of $\tau_0 c_T^2$, the viscosity due to the exchange of $\boldsymbol{v}_0$ is negligible. Since $\boldsymbol{v} = \boldsymbol{v}_0 + \frac{\partial \boldsymbol{u}}{\partial t}$, we interpolate between Eqs. (7.6.20) and (7.6.29) so that we obtain the generalized Navier-Stokes equation for $q_0\tau_0 > 1$:

$$M\rho\left(\frac{\partial}{\partial t} + \tau_M \frac{\partial^2}{\partial t^2}\right)\boldsymbol{v} = -\boldsymbol{\nabla}p + \eta\nabla^2\boldsymbol{v} + \left(\frac{1}{3}\eta + \zeta\right)\boldsymbol{\nabla\nabla}\cdot\boldsymbol{v}. \tag{7.6.30}$$

Using the relation $U_J^1 > |U_J^{01}| > U_J^0$ in Eq. (7.3.6), we can expect

$$1/\tau_M > 1/\tau_0. \tag{7.6.31}$$

The generalized Navier-Stokes equation must reduce to the Navier-Stokes equation for fluid mechanics in the fluid state and to the dynamical equation of elastic waves in the glassy state. Here we investigate Eq. (7.6.30) for two cases:

1. $1/\tau_M > q_0 > 1/\tau_0$:
Equation (7.6.30) reduces to the Navier-Stokes equation [30]:

$$M\rho\frac{\partial \boldsymbol{v}}{\partial t} = -\boldsymbol{\nabla}p + \eta\nabla^2\boldsymbol{v} + \left(\zeta + \frac{1}{3}\eta\right)\boldsymbol{\nabla\nabla}\cdot\boldsymbol{v}. \tag{7.6.32}$$

The Navier-Stokes equation is the dynamical equation of fluid in the classical fluid mechanics.

2. $q_0 > 1/\tau_M$:
Equation (7.6.30) leads to

$$M\rho\tau_M \frac{\partial^2}{\partial t^2}\boldsymbol{v} = -\nabla p + \eta\nabla^2\boldsymbol{v} + \left(\frac{1}{3}\eta + \zeta\right)\nabla\nabla\cdot\boldsymbol{v}.$$

There are relations:

$$\nu_t/(M\rho) = \eta/(M\rho) = \tau_M c_t^2, \tag{7.6.33}$$

$$\nu_l/(M\rho) = \left(\zeta + \frac{4}{3}\eta\right)/(M\rho) = \tau_M c_l^2, \tag{7.6.34}$$

where c_l and c_t are a longitudinal and two transverse phonon or elastic wave velocities, respectively. Near the liquid-glass transition, $\tau_M \propto J^{-1}$ and $\boldsymbol{v} \to \frac{\partial \boldsymbol{u}}{\partial t}$. Thus Eq. (7.6.30) reduces to the dynamical equation of elastic waves [32]:

$$\frac{\partial^2 \boldsymbol{u}}{\partial t^2} = c_t^2\nabla^2\boldsymbol{u} + (c_l^2 - c_t^2)\nabla\nabla\cdot\boldsymbol{u}. \tag{7.6.35}$$

The dynamical equation of the elastic waves plays an essential role in the solid state. Elastic waves correspond to phonons in the long wavelength regime.

7.6.4 Viscoelastic Theory

The dynamical equations of phonons are given by (7.6.21):

$$q_0\left(q_0 + \frac{\mathrm{i}}{\tau_M}\right) - \omega_{\lambda\boldsymbol{q}}^2 = 0. \tag{7.6.21}$$

Equation (7.6.21) in the long wavelength regime, $aq \ll 1$, explains the conventional viscoelastic theory. We show this.

We consider the displacement $\boldsymbol{u}$ depends only on z. Then the instantaneous displacement corresponds to the dynamical equation of elastic waves for $q_0\tau_M \gg 1$:

$$q_0^2 - \omega_{\lambda\boldsymbol{q}}^2 \cong q_0^2 - c_\lambda^2\boldsymbol{q}^2 = 0.$$

Then we have

$$\left\{M\rho\frac{\partial^2}{\partial t^2} - G_{\infty\lambda}\frac{\partial^2}{\partial x_\lambda^2}\right\}u_\lambda = 0, \tag{7.6.36}$$

$$G_{\infty\lambda} = M\rho c_\lambda^2. \tag{7.6.37}$$

where for λ = transverse, $\lambda = x$ or y, $G_{\infty t}$ is a high frequency shear modulus and for λ = longitudinal, $\lambda = z$, $G_{\infty l}$ a high frequency bulk modulus.

The viscous equation corresponds to the dynamical equation of viscosity for $q_0\tau_M \ll 1$:

$$q_0 + \mathrm{i}\tau_M\omega_{\lambda\boldsymbol{q}}^2 \cong q_0 + \mathrm{i}\tau_M c_\lambda^2\boldsymbol{q}^2 = q_0 + \mathrm{i}\frac{\nu_\lambda}{M\rho}\boldsymbol{q}^2 = 0.$$

Then we have

$$\left\{ M\rho \frac{\partial}{\partial t} - \nu_\lambda \frac{\partial^2}{\partial x_\lambda^2} \right\} \frac{\partial u_\lambda}{\partial t} = 0, \tag{7.6.38}$$

where

$$\nu_\lambda / M\rho = \tau_M c_\lambda^2. \tag{7.6.39}$$

The viscoelastic equation is obtained by interpolating between Eqs. (7.6.36) and (7.6.38):

$$\left\{ M\rho \left(\frac{1}{\nu_\lambda} + \frac{1}{G_{\infty\lambda}} \frac{\partial}{\partial t} \right) \frac{\partial}{\partial t} - \frac{\partial^2}{\partial x_\lambda^2} \right\} \frac{\partial u_\lambda}{\partial t} = 0. \tag{7.6.40}$$

The Fourier transformation of Eq. (7.6.40) leads to

$$q_0 \left(q_0 + \mathrm{i} \frac{G_{\infty\lambda}}{\nu_\lambda} \right) - \frac{G_{\infty\lambda}}{M\rho} q^2 = 0. \tag{7.6.41}$$

Equations (7.6.37) and (7.6.39) lead to

$$\frac{G_{\infty\lambda}}{M\rho} = c_\lambda^2, \quad \frac{G_{\infty\lambda}}{\nu_\lambda} = \frac{1}{\tau_M}. \tag{7.6.42}$$

Thus the viscoelastic Eq. (7.6.40) is just the dynamical Eq. (7.6.21) in the long wavelength regime.

References

[1] T. Kitamura, Physica A 245 (1997) 543.
[2] T.Kitamura, Phys. Lett. A 235 (1997) 515.
[3] T. Kitamura, Phys. Lett. A 243 (1998) 325.
[4] T. Kitamura, Physica A 262 (1999) 16.
[5] T.Kitamura, Physica A 285 (2001) 234.
[6] T. Kitamura, Phys. Rep. 383 (2003) 1.
[7] D. Vollhandt, P. Wölfle, Phys. Rev. B22 (1980) 4666.
[8] P. Shen, Introduction to Wave Scatterng, Localization, and Mesoscopic Phenomena, Academic Press, New York, 1995.
[9] T. Schneider, R. Brout, H. Thomas, J. Feder, Phys. Rev. Lett. 25 (1970) 1423.
[10] M.C. Phillips, A.J. Barlow, J. Lamb. Proc. R. Soc. London A 329 (1972) 193.
[11] A.A. Abrikosov, L.P. Gorkov, I.E. Dzyaloshinski, Method of Quantum Field Theory in Statistical Physics, Prince-Hall, Englewood Cliffs, NJ, 1963.
[12] A.L. Fetter, J.D. Walecka, Quantum Field Theory of Many-Particle Systems, McGraw-Hill, New York, 1971.
[13] R.C. Zeller, R.O. Pohl, Phys. Rev. B 4 (1971) 2029.
[14] P.W. Anderson, B.I. Halperin, C.M. Varma, Phil. Mag. 25 (1972) 1.

[15] A.J. Sievers, S. Takeno, Phys. Rev. B 39 (1989) 3374.
[16] T. kitamura, S. Takeno, Phys. Lett. A 266 (2000) 198.
[17] T. Kitamura, H. Umezawa, Physica A 121 (1983) 38.
[18] T. Kitamura, Physica A 135 (1986) 21.
[19] T. Kitamura, Phys. Lett. A 117 (1986) 81.
[20] T. Kitamura, Phys. Lett. A118 (1986) 341.
[21] T. Kitamura, Physica A 144 (1987) 29.
[22] T. Kitamura, Physica A 151 (1988) 303.
[23] T. Kitamura, Il Nuovo Cimento D 11 (1989) 1441.
[24] T. Kitamura, Physica A 160 (1989) 181.
[25] F.A. Lindemann, Z. Phys. 11 (1910) 609.
[26] G. Adam, J.H. Gibbs, J. Chem. Phys. 43 (1965) 139.
[27] T. Kitamura, Phys. Lett. A 297 (2002) 253.
[28] S.W. Loversey, Condensed Matter Physics: Dynamical Correlations (Benjamin, 1980).
[29] J.P. Hansen, I.R. MacDonald, Theory of Simple Liquids, Academic Press, London, 1990.
[30] L.D. Landau, E.M. Lifshitz, Fluid Mechanics, Pergamon Press, London, 1963.
[31] T. Kitamura, Phys. Lett. A 282 (2001) 59.
[32] L.D. Landau, E.M. Lifshitz, Theory of Elasticity, Pergamon Press, London, 1970.

8 Phonon Operators in Nonlinear Interaction Potentials

So far we have investigated phonon operators in a linear interaction potential. There we have investigated the phonons and the viscosities. The viscosities are the dissipative processes of the phonon modes. If we want thermal conductivity, since the thermal conductivity is the dissipative process of the heat fluctuation modes, i.e. the entropy fluctuation modes, we must investigate the entropy fluctuation modes. The phonon density fluctuations correspond to the fluctuations of the phonon frequencies so that the phonon density fluctuations correspond to the entropy fluctuations due to phonons. So we must investigate the phonon density fluctuation modes. Since the phonon density fluctuations are the bilinear form of the phonon operators, we need the nonlinear interaction potentials. Therefore we first construct the nonlinear interaction potentials, which originate from the repetition of the up and down transitions of particles between the two levels in their respective localized harmonic potentials. We construct the nonlinear interaction potentials such that they satisfy the WT relations and reduce to the classical nonlinear interaction potentials. Using the nonlinear interaction potentials, we establish the dynamical equations of phonon operators.

Next we exemplify that the established dynamical equations of phonon operators in the nonlinear interaction potentials can explain solitons and localized modes in the classical nonlinear theory. We also show that we can derive a new aspect of bound states in moving solitons. The conventional theories have not been able to treat the temperature dependence of the localized modes, but our theory can explain the temperature dependence behaviors. Such situations suggest that the present theory is a potentially powerful method in developing the thermal conductivity near the liquid–glass transition. In the next chapter, using the nonlinear interaction potentials, we deal with thermal conductivity due to phonon density fluctuations.

8.1 The Dynamical Equation for Phonon Operators in Nonlinear Interaction Potentials

The two body interaction potential in the Hamiltonian is essentially nonlinear. But up to now we have made the linear approximation to the two body interaction potential with respect to inter-band density fluctuations (phonon operators). To extend the dynamics of phonon operators to the nonlinear ones, we have to construct the nonlinear interaction potentials for phonon operators. In constructing the nonlinear interaction potentials for phonon operators [1–6], we consider the nonlinear interaction potentials satisfy the WT

Liquid Glass Transition. DOI: http://dx.doi.org/10.1016/B978-0-12-407177-3.00008-8

relation and reduce to the classical nonlinear interaction potentials. Using the nonlinear interaction potentials we establish the dynamical equation for phonon operators.

8.1.1 Nonlinear Interaction Potentials

We start with the following model Hamiltonian:

$$H = \sum_{\mu m}\{\hbar\tilde{\omega}_\mu b^\dagger_{\mu m}b_{\mu m} + \hbar J_\mu(\boldsymbol{R}_m - \boldsymbol{R}_n)b^\dagger_{\mu m}b_{\mu n}\} + \frac{1}{2}\int d^3x\, d^3y n(x)V(\boldsymbol{x}-\boldsymbol{y})n(y), \tag{8.1.1}$$

where we have considered the averaged harmonic frequencies and hopping magnitudes:

$$\tilde{\omega}_\mu = \langle\tilde{\omega}_{m\mu}\rangle_c, \qquad J_\mu(\boldsymbol{R}_m - \boldsymbol{R}_n) = \langle J_{\mu mn}\rangle_c, \tag{8.1.2}$$

The randomness of the system are taken into account by the simultaneously scattering processes of two particles by the same random scattering potentials; the elementary random scattering magnitudes, $U^{\mu\nu}_\omega$ due to the random frequencies and $U^{\mu\nu}_J$ due to the random hopping magnitudes.

In order to construct the nonlinear interaction potentials we introduce the phonon operators as

$$B_{mi} \equiv b^\dagger_{mi}b_{m0} + b^\dagger_{m0}b_{mi}, \tag{8.1.3}$$

where $b_{m\mu}$ is the annihilation operator of a particle with μ-state at a position $\boldsymbol{R}_m$. The effective interaction Hamiltonian is established as follows:

1. Since particles are well localized, the width of energy bands of a particle $|J_\mu|$ is very much smaller than the harmonic frequency ω:

$$|J_\mu| \ll \omega. \tag{8.1.4}$$

   From the uncertainty principle, $|J_\mu|$ is nearly equal to $\frac{1}{T}$, where T is a hopping time, while ω is nearly equal to $\frac{1}{\tau}$, where τ is a time of an up and down transition of a particle, we have

$$T \gg \tau, \qquad |J_\mu| \sim \frac{1}{T}, \quad \omega \sim \frac{1}{\tau}. \tag{8.1.5}$$

   In an interval T, many repetition of up and down transitions of a particle between the two bands can occur at the same site. Thus we can include higher powers of the phonon operator $\boldsymbol{B}_m$ in the number operator $n(x)$ and we can neglect the second term, the hopping term $\hbar J_\mu(\boldsymbol{R}_m - \boldsymbol{R}_n)b^\dagger_{\mu m}b_{\mu n}$ in Eq. (8.1.1) because of Eq. (8.1.5) and $\beta\hbar|J_\mu| \ll 1$, which means that the bandwidth can be neglected.
2. The expansion form of $n(x)$ with respect to $\boldsymbol{B}_m$ is determined by the WT relation (6.1.13):

$$\hbar\nabla\langle n(x)\rangle_\epsilon = \epsilon\int d^4y\langle T_\tau n(x)n(y)\rangle_\epsilon\nabla g(y), \tag{6.1.13}$$

An up and down transition of a particle according to the phonon operators $\boldsymbol{B}_m$ yields the spread of wave function from the ground state wave function, the magnitude of which is nearly $\zeta \boldsymbol{B}$, where

$$\zeta = \sqrt{\frac{\hbar}{2M\omega}}, \tag{8.1.6}$$

which is the half width of the zero point motion. The repetition of the up and down transitions also yields the excited state wave functions $\nabla \tilde{w}_0(\boldsymbol{x} - \boldsymbol{R_m})$. Thus we can put

$$n(x) = \sum_m \left[\tilde{w}_0^2(\boldsymbol{x} - \boldsymbol{R}_m) b_{m0}^\dagger b_{m0} + \cdots + \sum_\alpha c_\alpha \frac{(-\zeta \boldsymbol{B}_m \cdot \nabla)^\alpha}{\alpha!} \right.$$
$$\left. \tilde{w}_0^2(\boldsymbol{x} - \boldsymbol{R}_m) + \cdots \right]. \tag{8.1.7}$$

If we substitute $n(x)$ in Eq. (8.1.7) into both the sides of Eq. (6.1.13), the term of the nth spatial derivative in $n(x)$ becomes to the $(n+1)$th derivative on the left-hand side of Eq. (6.1.13), which corresponds to the term of the $(n+1)$th derivative of $n(x)$ on the right-hand side of Eq. (6.1.13). Thus we can verify the form of the expansion in $n(x)$.

3. We compare the interaction Hamiltonian with $n(x)$ in Eq. (8.1.7) with the classical nonlinear potential,

$$H_I = \frac{1}{2} \sum_{m \neq n, \alpha\beta} \left[\frac{1}{\alpha!} (\boldsymbol{u}_m \cdot \nabla_{\boldsymbol{R}_m})^\alpha V(\boldsymbol{R}_m - \boldsymbol{R}_n) \frac{1}{\beta!} (\boldsymbol{u}_n \cdot \overleftarrow{\nabla}_{\boldsymbol{R}_n})^\beta \right], \tag{8.1.8}$$

where $\boldsymbol{u}_m$ is a displacement of the particle from the site $\boldsymbol{R}_m$. In the classical limit, we can regard $\zeta \boldsymbol{B}_m$ as a displacement $\boldsymbol{u}_m$ and $\tilde{w}_{m0}^2$ as a delta function:

$$\boldsymbol{u}_m = -\zeta \boldsymbol{B}, \tag{8.1.9}$$
$$\tilde{w}_0^2(\boldsymbol{x} - \boldsymbol{R}_m) = \delta(\boldsymbol{x} - \boldsymbol{R}_m). \tag{8.1.10}$$

If we partially integrate H_I in Eq. (8.1.1) in x and y, and compare Eqs. (8.1.1) and (8.1.8), we obtain $c_\alpha = 1$.

Thus we obtain the effective interaction Hamiltonian:

$$H_I = \frac{1}{2} \sum_{m \neq n, \alpha\beta} \int d^3x d^3y \frac{(-\zeta \boldsymbol{B}_m \cdot \nabla)^\alpha}{\alpha!} \tilde{w}_0^2(\boldsymbol{x} - \boldsymbol{R}_m) V(\boldsymbol{x} - \boldsymbol{y})$$
$$\frac{(-\zeta \boldsymbol{B}_n \cdot \nabla)^\beta}{\beta!} \tilde{w}_0^2(\boldsymbol{y} - \boldsymbol{R}_n)$$
$$\equiv \frac{1}{2} \sum_{m \neq n, \alpha\beta} (-\zeta)^\alpha \underbrace{(\boldsymbol{B}_m \cdots \boldsymbol{B}_m)}_{\alpha} \cdot \overleftrightarrow{V}_{mn}^{(\alpha\beta)} \cdot (-\zeta)^\beta \underbrace{(\boldsymbol{B}_n \cdots \boldsymbol{B}_n)}_{\beta}, \tag{8.1.11}$$

where we abbreviate the first equation as the second one. H_I is the same form as that in crystals, but the sites $\{\boldsymbol{R}_l\}$ distribute randomly.

8.1.2 The Heisenberg Equation of the Phonon Operators B_m

Using the interaction Hamiltonian (8.1.11), we derive a Heisenberg equation for phonon operators $\{\boldsymbol{B}_m\}$. To do so, we first derive a Heisenberg equation for particle operators $\{b_{m\mu}\}$:

$$\begin{aligned}
&\left\{-\hbar\frac{\partial}{\partial\tau} - \hbar\tilde{\omega}_\mu\right\} b_{m\mu} \\
&= b_{m\bar{\mu}} \sum_{n(\neq m)} \left\{(-\zeta)\boldsymbol{e}_i \cdot \overleftrightarrow{V}^{(11)}_{mn} \cdot (-\zeta)\boldsymbol{B}_n \right. \\
&\quad \left. + \sum_{\alpha+\beta>2} \alpha(-\zeta)^\alpha(\boldsymbol{e}_i\boldsymbol{B}_m \cdots \boldsymbol{B}_m) \cdot \overleftrightarrow{V}^{(\alpha\beta)}_{mn} \cdot (-\zeta)^\beta(\boldsymbol{B}_n \cdots \boldsymbol{B}_n)\right\} \\
&\equiv b_{m\bar{\mu}} \sum_{n(\neq m)} \left\{(-\zeta)\boldsymbol{e}_i \cdot \overleftrightarrow{V}^{(11)}_{mn} \cdot (-\zeta)\boldsymbol{B}_n + \sum_{\alpha+\beta>2} \cdots\right\},
\end{aligned} \tag{8.1.12}$$

where we have been concerned with the ith Cartesian coordinate. In the last equation we abbreviate the second term in the bracket as .... Thus $\bar{\mu}$ is 0 or i for $\mu = i$ or 0, respectively, $\boldsymbol{e}_i$ is a unit vector. Note that in Eq. (8.1.12) the terms with $\overleftrightarrow{V}^{(20)}$ disappear, because these terms are taken into account in the unperturbed Hamiltonian.

Since a phonon operator $\boldsymbol{B}_m$ is a particle-hole pair, we should consider the internal freedom. So we put

$$B_{mi} \equiv \pm \lim_{\tau\to\tau'} B_{mi}(\tau, \tau') \tag{8.1.13}$$

$$B_{mi}(\tau, \tau') \equiv T_\tau\{b_{m0}(\tau)b^\dagger_{mi}(\tau') + b_{mi}(\tau)b_{m0}(\tau')\}, \tag{8.1.14}$$

where T_τ is a chronological operator. Using Eq. (8.1.12), we obtain

$$\begin{aligned}
&\left(-\hbar\frac{\partial}{\partial\tau} - \hbar\tilde{\omega}_\mu\right)\{T_\tau b_{m\mu}(\tau)b^\dagger_{m\bar{\mu}}(\tau')\} \\
&= \left(-\hbar\frac{\partial}{\partial\tau} - \hbar\tilde{\omega}_\mu\right)\{\theta(\tau-\tau')b_{m\mu}(\tau)b^\dagger_{m\bar{\mu}}(\tau') \pm \theta(\tau'-\tau)b^\dagger_{m\bar{\mu}}(\tau')b_{m\mu}(\tau)\} \\
&= \theta(\tau-\tau')b_{m\bar{\mu}}(\tau) \sum_{n(\neq m)} \\
&\quad \left\{(-\zeta)\boldsymbol{e}_i \cdot \overleftrightarrow{V}^{(11)}_{mn} \cdot (-\zeta)\boldsymbol{B}_n + \sum_{\alpha+\beta>2} \cdots\right\}(\tau)b^\dagger_{m\bar{\mu}}(\tau') \\
&\quad \pm\theta(\tau'-\tau)b^\dagger_{m\bar{\mu}}(\tau')b_{m\bar{\mu}}(\tau) \sum_{n(\neq m)} \\
&\quad \left\{(-\zeta)\boldsymbol{e}_i \cdot \overleftrightarrow{V}^{(11)}_{mn} \cdot (-\zeta)\boldsymbol{B}_n + \sum_{\alpha+\beta>2} \cdots\right\}(\tau),
\end{aligned} \tag{8.1.15}$$

where in the last equation, the simultaneous commutation relation of $[b_{m\mu}, b^{\dagger}_{m\bar{\mu}}]$ coming from the τ-derivative of θ functions disappears. Then we obtain

$$\left(-\hbar\frac{\partial}{\partial\tau} - \hbar\tilde{\omega}_{\mu}\right)\left\{T_{\tau}b_{m\mu}(\tau)b^{\dagger}_{m\bar{\mu}}(\tau')\right\}\left(\hbar\frac{\overleftarrow{\partial}}{\partial\tau'} - \hbar\tilde{\omega}_{\bar{\mu}}\right)$$

$$= -\hbar\delta(\tau-\tau')\Bigg[b_{m\bar{\mu}}(\tau)\sum_{n(\neq m)}\Bigg\{(-\zeta)\boldsymbol{e}_i\cdot\overset{\leftrightarrow}{V}{}^{(11)}_{mn}\cdot(-\zeta)\boldsymbol{B}_n$$

$$+\sum_{\alpha+\beta>2}\cdots\Bigg\}(\tau), b^{\dagger}_{m\bar{\mu}}(\tau')\Bigg]_{\mp}$$

$$+T_{\tau}\Bigg\{b_{m\bar{\mu}}(\tau)\sum_{n(\neq m)}\Bigg\{(-\zeta)\boldsymbol{e}_i\cdot\overset{\leftrightarrow}{V}{}^{(11)}_{mn}\cdot(-\zeta)\boldsymbol{B}_n+\sum_{\alpha+\beta>2}\cdots\Bigg\}(\tau)$$

$$\times\, b^{\dagger}_{m\bar{\mu}}(\tau')\sum_{n(\neq m)}\Bigg\{(-\zeta)\boldsymbol{e}_i\cdot\overset{\leftrightarrow}{V}{}^{(11)}_{mn}\cdot(-\zeta)\boldsymbol{B}_n+\sum_{\alpha+\beta>2}\cdots\Bigg\}(\tau')\Bigg\},$$

(8.1.16)

where the arrow on the partial derivative means that it operates on the term on the left-hand side.

When we derive a dynamical equation for $B_{mi}(\tau,\tau')$, we make the random phase approximation; we only consider the term involving the $\delta(\tau-\tau')$ function and the commutation relation between particle operators at the same site. Thus we obtain

$$\left(-\hbar\frac{\partial}{\partial\tau} - \hbar\tilde{\omega}_{\mu}\right)\left\{T_{\tau}b_{m\mu}(\tau)b^{\dagger}_{m\bar{\mu}}(\tau')\right\}\left(\hbar\frac{\overleftarrow{\partial}}{\partial\tau'} - \hbar\tilde{\omega}_{\bar{\mu}}\right)$$

$$\cong \hbar\delta(\tau-\tau')\sum_{n(\neq m)}\Bigg\{\zeta^2\boldsymbol{e}_i\cdot\overset{\leftrightarrow}{V}{}^{(11)}_{mn}\cdot\boldsymbol{B}_n+\sum_{\alpha+\beta>2}\cdots\Bigg\}(\tau'), \tag{8.1.17}$$

This equation is a dynamical equation for $B_{mi}(\tau,\tau')$.

Now we introduce particle Green's functions and bubble diagrams:

$$\left(-\hbar\frac{\partial}{\partial\tau} - \hbar\tilde{\omega}_{\mu}\right)G_{m\mu}(\tau-\tau') = \hbar\delta(\tau-\tau') \tag{8.1.18}$$

$$Q_{mi}(\tau-\tau') = \mp\frac{1}{\hbar}[G_{m0}(\tau-\tau')G_{mi}(\tau'-\tau) + G_{mi}(\tau-\tau')G_{m0}(\tau'-\tau)], \tag{8.1.19}$$

where $G_{m\mu}$ and Q_{mi} do not depend on the site $\boldsymbol{R}_m$, since we have assumed the averaged harmonic frequencies in Eq. (8.1.1). Q_{mi} does not also depend on the Cartesian coordinate i because of the isotropic energy spectrum of particles. So we omit these subscripts below. The Fourier transformation of the Green's functions gives

$$
\begin{aligned}
G_\mu(\mathrm{i}\omega_n) &= \int_0^{\beta\hbar} d\tau e^{\mathrm{i}\omega_n\tau} G_\mu(\tau) \\
&= \frac{1}{\mathrm{i}\omega_n - \tilde{\omega}_\mu},
\end{aligned} \tag{8.1.20}
$$

$$
\begin{aligned}
Q(i\nu_n) &= \int_0^{\beta\hbar} d\tau e^{\mathrm{i}\nu_n\tau} Q(\tau) \\
&= \mp\frac{1}{\beta\hbar^2}\sum_{\mathrm{i}\omega_n}\{G_0(\mathrm{i}\omega_n)G_i(\mathrm{i}\omega_n - \mathrm{i}\nu_n) + G_i(\mathrm{i}\omega_n)G_0(\mathrm{i}\omega_n - \mathrm{i}\nu_n)\} \\
&= -\frac{2\omega/\hbar}{(\mathrm{i}\nu_n)^2 - \omega^2}\{f(\omega_i) - f(\omega_0)\},
\end{aligned} \tag{8.1.21}
$$

where f is a bose or fermi distribution function. At low temperatures, $f(\tilde{\omega}_0) \cong 1$ and $f(\tilde{\omega}_i) \cong 0$, we obtain

$$
Q(\mathrm{i}\nu_n) = \frac{2\omega/\hbar}{(\mathrm{i}\nu_n)^2 - \omega^2}, \tag{8.1.22}
$$

$$
Q(\mathrm{i}\nu_n) - Q(0) = \frac{(\mathrm{i}\nu_n)^2/\hbar}{2\omega} Q(\mathrm{i}\nu_n)Q(0), \tag{8.1.23}
$$

we will use these equations in deriving the Heisenberg equation for phonon operators.

From Eqs. (8.1.18), (8.1.17) leads to

$$
\begin{aligned}
\{T_\tau b_{m\mu}(\tau)b^\dagger_{m\bar{\mu}}(\tau')\} = \frac{1}{\hbar}\int_0^{\beta\hbar} d\tau_z G_\mu(\tau - \tau_z)G_{\bar{\mu}}(\tau_z - \tau') \sum_{n(\neq m)} \\
\left\{\zeta^2 \boldsymbol{e}_i \cdot \overset{\leftrightarrow}{V}{}^{(11)}_{mn} \cdot \boldsymbol{B}_n + \sum_{\alpha+\beta>2} \cdots\right\}(\tau_z).
\end{aligned} \tag{8.1.24}
$$

Considering Eqs. (8.1.13) and (8.1.14) in Eq. (8.1.24), we obtain

$$
\begin{aligned}
B_{mi}(\tau) = \int_0^{\beta\hbar} d\tau' Q(\tau - \tau') \sum_{n(\neq m)} \Big\{(-\zeta)\boldsymbol{e}_i \cdot \overset{\leftrightarrow}{V}{}^{(11)}_{mn} \cdot (-\zeta)\boldsymbol{B}_n(\tau') \\
+ \sum_{\alpha+\beta>2} \alpha(-\zeta)^\alpha(\boldsymbol{e}_i\boldsymbol{B}_m \cdots \boldsymbol{B}_m) \\
\cdot \overset{\leftrightarrow}{V}{}^{(\alpha\beta)}_{mn} \cdot (-\zeta^\beta)(\boldsymbol{B}_n \cdots \boldsymbol{B}_n)(\tau')\Big\}.
\end{aligned} \tag{8.1.25}
$$

The Fourier transformation of Eq. (8.1.25) leads to

$$0 = \sum_{nj} \left\{ \delta_{mn}\delta_{ij} - Q(\mathrm{i}\nu_n)\zeta^2 V^{(11)}_{mn;ij} \right\} B_{nj}(\mathrm{i}\nu_n)$$
$$- \sum_{\alpha+\beta>2} Q(\mathrm{i}\nu_n)\alpha(-\zeta)^{\alpha}(\boldsymbol{e}_i \boldsymbol{B}_m \cdots \boldsymbol{B}_m)$$
$$\cdot \overleftrightarrow{V}^{(\alpha\beta)}_{mn} \cdot (-\zeta)^{\beta}(\boldsymbol{B}_n \cdots \boldsymbol{B}_n)(\mathrm{i}\nu_n), \tag{8.1.26}$$

where $\overleftrightarrow{V}^{(\alpha\beta)}_{mm} = 0$. Hereafter, we also denote the interaction Hamiltonian H_I as

$$H_I = \frac{1}{2} \sum_{m\neq n,\alpha\beta} \frac{(-\zeta)^{\alpha}}{\alpha!} B_{mi_1} \cdots B_{mi_\alpha}$$
$$V^{(\alpha\beta)}_{mn;i_1\cdots i_\alpha, j_1\cdots j_\beta} \frac{(-\zeta)^{\beta}}{\beta!} B_{nj_1} \cdots B_{nj_\beta}, \tag{8.1.27}$$

$$V^{(\alpha\beta)}_{mn;i_1\cdots i_\alpha, j_1\cdots j_\beta} = \int d^3x d^3y \nabla_{i_1} \cdots \nabla_{i_\alpha} \tilde{w}_0^2(\boldsymbol{x} - \boldsymbol{R}_m)$$
$$V(\boldsymbol{x} - \boldsymbol{y}) \nabla_{j_1} \cdots \nabla_{j_\beta} \tilde{w}_0^2(\boldsymbol{y} - \boldsymbol{R}_n). \tag{8.1.28}$$

8.1.3 The Dynamical Equation of Phonon Green's Function

Now we introduce phonon Green's functions:

$$D_{minj}(\tau - \tau') = -\frac{1}{\hbar}\langle T_\tau B_{mi}(\tau) B_{nj}(\tau')\rangle_c$$
$$= \frac{1}{\beta\hbar N} \sum_{\mathrm{i}\nu_n \boldsymbol{q}} e^{\mathrm{i}\boldsymbol{q}\cdot(\boldsymbol{R}_m - \boldsymbol{R}_n) - \mathrm{i}\nu_n(\tau-\tau')} D_{ij}(\boldsymbol{q}, \mathrm{i}\nu_n). \tag{8.1.29}$$

The phonon Green's function $D_{minj}(\tau - \tau')$ in the random phase approximation is given by

$$D_{minj}(\tau - \tau') = Q_{mi}(\tau - \tau')\delta_{mn}\delta_{ij} + \int_0^{\beta\hbar} d\tau_z \sum_{lk} Q_{mi}(\tau - \tau_z)\zeta^2$$
$$\langle V^{(11)}_{ml;ik}\rangle_c D_{lknj}(\tau_z - \tau'), \tag{8.1.30}$$

The Fourier transformation of Eq. (8.1.30) in $\mathrm{i}\nu_n$ leads to

$$D_{minj}(\mathrm{i}\nu_n) = Q(\mathrm{i}\nu_n)\delta_{mn}\delta_{ij} + Q(\mathrm{i}\nu_n)\sum_{lk} \zeta^2 \langle V^{(11)}_{ml;ik}\rangle_c D_{lknj}(\mathrm{i}\nu_n). \tag{8.1.31}$$

The Fourier transformation of Eq. (8.1.31) in $\boldsymbol{q}$ leads to

$$D_{ij}(\boldsymbol{q}, \mathrm{i}\nu_n) = Q(\mathrm{i}\nu_n)\delta_{ij} + Q(\mathrm{i}\nu_n)\sum_{k} V_{ik}(\boldsymbol{q}) D_{kj}(\boldsymbol{q}, \mathrm{i}\nu_n), \tag{8.1.32}$$

where

$$V_{ij}(\boldsymbol{q}) = \zeta^2 \sum_{\boldsymbol{R}_m - \boldsymbol{R}_n} e^{-\mathrm{i}\boldsymbol{q}\cdot(\boldsymbol{R}_m - \boldsymbol{R}_n)} \langle V^{(11)}_{mn;ij}\rangle_c$$
$$= \zeta^2 \int d^3R g(R) e^{-\mathrm{i}\boldsymbol{q}\cdot\boldsymbol{R}} V_{ij}(\boldsymbol{R}), \tag{8.1.33}$$

where we have put $\langle V^{(11)}_{nni,j}\rangle_c = 0$.

Using the gap equation:

$$\delta_{ij} = Q(\mathrm{i}\nu_n = 0) V_{ij}(\boldsymbol{q} = 0)$$
$$= Q(0)\zeta^2 \sum_{\boldsymbol{R}_m - \boldsymbol{R}_n} \langle V^{(11)}_{mn;ij}\rangle_c = Q(0)\zeta^2 \sum_{\boldsymbol{R}_n} \langle V^{(11)}_{mn;ij}\rangle_c, \tag{8.1.34}$$

and Eq. (8.1.23), we obtain

$$\delta_{ij} - Q(i\nu) V_{ij}(\boldsymbol{q}) = Q(0) V_{ij}(0) - Q(\mathrm{i}\nu) V_{ij}(\boldsymbol{q})$$
$$= \{Q(0) - Q(\mathrm{i}\nu_n)\} V_{ij}(0) + Q(\mathrm{i}\nu_n)\{V_{ij}(0) - V_{ij}(\mathrm{q})\}$$
$$= \zeta^2 Q(\mathrm{i}\nu_n)\{M(\mathrm{i}\nu_n)^2 \delta_{ij} + \zeta^{-2}(V_{ij}(0) - V_{ij}(\boldsymbol{q}))\} \tag{8.1.35}$$

Equations (8.1.32) and (8.1.35) lead to

$$\sum_k \left[M(\mathrm{i}\nu_n)^2 \delta_{ik} + \zeta^{-2}\{V_{ik}(0) - V_{ik}(\boldsymbol{q})\}\right] D_{kj}(\boldsymbol{q}, \mathrm{i}\nu_n) = \frac{\delta_{ij}}{\zeta^2}. \tag{8.1.36}$$

This equation is the dynamical equation of phonon Green's function.

8.1.4 The Dynamical Equation of Phonon Operators

Equation (8.1.31) can be rewritten by

$$\sum_{lk} \left\{\delta_{ml}\delta_{kj} - Q(\mathrm{i}\nu_n)\zeta^2 \langle V^{(11)}_{ml;ik}\rangle_c\right\} D_{lknj}(\mathrm{i}\nu_n) = Q(\mathrm{i}\nu_n)\delta_{mn}\delta_{ij}. \tag{8.1.37}$$

Using the second equation in Eq. (8.1.34) and Eq. (8.1.23), we obtain

$$\delta_{mn}\delta_{ij} - Q(\mathrm{i}\nu_n)\zeta^2 \langle V^{(11)}_{mn;ij}\rangle_c$$
$$= \delta_{mn} Q(0)\zeta^2 \sum_l \langle V^{(11)}_{ml;ij}\rangle_c - Q(\mathrm{i}\nu_n)\zeta^2 \langle V^{(11)}_{mn;ij}\rangle_c$$
$$= \{Q(0) - Q(\mathrm{i}\nu_n)\}\zeta^2 \delta_{mn} \sum_l \langle V^{(11)}_{ml;ij}\rangle_c$$
$$+ Q(\mathrm{i}\nu_n)\zeta^2 \left\{\delta_{mn} \sum_l \langle V^{(11)}_{ml;ik}\rangle_c - \langle V^{(11)}_{mn;ij}\rangle_c\right\}$$
$$= \zeta^2 Q(\mathrm{i}\nu_n) \left\{M(\mathrm{i}\nu_n)^2 \delta_{mn}\delta_{ij} + \delta_{mn} \sum_l \langle V^{(11)}_{ml;ij}\rangle_c - \langle V^{(11)}_{ml;ij}\rangle_c\right\} \tag{8.1.38}$$

Thus Eqs. (8.1.37) and (8.1.38) lead to

$$\sum_{lk} \lambda_{milk} D_{lknj}(\tau - \tau') = \frac{1}{\zeta^2}\delta_{mn}\delta_{ij}\delta(\tau - \tau'), \tag{8.1.39}$$

$$\lambda_{minj} = M\delta_{mn}\delta_{ij}\frac{\partial^2}{\partial\tau^2} + \delta_{mn}\sum_{l}\langle V^{(11)}_{ml;ij}\rangle_c - \langle V^{(11)}_{ml;ij}\rangle_c, \tag{8.1.40}$$

Using Eqs. (8.1.26), (8.1.38), and (8.1.40) and taking the ensemble average of the position of particles, we obtain a Heisenberg equation for phonon operators:

$$\sum_{nj} \lambda_{minj} B_{nj}(\tau) = J_{mi}(\tau), \tag{8.1.41}$$

$$J_{mi}(\tau) = \frac{1}{\zeta^2}\sum_{\alpha+\beta>2} \alpha(-\zeta)^{\alpha}(\boldsymbol{e}_i\boldsymbol{B}_m \cdots \boldsymbol{B}_m) \cdot \left\langle \overleftrightarrow{V}^{(\alpha+\beta)}_{mn}\right\rangle_c \cdot (-\zeta)^{\beta}(\boldsymbol{B}_n \cdots \boldsymbol{B}_n). \tag{8.1.42}$$

From Eqs. (8.1.39) and (8.1.40), we can write a dynamical Eq. (8.1.41) as:

$$\boldsymbol{B}_m(\tau) = \sum_{n} \zeta^2 \int_0^{\beta\hbar} d\tau' D_{mn}(\tau - \tau')\boldsymbol{J}_n(\tau'), \tag{8.1.43}$$

where we have written in a matrix form.

8.1.5 The Nonlinear Interaction Potentials

Now we investigate the nonlinear interaction potential J_{mi} in Eq. (8.1.42), which can be rewritten as

$$J_{mi} = \frac{1}{\zeta^2}\sum_{m\neq n,\alpha+\beta>2} \frac{(-\zeta)^{\alpha}}{(\alpha-1)!}B_{mi_1}\cdots B_{mi_\alpha} V^{(\alpha\beta)}_{mn;i_1\cdots i_{\alpha-1},j_1\cdots j_\beta}\frac{(-\zeta)^{\beta}}{\beta!}B_{nj_1}\cdots B_{nj_\beta}, \tag{8.1.44}$$

To make the nonlinear interaction potentials in a transparent form, we make the following approximations:

1. We make the nearest neighbor approximation for couplings between particles.
2. Since the overlappings between the same excited state wave functions contribute mainly to the couplings of particles, we only consider the potentials consisting of the same excited state wave functions:

$$V^{(\alpha\beta)}_{mn;i_1\cdots i_\alpha,j_1\cdots j_\beta} \longrightarrow V^{(\alpha\beta)}_{mn;\underbrace{i\cdots i}_{\alpha},\underbrace{i\cdots i}_{\beta}}.$$

Using the relation:

$$V_{mn;\underbrace{i\cdots i}_{\alpha},\underbrace{i\ldots i}_{\beta}} = -V_{mn;\underbrace{i\cdots i}_{\alpha\pm 1},\underbrace{i\cdots i}_{\beta\mp 1}} \tag{8.1.45}$$

and introducing the notation:

$$V^{(\alpha+\beta,0)}_{mn;\underbrace{i\cdots i}_{\alpha+\beta},0} \equiv -(\alpha+\beta-1)!W^{(\alpha+\beta)}_{mn;i}, \tag{8.1.46}$$

we obtain

$$\begin{aligned} H_I &= \frac{1}{2}\sum_{\{\boldsymbol{R}_m\}_n\alpha,\beta} V^{(\alpha+\beta)}_{mn;\underbrace{i\cdots i}_{\alpha},\underbrace{i\cdots i}_{\beta}} \frac{(-\zeta)^\alpha}{\alpha!} B_{mi}\cdots B_{mi}\frac{(-\zeta)^\beta}{\beta!}B_{ni}\cdots B_{ni} \\ &= -\frac{1}{2}\sum_{\{\boldsymbol{R}_m\}_n\alpha+\beta}\frac{W^{(\alpha+\beta)}_{mn;i}}{\alpha+\beta}(-\zeta)^{\alpha+\beta}\sum_{\gamma}\frac{(\alpha+\beta)!}{(\alpha+\beta-\gamma)!\gamma!}(B_{mi})^{\alpha+\beta-\gamma}(-B_{ni})^{\gamma} \\ &= -\frac{1}{2}\sum_{\{\boldsymbol{R}_m\}_n\alpha+\beta}\frac{W^{(\alpha+\beta)}_{mn;i}}{\alpha+\beta}(-\zeta)^{\alpha+\beta}(B_{mi}-B_{ni})^{(\alpha+\beta)}. \end{aligned} \tag{8.1.47}$$

Then we have

$$\begin{aligned} J_{mi}(\tau) &= \sum_{\{\boldsymbol{R}_m\}_n\alpha+\beta>2} V^{(\alpha+\beta)}_{mn;\underbrace{i\cdots i}_{\alpha},\underbrace{i\cdots i}_{\beta}}\frac{(-\zeta)^\alpha}{(\alpha-1)!}B_{mi}\cdots B_{mi}\frac{(-\zeta)^\beta}{\beta!}B_{ni}\cdots B_{ni} \\ &= \sum_{\{\boldsymbol{R}_n\}_m,\alpha+\beta>2)}\frac{1}{\zeta}\langle W^{\alpha+\beta}_{mn:i}\rangle_c\{-\zeta(B_{mi}-B_{ni})\}^{\alpha+\beta-1} \\ &= -\sum_{\{\boldsymbol{R}_n\}_m}\Big\{\langle W^{(3)}_{mn;i}\rangle_c(-\zeta)(B_{mi}-B_{ni})^2 \\ &\qquad + \langle W^{(4)}_{mn;i}\rangle_c(-\zeta)^2(B_{mi}-B_{ni})^3+\cdots\Big\}, \end{aligned} \tag{8.1.48}$$

where $\{\boldsymbol{R}_n\}_m$ means the positions of the nearest neighbors of the atom at $\boldsymbol{R}_m$. We use the nonlinear dynamical equation for phonon operators (8.1.41) with the nonlinear interaction potential (8.1.48).

8.2 Solitons and Bound States of the Self-Consistent Potential by the Boson Transformation Method

In the previous section, we have established the nonlinear interaction Hamiltonian for phonon operators and the dynamical equation for phonon operators. In the next chapter, we use the nonlinear interaction Hamiltonian to derive the entropy fluctuation modes and thermal conductivity for phonons. In this section, we investigate the efficiency of the nonlinear interaction potentials by applying the nonlinear dynamical equation for phonon operators to a one-dimensional system [7]. We can show how significant the nonlinear dynamical equation for phonon operators is.

Applying the boson transformation method to the nonlinear dynamical equation for phonon operators in the long wavelength regions, we derive the conventional soliton equations in the classical system [8,9]. We also derive a new aspect of the quantum field theory: there appears quantum bound states in the potential created by the soliton solutions. Thus we can see how powerful the present method of the nonlinear dynamical equation for phonon operator is. This is because the present method is developed from the quantum field theory.

8.2.1 The Boson Transformation Method and the Dynamical Equations

Now we investigate the nonlinear dynamical equation for phonon operators (8.1.41) with the nonlinear interaction potentials (8.1.48). To do so, we apply the boson transformation method to Eq. (8.1.41) [10,11]. The boson transformed B_m is denoted by B_m^f, which satisfies Eq. (8.1.41). We put

$$B_m^f = B_m^{(0)} + B_m^{(1)} + \cdots , \tag{8.2.1}$$

where $B_m^{(0)}$ is the configurationally averaged value in the condensed state:

$$B_m^{(0)} = \langle B_m^f \rangle_c, \tag{8.2.2}$$

and $B_m^{(1)}$ is a linear term of free phonon fields. Analytically continue: $\tau \to it$ and making the nearest neighbor approximation in Eq. (8.1.41) with Eq. (8.1.48), we obtain

$$M\frac{d^2 B_m^f}{dt^2} = -\phi'(B_m^f - B_{m-1}^f) + \phi'(B_{m+1}^f - B_m^f), \tag{8.2.3}$$

where we put

$$\phi(r) \equiv \frac{1}{2}Ur^2 + \frac{1}{3}Vr^3 + \frac{1}{4}Wr^4, \tag{8.2.4}$$

and the prime on ϕ means the derivative. $\phi(r)$ in Eq. (8.2.4) corresponds to the terms $\alpha + \beta = 2, 3, 4$ in the interaction Hamiltonian H_I in Eq. (8.1.47). We have considered the following:

$$\begin{aligned}\sum_n \lambda_{mn} B_n(t) &= \sum_n \left\{ M\delta_{mn}\frac{d^2}{dt^2} + \delta_{mn}\left\langle \sum_l V_{ml}^{(11)} \right\rangle_c - \langle V_{mn}^{(11)} \rangle_c \right\} B_n(t) \\ &= M\frac{d^2}{dt^2}B_m(t) - U(B_{m+1} + B_{m-1} - 2B_m)(t),\end{aligned}$$

where $V_{m,m\pm 1}^{(11)} = -U$.

Configurationally averaged Eq. (8.2.3) leads to

$$M\frac{d^2B_m^{(0)}}{dt^2} = -\phi'(B_m^{(0)} - B_{m-1}^{(0)}) + \phi'(B_{m+1}^{(0)} - B_m^{(0)}), \tag{8.2.5}$$

Eq. (8.2.5) is nothing but the dynamical equation for the displacement fields in the conventional nonlinear lattice theory but randomly distributed particles. The dynamical equation of the phonon field $B_m^{(1)}$ is obtained by substituting Eq. (8.2.1) into Eq. (8.2.3) and taking the linear terms of $B_m^{(1)}$ as

$$M\frac{d^2B_m^{(1)}}{dt^2} - U(B_{m-1}^{(1)} + B_{m+1}^{(1)} - 2B_m^{(1)}) - \sum_n V_{mn}^{(1)}B_n^{(1)} = 0, \tag{8.2.6}$$

$$V_{mn}^{(1)} = \frac{\partial}{\partial B_n^{(0)}}\left[-\phi_N{}'(B_m^{(0)} - B_{m-1}^{(0)}) + \phi_N{}'(B_{m+1}^{(0)} - B_m^{(0)})\right], \tag{8.2.7}$$

where ϕ_N is the nonlinear potential in Eq. (8.2.4) and $n = m,\ m \pm 1$. Eq. (8.2.6) determines the behavior of a single quantum under the self-consistent potential $V_{mn}^{(1)}$ in Eq. (8.2.7) created by the condensed deviations $B_m^{(0)}$. There appear bound states of the physical quantum due to the self-consistent potential.

8.2.2 The Boussinesq Equation and the Dynamical Equation of Phonon Fields in the Long Wavelength Limit

Now we investigate Eqs. (8.2.5) and (8.2.6) in the long wavelength limit. First we investigate B_m^f. If we put

$$B_m^f = B^f(x,t), \qquad x = ma, \tag{8.2.8}$$

and we consider $g(x) \cong \delta(x-a)$, we obtain

$$B_{m\pm1} = B \pm aB_x + \frac{a^2}{2!}B_{xx} \pm \frac{a^3}{3!}B_{xxx} + \frac{a^4}{4!}B_{xxxx} + \cdots, \tag{8.2.9}$$

where we have abbreviated $B_m^f = B^f(x,t)$ as B. Considering up to the order of a^4, we have

$$\begin{aligned}
&-\phi'(B_m - B_{m-1}) + \phi'(B_{m+1} - B_m)\\
&= U\left(a^2B_{xx} + \frac{a^4}{12}B_{xxxx}\right) - \phi_N'\left(aB_x - \frac{a^2}{2}B_{xx}+\right) + \phi_N'\left(aB_x + \frac{a^2}{2}B_{xx}\right)\\
&= U\left(a^2B_{xx} + \frac{a^4}{12}B_{xxxx}\right) + 2a^3VB_xB_{xx} + 3a^4WB_xB_{xx} + \cdots\\
&= U\left(a^2B_{xx} + \frac{a^4}{12}B_{xxxx}\right) + a^2V\frac{\partial}{\partial x}B_x^2 + a^4W\frac{\partial}{\partial x}B_x^3.
\end{aligned} \tag{8.2.10}$$

Substituting Eq. (8.2.10) into Eq. (8.2.3), we obtain

$$MB_{tt}^f = U\left(a^2B_{xx}^f + \frac{a^4}{12}B_{xxxx}^f\right) + a^2V\frac{\partial}{\partial x}B_x^{f\,2} + a^4W\frac{\partial}{\partial x}B_x^{f\,3}. \tag{8.2.11}$$

If we define

$$\varphi^f(x,t) \equiv \frac{\partial}{\partial x} B_x^f, \tag{8.2.12}$$

and

$$c_0^{-2} = \frac{m}{Ua^2}, \qquad \alpha = \frac{aV}{U}, \qquad \beta = \frac{a^2 W}{U}, \tag{8.2.13}$$

we obtain

$$c_0^{-2}\varphi_{tt}^f - \varphi_{xx}^f - \frac{a^2}{12}\varphi_{xxxx}^f = \alpha\left(\varphi^{f2}\right)_{xx} + \beta\left(\varphi^{f3}\right)_{xx}. \tag{8.2.14}$$

Here we carry out the boson transformation:

$$\varphi^f(x,t) = \varphi^{(0)}(x,t) + \varphi^{(1)}(x,t) + \cdots. \tag{8.2.15}$$

we obtain

$$c_0^{-2}\varphi_{tt}^{(0)} - \varphi_{xx}^{(0)} - \frac{a^2}{12}\varphi_{xxxx}^{(0)} = \alpha\left(\varphi^{(0)2}\right)_{xx} + \beta\left(\varphi^{(0)3}\right)_{xx}, \tag{8.2.16}$$

$$c_0^{-2}\varphi_{tt}^{(1)} - \varphi_{xx}^{(1)} - \frac{a^2}{12}\varphi_{xxxx}^{(1)} = 2\alpha\left(\varphi^{(0)}\varphi^{(1)}\right)_{xx} + 3\beta\left(\varphi^{(0)2}\varphi^{(1)}\right)_{xx}, \tag{8.2.17}$$

Eq. (8.2.16) is the Boussinesq equation and Eq. (8.2.17) is the dynamical equation of phonon fields.

The Boussinesq equation has well-known soliton solutions:

1. The cubic potential:

$$c_0^{-2}\varphi_{tt}^{(0)} - \varphi_{xx}^{(0)} - \frac{a^2}{12}\varphi_{xxxx}^{(0)} = \alpha\left(\varphi^{(0)2}\right)_{xx}, \tag{8.2.18}$$

$$\varphi^{(0)}(x,t) = \frac{k^2a^2}{2\alpha}\text{sech}^2 k\xi, \qquad \left(\frac{c}{c_0}\right)^2 - 1 = \frac{(2ka)^2}{12}. \tag{8.2.19}$$

2. The quartic potential:

$$c_0^{-2}\varphi_{tt}^{(0)} - \varphi_{xx}^{(0)} - \frac{a^2}{12}\varphi_{xxxx}^{(0)} = \beta\left(\varphi^{(0)3}\right)_{xx}, \tag{8.2.20}$$

$$\varphi^{(0)}(x,t) = \frac{ak}{\sqrt{6\beta}}\text{sech} k\xi, \qquad \left(\frac{c}{c_0}\right)^2 - 1 = \frac{(ak)^2}{12}, \tag{8.2.21}$$

where $\xi = x - ct$ and k is a wavenumber.

The dynamical equation of phonon fields $\phi^{(1)}$ with the self-consistent potential $\tilde{\varphi}(x)$ created by the soliton solutions of the Boussinesq equation is given by

$$c_0^{-2}\varphi_{tt}^{(1)} - \varphi_{xx}^{(1)} - \frac{a^2}{12}\varphi_{xxxx}^{(1)} = \left(\tilde{\varphi}(x)\varphi^{(1)}\right)_{xx}. \tag{8.2.22}$$

1. The cubic potential:

$$\tilde{\varphi}(x) = 2\alpha\varphi^{(0)} = k^2a^2\text{sech}^2 k\xi. \tag{8.2.23}$$

2. The quartic potential:

$$\tilde{\varphi}(x) = 3\beta\varphi^{(0)2} = \frac{k^2a^2}{2}\text{sech}^2 k\xi. \tag{8.2.24}$$

Note that $\tilde{\varphi}(x)$ does not depend on the coupling constants α and β and has the same $\text{sech}^2 k\xi$-type potential.

8.2.3 Bound States in the Dynamical Equation of Phonon Fields with the Self-Consistent Fields Due to Solitons

The dynamical equation of phonon fields $B_m^{(1)}$ (8.2.6) with the self-consistent potential $V_{mn}^{(1)}$ determines the scattering states, the bound states, and the translational modes. Therefore, the dynamical equation of phonon fields $\phi^{(1)}$ (8.2.22) with the self-consistent potential $\tilde{\varphi}(x)$ (8.2.23) for cubic potential and (8.2.24) for the quartic potential determines the scattering states, the bound states, and the translational modes. Now we investigate solutions of the dynamical equation of phonon fields $\phi^{(1)}$ with the self-consistent potential $\tilde{\varphi}(x)$.

To do so, we seek solutions to Eq. (8.2.22) in the form [12,13]:

$$\varphi^{(1)}(x,t) = \cos(Kx - \Omega t)D(\xi). \tag{8.2.25}$$

Substituting Eq. (8.2.25) into Eq. (8.2.22), we obtain

$$\begin{aligned}&\left[\left(\frac{c^2}{c_0^2} - 1\right)D - \frac{a^2}{12}D'' - \tilde{\varphi}D\right]'' \\ &\quad - \left[\left(\frac{\Omega^2}{c_0^2K^2} - 1 + \frac{a^2K^2}{12}\right)D - \frac{a^2}{2}D'' - \tilde{\varphi}D\right]K^2 = 0,\end{aligned} \tag{8.2.26}$$

$$\left[\left(\frac{c\Omega}{c_0^2K} - 1 + \frac{a^2K^2}{6}\right)D - \frac{a^2}{6}D'' - \tilde{\varphi}D\right]' K = 0. \tag{8.2.27}$$

In order to solve Eqs. (8.2.26) and (8.2.27), we first consider the properties of Eqs. (8.2.26) and (8.2.27). To do so, as a matter of convenience, we introduce the following equations relating with the first and second brackets in Eq. (8.26):

$$\left[\left(\frac{c^2}{c_0^2} - 1\right)D - \frac{a^2}{12}D'' - \tilde{\varphi}D\right] = 0, \tag{8.2.26a}$$

$$\left[\left(\frac{\Omega^2}{c_0^2K^2} - 1 + \frac{a^2K^2}{12}\right)D - \frac{a^2}{2}D'' - \tilde{\varphi}D\right] = 0. \tag{8.2.26b}$$

Then we investigate the properties of Eqs. (8.2.26) and (8.2.27).

1. $K = 0,\ \Omega = 0$:
   The first bracket (8.2.26a) remains. This equation has a solution $D(x) = \frac{d\varphi^{(0)}(x)}{dx}$, which is a translational mode.
2. $K \neq 0,\ \Omega \neq 0$:
   Eq. (8.2.27) always holds. There exists a bound state, if a solution of Eq. (8.2.27) is not orthogonal to a solution of Eq. (8.2.26).
3. $K > k,\ \Omega > kc$:
   In Eq. (8.2.26) the second bracket (8.2.26b) dominates. If a solution of the second bracket (8.2.26b) is not orthogonal to a solution of Eq. (8.2.27), it is a bound state.

The Equations (8.2.26a, 8.2.26b) and Equation (8.2.27) reduce to the Schrödinger equation:

$$D'' + U_0 \operatorname{sech}^2 k\xi\, D + ED = 0. \tag{8.2.28}$$

Here we have

1. $\left(\frac{c^2}{c_0^2} - 1\right) D - \frac{a^2}{12} D'' - \tilde{\varphi} D = 0$: in Eq. (8.2.26a).

$$U_0 = \begin{cases} 12k^2 & \text{for cubic} \\ 6k^2 & \text{for quartic,} \end{cases} \qquad E = \frac{12}{a^2}\left(-\frac{c^2}{c_0^2} + 1\right) \equiv E_0. \tag{8.2.29}$$

2. $\left(\frac{\Omega^2}{c_0^2 K^2} - 1 + \frac{a^2 K^2}{12}\right) D - \frac{a^2}{2} D'' - \tilde{\varphi} D = 0$: in Eq. (8.2.26b).

$$U_0 = \begin{cases} 2k^2 & \text{for cubic} \\ k^2 & \text{for quartic,} \end{cases} \qquad E = \frac{2}{a^2}\left(-\frac{\Omega^2}{c_0^2 K^2} + 1 - \frac{a^2 K^2}{12}\right) \equiv E_I. \tag{8.2.30}$$

3. $\left(\frac{c\Omega}{c_0^2 K} - 1 + \frac{a^2 K^2}{6}\right) D - \frac{a^2}{6} D'' - \tilde{\varphi} D = 0$: in Eq. (8.2.27).

$$U_0 = \begin{cases} 6k^2 & \text{for cubic} \\ 3k^2 & \text{for quartic,} \end{cases} \qquad E = \frac{6}{a^2}\left(-\frac{\Omega^2}{c_0^2 K^2} + 1 - \frac{a^2 K^2}{6}\right) \equiv E_{II}. \tag{8.2.31}$$

If we put $u = \tanh k\xi$, Eq. (8.2.28) leads to the associated Legendre differential equation [14]:

$$\frac{d}{du}\left[(1-u^2)\frac{dD}{du}\right] + \left[\nu(\nu+1) - \frac{\mu^2}{1-u^2}\right] D = 0, \tag{8.2.32}$$

where

$$\mu^2 = -\frac{E}{k^2}, \qquad \nu(\nu+1) = \frac{U_0}{k^2}. \tag{8.2.33}$$

The solution finite for $u = 1$ is

$$D(u) = (1-u^2)^{\frac{\mu}{2}} F\left(\mu-\nu, \mu+\nu+1, \mu+1; \frac{1-u}{2}\right), \tag{8.2.34}$$

where F is the hypergeometric function. If D remains finite for $u = -1$,

$$\mu - \nu = -n, \tag{8.2.35}$$

where $n = 0, 1, 2, \ldots$; then F is a polynomial of degree n. From Eqs. (8.2.33) and (8.2.35) we obtain

$$\frac{E}{k^2} = -(n-\nu)^2 = -\frac{1}{4}\left\{-(1+2n) + \sqrt{1+\frac{4U_0}{k^2}}\right\}^2. \tag{8.2.36}$$

Now, we investigate solutions of Eqs. (8.2.26) and (8.2.27).

1. The cubic potential

a. $K = 0, \quad \Omega = 0.$
From Eq. (8.2.29) we have

$$\frac{U_0}{k^2} = 12: \qquad \nu = 3, \tag{8.2.37}$$

$$\frac{E_0}{k^2} = \frac{12}{(ka)^2}\left(-\frac{c^2}{c_0^2}+1\right) = -\mu^2, \qquad \mu = 1, 2, 3. \tag{8.2.38}$$

The Boussinesq equation for cubic potential has the solution (8.2.19):

$$\varphi^{(0)}(x,t) = \frac{k^2a^2}{2\alpha}\text{sech}^2 k\xi, \qquad \left(\frac{c}{c_0}\right)^2 - 1 = \frac{(2ka)^2}{12}. \tag{8.2.19}$$

The Eigenvalue of $\frac{E_0}{k^2}$ in Eq. (8.2.38) satisfying that of the Boussinesq Eq. (8.2.19) is of $\mu = 2$. Thus

$$D(\xi) = (1-u^2)F\left(-1, 6, 3, \frac{1-u}{2}\right) = (1-u^2)u = \text{sech}^2 k\xi \tanh k\xi$$
$$\propto \frac{\partial}{\partial x}\varphi^{(0)}(x.t). \tag{8.2.39}$$

This is just the translational mode.

b. $K \neq 0, \ \Omega \neq 0.$
From Eq. (8.2.31) we have

$$\frac{U_0}{k^2} = 6: \ \nu = 2, \tag{8.2.40}$$

$$\frac{E_{II}}{k^2} \equiv \frac{6}{a^2k^2}\left(-\frac{c\Omega}{c_0^2 K} + 1 - \frac{a^2K^2}{6}\right) = -\mu^2; \qquad \mu = 1, 2, \tag{8.2.41}$$

where
$\mu = 1$:

$$D(\xi) = (1-u^2)^{\frac{1}{2}} F\left(-1, 4, 2, \frac{1-u}{2}\right)$$
$$= (1-u^2)^{\frac{1}{2}} u = \text{sech} k\xi \tanh k\xi, \tag{8.2.42}$$

$\mu = 2$:

$$D(\xi) = \text{sech}^2 k\xi. \tag{8.2.43}$$

If a wave function is not orthogonal to that of Eq. (8.2.30), there is a bound state which satisfies both the equations.

c. $K > k,\ \Omega > kc$.
From Eq. (8.2.30) we have

$$\frac{U_0}{k^2} = 2: \quad \nu = 1, \tag{8.2.44}$$

$$\frac{E_I}{k^2} \equiv \frac{2}{a^2k^2}\left(-\frac{\Omega^2}{c_0^2K^2} + 1 - \frac{a^2K^2}{12}\right) = -1; \quad \mu = 1, \tag{8.2.45}$$

where

$$D(\xi) = (1-u^2)^{\frac{1}{2}} = \text{sech} k\xi, \tag{8.2.46}$$

The solution of Eq. (8.2.46) is not orthogonal to that of Eq. (8.2.42).

Thus Eq. (8.2.42) with $\frac{E_{II}}{k^2} = -1$ and Eq. (8.2.48) with $\frac{E_I}{k^2} = -1$ can be a bound state.

2. The quartic potential

a. $K = 0,\ \Omega = 0$.
From Eq. (8.2.29) we have

$$\frac{U_0}{k^2} = 6: \quad \nu = 2, \tag{8.2.47}$$

$$\frac{E_0}{k^2} = \frac{12}{(ka)^2}\left(-\frac{c^2}{c_0^2} + 1\right) = -\mu^2; \quad \mu = 1, 2. \tag{8.2.48}$$

The Boussinesq equation for quartic potential has the solution (8.2.21):

$$\varphi^{(0)}(x,t) = \frac{ak}{\sqrt{6\beta}} \text{sech} k\xi, \quad \left(\frac{c}{c_0}\right)^2 - 1 = \frac{(ak)^2}{12}. \tag{8.2.21}$$

The Eigenvalue of $\frac{E_0}{k^2}$ in Eq. (8.2.48) satisfying that of the Boussinesq Eq. (8.2.21) is of $\mu = 1$. Thus we obtain

$$\begin{aligned} D(\xi) &= (1-u^2)^{\frac{1}{2}} F\left(-1, 4, 2, \frac{1-u}{2}\right) \\ &= (1-u^2)^{\frac{1}{2}} u = \operatorname{sech} k\xi \tanh k\xi \\ &\propto \frac{\partial}{\partial x}\varphi^{(0)}(x,t), \end{aligned} \tag{8.2.49}$$

which is the translational mode.

b. $K \neq 0,\ \Omega \neq 0$.
From Eq. (8.2.31) we have

$$\frac{U_0}{k^2} = 3: \qquad \nu = \frac{1}{2}(-1+\sqrt{13}), \tag{8.2.50}$$

$$\frac{E_{II}}{k^2} \equiv \frac{6}{a^2k^2}\left(-\frac{c\Omega}{c_0^2 K} + 1 - \frac{a^2K^2}{6}\right) = -\mu^2;$$

$$\mu = \frac{1}{2}(-1+\sqrt{13}),\ \frac{1}{2}(-3+\sqrt{13}), \tag{8.2.51}$$

where
$\mu = \frac{1}{2}(-1+\sqrt{13})$:

$$D(\xi) = (\operatorname{sech} k\xi)^{\frac{1}{2}(-1+\sqrt{13})}. \tag{8.2.52}$$

$\mu = \frac{1}{2}(-3+\sqrt{13})$:

$$\begin{aligned} D(\xi) &= (1-u^2)^{\frac{1}{4}(-3+\sqrt{13})} F \\ &\quad \left(-1, -1+\sqrt{13}, \frac{1}{2}(-1+\sqrt{13}), \frac{1-u}{2}\right) \\ &= (\operatorname{sech} k\xi)^{\frac{1}{2}(-3+\sqrt{13})} \tanh k\xi. \end{aligned} \tag{8.2.53}$$

c. $K > k,\ \Omega > kc$.
From Eq. (8.2.30) we have

$$\frac{U_0}{k^2} = 1: \qquad \nu = \frac{1}{2}(-1+\sqrt{5}), \tag{8.2.54}$$

$$\frac{E_I}{k^2} = -\frac{1}{4}(-1+\sqrt{5})^2; \qquad \mu = \frac{1}{2}(-1+\sqrt{5}), \tag{8.2.55}$$

$$D(\xi) = (1-u^2)^{\frac{1}{4}(-1+\sqrt{5})} = (\operatorname{sech} k\xi)^{\frac{1}{2}(-1+\sqrt{5})}. \tag{8.2.56}$$

The solution of Eq. (8.2.56) is not orthogonal to that of Eq. (8.2.52). Thus Eq. (8.2.52) with $\frac{E_{II}}{k^2} = -\frac{1}{2}(3-\sqrt{5})$ and Eq. (8.2.56) with $\frac{E_I}{k^2} = -\frac{1}{2}(7-\sqrt{13})$ can be a bound state.

Starting with the Hamiltonian with higher powers of phonon operators $\{B_m\}$, we have derived the dynamical equation for the phonon operators, Eq. (8.1.41) with Eq. (8.1.48). Applying the boson transformation (8.2.1) to Eq. (8.1.41), we have obtained the dynamical Eq. (8.2.5) for the condensed displacements $B_m^{(0)} = \langle B_m \rangle_c$ and the dynamical Eq. (8.2.6) for phonon fields $B_m^{(1)}$ with the self-consistent potential (8.2.7) created by the condensed displacements. The dynamical equation for condensed displacements is the same equation as that of the classical nonlinear lattice theory.

In the long wavelength limit, Eq. (8.2.5) reduces to the Boussinesq equation (8.2.16), which yields to the soliton solutions (8.2.19) for a cubic potential and (8.2.21) for a quartic potential. In the long wavelength regime, the dynamical equation for phonon fields (8.2.6) reduces to the dynamical equation for phonon fields (8.2.22) with the self-consistent potential created by the cubic and quartic solitons. We have sought solutions to Eq. (8.2.22) in the form of Eq. (8.2.25). We have three different envelope equations: (8.2.26a), (8.2.26b) in Eq. (8.2.26), and (8.2.27). Each envelope function $D(\xi)$ satisfies the Schrödinger Eq. (8.2.28) with the same $-\text{sech}^2 k\xi$-type potential for both cubic and quartic potentials. There always exists a translational mode; $K = 0$, $\Omega = 0$. Eq. (8.2.27) always has bound states for $K \neq 0$, $\Omega \neq 0$. Actual bound states can exist if the wave function of Eq. (8.2.27) is not orthogonal to the wave function of Eq. (8.2.26). Under the condition $K > 0$, $\Omega > kc$, there exists a bound state for each nonlinear potential.

Takeno et al. [12, 13] have investigated the localized modes in the nonlinear dynamical Eq. (8.2.6). Our bound states come from the dynamical equation for phonon fields (8.2.6) with the potential (8.2.7) created by solitons in the long wavelength regime. The potential, which a single quantum feels, corresponds to the mean field in the nonlinear potential in Eq. (8.2.17). So the bound state can be regarded as the localized mode. This situation is explained as follows: The self-consistent potential $V_{mn}^{(1)}$ around a position in Eq. (8.2.6) causes the increase of the elastic constant around the position, which creates the localized mode around the position in the classical lattice theory. However, since the moving localized modes are not stable in lattices, the bound states or the localized modes are not stable for a long time. On the other hand, in the present theory, since there are stable solitons in the long wavelength regime, the stable bound states or the localized modes can exist.

8.3 Localized Modes for a Quartic Potential in the One Loop Approximation

Once a localized or a resonant mode appears around a site, the mode creates the self-consistent potential which ensures the existence of the mode; the self-consistent potential yields the change of the elastic constants around the site, which ensures the existence of the mode [1,2,4–6]. These localized and resonant modes first have been investigated by Takeno et al. [15–17]. This self-consistent localized potential behaves in a similar manner to the potential caused by an impurity in linear lattices. These localized modes caused by impurities have been investigated fully by Montroll and Potts [18,19].

In this section, we construct the self-consistent localized potential, making the one loop approximation in the dynamical equation for phonon operators in the nonlinear interaction potential. The one loop approximation corresponds to the mean field approximation, which takes the configuration average of the nonlinear potential. Under the mean field approximation, we show localized modes. Furthermore, we can show the temperature dependence of the localized modes. This is very significant because the conventional theories of localized modes have not treated with the temperature dependence. In our case, since we consider the nonlinear potential made up by the phonon fields at finite temperatures, we can investigate the temperature-dependent localized modes.

8.3.1 The Quantization of Phonon Operators

In order to investigate localized mode, we must quantize the phonon operators by using free phonon fields. The method of quantization of field operators is explained in Section 3.4. Here we apply the method of quantization in Section 3.4 to the phonon operators. To do so, we consider the dynamical equation of phonons in real time. First we define $\Lambda(q)$ as

$$\Lambda(\partial_t, \boldsymbol{q})e^{\mathrm{i}(\boldsymbol{q}\cdot\boldsymbol{R}_m - q_0 t)} = \Lambda(q)e^{\mathrm{i}(\boldsymbol{q}\cdot\boldsymbol{R}_m - q_0 t)}. \tag{8.3.1}$$

The dynamical equation of phonons is given by

$$\Lambda(\partial_t, \boldsymbol{q})u_{\lambda\boldsymbol{q}}(t) = 0, \tag{8.3.2}$$

where

$$\Lambda_{ij}(\partial_t, \boldsymbol{q}) = -M\delta_{ij}\frac{\partial^2}{\partial t^2} + \zeta^{-2}\{V_{ij}(0) - V_{ij}(\boldsymbol{q})\}, \tag{8.3.3}$$

where $\Lambda(\partial_t, \boldsymbol{q})$ and $u_{\lambda\boldsymbol{q}}(t)$ are written in matrix form. We can regard $v_{\lambda\boldsymbol{q}} = u^*_{\lambda\boldsymbol{q}}$. From Eq. (3.4.18) since

$$\Lambda^\dagger_{ij}(\partial_t, \boldsymbol{q}) = \Lambda_{ij}(\partial_t, \boldsymbol{q}), \tag{8.3.4}$$

we obtain

$$\eta = 1. \tag{8.3.5}$$

From Eq. (3.4.26), we obtain

$$\overleftrightarrow{\Gamma} = -\mathrm{i}M\left(\frac{\partial}{\partial t} - \frac{\overleftarrow{\partial}}{\partial t}\right). \tag{8.3.6}$$

From Eq. (3.4.29), we define

$$u_{\lambda\boldsymbol{q}}(\boldsymbol{R}_m) = u_{\lambda\boldsymbol{q}}e^{\mathrm{i}(\boldsymbol{q}\cdot\boldsymbol{R}_m - \omega_{\lambda\boldsymbol{q}}t)}, \tag{8.3.7}$$

then from Eq. (3.4.35), we obtain

$$\begin{aligned}\sum_m u_{\lambda\boldsymbol{q}}(\boldsymbol{R}_m)\overleftrightarrow{\Gamma}u^*_{\lambda'\boldsymbol{q}}(\boldsymbol{R}_m) &= u_{\lambda\boldsymbol{q}}NM(\omega_{\lambda\boldsymbol{q}} + \omega_{\lambda'\boldsymbol{q}})u^*_{\lambda\boldsymbol{q}} \\ &= \hbar\delta_{\lambda\lambda'},\end{aligned} \tag{8.3.8}$$

and we have

$$u_{\lambda q} = \sqrt{\frac{\hbar}{2MN\omega_{\lambda q}}}. \tag{8.3.9}$$

Thus we can quantize phonon operators ζB_m by using free phonon fields $\beta_{\lambda q}$.

$$\begin{aligned} -\zeta B_{mi}(\tau) &= -\frac{\zeta}{N}\sum_{q} e^{-\mathrm{i}q\cdot R_m}\rho^{\dagger}_{iq}(\tau) \\ &= \sum_{\lambda q}\sqrt{\frac{\hbar}{2MN\omega_{\lambda q}}}e^{i}_{\lambda q} \\ &\qquad \left\{e^{\mathrm{i}q\cdot R_m - \omega_{\lambda q}\tau}\beta_{\lambda q} + e^{-\mathrm{i}q\cdot R_m + \omega_{\lambda q}\tau}\beta^{\dagger}_{\lambda q}\right\}, \end{aligned} \tag{8.3.10}$$

where $e^{i}_{\lambda q}$ is the ith Cartesian component of the polarization vector with the mode λq.

8.3.2 The One Loop Approximation

Here we investigate localized modes with a quartic potential. We solve the dynamical equation of phonon operators (8.1.41) with the nonlinear interaction potential (8.1.48). Then we investigate the following quartic potential:

$$J_{mi}(\tau) = -\sum_{\{R_n\}_m}\langle W^{(4)}_{mn;i}\rangle_c \zeta^2 (B_{mi} - B_{ni})^3, \tag{8.3.11}$$

where $\{R_n\}_m$ means the positions of the nearest neighbors of the atom at R_m. Here we make the one loop approximation:

$$\begin{aligned} J_{mi} &\cong \sum_{lj}\left\langle\frac{\partial J_{mi}}{\partial B_{lj}}\right\rangle_c B_{lj} \\ &= -3\sum_{\{R_n\}_m}\langle W^{(4)}_{mn;i}\zeta^2(B_{mi} - B_{ni})^2\rangle_c (B_{mi} - B_{ni}). \end{aligned} \tag{8.3.12}$$

In order to calculate the one loop diagrams, which represent the mean fields, we first quantize phonon operator. Here we consider amorphous solids containing spatially localized modes. To consider the localized mode, considering Eq. (8.3.10), we put

$$\begin{aligned} -\zeta B_m &= \sum_{l}\sqrt{\frac{\hbar}{2M\omega_l}}\left[v_m(l)\beta_l e^{-\omega_l\tau} + v^{*}_m(l)\beta^{\dagger}_l e^{\omega_l\tau}\right] \\ &\quad + \sum_{\lambda q}\sqrt{\frac{\hbar}{2M\omega_{\lambda q}}}\left[v_{\lambda m}(q)\beta_{\lambda q}e^{-\omega_{\lambda q}\tau} + v^{*}_{\lambda m}(q)\beta^{\dagger}_{\lambda q}e^{-\omega_{\lambda q}\tau}\right], \end{aligned} \tag{8.3.13}$$

$$\sum_{m} v^{\dagger}_m(l)\cdot v_m(l') = \delta_{l,l'}, \tag{8.3.14}$$

$$\sum_{m} v^{\dagger}_{\lambda m}(q)\cdot v_{\lambda m}(q) = \delta_{\lambda\lambda'}\delta_{qq'}, \tag{8.3.15}$$

where the subscripts l and $\lambda\boldsymbol{q}$ mean spatially localized modes and plane wave modes, respectively. The localized modes and the plane wave modes have 3 and $3N-3$ components, respectively. β_l and $\beta_{\lambda\boldsymbol{q}}$ are annihilation operators satisfying

$$[\beta_l, \beta_{l'}^\dagger] = \delta_{ll'}, \tag{8.3.16}$$

$$[\beta_{\lambda\boldsymbol{q}}, \beta_{\lambda'\boldsymbol{q}'}^\dagger] = \delta_{\lambda\lambda'}\delta_{\boldsymbol{q}\boldsymbol{q}'} \tag{8.3.17}$$

and other commutation relations are zero. With the localized modes at $\boldsymbol{R}_m = 0$:

$$\boldsymbol{v}_m^\dagger(l) \cdot \boldsymbol{v}_m(l) \cong 1 \qquad \text{for } \boldsymbol{R}_m = 0, \tag{8.3.18}$$

and

$$\boldsymbol{v}_m^\dagger(l) \cdot \boldsymbol{v}_m(l) \cong 0 \qquad \text{for } \boldsymbol{R}_m \neq 0. \tag{8.3.19}$$

$$\boldsymbol{v}_{\lambda m}^\dagger(\boldsymbol{q}) \cdot \boldsymbol{v}_{\lambda m}(\boldsymbol{q}) = \frac{1}{N} \tag{8.3.20}$$

Using Eq. (8.3.13), we obtain

$$\begin{aligned}
&\zeta^2\langle(B_{mi} - B_{ni})^2\rangle_c \\
&\quad = \sum_l \frac{\hbar}{2M\omega_l}\left\{|v_{mi}(l) - v_{ni}(l)|^2(\langle\beta_l\beta_l^\dagger\rangle + \langle\beta_l^\dagger\beta_l\rangle)\right\} \\
&\qquad + \sum_{\lambda\boldsymbol{q}} \frac{\hbar}{2M\omega_{\lambda\boldsymbol{q}}}\left\{|v_{\lambda mi}(\boldsymbol{q}) - v_{\lambda ni}(\boldsymbol{q})|^2(\langle\beta_{\lambda\boldsymbol{q}}\beta_{\lambda\boldsymbol{q}}^\dagger\rangle + \langle\beta_{\lambda\boldsymbol{q}}^\dagger\beta_{\lambda\boldsymbol{q}}\rangle)\right\} \\
&\quad = \sum_l |v_{mi}(l) - v_{ni}(l)|^2\alpha_l^2 + \sum_{\lambda\boldsymbol{q}} |v_{\lambda mi}(\boldsymbol{q}) - v_{\lambda ni}(\boldsymbol{q})|^2\alpha_{\lambda q}^2,
\end{aligned} \tag{8.3.21}$$

$$\alpha_l^2 \equiv \frac{\hbar}{2M\omega_l}\coth\frac{\hbar\omega_l}{2k_BT}, \qquad \alpha_{\lambda q}^2 \equiv \frac{\hbar}{2M\omega_{\lambda q}}\coth\frac{\hbar\omega_{\lambda q}}{2k_BT}. \tag{8.3.22}$$

This procedure corresponds to the rotating wave approximation in Refs.[15–17]. Since the contribution of the second term of Eq. (8.3.13) does not depend on $\boldsymbol{R}_m$ from Eq. (8.3.20), we focus our attention on the first term of Eq. (8.3.13).

Now we consider spatially well-localized modes located at the original point; we consider the amplitude of the localized modes exists only in the original point as the first approximation. So we can put

$$\begin{aligned}
J_{mi}(\tau) &= 3\alpha_l^2 \sum_{\{\boldsymbol{R}_n\}_m} \langle W_{mn;i}^{(4)}\rangle_c\{\delta_{m0} + \delta_{n0}\}|v_{0i}(l)|^2(B_{mi} - B_{ni}) \\
&= \alpha_l^2 \sum_{\{\boldsymbol{R}_n\}_m} \langle W_{mn;i}^{(4)}\rangle_c\{\delta_{m0} + \delta_{n0}\}(B_{mi} - B_{ni}),
\end{aligned} \tag{8.3.23}$$

where we have put $|v_{\mathbf{0}i}(l)|^2 \cong \frac{1}{3}$.

8.3.3 The Eigenvalue Equation for the Localized Modes

Substituting Eq. (8.3.23) into Eq. (8.1.43), putting

$$\boldsymbol{B}_m(\tau) \equiv \boldsymbol{B}_m e^{-\mathrm{i}\nu_n \tau}$$

and Fourier transforming the equation, we obtain the eigenvalue equation for localized modes:

$$\begin{aligned}\boldsymbol{B}_m &= \sum_n \zeta^2 D_{mn}(\mathrm{i}\nu_n)\boldsymbol{J}_n \\ &= \alpha_l^2 \sum_n \zeta^2 D_{mn}(\mathrm{i}\nu_n) \sum_{\{\boldsymbol{R}_l\}_n} \langle W_{nl}^{(4)}\rangle_c (\delta_{n0}+\delta_{l0})(\boldsymbol{B}_n - \boldsymbol{B}_l) \\ &= \alpha_l^2 \frac{1}{N}\sum_{nl}\zeta^2 \sum_{\lambda \boldsymbol{q}} e^{\mathrm{i}\boldsymbol{q}\cdot(\boldsymbol{R}_m-\boldsymbol{R}_n)} D_{\lambda\boldsymbol{q}}(\mathrm{i}\nu_n)\boldsymbol{e}_{\lambda\boldsymbol{q}}\boldsymbol{e}_{\lambda\boldsymbol{q}}^T \langle W_{nl}^{(4)}\rangle_c \\ &\quad (\delta_{n0}+\delta_{l0})(\boldsymbol{B}_n - \boldsymbol{B}_l),\end{aligned} \tag{8.3.24}$$

where $\boldsymbol{e}_{\lambda\boldsymbol{q}}$ is a unit polarization vector of a phonon mode λ. Thus we obtain

$$\begin{aligned}B_{mi} &= -\alpha_l^2 \frac{1}{N} \sum_{\lambda \boldsymbol{q} j} \zeta^2 D_{\lambda\boldsymbol{q}}(\mathrm{i}\nu_n) e_{\lambda\boldsymbol{q}}^i e_{\lambda\boldsymbol{q}}^j \sum_{\{\boldsymbol{R}_n\}_0} \\ &\quad \left\langle e^{\mathrm{i}\boldsymbol{q}\cdot\boldsymbol{R}_m}(1-e^{-\mathrm{i}\boldsymbol{q}\cdot\boldsymbol{R}_n}) W_{0nj}^{(4)}\right\rangle_c (B_{0j}-B_{nj}).\end{aligned} \tag{8.3.25}$$

Now we solve Eq. (8.3.25) analytically continued $\mathrm{i}\nu_n \to \omega_l + \mathrm{i}\delta$. Here we consider localized modes for the limiting case $\omega_l \gg \omega_{\lambda\boldsymbol{q}}$. Since for any function $F_i(\boldsymbol{q})$

$$\sum_{\lambda \boldsymbol{q} j} \zeta^2 D_{\lambda\boldsymbol{q}}(\omega_l) e_{\lambda\boldsymbol{q}}^i e_{\lambda\boldsymbol{q}}^j F_j(\boldsymbol{q}) \longrightarrow \frac{3}{M\omega_l^2}\sum_{\boldsymbol{q}} F_i(\boldsymbol{q}), \tag{8.3.26}$$

Eq. (8.3.25) can be written down in terms of only one Cartesian component of B_{mi}. If we introduce

$$V_{\boldsymbol{0}Re_i iii,0}^{(4,0)} \equiv \frac{1}{6}\zeta^2 W_\sigma(R), \qquad V_{\boldsymbol{0}Re_j iii,0}^{(4,0)} \equiv \frac{1}{6}\zeta^2 W_\pi(R). \tag{8.3.27}$$

Here we obtain

$$W_{\boldsymbol{0}Re_R i}^{(4)} = -W_\sigma(R)\cos^4\theta - W_\pi(R)\sin^4\theta, \qquad \boldsymbol{e}_R\cdot\boldsymbol{e}_i = \cos\theta, \tag{8.3.28}$$

where $\boldsymbol{e}_R$ is a unit vector in the direction of the vector $\boldsymbol{R}$. Here we consider $W_\sigma = W_\pi$. Further, since essential properties of motion does not change, we make the constant coupling approximation: $W_{\boldsymbol{0}Re_R i}^{(4)} \equiv W(R)$. Thus we have

$$\begin{aligned}B_{mi} &= -3\alpha_l^2 \frac{1}{N}\sum_{\boldsymbol{q}} \zeta^2 D_{\lambda\boldsymbol{q}}(\omega_l) \sum_{\{\boldsymbol{R}_n\}_0} \\ &\quad \left\langle e^{\mathrm{i}\boldsymbol{q}\cdot\boldsymbol{R}_m}(1-e^{-\mathrm{i}\boldsymbol{q}\cdot\boldsymbol{R}_n}) W(R)\right\rangle_c (B_{0i}-B_{ni}).\end{aligned} \tag{8.3.29}$$

Under these conditions, we can reduce Eq. (8.3.25) to the eigenvalue equation in two components; $B_{0i} \equiv B_0$ for the particle at the original point and $B_{R_{eR}i} \equiv B_1$ for the nearest neighbor particles. If we define

$$\xi = 3\alpha_l^2 \frac{1}{N} \sum_{\lambda q} \zeta^2 D_{\lambda q}(\omega_l) \sum_{\{R_n\}0} \langle (1 - e^{-iq \cdot R_n}) W(R) \rangle_c, \tag{8.3.30}$$

$$\eta = 3\alpha_l^2 \frac{1}{N} \sum_{\lambda q j} \zeta^2 D_{\lambda q}(\omega_l) \sum_{\{R_n\}0} \langle e^{iq \cdot R_m} (1 - e^{-iq \cdot R_n}) W(R) \rangle_c, \tag{8.3.31}$$

where R_m is a site of a nearest neighbor particle, then we obtain

$$B_0 = \xi(B_0 - B_1), \tag{8.3.32}$$

$$B_1 = \eta(B_0 - B_1). \tag{8.3.33}$$

Thus we obtain the eigenvalue equation:

$$\begin{pmatrix} 1-\xi & \xi \\ \eta & -1-\eta \end{pmatrix} \begin{pmatrix} B_0 \\ B_1 \end{pmatrix} = 0, \tag{8.3.34}$$

From the secular equation of (8.3.34), we obtain the frequency of the localized modes and the ratio of the amplitudes:

$$1 - \xi + \eta = 0, \tag{8.3.35}$$

$$\frac{B_1}{B_0} = \frac{\eta}{\xi}. \tag{8.3.36}$$

Next we calculate ξ and η in Eqs. (8.3.30) and (8.3.31):

$$\begin{aligned}
\xi &= 3\alpha_l^2 \frac{1}{NM\omega_l^2} \frac{V}{(2\pi)^3} \int_0^{q_M} d^3q \int d^3R g(R) W(R)(1 - e^{iq \cdot R}) \\
&= \frac{6\alpha_l^2 V}{M\omega_l^2 N a^3 \pi} \int_0^\infty R^2 dR g(R) W(R) \int_0^{\alpha_M} \alpha^2 d\alpha \left(1 - \frac{\sin\alpha}{\alpha}\right) \\
&= \frac{6\alpha_l^2 V}{M\omega_l^2 N a^3 \pi} \int_0^\infty R^2 dR g(R) W(R) \\
&\quad \left(\frac{\alpha_M^3}{3} + \alpha_M \cos\alpha_M - \sin\alpha_M\right),
\end{aligned} \tag{8.3.37}$$

$$\begin{aligned}
\eta &= 3\alpha_l^2 \frac{1}{NM\omega_l^2} \frac{V}{(2\pi)^3} \int_0^{q_M} d^3q e^{iq \cdot R_m} \int d^3R g(R) W(R)(1 - e^{iq \cdot R}) \\
&= \frac{6\alpha_l^2 V}{M\omega_l^2 N a^3 \pi} \int_0^\infty R^2 dR g(R) W(R) \int_0^{\alpha_M} \alpha^2 d\alpha \frac{\sin\alpha}{\alpha} \left(1 - \frac{\sin\alpha}{\alpha}\right) \\
&= \frac{6\alpha_l^2 \rho V}{M\omega_l^2 N a^3 \pi} \int_0^\infty R^2 dR g(R) W(R) \\
&\quad \left(-\frac{\alpha_M}{2} + \frac{\sin 2\alpha_M}{4} - \alpha_M \cos\alpha_M + \sin\alpha_M\right),
\end{aligned} \tag{8.3.38}$$

where we have considered the nearest neighbor particles with a constant distance from the original point $g(R) = \rho\delta(R - a)$, and $\alpha = aq$, $\alpha_M = aq_M$, q_M is a cut-off momentum, and V a volume of the system. In Eqs. (8.3.37) and (8.3.38) from the first to the second equation, we have first integrated in the angular part of $\boldsymbol{R}$, secondly integrated in the angular part of $\boldsymbol{q}$, and then put $R_m = R$. If we put $q_M a = \pi$, we obtain

$$\xi = \frac{6V}{\pi a^3 N}\frac{\alpha_l^2}{M\omega_l^2}\rho a^2 W(a)\left(\frac{\pi^3}{3} - 1\right), \tag{8.3.39}$$

$$\eta = \frac{6V}{\pi a^3 N}\frac{\alpha_l^2}{M\omega_l^2}\rho a^2 W(a)\left(-\frac{\pi}{2} + 1\right). \tag{8.3.40}$$

8.3.4 The Eigenvalue and the Amplitude of the Localized Modes

In order to compare ω_l with the phonon harmonic frequency ω and the coupling strength $W(a)$ with the phonon coupling $V_\sigma(a)$, we remember the relation of ω and V_σ for the case of $V_\sigma = V_\pi$:

$$M\omega^2 = 4\pi \int_0^\infty R^2 dR g(R) V_\sigma(R) = 4\pi\rho a^2 V_\sigma(a). \tag{8.3.41}$$

Introducing

$$f \equiv \frac{\omega_l}{\omega}, \tag{8.3.42}$$

and putting $V = \frac{4\pi a^3}{3}N$, we obtain

$$\xi = \frac{2}{\pi}\alpha_l^2 \frac{W(a)}{V_\sigma(a)}\frac{1}{f^2}\left(\frac{\pi^3}{3} - 1\right) \cong 5.77\alpha_l^2 \frac{W(a)}{V_\sigma(a)} f^2, \tag{8.3.43}$$

$$\eta = \frac{2}{\pi}\alpha_l^2 \frac{W(a)}{V_\sigma(a)}\frac{1}{f^2}\left(-\frac{\pi}{2} + 1\right) \cong -0.36\alpha_l^2 \frac{W(a)}{V_\sigma(a)} f^2, \tag{8.3.44}$$

where

$$\alpha_l^2 = \frac{\zeta^2}{f}\coth\left(\frac{\hbar\omega}{2k_B T} f\right). \tag{8.3.45}$$

First we calculate the frequency of the localized mode from Eq. (8.3.35) in two limiting cases:

1. $\dfrac{\hbar\omega_l}{k_B T} \gg 1: \quad \alpha_l^2 \cong \dfrac{\zeta^2}{f}.$

Since

$$\xi - \eta \cong 6.13 \frac{W(a)}{V_\sigma(a)}\frac{\zeta^2}{f^3},$$

we get

$$f \cong \left(6.13 \frac{W(a)}{V_\sigma(a)}\zeta_M^2\right)^{\frac{1}{3}}. \tag{8.3.46}$$

2. For $\frac{\hbar\omega_l}{k_BT} \ll 1$: $\quad \alpha_l^2 \cong \frac{\zeta^2 2k_\mathrm{B}T}{\hbar\omega f^2}$.

Since

$$\xi - \eta = 12.26\frac{W(a)}{V_\sigma(a)}\frac{\zeta^2 2k_\mathrm{B}T}{\hbar\omega f^4},$$

we get

$$f \cong \left(12.26\frac{W(a)}{V_\sigma(a)}\frac{\zeta^2 k_\mathrm{B}T}{\hbar\omega}\right)^{\frac{1}{4}}. \tag{8.3.47}$$

Next we calculate the ratio of the amplitudes of the particles. Eq. (8.3.36) leads to

$$\frac{B_1}{B_0} \cong -\frac{1}{16}. \tag{8.3.48}$$

The ratio is independent of temperature. This result is consistent with the assumption (8.3.23).

8.3.5 The Condition of the Existence of the Localized Modes

Now we estimate the condition of the existence of localized modes. For $V_\sigma = V_\pi$, since $\omega_{Lq} = \omega_{Tq}$, we can write Eq. (8.3.35) explicitly as

$$1 = -3\alpha_l^2\frac{1}{N}\sum_q \zeta^2 D_{Lq}(\omega_l)\sum_{\{\boldsymbol{R}_n\}_0}\langle(1 - e^{\mathrm{i}\boldsymbol{q}\cdot\boldsymbol{R}_m})(1 - e^{-\mathrm{i}\boldsymbol{q}\cdot\boldsymbol{R}_n})W(R)\rangle_c, \tag{8.3.49}$$

where $\boldsymbol{R}_m$ is a position of a nearest neighbor particle. The summing Eq. (8.3.49) over $\boldsymbol{q}$ and $\{\boldsymbol{R}_n\}_0$ leads to the equation independent of $\boldsymbol{R}_m$. Since α_l^2 and $D_{L\boldsymbol{q}}(\omega_l)$ are monotonically decreasing function with respect to $\omega_l (> \omega_M)$ and $\frac{1}{k_BT}$, the condition for the existence of localized modes is given by the equation replaced $=$ by $<$ in Eq. (8.3.49). For simple estimation, we put

$$\sum_{\boldsymbol{R}_n}\langle(1 - e^{\mathrm{i}\boldsymbol{q}\cdot\boldsymbol{R}_m})(1 - e^{-\mathrm{i}\boldsymbol{q}\cdot\boldsymbol{R}_n})W(R)\rangle_c \cong 4\pi\rho a^2 W(a) \tag{8.3.50}$$

and

$$\frac{1}{N}\sum_{\boldsymbol{q}} D_{L\boldsymbol{q}} \cong 3\frac{1}{M\omega^2} = \frac{3}{4\pi\rho a^2 V_\sigma(a)}. \tag{8.3.51}$$

Thus from Eqs. (8.3.50) and (8.3.51) the condition for the existence of localized modes is estimated as

$$1 \lesssim \frac{9W(a)\zeta^2}{V_\sigma(a)}\coth\frac{\hbar\omega}{2k_BT}. \tag{8.3.52}$$

The localized modes in glasses are the same as those in simple cubic lattices except the numerical factors [4,5]. This is because the dynamical equations of the localized modes in glasses correspond to those simple cubic lattices, if we consider $(0, 0, 0)$ as the original point and $(\pm a, 0, 0)$, $(0, \pm a, 0)$, $(0, 0, \pm a)$ in simple cubic lattices as the position of the original point and the positions of the nearest neighbor particles in glasses. Thus the Eigenfrequencies (8.3.46) and (8.3.47), the amplitude (8.3.48), and the condition of the existence of the localized modes (8.3.52) are the same equations as those in simple cubic lattices except the numerical factors and the frequency ω in the relative frequency $f = \frac{\omega_l}{\omega}$ in glasses replaced by the maximum frequency of phonons ω_M in simple cubic lattices.

For the spatially well-localized modes at the original point oscillate with the frequency ω_l, the surrounding particles oscillate in the opposite direction to the particle at the original point due to the conservation of the center of mass. This situation resembles the oscillation of a shell and its central particle which are bound by springs. For $\omega_l \gg \omega$, the particle feels a quartic potential, where the spring constant corresponds to the mean field of the quartic potential.

In actual glasses, since particles seldom hop, the phonons are stable so that the localized modes are stable. However, the phonon dispersion curves are vague due to the randomness of particle positions, so are the localized modes.

References

[1] T. Kitamura, Phys. Lett. A 186 (1994) 351.
[2] T. Kitamura, S. Takeno, Phys. Lett. A 190 (1994) 327.
[3] T. Kitamura, Physica A 214 (1995) 295.
[4] T. Kitamura, Physica A 213 (1995) 525.
[5] T. Kitamura, S. Takeno, Physica A 213 (1995) 539.
[6] T. Kitamura, Phys. Lett. A 203 (1995) 395.
[7] T. Kitamura, S. Takeno, Phys. Lett. A 172 (1992) 184.
[8] M. Toda, Theory of Nonlinear Lattices, Springer, Berlin, 1982.
[9] G.L. Lamb Jr., Elements of Soliton Theory, Wiley, New York, 1983.
[10] H. Umezawa, H. Matsumoto, M. Tachiki, Thermo Field Dynamics and Condensed States, North-Holland, Amsterdam, 1982.
[11] H. Matsumoto, P. Sadano, H. Umezawa, Phys. Rev. D 19 (1979) 511.
[12] S. Takeno, K. Hori, J. Phys. Soc. Jp, 59 (1990) 3037.
[13] S.R. Bickham, A.J. Sievers, S. Takeno, Phys. Rev.B45 (1992) 10344.
[14] L.D. Landau, E.M. Lifshitz, Quantum Mechanics, Pergamon Press, London, 1964 (See Problem 4 in Section 23).
[15] S. Takeno, A.J. Sievers, Solid State Commun. 67 (1988) 1023.
[16] A. J. Sievers, S. Takeno, Phys. Rev. Lett. 61 (1988) 970.
[17] S. Takeno, K. Kisoda, A.J. Sievers, Prog. Theor. Phys. Suppl. 94 (1988) 242.
[18] A.A. Maradudin, E.W. Montroll, G.H. Weiss, I.P. Ipatova, Solid State Physics, 2nd Edition, Suppl. 3, Academic Press, New York, 1971.
[19] E.W. Montroll, R.B. Potts, Phys. Rev. 100 (1955) 525.

9 Phonon and Sound Fluctuation Modes and Thermal Conductivities

The correlations of intra-band and inter-band particle density fluctuations yield the respective collective modes, sound and phonons at high frequencies, $q_0\tau_\mu \gg 1$, and the respective dissipative modes, viscosities and diffusivity at low frequencies, $q_0\tau_\mu \ll 1$. In calculating the correlation functions, we use the interaction potential of intra-band and inter-band particle density fluctuations for sound and phonons, respectively.

Sound and phonons have respective sound and phonon frequencies (energies) so that the sound and phonon density fluctuations relate with their respective frequency (energy) density fluctuations, i.e. the respective entropy density fluctuations. Therefore, the correlations of sound and phonon density fluctuations yield their respective entropy (heat) density fluctuation modes at high frequencies and their respective thermal conductivities at low frequencies [1,2]. In calculating the correlations of sound and phonon density fluctuations, we need the interaction potentials of sound and phonon density fluctuations, which are nonlinear interaction potentials for sound and phonon operators. In the previous chapter we have derived the nonlinear interaction potentials for phonons. For the sound case, we construct the nonlinear interaction potential in a perturbative method in this chapter.

The relaxation times of the entropy fluctuation modes due to sound and phonons are essentially the same as those of sound and phonons, but modified by the contributions of the corresponding inverse frequencies which appear in the self-energy part of sound and phonons, respectively. The powers of the VTF law J of the transport coefficients and the relaxation times due to sound density fluctuations are different from those due to sound, while those due to phonon density fluctuations are the same as those due to phonons as shown in Table 7.1. This is owing to that τ_0^{-1}, $c_T \propto J$ for sound, while $\tau_M^{-1} \propto J$, but c_λ = constant for phonons.

In Chapter 7, we have explained the specific anomaly in the low temperature glasses. Here we explain the thermal conductivity anomaly in the low temperature glasses.

9.1 The Effective Interaction Hamiltonian for Phonon Fields and the Elementary Scattering Processes of Phonons

First we investigate the nonlinear interaction potentials for phonon operators. In the previous chapter, under the two approximations: (1) the nearest neighbor approximation,

Liquid Glass Transition. DOI: http://dx.doi.org/10.1016/B978-0-12-407177-3.00009-X

and (2) the same state coupling approximation, we have constructed the nonlinear interaction Hamiltonian for phonon operators $\{B_m\}$ in the localized representation:

$$H_I = -\frac{1}{2} \sum_{\{\mathbf{R}_m\}n\alpha,\beta,i} \frac{W^{(\alpha+\beta)}_{mn;i}}{\alpha+\beta} (-\zeta)^{\alpha+\beta} (B_{mi} - B_{ni})^{(\alpha+\beta)}. \tag{8.1.47}$$

Here we derive the effective interaction Hamiltonian for cubic and quartic interaction potentials. For the convenience to calculate the collective modes and the dissipative processes, we rewrite the effective nonlinear interaction potentials by free sound and phonon fields explained in Section 3.4. In the relaxation and dissipative processes, we must consider the random scattering processes of sound and phonon density fluctuations due to the random harmonic frequencies and the random hopping magnitudes. In a similar manner to the inclusion of the elementary scattering processes of particles in sound and phonons systems, we include the elementary scattering processes of phonons, i.e., the simultaneously scattering processes of two phonons by the same random potentials due to the random harmonic frequencies and the random hopping magnitudes.

9.1.1 Quantization of Phonon Operators and the Free Phonon Green's Function

We have quantized phonon operators ζB_m by using free phonon fields $\beta_{\lambda\mathbf{q}}$ in the previous chapter. Here we show an alternative method to quantize the phonon operators by free phonon fields and free phonon field Green's functions, by which we can calculate the nonlinear dynamics more easily.

We start with the phonon Green's function in the previous chapter:

$$\begin{aligned} D_{mnij}(\tau - \tau') &\equiv -\frac{1}{\hbar} \langle T_\tau B_{mi}(\tau) B_{nj}(\tau') \rangle_c \\ &= \frac{1}{\beta\hbar N} \sum_{\mathrm{i}\nu_n \mathbf{q}} e^{\mathrm{i}\mathbf{q}\cdot(\mathbf{R}_m - \mathbf{R}_n) - \mathrm{i}\nu_n(\tau-\tau')} D_{ij}(q) \\ &= \frac{1}{\beta\hbar N} \sum_{\mathrm{i}\nu_n \mathbf{q}} e^{\mathrm{i}\mathbf{q}\cdot(\mathbf{R}_m - \mathbf{R}_n) - \mathrm{i}\nu_n(\tau-\tau')} e^i_{\lambda\mathbf{q}} D_\lambda(q) e^{j*}_{\lambda\mathbf{q}}, \end{aligned} \tag{8.1.29}$$

where we have the phonon Green's function in the extended representation:

$$\begin{aligned} D_\lambda(q) &= \frac{\zeta^{-2}}{M} \frac{1}{(\mathrm{i}\nu_n)^2 - \omega^2_{\lambda\mathbf{q}}} \\ &= \frac{\zeta^{-2}}{2\omega_{\lambda\mathbf{q}} M} \left\{ \frac{1}{\mathrm{i}\nu_n - \omega_{\lambda\mathbf{q}}} - \frac{1}{\mathrm{i}\nu_n + \omega_{\lambda\mathbf{q}}} \right\}. \end{aligned} \tag{9.1.1}$$

We introduce free phonon Hamiltonian:

$$H_0 = \sum_{\lambda\mathbf{q}} \hbar\omega_{\lambda\mathbf{q}} \beta^\dagger_{\lambda\mathbf{q}} \beta_{\lambda\mathbf{q}}, \tag{9.1.2}$$

$$[\beta_{\lambda\mathbf{q}}, \beta^\dagger_{\lambda'\mathbf{q}'}] = \delta_{\lambda\lambda'} \delta_{\mathbf{q}\mathbf{q}'}, \tag{9.1.3}$$

and free phonon Green's functions:

$$\begin{aligned}\mathcal{D}_{\lambda\mathbf{q}}(\tau-\tau') &\equiv -\langle T_\tau \beta_{\lambda\mathbf{q}}(\tau)\beta^\dagger_{\lambda\mathbf{q}}(\tau')\rangle_c \\ &= \frac{1}{\beta\hbar}\sum_{\mathrm{i}\nu_n} e^{\mathrm{i}\nu_n(\tau-\tau')}\mathcal{D}_\lambda(q). \end{aligned} \tag{9.1.4}$$

For $\tau > 0$, we obtain

$$\begin{aligned}\mathcal{D}_{\lambda\mathbf{q}}(\tau) &= -\langle T_\tau \beta_{\lambda\mathbf{q}}(\tau)\beta^\dagger_{\lambda\mathbf{q}}(0)\rangle \\ &= -e^{-\omega_{\lambda\mathbf{q}}\tau}\langle \beta_{\lambda\mathbf{q}}\beta^\dagger_{\lambda\mathbf{q}}\rangle \\ &= e^{-\omega_{\lambda\mathbf{q}}\tau}(1+f(\omega_{\lambda\mathbf{q}}) = -e^{-\omega_{\lambda\mathbf{q}}\tau}\frac{e^{\beta\hbar\omega_{\lambda\mathbf{q}}}}{e^{\beta\hbar\omega_{\lambda\mathbf{q}}}-1}. \end{aligned} \tag{9.1.5}$$

Therefore we obtain

$$\begin{aligned}\mathcal{D}_\lambda(q) &= \int_0^{\beta\hbar} d\tau e^{\mathrm{i}\nu_n\tau}\mathcal{D}_{\lambda\mathbf{q}}(\tau) \\ &= -\int_0^{\beta\hbar} d\tau e^{(\mathrm{i}\nu_n-\omega_{\lambda\mathbf{q}})\tau}\frac{e^{\beta\hbar\omega_{\lambda\mathbf{q}}}}{e^{\beta\hbar\omega_{\lambda\mathbf{q}}}-1} \\ &= \frac{1}{\mathrm{i}\nu_n-\omega_{\lambda\mathbf{q}}}. \end{aligned} \tag{9.1.6}$$

For the case:

$$\begin{aligned}\mathcal{D}^\dagger_{\lambda\mathbf{q}}(\tau) &= -\langle T_\tau \beta^\dagger_{\lambda\mathbf{q}}(\tau)\beta_{\lambda\mathbf{q}}(0)\rangle \\ &= \mathcal{D}_{\lambda\mathbf{q}}(-\tau) = \mathcal{D}_{\lambda\mathbf{q}}(\beta\hbar-\tau) \\ &= -e^{\omega_{\lambda\mathbf{q}}\tau}\frac{1}{e^{\beta\hbar\omega_{\lambda\mathbf{q}}}-1}, \end{aligned} \tag{9.1.7}$$

then we obtain

$$\mathcal{D}^\dagger_\lambda(q) = -\frac{1}{\mathrm{i}\nu_n+\omega_{\lambda\mathbf{q}}}. \tag{9.1.8}$$

Therefore we can write Eq. (9.1.1) as

$$\begin{aligned}D_\lambda(q) &= \frac{\zeta^{-2}}{2\omega_{\lambda\mathbf{q}}M}\left\{\frac{1}{\mathrm{i}\nu_n-\omega_{\lambda\mathbf{q}}}-\frac{1}{\mathrm{i}\nu_n+\omega_{\lambda\mathbf{q}}}\right\} \\ &= -\frac{\zeta^{-2}}{2\omega_{\lambda\mathbf{q}}M}\int_0^{\beta\hbar} d\tau e^{\mathrm{i}\nu_n\tau}\langle T_\tau\{\beta_{\lambda\mathbf{q}}(\tau)\beta^\dagger_{\lambda\mathbf{q}}(0)+\beta^\dagger_{\lambda\mathbf{q}}(\tau)\beta_{\lambda\mathbf{q}}(0)\rangle, \end{aligned} \tag{9.1.9}$$

where it should be noted that we should discriminate between the Green's function for phonon operators D and the Green's function for free phonon fields $\mathcal{D}$. Thus we can obtain the relation of phonon operators between the localized and extended representations.

$$
\begin{aligned}
-\zeta B_{mi}(\tau) &= -\frac{\zeta}{N}\sum_{\mathbf{q}} e^{-i\mathbf{q}\cdot\mathbf{R}_m}\rho^{\dagger}_{i\mathbf{q}}(\tau) \\
&= \sum_{\lambda\mathbf{q}}\sqrt{\frac{\hbar}{2\omega_{\lambda\mathbf{q}}MN}}e^{i}_{\lambda\mathbf{q}}\left\{e^{i\mathbf{q}\cdot\mathbf{R}_m-\omega_{\lambda\mathbf{q}}\tau}\beta_{\lambda\mathbf{q}}+e^{-i\mathbf{q}\cdot\mathbf{R}_m+\omega_{\lambda\mathbf{q}}\tau}\beta^{\dagger}_{\lambda\mathbf{q}}\right\},
\end{aligned}
\tag{9.1.10}
$$

where $e^{i}_{\lambda\mathbf{q}}$ is the ith Cartesian component of the polarization vector with the mode $\lambda\mathbf{q}$.

Next, in order to investigate the dynamical equations for free phonon fields, we establish the effective interaction Hamiltonian among free phonon fields. The effective interaction Hamiltonian consists of the nonlinear interaction between phonons and the random scattering processes through the bubble diagrams of particles. First we investigate the nonlinear interaction between phonon fields.

9.1.2 The Effective Interaction Hamiltonian for Phonon Fields

First, we consider the cubic interaction Hamiltonian in Eq. (8.1.47). Here we only take into account $B_m^2 B_n - B_m B_n^2$. Using Eq. (9.1.10), we obtain

$$
\begin{aligned}
H_3 &\equiv -\frac{1}{2}\sum_{m\neq n}\frac{1}{3}W^{(3)}_{mn;i}(-\zeta)^3(B^2_{mi}B_{ni}-B_{mi}B^2_{ni}) \\
&= -\sum_{m\neq n}\frac{1}{3}W^{(3)}_{mn;i}(-\zeta)^3B^2_{mi}B_{ni} \\
&= -\sum_{\lambda\lambda'\mathbf{q}\mathbf{q}'i}\Bigg[\left(\frac{\hbar}{2MN}\right)^{\frac{3}{2}}W^{(3)}(\mathbf{q}-\mathbf{q}')\frac{e^{i}_{\lambda\mathbf{q}}e^{i}_{\lambda\mathbf{q}'}e^{i}_{\lambda'\mathbf{q}-\mathbf{q}'}}{\sqrt{\omega_{\lambda\mathbf{q}}\omega_{\lambda\mathbf{q}'}\omega_{\lambda'\mathbf{q}-\mathbf{q}'}}} \\
&\qquad\left\{\beta_{\lambda\mathbf{q}}\beta^{\dagger}_{\lambda\mathbf{q}'}\beta_{\lambda'\mathbf{q}-\mathbf{q}'}+\beta_{\lambda\mathbf{q}}\beta_{\lambda\mathbf{q}'}\beta^{\dagger}_{\lambda'\mathbf{q}-\mathbf{q}'}\right\}+\text{h.c}\Bigg],
\end{aligned}
\tag{9.1.11}
$$

$$
W^{(3)}(\mathbf{q}) = \frac{N}{3}\sum_{\mathbf{R}_m-\mathbf{R}_n}W^{(3)}_{mn;i}e^{i\mathbf{q}\cdot(\mathbf{R}_m-\mathbf{R}_n)},
\tag{9.1.12}
$$

where we have considered spherical symmetry. From the first to the second equation, we have used the relation (8.1.45) in exchanging the indices m and n. H_3 is illustrated in Figure 9.1.

Next we consider the quartic interaction terms. We only take into account $B_m^2 B_n^2$ term. Thus we obtain

$$H_4 = -\sum_{m\neq n} \frac{3}{4} W^{(4)}_{mn;i} (-\zeta B_m)^2 (-\zeta B_n)^2$$

$$= -\sum_{\lambda \mathbf{q}\mathbf{q}'\mathbf{q}_1\mathbf{q}_1'} 3\left(\frac{\hbar}{2MN}\right)^2 \left[W^{(4)}(\mathbf{q}-\mathbf{q}') \frac{e^i_{\lambda\mathbf{q}} e^i_{\lambda\mathbf{q}'} e^i_{\lambda\mathbf{q}_1} e^i_{\lambda\mathbf{q}_1'}}{\sqrt{\omega_{\lambda\mathbf{q}}\omega_{\lambda\mathbf{q}'}\omega_{\lambda\mathbf{q}_1}\omega_{\lambda\mathbf{q}_1'}}} \right.$$

$$\left\{ \beta_{\lambda\mathbf{q}} \beta^\dagger_{\lambda\mathbf{q}'} \beta_{\lambda\mathbf{q}_1} \beta^\dagger_{\lambda\mathbf{q}_1'} \delta_{\mathbf{q}-\mathbf{q}',\mathbf{q}_1'-\mathbf{q}_1} \right.$$

$$\left. \left. + \beta_{\lambda\mathbf{q}} \beta^\dagger_{\lambda\mathbf{q}'} \beta^\dagger_{\lambda\mathbf{q}_1} \beta_{\lambda\mathbf{q}_1'} \delta_{\mathbf{q}-\mathbf{q}',\mathbf{q}_1-\mathbf{q}_1'} \right\} + \text{h.c.} \right], \tag{9.1.13}$$

$$W^{(4)}(\mathbf{q}) \equiv \frac{N}{4} \sum_{\mathbf{R}_m-\mathbf{R}_n} e^{i\mathbf{q}\cdot(\mathbf{R}_m-\mathbf{R}_n)} W^{(4)}_{mn}$$

$$= W^{(4)*}(\mathbf{q}). \tag{9.1.14}$$

H_4 is illustrated in Figure 9.2.

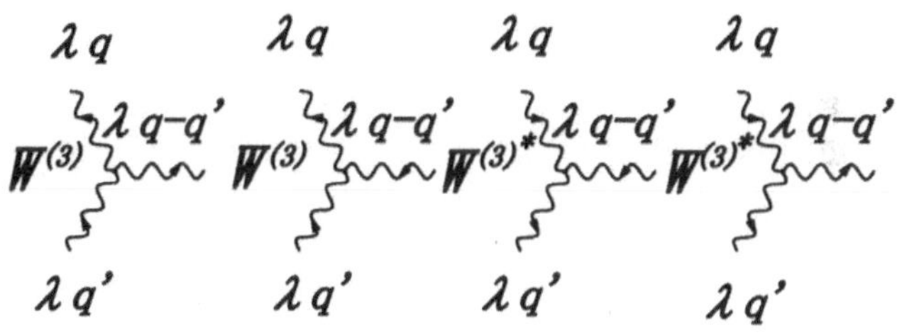

Figure 9.1 The cubic interaction Hamiltonian H_3 for phonon field operators is diagrammatically illustrated.

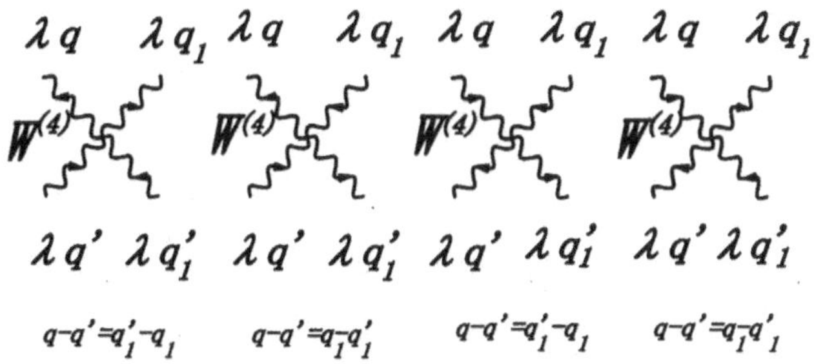

Figure 9.2 The quartic interaction Hamiltonian H_4 for phonon field operators is diagrammatically illustrated.

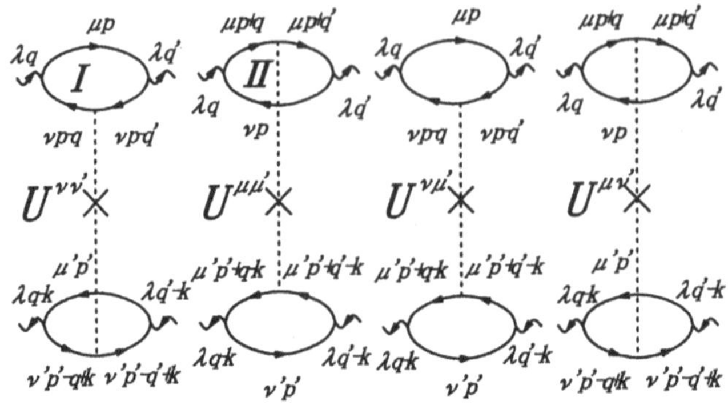

Figure 9.3 The elementary scattering process of phonons due to random scattering $U^{\mu\nu} = U_{\omega}^{\mu\nu} + U_{J}^{\mu\nu}$ are shown. The effect of random scatterings comes through the bubble diagrams of particle-hole pairs.

9.1.3 The Elementary Scattering Processes of Phonons

When we have investigated the relaxation and dissipative processes for phonons, we have considered the random scattering processes due to random harmonic frequencies and random hopping magnitudes. In order to take into account the simultaneously scattering processes of two phonons by the same random potentials due to random harmonic frequencies and random hopping magnitudes, we have introduced the elementary scattering processes due to random harmonic frequencies, $U_{\omega}^{\mu\nu}$, and due to random hopping magnitudes, $U_{J}^{\mu\nu}$. By using the elementary scattering processes $U_{\omega,J}^{\mu\nu}$ between particles, we have taken into account the randomness of the systems in calculating the dynamical equations of inter-band density fluctuations.

Therefore, when we develop the dynamical equations of phonon density fluctuations, we must consider the elementary scattering processes of phonons, i.e. the simultaneously scattering processes of two phonons by the same random scattering potentials due to random harmonic frequencies and random hopping magnitudes. The simultaneously scattering processes of the two phonons by the same random potentials only occur through two particles. The phonon is the sum of an infinite series of interacting bubbles of inter-band particle-hole pairs. Therefore the simultaneously scattering processes of phonons by the same random potentials occur through the scattering processes of two particles in inter-bubbles. Thus the elementary scattering process of phonons due to random scatterings potentials is diagrammatically shown in Figure 9.3. Taking into account the elementary scattering processes, we calculate the dynamical equations of phonon density fluctuations.

Now we calculate the bubble diagrams *I* and *II* in Figure9.3. A bubble diagram denoted as *I* is calculated as:

$$
\begin{aligned}
I : &\frac{1}{\hbar N}\rho_{i\mathbf{q}}^{\dagger} V_{ij}(\mathbf{q})\rho_{j\mathbf{q}} a_{\nu\mathbf{p}-\mathbf{q}'}^{\dagger} a_{\nu\mathbf{p}-\mathbf{q}} \frac{1}{\hbar N}\rho_{j\mathbf{q}'}^{\dagger} V_{jl}(\mathbf{q}')\rho_{l\mathbf{q}'} \\
&\to \frac{1}{\hbar}\sqrt{\frac{\hbar}{2\omega_{\lambda\mathbf{q}} M N}}\, e_{\lambda\mathbf{q}}^{i} \beta_{\lambda\mathbf{q}}^{\dagger} \widetilde{V}_{ij}(\mathbf{q})\rho_{j\mathbf{q}} a_{\nu\mathbf{p}-\mathbf{q}'}^{\dagger} a_{\nu\mathbf{p}-\mathbf{q}} \rho_{j\mathbf{q}'}^{\dagger} \widetilde{V}_{jl}(\mathbf{q}')
\end{aligned}
$$

$$\times \frac{1}{\hbar}\sqrt{\frac{\hbar}{2\omega_{\lambda\mathbf{q}'}MN}}e^{l}_{\lambda\mathbf{q}'}\beta_{\lambda\mathbf{q}'}$$
$$= \beta^{\dagger}_{\lambda\mathbf{q}}\widetilde{I}\beta_{\lambda\mathbf{q}'}, \tag{9.1.15}$$

$$\widetilde{I} = \frac{1}{2\hbar M}\frac{e^{i}_{\lambda\mathbf{q}}}{\sqrt{\omega_{\lambda\mathbf{q}}}}\widetilde{V}_{ij}(\mathbf{q})\widetilde{V}_{jl}(\mathbf{q}')\frac{e^{l}_{\lambda\mathbf{q}'}}{\sqrt{\omega_{\lambda\mathbf{q}'}}}$$
$$\times\left[\mp\frac{1}{(\beta\hbar)^2N}\sum_{p}G^{\mu}(i\omega_n,\mathbf{p})G^{\nu}(\mathrm{i}\omega_n-\mathrm{i}\nu_n,\mathbf{p}-\mathbf{q})\right.$$
$$\left.G^{\nu}(\mathrm{i}\omega_n-\mathrm{i}\nu_n,\mathbf{p}-\mathbf{q}')\right], \tag{9.1.16}$$

where $\widetilde{V}_{ij}(\mathbf{q}) \equiv V_{ij}(\mathbf{q})/\zeta^2$. Putting $\epsilon_{\mu\mathbf{p}} \cong \omega_\mu$ and $\mathrm{i}\nu_n = 0$ because of $\beta\hbar\omega \gg 1$, the term in the bracket leads to

$$[\cdots] \cong \mp\frac{1}{\beta\hbar N}\sum_{\mathrm{i}\omega_n,\mathbf{p}}\frac{1}{\mathrm{i}\omega_n-\tilde{\omega}_\mu}\frac{1}{(\mathrm{i}\omega_n-\mathrm{i}\nu_n-\tilde{\omega}_\nu)^2}$$
$$= \mp\frac{1}{\beta\hbar N}\frac{\partial}{\partial\tilde{\omega}_\nu}\sum_{p}\frac{1}{\mathrm{i}\omega_n-\tilde{\omega}_\mu}\frac{1}{\mathrm{i}\omega_n-\mathrm{i}\nu_n-\tilde{\omega}_\nu}$$
$$= \mp\frac{1}{\beta\hbar N}\frac{\partial}{\partial\tilde{\omega}_\nu}\sum_{p}\frac{1}{\mathrm{i}\nu_n-\tilde{\omega}_\mu+\tilde{\omega}_\nu}\left(\frac{1}{\mathrm{i}\omega_n-\mathrm{i}\nu_n-\tilde{\omega}_\nu}-\frac{1}{\mathrm{i}\omega_n-\tilde{\omega}_\mu}\right)$$
$$\cong -\frac{1}{\beta\hbar}\frac{f(\tilde{\omega}_\nu)-f(\tilde{\omega}_\mu)}{\omega^2}, \tag{9.1.17}$$

where we have considered $f'(\omega_\nu) \cong 0$. Note that $f(\omega_\mu) \cong \delta_{\mu,0}$ and $|\tilde{\omega}_\mu - \tilde{\omega}_\nu| = \omega$. Thus we obtain

$$\widetilde{I} = -\frac{1}{2M\beta\hbar^2\omega^2}\frac{e^{i}_{\lambda\mathbf{q}}}{\sqrt{\omega_{\lambda_\mathbf{q}}}}\widetilde{V}_{ij}\widetilde{V}_{jl}\frac{e^{l}_{\lambda\mathbf{q}'}}{\sqrt{\omega_{\lambda\mathbf{q}'}}}\{f(\tilde{\omega}_\nu)-f(\tilde{\omega}_\mu)\}. \tag{9.1.18}$$

A bubble diagram denoted as *II* is given in a similar manner to *I*:

$$II = \beta^{\dagger}_{\lambda\mathbf{q}}\widetilde{II}\beta_{\lambda\mathbf{q}'}, \tag{9.1.19}$$

$$\widetilde{II} = -\widetilde{I}. \tag{9.1.20}$$

By considering the calculation of *I* and *II*, the summation of the elementary scattering processes due to random scatterings U_ω and U_J is denoted as $U^{\lambda}_{R}(q,q';k)$, which is diagrammatically illustrated in Figure 9.4. We call $U^{\lambda}_{R}(q,q';k)$ the elementary

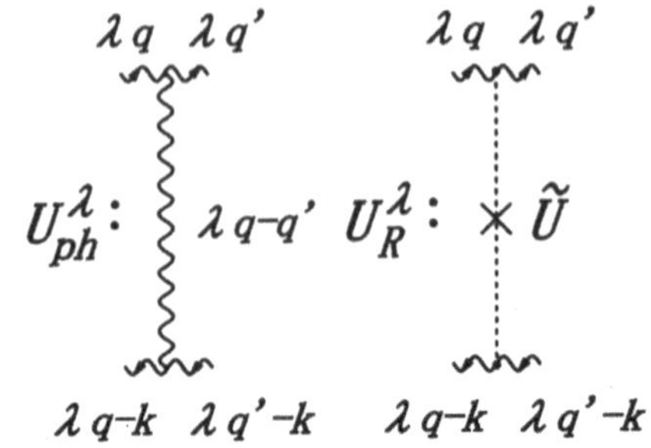

Figure 9.4 The elementary scattering process of phonons due to a phonon, $U^{\lambda}_{\mathrm{ph}}(q, q'; k)$ and the sum of the elementary scattering processes due to the random scattering shown in Figure 9.3, $U^{\lambda}_{R}(q, q'; k)$ are diagrammatically illustrated.

scattering magnitude of phonons due to the random scattering potentials.

$$U^{\lambda}_{R}(q, q'; k) = \frac{1}{(M\beta\hbar^2\omega^2)^2}\sum_{ij}\frac{e^{i}_{\lambda\mathbf{q}}e^{l}_{\lambda\mathbf{q}'}}{\sqrt{\omega_{\lambda\mathbf{q}}\omega_{\lambda\mathbf{q}'}}}\widetilde{V}_{ij}(\mathbf{q})\widetilde{V}_{jl}(\mathbf{q}')\widetilde{U}\widetilde{V}_{i'j'}(\mathbf{q}-\mathbf{k})$$
$$\widetilde{V}_{j'l'}(\mathbf{q}'-\mathbf{k})\frac{e^{i'}_{\lambda\mathbf{q}-\mathbf{k}}e^{l'}_{\lambda\mathbf{q}'-\mathbf{k}}}{\sqrt{\omega_{\lambda\mathbf{q}-\mathbf{k}}\omega_{\lambda\mathbf{q}'-\mathbf{k}}}}, \tag{9.1.21}$$

$$\widetilde{U} \equiv \frac{1}{N}\sum_{\mathbf{q}}\{U^{00}(\mathbf{q}) + U^{11}(\mathbf{q}) - U^{01}(\mathbf{q}) - U^{10}(\mathbf{q})\}, \tag{9.1.22}$$

$$U^{\mu\nu}(\mathbf{q}) \equiv U^{\mu\nu}_{\omega} + U^{\mu\nu}_{J}(\mathbf{q}). \tag{9.1.23}$$

Next we include the elementary scattering processes due to a phonon via the cubic interaction H_3, denoted as $U^{\lambda}_{\mathrm{ph}}(q, q'; k)$, which is also shown in Figure 9.4. We call $U^{\lambda}_{\mathrm{ph}}(q, q'; k)$ the elementary scattering magnitude due to phonon scatterings.

$$U^{\lambda}_{\mathrm{ph}}(q, q'; k) = -\frac{\left|\widetilde{W}_3\right|^2}{\hbar^2}\frac{1}{\beta\hbar}\sum_{\lambda' ii'}\frac{e^{i}_{\lambda\mathbf{q}}e^{i}_{\lambda\mathbf{q}'}e^{i}_{\lambda\mathbf{q}-\mathbf{k}}e^{i}_{\lambda\mathbf{q}'-\mathbf{k}}}{\sqrt{\omega_{\lambda\mathbf{q}}\omega_{\lambda\mathbf{q}'}\omega_{\lambda\mathbf{q}-\mathbf{k}}\omega_{\lambda\mathbf{q}'-\mathbf{k}}}}\frac{e^{i}_{\lambda'\mathbf{q}-\mathbf{q}'}{}^{2}}{\omega_{\lambda'\mathbf{q}-\mathbf{q}'}}$$
$$\times\left\{D_{\lambda}\left(q-q'\right) + D_{\lambda}\left(q'-q\right)\right\}, \tag{9.1.24}$$

$$\widetilde{W}_3(\mathbf{q}) \equiv \left(\frac{\hbar}{2MN}\right)^{\frac{3}{2}} W^{(3)}(\mathbf{q}). \tag{9.1.25}$$

It should be noted that the frequency of a phonon in $U^{\lambda}_{\mathrm{ph}}$ changes as $q = (\mathbf{q}, i\nu_n) \rightarrow q' = (\mathbf{q}', i\nu'_n)$, but that in U^{λ}_{R} does not change as $(\mathbf{q}, i\nu_n) \rightarrow (\boldsymbol{q}', i\nu_n)$. U^{λ}_{R} and $U^{\lambda}_{\mathrm{ph}}$ are used in calculating the vertex corrected bubble diagrams for phonon density fluctuations in a similar manner to those for inter-band density fluctuations.

9.2 Phonon Density Fluctuations: Phonon Entropy Fluctuation Modes and Thermal Conductivities

In a similar manner to the calculation procedure of the dynamical processes of intra-band and inter-band density fluctuations in Chapter 6, we calculate the dynamical equation of phonon density fluctuations. The dynamical equations of phonon density fluctuations are calculated by summing up the interacting vertex corrected bubble diagrams in the random phase approximation. The vertex corrected bubble diagrams are calculated by taking into account the elementary scattering amplitudes $U^{\lambda}_{\mathrm{ph}}$ due to the multi-phonon scattering processes and U^{λ}_{R} due to the random scattering processes in the ladder approximation. The self-energy part of Green's function for phonon density fluctuations is determined according to the approximate WT relations. We obtain the phonon entropy fluctuation modes at high frequencies and phonon thermal conductivities at low frequencies.

9.2.1 The Correlation Functions of Phonon Density Fluctuations

We define phonon density fluctuations with a phonon mode λ as

$$\eta^{\dagger}_{\lambda \boldsymbol{k}} \equiv \sum_{\boldsymbol{q}} \beta^{\dagger}_{\lambda \boldsymbol{q}} \beta_{\lambda \boldsymbol{q}-\boldsymbol{k}}. \tag{9.2.1}$$

The correlation functions for phonon density fluctuations are defined by

$$\mathcal{B}_{\lambda \boldsymbol{k}}(\tau - \tau') \equiv -\frac{1}{\hbar}\langle T_{\tau}\eta_{\lambda \boldsymbol{k}}(\tau)\eta^{\dagger}_{\lambda \boldsymbol{k}}(\tau')\rangle_c = \frac{1}{\beta\hbar}\sum_{\mathrm{i}\kappa_n} e^{\mathrm{i}\kappa_n(\tau-\tau')}\mathcal{B}_{\lambda}(k), \tag{9.2.2}$$

where $k = (\mathrm{i}\kappa_n, \boldsymbol{k})$. We calculate Eq. (9.2.2) in the same manner as the correlation functions for inter-band density fluctuations. Using the interaction Hamiltonian (9.1.12) in the random phase approximation, we obtain

$$\mathcal{B}_{\lambda}(k) = R_{\lambda}(k) + R_{\lambda}(k) W_4(\boldsymbol{k})\mathcal{B}_{\lambda}(k), \tag{9.2.3}$$

$$R_{\lambda}(k) \equiv -\frac{1}{\beta\hbar^2}\sum_{q}\varphi^{\lambda}(q, q-k) \tag{9.2.4}$$

$$W_4(\boldsymbol{k}) = -3\left(\frac{\hbar}{2MN}\right)^2 W^{(4)}(\boldsymbol{k}), \tag{9.2.5}$$

where $R_{\lambda}(k)$ is the vertex corrected bubble diagrams for phonon density fluctuations similar to sound and phonons.

9.2.2 The Vertex Corrected Bubble Diagrams for Phonon Density Fluctuations

Using the elementary scattering amplitudes U_R^λ and $U_{\rm ph}^\lambda$, making the ladder approximation, we obtain the vertex corrected bubble diagrams for phonon density fluctuations:

$$\varphi^\lambda(q,q-k) = \mathcal{D}_\lambda(q)\mathcal{D}_\lambda(q-k)\left[1+\sum_{q_1}\left\{U_{\rm ph}^\lambda(q,q_1;k)+U_R^\lambda(q,q_1;k)\right\}\times\varphi^\lambda(q_1,q_1-k)\right], \tag{9.2.6}$$

where

$$\begin{aligned}
&1+\sum_{q_1}\left\{U_{\rm ph}^\lambda(q,q_1;k)+U_R^\lambda(q,q_1;k)\right\}\varphi^\lambda(q_1,q_1-k)\\
&=1-\frac{|W_3|^2}{\hbar^2}\frac{1}{\beta\hbar}\sum_{ll'q_1}\frac{e^l_{\lambda\boldsymbol{q}}e^{l'}_{\lambda\boldsymbol{q}-\boldsymbol{k}}}{\sqrt{\omega_{\lambda\boldsymbol{q}}\omega_{\lambda\boldsymbol{q}-\boldsymbol{k}}}}\varphi^{\lambda ll'}(q-q_1)\frac{e^l_{\lambda\boldsymbol{q}_1}e^{l'}_{\lambda\boldsymbol{q}_1-\boldsymbol{k}}}{\sqrt{\omega_{\lambda\boldsymbol{q}_1}\omega_{\lambda\boldsymbol{q}_1-\boldsymbol{k}}}}\varphi^\lambda(q_1,q_1-k)\\
&\quad+\frac{\widetilde{U}}{(M\beta\hbar^2\omega^2)^2}\sum_{mm'q_1}\frac{\Pi^\lambda_{mm'}(\boldsymbol{q},\boldsymbol{q}-\boldsymbol{k})}{\sqrt{\omega_{\lambda\boldsymbol{q}}\omega_{\lambda\boldsymbol{q}-\boldsymbol{k}}}}\frac{\Pi^\lambda_{mm'}(\boldsymbol{q}_1,\boldsymbol{q}_1-\boldsymbol{k})}{\sqrt{\omega_{\lambda\boldsymbol{q}_1}\omega_{\lambda\boldsymbol{q}_1-\boldsymbol{k}}}}\\
&\quad\times\delta_{\nu_n\nu_{n_1}}\varphi^\lambda(q_1,q_1-k),
\end{aligned} \tag{9.2.7}$$

and

$$\varphi^{\lambda ll'}(q-q_1) = \frac{e^l_{\lambda\boldsymbol{q}-\boldsymbol{q}_1}e^{l'}_{\lambda\boldsymbol{q}-\boldsymbol{q}_1}}{\omega_{\lambda\boldsymbol{q}-\boldsymbol{q}_1}}\{\mathcal{D}_\lambda(q-q_1)+\mathcal{D}_\lambda(q_1-q)\}, \tag{9.2.8}$$

$$\Pi^\lambda_{mm'}(\boldsymbol{q},\boldsymbol{q}-\boldsymbol{k}) \equiv \sum_{ll'} e^l_{\lambda\boldsymbol{q}}e^{l'}_{\lambda\boldsymbol{q}-\boldsymbol{k}}V_{lm}(\boldsymbol{q})V_{l'm'}(\boldsymbol{q}-\boldsymbol{k}). \tag{9.2.9}$$

The phonon Green's function and its self-energy part are given by

$$\mathcal{D}_\lambda(q) \equiv \frac{1}{\mathrm{i}\nu_n-\omega_{\lambda\boldsymbol{q}}-\Sigma_\lambda(q)}, \tag{9.2.10}$$

and

$$\Sigma^\lambda(q) = \sum_{q_1}\left\{U_{\rm ph}^\lambda(q,q_1;0)+U_R^\lambda(q,q_1;0)\right\}\mathcal{D}_\lambda(q_1), \tag{9.2.11}$$

$$\begin{aligned}
&\sum_{q_1}\left\{U_{\rm ph}^\lambda(q,q_1;0)+U_R^\lambda(q,q_1;0)\right\}\mathcal{D}_\lambda(q_1)\\
&=-\frac{|W_3|^2}{\hbar^2}\frac{1}{\omega_{\lambda\boldsymbol{q}}}\phi^\lambda(q)+\frac{\widetilde{U}}{(M\beta\hbar^2\omega^2)^2}\frac{1}{\omega_{\lambda\boldsymbol{q}}}\\
&\quad\times\sum_{q_1}\frac{\Pi^\lambda(\boldsymbol{q},\boldsymbol{q}_1)}{\omega_{\lambda\boldsymbol{q}_1}}\mathcal{D}_\lambda(q_1)\delta_{\nu_n\nu_{n1}},
\end{aligned} \tag{9.2.12}$$

$$\phi^{\lambda}(q) \equiv \frac{1}{\beta\hbar}\sum_{q_1}\frac{\phi^{\lambda}(q,q_1)}{\omega_{\lambda \boldsymbol{q}_1}}\mathcal{D}^{\lambda}(q_1), \tag{9.2.13}$$

$$\phi^{\lambda}(q,q_1) \equiv \sum_{ll'} e^{l}_{\lambda \boldsymbol{q}} e^{l'}_{\lambda \boldsymbol{q}} \varphi^{\lambda ll'}(q-q_1) e^{l}_{\lambda \boldsymbol{q}_1} e^{l'}_{\lambda \boldsymbol{q}_1}, \tag{9.2.14}$$

$$\Pi^{\lambda}(\boldsymbol{q},\boldsymbol{q}_1) = \sum_{mm'} \Pi^{\lambda}_{mm'}(\boldsymbol{q},\boldsymbol{q})\Pi^{\lambda}_{mm'}(\boldsymbol{q}_1,\boldsymbol{q}_1). \tag{9.2.15}$$

In $\mathcal{D}_{\lambda}(q-q_1)$ and $\mathcal{D}_{\lambda}(q_1-q)$, since analytic continuation is done as $\mathrm{i}\nu_n \to q_0 + \mathrm{i}\delta$, $\mathrm{i}\nu_{n_1} \to q_{0_1} + \mathrm{i}\delta$ and $\mathrm{i}\nu_n - \mathrm{i}\nu_{n_1} \to q_0 - q_{0_1} - \mathrm{i}\delta$, $\mathrm{i}\nu_{n_1} - \mathrm{i}\nu_n \to q_{0_1} - q_0 - \mathrm{i}\delta$, $\mathcal{D}_{\lambda}(q-q_1)$ and $\mathcal{D}_{\lambda}(q_1-q_0)$ have the poles on the opposite side complex plane to $\mathcal{D}_{\lambda}(q)$.

Multiplying both the sides of Eq. (9.2.6) by $\mathcal{D}^{\lambda}(q-k)^{-1} - \mathcal{D}^{\lambda}(q)^{-1}$, we obtain the basic equation for the vertex corrected bubble diagram for phonon density fluctuation:

$$\begin{aligned}&\left\{-\mathrm{i}\kappa_n + \omega_{\lambda \boldsymbol{q}} - \omega_{\lambda \boldsymbol{q}-k} + \Delta\Sigma^{\lambda}(q,k)\right\}\varphi^{\lambda}(q,q-k)\\ &\quad = \Delta\mathcal{D}^{\lambda}(q,k)\left[1 + \sum_{q_1}\left\{U^{\lambda}_{\mathrm{ph}}(q,q_1;k) + U^{\lambda}_{R}(q,q_1;k)\right\}\varphi^{\lambda}(q_1,q_1-k)\right],\end{aligned} \tag{9.2.16}$$

where

$$\Delta\Sigma^{\lambda}(q,k) \equiv \Sigma^{\lambda}(q) - \Sigma^{\lambda}(q-k), \tag{9.2.17}$$

$$\Delta\mathcal{D}(q,k) \equiv \mathcal{D}^{\lambda}(q) - \mathcal{D}^{\lambda}(q-k), \tag{9.2.18}$$

We use Eq. (9.2.16) to obtain the vertex corrected bubble diagrams for phonon density fluctuations. There is the WT relation similar to that for sound:

$$\Delta\Sigma^{\lambda}(q,k) = \sum_{q_1}\left\{U^{\lambda}_{\mathrm{ph}}(q,q_1;k) + U^{\lambda}_{R}(q,q_1;k)\right\}\Delta\mathcal{D}^{\lambda}(q,k). \tag{9.2.19}$$

Therefore summing Eq. (9.1.16) over q, we obtain the following exact equation:

$$\sum_{q}\left\{-\mathrm{i}\kappa_n + \omega_{\lambda \boldsymbol{q}} - \omega_{\lambda \boldsymbol{q}-k}\right\}\varphi^{\lambda}(q,k) = \sum_{q}\Delta\mathcal{D}^{\lambda}(q,k). \tag{9.2.20}$$

In solving Eq. (9.2.16), we use this equation.

9.2.3 The Approximation Procedure of the Phonon Self-Energy Part

Now we investigate the self-energy part due to the random scattering process, which plays an important role in the VTF law. We denote the second term of the self-energy part in Eq. (9.2.11) or Eq. (9.2.12) as $\Sigma^{\lambda}_{R}(q)$:

$$\Sigma^{\lambda}_{R}(q) \equiv \frac{\widetilde{U}}{(M\beta\hbar^2\omega^2)^2}\sum_{q_1}\frac{\Pi^{\lambda}(\boldsymbol{q},\boldsymbol{q}_1)}{\omega_{\lambda \boldsymbol{q}}\omega_{\lambda \boldsymbol{q}_1}}\mathcal{D}_{\lambda}(q_1)\delta_{\nu_n\nu_{n1}}. \tag{9.2.21}$$

1. Since the self-energy part $\Sigma_R^\lambda(q)$ and the phonon Green's function $\mathcal{D}_\lambda(q_1)$ have the same frequency $i\nu_n$, we analytically continues $i\nu_n \to q_0 + i\delta$. In summing the phonon Green's function with respect to $\boldsymbol{q}$, since we are interested in the lifetime of phonons and the main contribution to the self-energy part comes from the mass shell, we put

$$\mathcal{D}^\lambda(q_1) \cong -\frac{1}{\Sigma_R^\lambda(q_0, \boldsymbol{q}_1)} \cong -\frac{1}{\Sigma_R^\lambda(0, \boldsymbol{q}_1)}. \tag{9.2.22}$$

2. We consider only the imaginary part of the self-energy part. The self-energy part $\Sigma_R^\lambda(0, \boldsymbol{q})$ has the factor $1/\omega_{\lambda \boldsymbol{q}}$. Since we can expect that $\Pi^\lambda(\boldsymbol{q}, q_1)$ is more smooth behavior than $1/\omega_{\lambda \boldsymbol{q}}$ with respect to $\boldsymbol{q}$, we approximate

$$\Pi^\lambda(\boldsymbol{q}, \boldsymbol{q}_1) \cong (\Pi^\lambda)^2 = \text{constant}. \tag{9.2.23}$$

3. We further make the approximation in the self-energy part as

$$\Sigma_R^\lambda(0, \boldsymbol{q}) = -i\frac{1}{\omega_{\lambda \boldsymbol{q}}}\alpha, \quad \alpha = \text{constant}. \tag{9.2.24}$$

Combining Eqs. (9.2.21)–(9.2.24), we have

$$\begin{aligned}
\Sigma_R^\lambda(0, \boldsymbol{q}) &= -i\frac{1}{\omega_{\lambda \boldsymbol{q}}}\alpha \\
&= \frac{1}{\omega_{\lambda \boldsymbol{q}}}\left(\frac{A^\lambda}{\beta}\right)^2 \widetilde{U} \sum_{\boldsymbol{q}_1} \frac{1}{\omega_{\lambda \boldsymbol{q}_1}} \frac{1}{\Sigma^\lambda(0, \boldsymbol{q}_1)} \\
&= -\frac{1}{\omega_{\lambda \boldsymbol{q}}}\left(\frac{A^\lambda}{2\beta\tau_M}\right)^2 \frac{1}{-i\alpha},
\end{aligned} \tag{9.2.25}$$

where

$$\frac{1}{2\tau_M} = \sqrt{N\widetilde{U}}, \qquad A^\lambda \equiv \frac{\Pi^\lambda}{M\hbar^2\omega^2}. \tag{9.2.26}$$

Note that τ_M is the Maxwell relaxation time in phonons. Thus we obtain

$$\Sigma_R^\lambda(0, \boldsymbol{q}) \cong -i\frac{1}{\omega_{\lambda \boldsymbol{q}}} \frac{A^\lambda}{2\beta\tau_M}. \tag{9.2.27}$$

Now we investigate the self-energy part due to the phonon scattering process, the first term of Eq. (9.2.11) or Eq. (9.2.12). We denote it as $\Sigma_{\text{ph}}(q)$:

$$\Sigma_{\text{ph}}^\lambda(q) = -\frac{|W_3|^2}{\hbar^2}\frac{1}{\omega_{\lambda \boldsymbol{q}}}\phi^\lambda(q), \tag{9.2.28}$$

We first calculate $\phi^\lambda(q)$ under the assumption: all $e^i_{\lambda \boldsymbol{q}} \cong 1$. Then we obtain

$$\phi^\lambda(q) = \frac{1}{\beta\hbar}\sum_{q_1}\frac{1}{\omega_{\lambda \boldsymbol{q}_1}\omega_{\lambda \boldsymbol{q}_1-\boldsymbol{q}}}\mathcal{D}^\lambda(q_1)[\mathcal{D}^\lambda(q-q_1)+\mathcal{D}^\lambda(q_1-q)]$$
$$\overset{\mathrm{i}\nu_n\to q_0+\mathrm{i}\delta}{\longrightarrow}\sum_{q_1}\frac{1}{\omega_{\lambda \boldsymbol{q}_1}\omega_{\lambda \boldsymbol{q}_1-\boldsymbol{q}}}\left[-\frac{1+f(\omega_{\lambda \boldsymbol{q}_1})+f(\omega_{\lambda \boldsymbol{q}-\boldsymbol{q}_1})}{q_0+\mathrm{i}\delta-\omega_{\lambda \boldsymbol{q}_1}-\omega_{\lambda \boldsymbol{q}-\boldsymbol{q}_1}}\right.$$
$$\left.+\frac{f(\omega_{\lambda \boldsymbol{q}_1})-f(\omega_{\lambda \boldsymbol{q}-\boldsymbol{q}_1})}{q_0+\mathrm{i}\delta-\omega_{\lambda \boldsymbol{q}_1}+\omega_{\lambda \boldsymbol{q}-\boldsymbol{q}_1}}\right]. \tag{9.2.29}$$

We are concerned with the imaginary part of the self-energy part, the imaginary part $\phi^\lambda_2(q)$ of $\phi^\lambda(q)$, and low frequencies $q_0 \sim 0$. Since ϕ_2 comes from the second term $\propto \delta(\omega_{\lambda \boldsymbol{q}_1}-\omega_{\lambda \boldsymbol{q}-\boldsymbol{q}_1})(f(\omega_{\lambda \boldsymbol{q}_1})-f(\omega_{\lambda \boldsymbol{q}-\boldsymbol{q}_1}))=0$, we can put

$$\Sigma^\lambda_{\mathrm{ph}} \cong -\mathrm{i}\frac{|W_3|^2}{\hbar^2}\frac{1}{\omega_{\lambda \boldsymbol{q}}}\phi^\lambda_2(q) \cong 0 \tag{9.2.30}$$

Thus we define

$$\Sigma^\lambda(q) \equiv -\mathrm{i}\frac{1}{2\omega_{\lambda \boldsymbol{q}}\tau_\lambda}, \tag{9.2.31}$$

where

$$\frac{1}{\tau_\lambda} \cong \frac{A^\lambda}{2\beta\tau_M}. \tag{9.2.32}$$

Hereafter in this section we do not consider the phonon scattering process.

9.2.4 The Approximate Basic Equation of the Vertex Corrected Bubble Diagrams for Phonon Density Fluctuations

Next we investigate Eq. (9.2.6). First we approximate the square bracket in Eq. (9.2.6) as:

$$1+\sum_{q_1}U^\lambda_R(q,q_1;k)\varphi^\lambda(q_1,q_1-k)$$
$$\cong 1+\frac{\widetilde{U}}{(M\beta\hbar^2\omega)^2}\sum_{q_1}\frac{\Pi^\lambda(\boldsymbol{q},\boldsymbol{q}_1)}{\omega_{\lambda \boldsymbol{q}}\omega_{\lambda \boldsymbol{q}_1}}\delta_{\nu_n\nu_{n_1}}\varphi^\lambda(q_1,q_1-k)$$
$$\cong 1+\frac{1}{\beta^2}\left(\frac{A^\lambda}{2\tau_M}\right)^2\frac{1}{N}\sum_{q_1}\frac{1}{\omega_{\lambda \boldsymbol{q}}\omega_{\lambda \boldsymbol{q}_1}}\delta_{\nu_n\nu_{n_1}}\varphi^\lambda(q_1,q_1-k). \tag{9.2.33}$$

Then we obtain the approximate basic equation of the vertex corrected bubble diagrams for phonon density fluctuations:

$$\left\{-\mathrm{i}\kappa_n + \omega_{\lambda \boldsymbol{q}} - \omega_{\lambda \boldsymbol{q}-k} + \Delta\Sigma^{\lambda}(q,k)\right\} \varphi^{\lambda}(q, q-k)$$
$$\cong \Delta\mathcal{D}^{\lambda}(q,k) + \frac{1}{\beta^2}\left(\frac{A^{\lambda}}{2\tau_M}\right)^2 \frac{1}{\omega_{\lambda \boldsymbol{q}}}\Delta\mathcal{D}^{\lambda}(q,k)\frac{1}{N}$$
$$\times \sum_{q_1} \delta_{\nu_n \nu_{n_1}} \frac{1}{\omega_{\lambda \boldsymbol{q}_1}} \varphi^{\lambda}(q_1, q_1 - k). \tag{9.2.34}$$

Hereafter we confine ourselves to the long wavelength regions.

Before we solve the approximate basic equation, we make some approximations for the terms necessary for its calculation:

1. We make the approximation of the difference of the self-energy part in Eq. (9.2.31) as

$$\Delta\Sigma^{\lambda}(q,k) \cong -\mathrm{i}\frac{1}{\omega_{\lambda \boldsymbol{q}}\tau_{\lambda}}, \tag{9.2.35}$$

2. The difference of phonon Green's function is given by

$$\sum_q \Delta\mathcal{D}^{\lambda}(q,k)$$
$$= \sum_{i\nu_n, \boldsymbol{q}} \left\{ \frac{1}{\mathrm{i}\nu_n - \omega_{\lambda \boldsymbol{q}} + \frac{\mathrm{i}}{2\omega_{\lambda \boldsymbol{q}}\tau_{\lambda}}} - \frac{1}{\mathrm{i}(\nu_n - \kappa_n) - \omega_{\lambda \boldsymbol{q}-\boldsymbol{k}} - \frac{\mathrm{i}}{2\omega_{\lambda \boldsymbol{q}-\boldsymbol{k}}\tau_{\lambda}}} \right\}$$
$$= \sum_{\boldsymbol{q}} \beta\hbar \left\{ -f(\omega_{\lambda \boldsymbol{q}} - \frac{\mathrm{i}}{2\omega_{\lambda \boldsymbol{q}}\tau_{\lambda}}) + f(\omega_{\lambda \boldsymbol{q}-\boldsymbol{k}} + \frac{\mathrm{i}}{2\omega_{\lambda \boldsymbol{q}-\boldsymbol{k}}\tau_{\lambda}}) \right\}$$
$$\cong \beta\hbar \sum_{\boldsymbol{q}} \left\{ (\beta\hbar\frac{\mathrm{i}}{2\omega_{\lambda \boldsymbol{q}}\tau_{\lambda}}) - \beta\hbar(\boldsymbol{k}\cdot\nabla\omega_{\lambda \boldsymbol{q}} + \frac{\mathrm{i}}{2\omega_{\lambda \boldsymbol{q}}\tau_{\lambda}}) \right\} \left.\frac{\partial f(x)}{\partial x}\right|_{x=\beta\hbar\omega_{\lambda \boldsymbol{q}}}$$
$$= (\beta\hbar)^2 \sum_{\boldsymbol{q}} \left\{ -\boldsymbol{k}\cdot\nabla\omega_{\lambda \boldsymbol{q}} + \frac{\mathrm{i}}{\omega_{\lambda \boldsymbol{q}}\tau_{\lambda}}) \right\} \left.\frac{\partial f(x)}{\partial x}\right|_{x=\beta\hbar\omega_{\lambda \boldsymbol{q}}}$$
$$= -(\beta\hbar)^2 \frac{ib_{\lambda}}{\tau_{\lambda}} B_{\lambda}, \tag{9.2.36}$$

where in the last equation, we have replaced $1/\omega_{\lambda \boldsymbol{q}}$ by the average value:

$$b_{\lambda} = \frac{1}{N}\sum_{\boldsymbol{q}} \frac{1}{\omega_{\lambda \boldsymbol{q}}}, \tag{9.2.37}$$

and

$$B_{\lambda} \equiv -\sum_{\boldsymbol{q}} \left.\frac{\partial f(x)}{\partial x}\right|_{x=\beta\hbar\omega_{\lambda \boldsymbol{q}}}. \tag{9.2.38}$$

3. On the other hand, we obtain

$$\sum_q \boldsymbol{k}\cdot\nabla\omega_{\lambda\boldsymbol{q}}\Delta\mathcal{D}^{\lambda}(q,k)$$
$$= (\beta\hbar)^2\sum_{\boldsymbol{q}}\left\{-(\boldsymbol{k}\cdot\nabla\omega_{\lambda\boldsymbol{q}})^2+\boldsymbol{k}\cdot\nabla\omega_{\lambda\boldsymbol{q}}\frac{ib_\lambda}{\tau_\lambda}\right\}\left.\frac{\partial f(x)}{\partial x}\right|_{x=\beta\hbar\omega_{\lambda\boldsymbol{q}}}$$
$$= -(\beta\hbar)^2\frac{k^2}{3}\sum_{\boldsymbol{q}}(\nabla\omega_{\lambda\boldsymbol{q}})^2\left.\frac{\partial f(x)}{\partial x}\right|_{x=\beta\hbar\omega_{\lambda\boldsymbol{q}}}$$
$$= (\beta\hbar)^2\frac{k^2c_\lambda^2}{3}B_\lambda, \tag{9.2.39}$$

where in deriving the last equation, we have replaced $\nabla\omega_{\lambda\boldsymbol{q}}$ by the average velocity of phonon with the mode λ, c_λ:

$$c_\lambda = \frac{1}{N}\sum_q \frac{\partial\omega_{\lambda\boldsymbol{q}}}{\partial q}. \tag{9.2.40}$$

We use Eq. (9.2.39) in solving the dynamical equation of phonon density fluctuations.

9.2.5 *The Calculation of the Vertex Corrected Bubble Diagrams for Phonon Density Fluctuations*

To obtain the vertex corrected bubble diagrams for phonon density fluctuations, we solve Eq. (9.2.34), which are derived approximately from the exact equation (9.2.16). In solving Eq. (9.2.34), we follow the similar way to solve the correlation functions for the intra-band density fluctuations and only consider the terms:

$$\sum_q \varphi^{\lambda}(q,q-k) \equiv \varphi^{\lambda}, \tag{9.2.41}$$

$$\sum_q \boldsymbol{k}\cdot\nabla\omega_{\lambda\boldsymbol{q}}\varphi^{\lambda}(q,q-k) \equiv \boldsymbol{k}\cdot\boldsymbol{\Phi}^{\lambda}. \tag{9.2.42}$$

At the first step, we sum Eq. (9.2.34) over q to get an equation with Eqs. (9.2.41) and (9.2.42). At the next step, we need another equation with the first step equation to obtain a closed set of the equation with Eqs. (9.2.41) and (9.2.42). Therefore we multiply $\boldsymbol{k}\cdot\nabla\omega_{\lambda\boldsymbol{q}}$ on both the sides of Eq. (9.2.34), and then sum the multiplied equation over q.

At the first step, we remember the role of the WT relation (9.2.19) in the procedure deriving Eq. (9.2.20) from the exact Eq. (9.2.16). Therefore, instead of summing both the sides of the approximate Eq. (9.2.34) over q, we sum the exact result (9.2.20) over q. Analytically continuing $i\kappa_n \to k_0 + i\delta$ and using Eq. (9.2.36), we obtain

$$-k_0\varphi^{\lambda} + \boldsymbol{k}\cdot\Phi^{\lambda} == -\mathrm{i}(\beta\hbar)^2\frac{b_\lambda}{\tau_\lambda}B_\lambda. \tag{9.2.43}$$

At the second step, multiplying $\boldsymbol{k}\cdot\nabla\omega_{\lambda\boldsymbol{q}}$ on both the sides of Eq. (9.2.34) with Eq. (9.2.35) and summing over q, we obtain

$$\sum_q \boldsymbol{k}\cdot\nabla\omega_{\lambda\boldsymbol{q}}\left\{-\mathrm{i}\kappa_n+\boldsymbol{k}\cdot\nabla\omega_{\lambda\boldsymbol{q}}-\frac{\mathrm{i}}{\omega_{\lambda\boldsymbol{q}}\tau_\lambda}\right\}\varphi^\lambda(q,q-k)$$
$$=\sum_q \boldsymbol{k}\cdot\nabla\omega_{\lambda\boldsymbol{q}}\Delta\mathcal{D}^\lambda(q,k)$$
$$+\sum_q \boldsymbol{k}\cdot\nabla\omega_{\lambda\boldsymbol{q}}\frac{1}{\beta^2}\left(\frac{A^\lambda}{2\tau_M}\right)^2\frac{1}{\omega_{\lambda\boldsymbol{q}}}\Delta\mathcal{D}^\lambda(q,k)\frac{1}{N}$$
$$\times\sum_{q_1}\delta_{\nu_n\nu_{n1}}\frac{1}{\omega_{\lambda\boldsymbol{q}_1}}\varphi^\lambda(q_1,q_1-k). \tag{9.2.44}$$

Here we denote every term on the left- and right-hand sides as L1, L2, L3, and R1, R2, in order, respectively. Now, we investigate every term:

$$\mathrm{L1}=-k_0\boldsymbol{k}\cdot\boldsymbol{\Phi}^\lambda. \tag{9.2.45}$$

$$\mathrm{L2}\cong\sum_{\boldsymbol{q}}(\boldsymbol{k}\cdot\nabla\omega_{\lambda\boldsymbol{q}})^2\varphi^\lambda\equiv\frac{k^2c_\lambda^2}{3}\varphi^\lambda, \tag{9.2.46}$$

where we have used $c_\lambda=\frac{1}{N}\sum_{\boldsymbol{q}}\frac{\partial\omega_{\lambda\boldsymbol{q}}}{\partial q}$.Using Eq. (9.2.37), we obtain

$$\mathrm{L3}=-\mathrm{i}\frac{b_\lambda}{\tau_\lambda}\boldsymbol{k}\cdot\boldsymbol{\Phi}^\lambda. \tag{9.2.47}$$

From Eq. (9.2.39), we have

$$\mathrm{R1}\cong(\beta\hbar)^2\frac{k^2c_\lambda^2}{3}B_\lambda. \tag{9.2.48}$$

Now we investigate R2. Considering the approximation procedure of the self-energy part from Eq. (9.2.22) to Eq. (9.2.25); the dominant contribution of $\Delta\mathcal{D}^\lambda(q,k)$, comes from the mass shell and the imaginary part self-energy part, we can approximate the integrand $\propto\boldsymbol{k}\cdot\nabla\omega_{\lambda\boldsymbol{q}}\frac{1}{\omega_{\lambda\boldsymbol{q}}}\Delta\mathcal{D}^\lambda(q,k)\propto\boldsymbol{k}\cdot\nabla\omega_{\lambda\boldsymbol{q}}$. Thus summing over $\boldsymbol{q}$ leads to

$$\mathrm{R2}\cong 0. \tag{9.2.49}$$

Summarizing the above, we obtain

$$-k_0\boldsymbol{k}\cdot\boldsymbol{\Phi}^\lambda+\frac{k^2c_\lambda^2}{3}\varphi^\lambda-\mathrm{i}\frac{b_\lambda}{\tau_\lambda}\boldsymbol{k}\cdot\boldsymbol{\Phi}^\lambda\cong(\beta\hbar)^2\frac{k^2v_\lambda^2}{3}B_\lambda. \tag{9.2.50}$$

Neglecting the right-hand side term in Eq. (9.2.43) because of $\frac{\beta\hbar}{\tau_\lambda}\ll 1$, we obtain

$$\varphi^\lambda\cong-\frac{(\beta\hbar)^2\frac{c_\lambda^2k^2}{3}B_\lambda}{k_0(k_0+\frac{\mathrm{i}}{\tau_{e\lambda}})-\frac{c_\lambda^2k^2}{3}}, \tag{9.2.51}$$

where we have put

$$\frac{1}{\tau_{e\lambda}} \equiv \frac{b_\lambda}{\tau_\lambda}. \tag{9.2.52}$$

From the definition (9.2.4), we obtain the vertex corrected bubble diagrams for phonon density fluctuations:

$$R_\lambda(k) \cong \frac{\beta B_\lambda k^2 c_\lambda^2/3}{k_0(k_0 + \frac{\mathrm{i}}{\tau_{e\lambda}}) - \frac{k^2 c_\lambda^2}{3}}. \tag{9.2.53}$$

Substituting Eq. (9.2.53) into Eq. (9.2.3), we obtain the Green's function for phonon density fluctuations:

$$\mathcal{B}_\lambda(k) = \frac{\beta B_\lambda c_\lambda^2 k^2/3}{k_0(k_0 + \frac{\mathrm{i}}{\tau_{e\lambda}}) - u_\lambda^2 k^2}, \tag{9.2.54}$$

where

$$u_\lambda^2 \equiv \frac{(1 + \beta W_4(k) B_\lambda)}{3} c_\lambda^2. \tag{9.2.55}$$

u_λ explicitly depends on the temperature different from the phonon velocities c_λ.

9.2.6 Phonon Entropy Fluctuation Modes and Phonon Thermal Conductivities

The denominator of Eq. (9.2.54) represents the dynamical equations for phonon density fluctuations. The phonon density fluctuations express the frequency fluctuations which represent the entropy fluctuations. Thus the dynamical equation for phonon density fluctuations represents that for the phonon entropy fluctuations. We denote the phonon entropy fluctuations with a mode λ as $S^\lambda(k)$. Then we have

$$\left\{ k_0 - \frac{k^2 u_\lambda^2}{k_0 + \frac{\mathrm{i}}{\tau_{e\lambda}}} \right\} S^\lambda(k) = 0. \tag{9.2.56}$$

Here we investigate the dynamical equation for phonon entropy fluctuations (9.2.56) in the two limiting cases:

1. $\tau_{e\lambda} k_0 \gg 1$:
   There are three entropy fluctuation modes corresponding to a longitudinal and two transverse phonon modes:

$$\{k_0^2 - \bar{\omega}_{\lambda k}^2\} S^\lambda(k) = 0, \tag{9.2.57}$$

$$\bar{\omega}_{\lambda k}^2 = u_\lambda^2 k^2, \tag{9.2.58}$$

   where $\bar{\omega}_{\lambda k}$ is the frequency of an entropy fluctuation mode with λ and u_λ is its velocity.

2. $\tau_{e\lambda} k_0 \ll 1$:
We get the dissipative modes.

$$\left\{k_0 + \mathrm{i}\tau_{e\lambda} k^2 u_\lambda^2\right\} S^\lambda(k) = 0. \tag{9.2.59}$$

Since these modes relate with the dissipation of the heat (entropy), they are thermal conductivity modes associated with the longitudinal and transverse phonon density fluctuation modes. To see more explicitly, we relate the gradient of the local entropy to the gradient of the local temperature:

$$\nabla S^\lambda(\boldsymbol{r}, t) = \frac{C_\lambda}{T} \nabla T(\boldsymbol{r}, t), \tag{9.2.60}$$

where C_λ is the specific heat of phonons with a mode λ. Here we identify the local entropy with the Fourier transformed entropy fluctuations in Eq. (9.2.60). Considering the relation:

$$(k_0, \boldsymbol{k}) \longleftrightarrow \left(\mathrm{i}\frac{\partial}{\partial t}, -\mathrm{i}\nabla\right),$$

we obtain the Fourier equation for phonon thermal conductivity:

$$\frac{\partial S^\lambda}{\partial t} - \frac{\kappa_\lambda}{T} \nabla^2 T = 0. \tag{9.2.61}$$

The thermal conductivity is given by

$$\kappa_\lambda = C_\lambda \tau_{e\lambda} u_\lambda^2. \tag{9.2.62}$$

The specific heat with the mode λ is given by

$$C_\lambda = \frac{1}{4} k_B \sum_q \frac{(\beta \hbar \omega_{\lambda \boldsymbol{q}})^2}{\sinh^2(\frac{\beta \hbar \omega_{\lambda \boldsymbol{q}}}{2})}. \tag{9.2.63}$$

This is just the ordinary specific heat due to phonons at low temperatures in the form of T^3:

$$C_\lambda \propto T^3, \quad \text{for } k_{\mathrm{B}} T / \hbar\omega \ll 1. \tag{9.2.64}$$

At low temperatures, substituting Eqs. (9.2.32), (9.2.52), and (9.2.55) into Eq. (9.2.62), we obtain

$$\begin{aligned} \kappa_\lambda &= C_\lambda \tau_{e\lambda} u_\lambda^2 \\ &= \frac{1 + \beta W_4(0) B_\lambda}{3 b_\lambda A^\lambda} \beta C_\lambda \tau_M c_\lambda^2 \\ &= \frac{1 + \beta W_4(0) B_\lambda}{3 b_\lambda A^\lambda} \beta C_\lambda \frac{\nu_\lambda}{M \rho}, \end{aligned} \tag{9.2.65}$$

where we have used the definition of viscosities:

$$\frac{\nu_\lambda}{M\rho} \equiv \tau_M c_\lambda^2, \tag{9.2.66}$$

where $\nu_t = \eta$ and $\nu_l = \frac{4}{3}\eta + \zeta$ are viscosity.

The thermal conductivity obeys the VTF law in a similar manner to viscosity:

1. $T_g < T$:
   Since $\frac{1}{\tau_M} \propto J$, $C_\lambda \sim$ constant, we obtain $\nu_\lambda \propto J^{-1}$,

$$\kappa_\lambda \propto C_\lambda J^{-1}, \quad u_\lambda \sim \text{constant}, \quad \frac{1}{\tau_{e\lambda}} \propto J, \tag{9.2.67}$$

   It should be noted that C_λ, u_λ $\sim$ constant compared with the drastic change of J near the glass transition.
2. $T < T_g$:
   Since $\frac{1}{\tau_M} \propto \Omega$, $c_\lambda =$ constant, we obtain $\nu_\lambda \propto \Omega^{-1}$,

$$\kappa_\lambda \propto C_\lambda \Omega^{-1}, \quad u_\lambda \sim \text{constant}, \quad \frac{1}{\tau_{e\lambda}} \propto \Omega. \tag{9.2.68}$$

These results are summarized in Table 7.1.

9.3 The Effective Interaction for Sound Fields

In the previous section, we calculated the phonon entropy fluctuations due to phonon density fluctuations, which relate with inter-band density fluctuations. The dynamical equations of phonon density fluctuations yield the phonon entropy fluctuation modes at high frequencies and the phonon thermal conductivity at low frequencies. There are also the sound entropy fluctuations due to sound density fluctuations, which relate with intra-band density fluctuations. The dynamical equation of sound density fluctuations also yields sound entropy modes at high frequencies and sound thermal conductivity at low frequencies. In order to investigate the dynamical equation of sound density fluctuations, we first quantize the intra-density fluctuations by using sound free fields. Next we introduce an effective nonlinear interaction Hamiltonian of sound modes. Contrary to the phonon nonlinear interaction Hamiltonian, there is no requirement for the sound nonlinear interaction Hamiltonian. Therefore we introduce the sound nonlinear interaction Hamiltonian in a perturbative manner.

9.3.1 Sound Free Fields

Since sound frequency depends on temperature, the quantization procedure explained in Chapter 3 is very difficult. Therefore we take the same quantization procedure of sound as that of phonons in Section 9.1. We start with the Green's function of sound:

$$F_{\mathbf{q}}(\tau - \tau') = -\frac{1}{\hbar N}\langle T_\tau \rho_{\mathbf{q}}(\tau)\rho_{\mathbf{q}}^\dagger(\tau')\rangle_c$$

$$= \frac{1}{\beta\hbar}\sum_{i\nu_n} e^{-\mathrm{i}\nu_n(\tau-\tau')}F(q), \tag{7.2.2}$$

and

$$\begin{aligned} F(q) &= \frac{\beta(\omega_{\boldsymbol{q}})^2}{q_0^2-\omega_{s\boldsymbol{q}}^2} \\ &= \frac{\beta\sigma_q}{2}\left\{\frac{1}{q_0-\omega_{s\boldsymbol{q}}}-\frac{1}{q_0+\omega_{s\boldsymbol{q}}}\right\}, \end{aligned} \tag{9.3.1}$$

where

$$\sigma_q \equiv \sqrt{S(q)}\omega_q. \tag{9.3.2}$$

In order to develop the dynamical equation of sound density fluctuations transparently, we introduce free sound fields $\alpha_{\boldsymbol{q}}$ satisfying

$$H_0 = \sum_{\boldsymbol{q}} \hbar\omega_{sq}\alpha_{\boldsymbol{q}}^{\dagger}\alpha_{\boldsymbol{q}}, \tag{9.3.3}$$

$$[\alpha_{\boldsymbol{q}},\alpha_{\boldsymbol{q}'}^{\dagger}] = \delta_{\boldsymbol{q},\boldsymbol{q}'}. \tag{9.3.4}$$

We introduce the sound Green's function as

$$\begin{aligned} \mathcal{F}_{\boldsymbol{q}}(\tau-\tau') &= -\langle T_\tau\alpha_{\boldsymbol{q}}(\tau)\alpha_{\boldsymbol{q}}^{\dagger}(\tau')\rangle \\ &= \frac{1}{\beta\hbar}\sum_{\mathrm{i}\nu_n} e^{\mathrm{i}\nu_n(\tau-\tau')}\mathcal{F}(q). \end{aligned} \tag{9.3.5}$$

In a similar manner to free phonon field case, we obtain:

$$\mathcal{F}(q) = \frac{1}{\mathrm{i}\nu_n-\omega_{sq}}, \tag{9.3.6}$$

Therefore we can write Eq. (9.3.1) as

$$\begin{aligned} F(q) &= \frac{\beta\sigma_q}{2}\left\{\frac{1}{q_0-\omega_{s\boldsymbol{q}}}-\frac{1}{q_0+\omega_{s\boldsymbol{q}}}\right\} \\ &= -\frac{\beta\sigma_q}{2}\int_0^{\beta\hbar}\left\langle T_\tau\{\alpha_{\boldsymbol{q}}(\tau)\alpha_{\boldsymbol{q}}^{\dagger}(0)+\alpha_{\boldsymbol{q}}^{\dagger}(\tau)\alpha_{\boldsymbol{q}}(0)\right\rangle. \end{aligned} \tag{9.3.7}$$

Thus, we obtain

$$-\sqrt{\frac{1}{\hbar N}}\rho_{\boldsymbol{q}}(\tau) = \sqrt{\frac{\beta\sigma_q}{2}}\left\{\alpha_{\boldsymbol{q}}(\tau)+\alpha_{\boldsymbol{q}}^{\dagger}(\tau)\right\}, \tag{9.3.8}$$

Before we investigate the Green's function of sound density fluctuations, we construct an effective interaction for free sound fields.

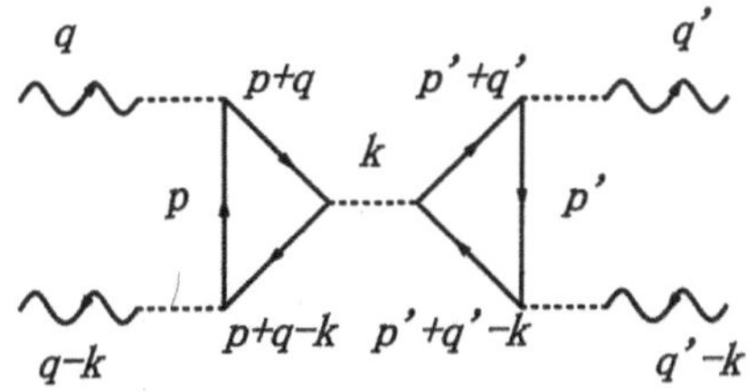

Figure 9.5 The effective interaction of free sound fields is illustrated. The wavy lines represent sound free fields. The dotted lines represent $V_{00}(\boldsymbol{q})$.

9.3.2 The Effective Nonlinear Interaction Hamiltonian for Free Sound Fields

The effective interaction Hamiltonian for phonon fields (8.1.47) is derived by considering the three points; (1) many repetitions of up and down transitions of particles between two levels in a hopping time, (2) the requirement of the WT relation, (3) the possibility of its reduction to the classical nonlinear potential. However, there is no requirement in constructing the effective nonlinear interaction Hamiltonian for sound fields. Therefore we construct the effective nonlinear interaction in a perturbative manner. Sound is the sum of an infinite series of interacting bubbles of intra-band particle-hole pairs. The interactions between sound modes come through the interaction Hamiltonian between the intra-band density fluctuations (intra-band particle-hole pairs), V_{00}. Therefore the simplest nonlinear interaction between sound modes comes through the interaction V_{00} between two inter-bubbles of the intra-band particle-hole pairs as illustrated in Figure 9.5:

$$\begin{aligned}
\text{Figure 9.5 :}&\frac{1}{(\hbar N)^5}\rho_{\boldsymbol{q}}^{\dagger}V_{00}(\boldsymbol{q})\rho_{\boldsymbol{q}}\rho_{\boldsymbol{q}-\boldsymbol{k}}V_{00}(\boldsymbol{q}-\boldsymbol{k})\rho_{\boldsymbol{q}-\boldsymbol{k}}^{\dagger}\rho_{\boldsymbol{k}}^{\dagger}V_{00}(\boldsymbol{k})\rho_{\boldsymbol{k}}\rho_{\boldsymbol{q}'}^{\dagger}\\
&\times V_{00}(\boldsymbol{q}')\rho_{\boldsymbol{q}'}\rho_{\boldsymbol{q}'-\boldsymbol{k}}V_{00}(\boldsymbol{q}'-\boldsymbol{k})\rho_{\boldsymbol{q}'-\boldsymbol{k}}^{\dagger}\\
&=\frac{1}{(\hbar N)^5}\rho_{\boldsymbol{q}}^{\dagger}\rho_{\boldsymbol{q}-\boldsymbol{k}}\Pi^{s}(\boldsymbol{q},\boldsymbol{q}-\boldsymbol{k})\rho_{\boldsymbol{q}}\rho_{\boldsymbol{q}-\boldsymbol{k}}^{\dagger}\rho_{\boldsymbol{k}}^{\dagger}V_{00}(\boldsymbol{k})\rho_{\boldsymbol{k}}\rho_{\boldsymbol{q}'}^{\dagger}\rho_{\boldsymbol{q}'-\boldsymbol{k}}\\
&\times\Pi^{s}(\boldsymbol{q}',\boldsymbol{q}'-\boldsymbol{k})\rho_{\boldsymbol{q}'}\rho_{\boldsymbol{q}'-\boldsymbol{k}}^{\dagger}\\
&\rightarrow\frac{1}{(\hbar N)^4}\rho_{\boldsymbol{q}}^{\dagger}\rho_{\boldsymbol{q}-\boldsymbol{k}}\Pi^{s}(\boldsymbol{q},\boldsymbol{q}-\boldsymbol{k})L(q,k)\\
&\times\frac{1}{\hbar N}V_{00}(\boldsymbol{k})L^{*}(q',k)\Pi^{s}(\boldsymbol{q}',\boldsymbol{q}'-\boldsymbol{k})\rho_{\boldsymbol{q}'}\rho_{\boldsymbol{q}'-\boldsymbol{k}}^{\dagger},
\end{aligned}\tag{9.3.9}$$

where we have put

$$\Pi^{s}(\boldsymbol{q},\boldsymbol{q}-\boldsymbol{k})=V_{00}(\boldsymbol{q})V_{00}(\boldsymbol{q}-\boldsymbol{k}).\tag{9.3.10}$$

The last equation in Eq. (9.3.9) shows the interaction between bubbles. The bubble on the right hand side is written as

$$L(q,k)=\mp\frac{1}{(\beta\hbar)^2N}\sum_{p}G^{0}(\mathrm{i}\omega_n,\boldsymbol{p})G^{0}(\mathrm{i}\omega_n+\mathrm{i}\nu_n,\boldsymbol{p}+\boldsymbol{q})$$

$$\times G^0(\mathrm{i}\omega_n + \mathrm{i}\nu_n - \mathrm{i}\kappa_n, \boldsymbol{p}+\boldsymbol{q}-\boldsymbol{k}). \tag{9.3.11}$$

In calculating Eq. (9.3.11), we put the energy dispersion $\epsilon_{\boldsymbol{p}} \cong \tilde{\omega}_0$. Thus, we have

$$L(q,k) \cong \frac{1}{(\beta\hbar)^2}\frac{1}{2}\frac{\partial^2}{\partial\tilde{\omega}_0^2}\sum_p \frac{1}{\mathrm{i}\omega_n - \tilde{\omega}_0} \cong \frac{N}{2\beta\hbar}\frac{\partial^2}{\partial\tilde{\omega}_0^2} f(\tilde{\omega}_0) \cong \frac{\beta\hbar N}{2}, \tag{9.3.12}$$

where we have considered the Boltzmann distribution. For simplicity, we make an approximation:

$$\Pi^s(\boldsymbol{q}, \boldsymbol{q}-\boldsymbol{k}) \cong \Pi^s = \text{constant}. \tag{9.3.13}$$

Then we can write Eq. (9.3.9) as follows:

$$\begin{aligned}\text{Figure 9.5} &\cong \frac{1}{(\hbar N)^4}\rho_{\boldsymbol{q}}^\dagger \rho_{\boldsymbol{q}-\boldsymbol{k}}\Pi^s\frac{\beta\hbar N}{2}\frac{1}{\hbar N}V_{00}(\boldsymbol{k})\frac{\beta\hbar N}{2}\Pi^s\rho_{\boldsymbol{q}'}\rho_{\boldsymbol{q}'-\boldsymbol{k}}^\dagger \\ &\to \alpha_{\boldsymbol{q}}^\dagger\alpha_{\boldsymbol{q}-\boldsymbol{k}}\frac{\beta^4\sqrt{\sigma_q\sigma_{|\boldsymbol{q}-\boldsymbol{k}|}\sigma_{q'}\sigma_{|\boldsymbol{q}'-\boldsymbol{k}'|}}(\Pi^s)^2}{16\hbar N}V_{00}(\boldsymbol{k})\alpha_{\boldsymbol{q}'}\alpha_{\boldsymbol{q}'-\boldsymbol{k}}^\dagger \\ &\to \alpha_{\boldsymbol{q}}^\dagger\alpha_{\boldsymbol{q}-\boldsymbol{k}}W^s(\boldsymbol{k})\alpha_{\boldsymbol{q}'}\alpha_{\boldsymbol{q}'-\boldsymbol{k}}^\dagger\end{aligned} \tag{9.3.14}$$

where since we consider the long wavelength regime of $\boldsymbol{k}$, we put $\sqrt{\sigma_q\sigma_{|\boldsymbol{q}-\boldsymbol{k}|}\sigma_{q'}\sigma_{|\boldsymbol{q}'-\boldsymbol{k}|}} \cong \sigma_q\sigma_{q'}$and we replace σ_q by the average $\langle\sigma\rangle$ over $\boldsymbol{q}$. Thus we obtain the interaction potential between sound density fluctuations:

$$W^s(\boldsymbol{k}) = \frac{\beta^4\langle\sigma\rangle^2(\Pi^s)^2}{16\hbar N}V_{00}(\boldsymbol{k}). \tag{9.3.15}$$

We use this interaction in calculating the dynamical equation for sound density fluctuations.

9.3.3 The Elementary Random Scattering Magnitude Due to the Random Scatterings

Next we investigate the elementary random scattering processes due to random scatterings: $U^{00}(\boldsymbol{q}) = U_\omega^{00} + U_J^{00}(\boldsymbol{q})$ in a similar manner to the phonon fields as illustrated in Figure 9.3. In this case, $\mu = \nu = 0$. The diagram corresponding to I in Figure 9.3 denoted as I^s is given by

$$\begin{aligned} I^s &= \frac{1}{\hbar}\sqrt{\frac{\beta\sigma_q}{2}}V_{00}(q)V_{00}(q')\sqrt{\frac{\beta\sigma_{q'}}{2}} \\ &\times\left[\mp\frac{1}{(\beta\hbar)^2 N}\sum_p G^0(\mathrm{i}\omega_n, \boldsymbol{p})G^0(\mathrm{i}\omega_n - \mathrm{i}\nu_n, \boldsymbol{p}-\boldsymbol{q})\right. \\ &\qquad \left. G^0(\mathrm{i}\omega_n - \mathrm{i}\nu_n, \boldsymbol{p}-\boldsymbol{q}')\right].\end{aligned} \tag{9.3.16}$$

Using the same approximation as Eq. (9.3.11), we obtain the elementary random scattering magnitude corresponding $U_R^\lambda(q, q'; k)$, which we denote as $U_R^s(q, q'; k)$:

$$U_R^s(q, q'; k) = \frac{\beta^4 \widetilde{U}^{00} (\Pi^s)^2}{4} \sqrt{\sigma_q \sigma_{|\boldsymbol{q}-\boldsymbol{k}|} \sigma_{q'} \sigma_{|\boldsymbol{q}'-\boldsymbol{k}|}} \tag{9.3.17}$$

$$\widetilde{U}^{00} \equiv \frac{1}{N} \sum_{\boldsymbol{q}} U^{00}(\boldsymbol{q}) = \frac{1}{N} \sum_{\boldsymbol{q}} \left[U_\omega^{00} + U_J^{00}(\boldsymbol{q}) \right]. \tag{9.3.18}$$

In a similar manner to the phonon density fluctuations, we approximate

$$U_R^s(q, q'; k) \cong \frac{\beta^4 \widetilde{U}^{00}}{4} (\Pi^s)^2 \sigma_q \sigma_{q'} = \frac{\beta^2 \widetilde{U}^{00}}{4} (\Pi^s)^2 \beta \sigma_q \beta \sigma_{q'}. \tag{9.3.19}$$

We use this elementary scattering magnitude in calculating the scattering processes of sound density fluctuations.

9.4 Sound Density Fluctuations; Sound Entropy Mode and Sound Thermal Conductivity

We have constructed the effective interaction potentials of free sound fields and the elementary random scattering magnitude due to random scatterings. Now we are ready to construct the dynamical equation of sound density fluctuations. We construct it in a similar manner to phonon density fluctuations.

9.4.1 *The Correlation Functions of Sound Density Fluctuations*

We introduce sound density fluctuations as

$$\xi_{\boldsymbol{k}}^\dagger \equiv \sum_{\boldsymbol{q}} \alpha_{\boldsymbol{q}}^\dagger \alpha_{\boldsymbol{q}-\boldsymbol{k}}. \tag{9.4.1}$$

The correlation functions for sound density fluctuations are defined by

$$\mathcal{A}_k(\tau - \tau') \equiv -\frac{1}{\hbar} \langle T_\tau \xi_k(\tau) \xi_k^\dagger(\tau') \rangle_c = \frac{1}{\beta\hbar} \sum_{i\nu_n} e^{i\kappa_n(\tau-\tau')} \mathcal{A}(k). \tag{9.4.2}$$

Using the effective interaction potential, (9.3.15), in the random phase approximation we obtain

$$\mathcal{A}(k) = R^s(k) + R^s(k) W^s(\boldsymbol{k}) \mathcal{A}(k), \tag{9.4.3}$$

$$R^s(k) = -\frac{1}{\beta\hbar^2} \sum_q \varphi^s(q, q-k), \tag{9.4.4}$$

where $R^s(q)$ is the vertex corrected bubble diagrams for sound density fluctuations.

9.4.2 The Vertex Corrected Bubble Diagrams of Sound Density Fluctuations

In the ladder approximation we obtain

$$\varphi^s(q,q-k) = \mathcal{F}(q)\mathcal{F}(q-k) \left[1 + \sum_{q_1} U_R^s(q,q_1;k)\delta_{\nu_n\nu_{n_1}}\varphi^s(q_1,q_1-k)\right] \tag{9.4.5}$$

$$\sum_{q_1} U_R^s(q,q_1;k)\delta_{\nu_n\nu_{n_1}}\varphi^s(q_1,q_1-k)] \cong \frac{\beta^2\widetilde{U}^{00}(\Pi^s)^2}{4}\beta\sigma_q \sum_{q_1}\beta\sigma_{q_1}\delta_{\nu_n\nu_{n_1}}\varphi^s(q_1,q_1-k). \tag{9.4.6}$$

The sound Green's function and the self-energy part are given by

$$\mathcal{F}(q) = \frac{1}{\mathrm{i}\nu_n - \omega_{sq} - \Sigma^s(q)}, \tag{9.4.7}$$

$$\Sigma^s(q) = \sum_{q_1} U_R^s(q,q_1;0)\delta_{\nu_n\nu_{n_1}}\mathcal{F}(q_1) \tag{9.4.8}$$

$$\sum_{q_1} U_R^s(q,q_1;0)\delta_{\nu_n\nu_{n_1}}\mathcal{F}(q_1) \cong \frac{\beta^2\widetilde{U}^{00}(\Pi^s)^2}{4}\beta\sigma_q\sum_{q_1}\beta\sigma_{q_1}\delta_{\nu_n\nu_{n1}}\mathcal{F}(q_1), \tag{9.4.9}$$

where $\Sigma^s(q)$ is the self-energy part of $\mathcal{F}(q)$ in the corresponding ladder approximation satisfying the WT relation.

Multiplying both the sides of Eq. (9.4.5) by $\mathcal{F}(q-k)^{-1} - \mathcal{F}(q)^{-1}$, we obtain

$$\left\{-\mathrm{i}\kappa_n + \omega_{s\mathbf{q}} - \omega_{s\mathbf{q}-k} + \Delta\Sigma^s(q,k)\right\}\varphi^s(q,q-k) = \Delta\mathcal{F}(q,k) + \Delta\mathcal{F}(q,k)\sum_{q_1} U_R^s(q,q_1:k)\delta_{\nu_n\nu_{n_1}}\varphi^s(q_1,q_1-k), \tag{9.4.10}$$

where

$$\Delta\Sigma^s(q,k) \equiv \Sigma^s(q) - \Sigma^s(q-k), \tag{9.4.11}$$

$$\Delta\mathcal{F}(q,k) \equiv \mathcal{F}(q) - \mathcal{F}(q-k). \tag{9.4.12}$$

Using the WT relation:

$$\Delta\Sigma^s(q,k) = \sum_{q_1} U_R^s(q,q_1;k)\delta_{\nu_n\nu_{n_1}}\Delta\mathcal{F}(q_1,k), \tag{9.4.13}$$

we obtain the following exact equation:

$$\sum_q\{-\mathrm{i}\kappa_n + \omega_{s\mathbf{q}} - \omega_{s\mathbf{q}-k}\}\varphi^s(q,q-k) = \sum_q \mathcal{F}(q,k). \tag{9.4.14}$$

We use this equation in calculating the dynamical equation of sound density fluctuations.

9.4.3 *The Approximation Procedure of the Self-Energy Part of Free Sound Fields*

First we investigate the self-energy part $\Sigma^s(q)$:

$$\Sigma^s(q) = \frac{\beta^2 \widetilde{U}^{00}(\Pi^s)^2}{4} \beta\sigma_q \sum_{q_1} \beta\sigma_{q_1} \delta_{\nu_n \nu_{n_1}} \mathcal{F}(q_1). \tag{9.4.15}$$

We follow the same approximation procedure to that of phonon density fluctuations in Eqs. (9.2.22)–(9.2.24):

1. We put $\mathcal{F}(q_1)$ as

$$\mathcal{F}(q_1) \cong -\frac{1}{\Sigma^s(q_0, \boldsymbol{q}_1)} \cong -\frac{1}{\Sigma^s(0, \boldsymbol{q}_1)}. \tag{9.4.16}$$

2. We further put

$$\Sigma^s(0, \boldsymbol{q}) \cong -\mathrm{i}\beta\sigma_q\alpha. \quad \alpha : \text{constant}. \tag{9.4.17}$$

Then we obtain

$$\begin{aligned}\Sigma^s(0, \boldsymbol{q}) &= -\mathrm{i}\beta\sigma_q\alpha \\ &= -\frac{\beta^4 \widetilde{U}^{00}(\Pi^s)^2}{4} \sigma_q \sum_{\boldsymbol{q}_1} \sigma_{q_1} \frac{1}{-\mathrm{i}\beta\sigma_{q_1}\alpha} \\ &= -\beta\sigma_q \left\{ \frac{\beta(\Pi^s)}{4\tau_0} \right\}^2 \frac{1}{-\mathrm{i}\alpha},\end{aligned} \tag{9.4.18}$$

where we have used the relaxations of sound.

$$\frac{1}{2\tau_0} = \sqrt{N\widetilde{U}^{00}}. \tag{9.4.19}$$

Thus we obtain

$$\Sigma^s(q) \equiv -\mathrm{i}\frac{\beta\sigma_q}{2\tau_s}, \tag{9.4.20}$$

where

$$\frac{1}{2\tau_s} = \frac{\beta\Pi^s}{4\tau_0}. \tag{9.4.21}$$

$\frac{\tau^s}{\beta\sigma_q}$ is the relaxation time of the sound density fluctuation mode with $\boldsymbol{q}$.

9.4.4 The Approximation Procedure of the Vertex Corrected Bubble Diagrams of Sound Density Fluctuations

Using Eq. (9.4.20), we rewrite the basic Eq. (9.4.10) as

$$\left\{-\mathrm{i}\kappa_n + \omega_{s\boldsymbol{q}} - \omega_{s\boldsymbol{q}-k} - \mathrm{i}\frac{\beta\sigma_q}{2\tau_s}\right\}\varphi^s(q, q-k)$$
$$= \Delta\mathcal{F}(q,k) + \Delta\mathcal{F}(q,k)\frac{1}{N}\left\{\frac{\beta\Pi^s}{4\tau_0}\right\}^2 \beta\sigma_q \sum_{q_1} \beta\sigma_{q_1}\delta_{\nu_n\nu_{n_1}}\varphi^s(q_1, q_1-k). \tag{9.4.22}$$

Following the approximation procedure of phonon density fluctuations, we consider only two terms:

$$\sum_q \varphi^s(q, q-k) \equiv \varphi^s, \tag{9.4.23}$$

$$\sum_q \boldsymbol{k}\cdot\nabla\omega_{sq}\varphi^s(q, q-k) \equiv \boldsymbol{k}\cdot\boldsymbol{\Phi}^s. \tag{9.4.24}$$

At the first step, we sum Eq. (3.4.22) over q_1, but since we have the exact Eq. (9.4.14), instead of summing Eq. (3.4.22) we employ Eq. (9.4.14). Making the same approximation as Eq. (9.2.36):

$$\sum_q \Delta\mathcal{F}(q,k) \cong -\mathrm{i}\frac{(\beta\hbar)^2}{\tau_{es}}E, \tag{9.4.25}$$

where we have used

$$\frac{1}{\tau_{es}} \cong \frac{\beta\langle\sigma\rangle}{\tau_s}, \quad \langle\sigma\rangle = \frac{1}{N}\sum_{\boldsymbol{q}}\sigma_{\boldsymbol{q}}, \tag{9.4.26}$$

$$E \equiv -\sum_{\boldsymbol{q}} \left.\frac{\partial f}{\partial x}\right|_{x=\beta\hbar\omega_{sq}}, \tag{9.4.27}$$

we obtain the first equation:

$$-k_0\varphi^s + \boldsymbol{k}\cdot\boldsymbol{\Phi}^s = -\mathrm{i}\frac{(\beta\hbar)^2}{\tau_{es}}E. \tag{9.4.28}$$

τ_{es} is the average of the relaxation time of sound entropy fluctuation mode.

At the next step, multiplying the term $\boldsymbol{k}\cdot\nabla\omega_{sq}$ on both the sides of Eq. (9.4.22), making the same approximation as Eq. (9.2.38)

$$\sum_q \boldsymbol{k}\cdot\nabla\omega_{sq}\Delta\mathcal{F}(q,k) \cong (\beta\hbar)^2\frac{k^2c_T^2}{3}E, \tag{9.4.29}$$

where c_T is the isothermal sound velocity:

$$c_T = \frac{1}{N}\sum_{q}\frac{\partial\omega_{sq}}{\partial q}, \tag{9.4.30}$$

and considering the same approximation procedure as the case of phonon density fluctuations we have the second equation:

$$-k_0\boldsymbol{k}\cdot\boldsymbol{\Phi}^s + \frac{k^2c_T^2}{3}\varphi^s - \frac{\mathrm{i}}{\tau_{es}}\boldsymbol{k}\cdot\boldsymbol{\Phi}^s \cong (\beta\hbar)^2\frac{k^2c_T^2}{3}E, \tag{9.4.31}$$

Combining Eqs. (9.4.28) and (9.4.31) and neglecting the right-hand side term, we obtain

$$\varphi^s = -\frac{(\beta\hbar)^2Ec_T^2k^2/3}{k_0(k_0+\frac{\mathrm{i}}{\tau_{es}}) - \frac{c_T^2k^2}{3}}. \tag{9.4.32}$$

From Eqs. (9.4.4) and (9.4.23), we obtain

$$R^s(k) = \frac{\beta Ec_T^2k^2/3}{k_0(k_0+\frac{\mathrm{i}}{\tau_{es}}) - \frac{c_T^2k^2}{3}}. \tag{9.4.33}$$

Thus we obtain

$$\mathcal{A}(k) = \frac{\beta Ec_T^2k^2}{k_0(k_0+\frac{\mathrm{i}}{\tau_{es}}) - u_s^2k^2}, \tag{9.4.34}$$

$$u_s^2 = \frac{1+\beta EW^s(k)}{3}c_T^2. \tag{9.4.35}$$

$\mathcal{A}(k)$ is the Green's function for sound density fluctuations.

9.4.5 Sound Entropy Fluctuation Mode and Sound Thermal Conductivity

The denominator of Eq. (9.4.34) yields the dynamical equation of sound entropy fluctuations:

$$\left\{k_0 - \frac{u_s^2k^2}{k_0+\frac{\mathrm{i}}{\tau_{es}}}\right\}S^s(k) = 0. \tag{9.4.36}$$

Here we investigate Eq. (9.4.36) in the two limiting cases.

1. $\tau_{es}k_0 \gg 1$:
   We obtain the equation of the sound entropy fluctuation mode:

$$\left\{k_0^2 - \bar{\omega}_{sk}^2\right\}S^s(k) = 0, \tag{9.4.37}$$

$$\bar{\omega}_{sk}^2 = u_s^2k^2. \tag{9.4.38}$$

   where $\bar{\omega}_{sk}$ is its dispersion curve. u_s is the velocity of the mode.

2. $\tau_{es}k_0 \ll 1$:
We obtain a dissipative mode:

$$\{k_0 + i\tau_{es}u_s^2k^2\}S^s(k) = 0. \tag{9.4.39}$$

Since this mode relates with the dissipation of the heat (entropy), this is thermal conductivity mode associated with sound entropy fluctuation mode. To see more explicitly, we relate the gradient of the local entropy to that of the local temperature:

$$\nabla S^s(\mathbf{r}, t) = \frac{C_{\text{intra}}}{T}\nabla T(\mathbf{r}, t), \tag{9.4.40}$$

where C_{intra} is the specific heat of the intra-band density fluctuations $C_{\text{intra}} = C_K + C_s$ in Chapter 6:

$$C_{\text{intra}} \cong \begin{cases} \frac{N_0k_BT}{2}\frac{T_0}{(T-T_0)^2} + k_B(N - N_0) & \text{for } T_g < T, \\ k_B(N - N_0) & \text{for } T < T_g. \end{cases} \tag{9.4.41}$$

Here we identify the local entropy with the Fourier transformed entropy fluctuation of Eq. (9.4.39). Thus we obtain the Fourier equation for the thermal conductivity:

$$\frac{\partial S^s(\mathbf{r}, t)}{\partial t} + \frac{\kappa_s}{T}\nabla^2 T = 0, \tag{9.4.42}$$

$$\kappa_s = C_{\text{intra}}\tau_{es}u_s^2 \tag{9.4.43}$$

where $\kappa_s$is the corresponding thermal conductivity.

Using the diffusivity:

$$D = \tau_0 c_T^2, \tag{9.4.44}$$

we can rewrite the sound thermal conductivity as

$$\begin{aligned} \kappa_s &= C_{\text{intra}}\tau_{es}u_s^2 \\ &= C_{\text{intra}}\frac{\tau_0}{\beta^2\langle\sigma\rangle\Pi^s}\frac{1 + \beta EW^s(0)}{3}c_T^2 \\ &= \frac{C_{\text{intra}}(1 + \beta EW^s(0))}{3\beta^2\langle\sigma\rangle\Pi^s}D. \end{aligned} \tag{9.4.45}$$

Next we investigate the VTF law. Since $\langle\sigma\rangle \propto J$, the VTF law governs as

1. $T_g < T$:
Since $\frac{1}{\tau_0} \propto J$, $c_T \propto J$, we obtain $D \propto J$,

$$\kappa_s \propto C_{\text{intra}}, \quad u_s \propto J, \quad \frac{1}{\tau_{es}} \propto J^2. \tag{9.4.46}$$

2. $T < T_g$:
Since $\frac{1}{\tau_0} \propto \Omega$, $c_T \propto J$, we obtain $D \propto J^2$

$$\kappa_s \propto C_{\text{intra}} J, \quad u_s \propto J, \quad \frac{1}{\tau_{es}} \propto J. \tag{9.4.47}$$

These results are summarized in Table 7.1.

Sound mode and diffusion disappear below the glass transition, so do corresponding entropy fluctuation mode and thermal conductivity. Since $\langle \sigma \rangle \propto J$, the powers of J in κ_s and τ_{es} are different from those of D and τ_0. On the other hand, phonon modes and viscosity exist in both phases, because these modes are the NG modes associated with the condensed states. The corresponding entropy fluctuation modes and thermal conductivity exist, too. Phonon modes and the corresponding entropy fluctuation modes do not obey the VTF law. Since the factor b_λ in Eq. (9.2.37) is almost constant, phonon modes and viscosity, and corresponding entropy fluctuation modes and thermal conductivity obey the same powers of J, respectively.

For $T_g < T$, the Stokes law, $D\eta \cong$ constant satisfies, but for $T < T_g$, the Stokes law breaks. This is because the relaxation time of sound and that of phonons obey the same power of J. This fact also reflects $C_t \eta / \kappa_t \cong$ constant.

9.5 The Anomaly of Thermal Conductivity and Specific Heat in Low Temperature Glasses

Zeller and Pohl's experimental results [3] showed that in the low temperature glasses, the specific heat C and the thermal conductivity κ behave as

$$C \propto T, \tag{9.5.1}$$
$$\kappa \propto T^2. \tag{9.5.2}$$

This is well known as the anomaly in low temperature glasses [3–6,2]. The specific heat anomaly has been explained in Section 7.4. Here we explain the low temperature thermal conductivity anomaly.

9.5.1 *The Specific Heat Anomaly*

As discussed in Chapter 7, below the glass transition $T < T_g$, the remaining entropy of sound yields the specific heat $C_s \cong k_B(N - N_0)$ in Eq. (7.4.33) around the glass transition temperature. At the lower temperature, the bandwidth vanishes as J. But from the uncertainty principle, there remains the bandwidth of the lower band $\sim \hbar/\tau_0$. Intra-band elementary excitations between random eigenfrequencies yield the specific heat $C_s \cong k_B(N - N_0)$ at higher temperatures, but at lower temperature $T \sim \hbar/(k_B \tau_0) \sim 1K$ yield

$$C_s \propto T \quad \text{for} \quad T \sim \hbar/(k_B \tau_0) \sim 1K. \tag{9.5.3}$$

This clarifies the specific heat anomaly in the low temperature glasses (9.5.1). However, since velocities of the particles, sound and sound density fluctuation mode vanish as $v_p, c_T, u_s \propto J$, thermal conductivity due to sound also vanishes:

$$\kappa_s \sim 0, \quad \text{below} \quad T_g. \tag{9.5.4}$$

Therefore we need to explain the thermal conductivity anomaly in the low temperature glasses.

9.5.2 The Thermal Conductivity Anomaly

On the other hand, at low temperatures $k_B T/\hbar\omega \ll 1$ phonons yield the specific heat,

$$C_\lambda \propto T^3, \quad \text{for} \quad k_B T \ll \hbar\omega. \tag{9.5.5}$$

Below T_g, we have

$$\tau_M \propto \Omega \quad \text{for} \quad T < T_g \tag{9.5.6}$$

and carrying out the same calculation as C_λ putting $\alpha = aq$ and $\omega_{\lambda \boldsymbol{q}} = c_\lambda q$, we have

$$\begin{aligned} B_\lambda &= -\sum_{\boldsymbol{q}} \frac{\partial f}{\partial x}\bigg|_{x=\beta\hbar\omega_{\lambda\boldsymbol{q}}} = -\frac{1}{4}\sum_{\boldsymbol{q}} \frac{1}{\sinh^2 \beta\hbar\omega_{\lambda\boldsymbol{q}}} \\ &= -\frac{N}{8\pi^2}\int_0^\infty \alpha^2 d\alpha \frac{1}{\sinh^2 \frac{\beta\hbar c_\lambda \alpha}{\alpha}} \propto -T^3 \end{aligned} \tag{9.5.7}$$

then Eq. (9.2.65) yields

$$\begin{aligned} \kappa_\lambda &\cong \frac{1 + \beta W_4(0) B_\lambda}{3 b_\lambda A^\lambda} \tau_M c_\lambda^2 \beta C_\lambda \\ &\propto T^2, \quad \text{for} \quad k_B T \ll \hbar\omega, \end{aligned} \tag{9.5.8}$$

where we have considered that $\beta W_4(0) B_\lambda (\propto T^2) \ll 1$ at the low temperature. Eq. (9.5.8) clarifies the thermal conductivity anomaly in the low temperature glasses (9.5.2). The specific heat anomaly originates from the intra-band density fluctuations, while the thermal conductivity anomaly from the phonon density fluctuations.

References

[1] T. Kitamura, Physica A 272 (1999) 330.
[2] T. Kitamura, Phys. Rep. 383 (2003) 1.
[3] P.C. Zeller, R.O. Pohl, Phys. Rev. B 4 (1971) 2029.
[4] P.W. Anderson, B.I. Halperlin, C.M Verma, Phil. Mag. 25 (1972) 1.
[5] A.J. Siever, S. Takeno, Phys. Rev. B 39 (1989) 3374.
[6] T. Kitamura, S. Takeno, Phys. Lett. A 266 (2000) 198.

10 The Liquid-Glass Transition in Multi-Component Liquids

So far, we have established the unified theory of the liquid-glass transition in one-component liquids in the framework of the two band model [1,2]. Phonons are the NG bosons, inter-band density fluctuation modes essential in glasses, while sound is intra-band density fluctuation mode essential in liquids. The sum of an infinite series of interacting bubbles of intra-band and inter-band particle-hole pairs constitutes respective modes, but the excess bubbles not constituting modes yield respective fluctuation entropies, which are compensated with the respective dissipations. In a supercooling process, enormous bubbles are created in the liquid and the fluctuation entropy mixing with the sound entropy constitutes the Kauzmann entropy, which plays an essential role in the liquid-glass transition. The hopping amplitude is the probability corresponding to the Kauzmann entropy per particle from the Einstein's relation on probability and entropy, and manifests the VTF law on the relaxation and dissipative processes.

Constituent particles in a one-component liquid can be particles or molecules. But in case of molecules, a molecule consists of subordinate particles. The subordinate particles not only form the structure of the molecule, but they also have the individual freedoms. In order to take into account the structure and freedoms, we must start with the subordinate particles so that we must develop a theory in a multi-component liquid [3–6].

A glass is normally formed by a supercooling process. However, a glass may also be formed by chemical vitrification, the process of which involves progressive polymerization of the constituent molecules via the formation of irreversible chemical bonds. In both physical and chemical vitrification, the slowing down of the dynamics due to the reduction of the configuration entropy, corresponding to the Kauzmann entropy, is commonly observed [7,8]. The polymerization of liquids is a process that modifies molecular motions in a reacting system, introducing configuration restriction which progressively reduce the accessible phase space, leading to a glassy behavior. This leads to a strong decrease of the configuration entropy [9].

In extending the theory to a multi-component liquid [6], we must clarify the structure of the liquid [10,11]. The structure is expressed by the pair distribution functions which depend on distances and the Euler angles. But it is very difficult to include realistic pair distribution functions at the first stage, so we makes the mean field approximation (the random phase approximation). This approximation is simple, but we can establish a unified theory in any component liquid. Once a unified theory has been established,

Liquid Glass Transition. DOI: http://dx.doi.org/10.1016/B978-0-12-407177-3.00010-6

the unified theory plays a crucial role in arranging experimental results and developing a more precise theory.

In multi-component liquids, there exists the electric polarization. The electric polarization relates to the p-state of particles. An excited state of a particle corresponds to a p-state, while a ground state to an s-state. Thus we need not take into account the polarization effect in intra-band density fluctuations, which annihilate and create the same states. However we must take into account the effect in inter-band density fluctuations, which annihilate an s-state and create a p-state, and vice versa. We take into account the electric polarization in the rigid ion approximation, the Ewald method [12–14].

Following the formulation of random systems in the previous chapters, at the first stage we construct the configurationally averaged model Hamiltonian. Then we construct the interaction Hamiltonian for intra-band and inter-band density fluctuations, which lead to the sound and the phonon modes. At the second stage, we introduce the random potentials due to random frequencies and random hopping, which yields random scattering processes, the relaxation, and dissipative processes. We can easily extend sound and diffusivity in one-component systems to multi-component systems, but in case of phonons and viscosities we must take into account the polarization due to the electric polarization in the extension. We also show the boson peaks. At the final stage, we investigate the liquid-glass transition in multi-component liquids. We show the Kauzmann entropy crisis and the VTF law on the relaxation and transport coefficients.

10.1 The Model Hamiltonian and the Random Scattering Hamiltonian

We extend the unified theory of the liquid-glass transition in one-component systems to that in multi-component systems. We first construct the model Hamiltonian of randomly distributed multi-component particles; this model Hamiltonian states that each particle stays in a random harmonic potential making up and down transitions for a while, then hops to a neighboring vacancy. Starting with the model Hamiltonian, we construct the two Hamiltonians corresponding to the two stages: (1) the configurationally averaged model Hamiltonian to develop the dynamical and thermodynamical properties at the first stage. (2) The random scattering interaction Hamiltonian to develop the relaxation and dissipative processes at the second stage.

10.1.1 The Configurationally Averaged Model Hamiltonian

We start with the following Hamiltonian [1,3–6,15,16]:

$$\begin{aligned} H = &\sum_{\alpha} \int d^3x \psi_{\alpha}^{\dagger}(x) \hbar\epsilon_{\alpha}^{0}(-\mathrm{i}\nabla)\psi_{\alpha}(x) \\ &+ \frac{1}{2}\sum_{\alpha\beta} \int d^3x d^3y n_{\alpha}(x) V_{\alpha\beta}(\boldsymbol{x}-\boldsymbol{y}) n_{\beta}(y), \end{aligned} \tag{10.1.1}$$

where $\hbar\epsilon_\alpha^0(-\mathrm{i}\nabla)$ is the energy operator of a free α-particle and $n_\alpha(x) = \psi_\alpha^\dagger(x)\psi_\alpha(x)$. We write an annihilation operator $\psi_\alpha(x)$ by using the localized operators, which are rewritten by the extended operators.

$$\psi_\alpha(x) = \sum_{m\mu} \tilde{w}_{\alpha\mu}(\boldsymbol{x} - \boldsymbol{R}_{\alpha m}) b_{\alpha m\mu}, \tag{10.1.2}$$

$$b_{\alpha m\mu} = \frac{1}{\sqrt{N_\alpha}} \sum_{\boldsymbol{p}} e^{\mathrm{i}\boldsymbol{p}\cdot\boldsymbol{R}_{\alpha m}} a_{\alpha\mu\boldsymbol{p}}, \tag{10.1.3}$$

where N_α is a number of α-particles.

The Heisenberg equation for $b_{\alpha m\mu}$ is given by

$$-\hbar\frac{\partial}{\partial\tau} b_{\alpha m\mu} = \sum_{n\nu} \int d^3x \tilde{w}_{\alpha m\mu}(\boldsymbol{x} - \boldsymbol{R}_{\alpha m})[\hbar\epsilon_\alpha^0(-\mathrm{i}\nabla) + \Phi_{\boldsymbol{R}_{\alpha n}}(\boldsymbol{x})]\tilde{w}_{\alpha n\nu}(\boldsymbol{x} - \boldsymbol{R}_{\alpha n}) b_{\alpha n\nu}, \tag{10.1.4}$$

$$\Phi_{\boldsymbol{R}_{\alpha n}}(\boldsymbol{x}) = \sum_\beta \int_{\{\boldsymbol{R}_{\alpha n}\}} d^3y V_{\alpha\beta}(\boldsymbol{x} - \boldsymbol{y}) n_\beta(y), \tag{10.1.5}$$

where $\{\boldsymbol{R}_{\alpha n}\}$ means summing the surrounding sites of the particle at $\boldsymbol{R}_{\alpha n}$ in $n_\beta(y)$. The term $\Phi_{\boldsymbol{R}_{\alpha n}}$ means that a potential is determined by the surrounding particles of the site $\boldsymbol{R}_{\alpha n}$. Thus we obtain the Schrödinger equation for a well-localized α-particle:

$$\left\{\hbar\epsilon_\alpha^0(-\mathrm{i}\nabla) + \Phi_{\boldsymbol{R}_{\alpha n}}(\boldsymbol{x})\right\} \tilde{w}_{\alpha n\nu}(\boldsymbol{x} - \boldsymbol{R}_{\alpha n}) = \hbar\tilde{\omega}_{\alpha n\nu}\tilde{w}_{\alpha n\nu}(\boldsymbol{x} - \boldsymbol{R}_{\alpha n}), \tag{10.1.6}$$

where $\tilde{\omega}_{\alpha n\mu}$ and $\tilde{w}_{\alpha n\mu}$ are the eigenfrequency and the eigenfunction of the μth state at the site $\boldsymbol{R}_{\alpha n}$ in a similar manner to one-component liquids in the harmonic approximation. We consider the following relations:

$$w_{n\alpha 1}(x_i) = -2\zeta_{n\alpha}\nabla_i w_{n\alpha 0}(x_i), \qquad w_{n\alpha 0}(x_i) = 2\zeta_{n\alpha}\nabla_i w_{n\alpha 1}(x_i), \tag{10.1.7}$$

$$\zeta_{n\alpha} = \sqrt{\frac{\hbar}{2M_\alpha\omega_{n\alpha}}}, \tag{10.1.8}$$

where $\zeta_{n\alpha}$ is the mean width of the zero point motion, M_α is the mass of the α-particle and $\omega_{n\alpha}$ is the harmonic frequency; $\hbar\tilde{\omega}_{n\alpha 0} = \frac{3}{2}\hbar\omega_{n\alpha}$ and $\hbar\tilde{\omega}_{n\alpha i} = \frac{5}{2}\hbar\omega_{n\alpha}$. At this stage, we can take the following model Hamiltonian in terms of localized operators:

$$H = \sum_{\alpha m\mu} \hbar\tilde{\omega}_{\alpha m\mu} b_{\alpha m\mu}^\dagger b_{\alpha m\mu} + \sum_{\alpha m\mu n\nu} \hbar J_{\alpha m\mu\alpha n\nu} b_{\alpha m\mu}^\dagger b_{\alpha n\nu}, \tag{10.1.9}$$

where the hopping matrix is given by

$$\hbar J_{\alpha m\mu\alpha n\nu} = \int d^3x \tilde{w}_{\alpha m\mu}(\boldsymbol{x} - \boldsymbol{R}_{\alpha m})\Phi_{\boldsymbol{R}_{\alpha n}}(\boldsymbol{x})\tilde{w}_{\alpha n\nu}(\boldsymbol{x} - \boldsymbol{R}_{\alpha n}). \tag{10.1.10}$$

Note that the interaction Hamiltonian is nonlinear through the term $n_\beta(y)$ in $\Phi_{\boldsymbol{R}_{\alpha n}}$ and the potential $\Phi_{\boldsymbol{R}_{\alpha n}}$ permits a particle to hop only to a vacancy.

Now we start with the new unperturbed Hamiltonian which is the configuration average of the model Hamiltonian (10.1.9):

$$H_0 = \langle H \rangle_c = \sum_{\alpha\mu m} \hbar\tilde{\omega}_{\alpha\mu} b^{\dagger}_{\alpha m\mu} b_{\alpha m\mu} + \sum_{\alpha\mu mn} \hbar J_{\alpha\mu}(\boldsymbol{R}_{\alpha m} - \boldsymbol{R}_{\alpha n}) b^{\dagger}_{\alpha m\mu} b_{\alpha n\mu}, \tag{10.1.11}$$

where $J_{\alpha\mu}$ is the hopping magnitude. $\langle\rangle_c$ means the configuration average. The configurationally averaged values do not include the indices of the positions. Here for simplicity we limit ourselves to the hopping between like particles and the hopping to those between the same levels. Using Eq. (10.1.3), we obtain

$$H_0 = \sum_{\alpha\mu\boldsymbol{p}} \hbar\epsilon_{\alpha\mu\boldsymbol{p}} a^{\dagger}_{\alpha\mu\boldsymbol{p}} a_{\alpha\mu\boldsymbol{p}}, \tag{10.1.12}$$

$$\epsilon_{\alpha\mu\boldsymbol{p}} = \tilde{\omega}_{\alpha\mu} + J_{\alpha\mu}(\boldsymbol{p}), \tag{10.1.13}$$

$$J_{\alpha\mu}(\boldsymbol{p}) \equiv \frac{1}{N_\alpha} \sum_m e^{i\boldsymbol{p}\cdot\boldsymbol{R}_{\alpha m}} J_{\alpha\mu}(\boldsymbol{R}_{\alpha m}), \tag{10.1.14}$$

which represent the two band model: the dispersion curves of the two bands are described by Eq. (10.1.13) with (10.1.14). We also obtain the number density operator:

$$\begin{aligned} n_\alpha(x) &= \frac{1}{N_\alpha} \sum_{m\boldsymbol{q}} e^{i\boldsymbol{q}\cdot\boldsymbol{R}_{\alpha m}} \Big[\sum_{\mu} \tilde{w}^2_{\alpha\mu}(\boldsymbol{x} - \boldsymbol{R}_{\alpha m}) \rho_{\alpha d\mu\boldsymbol{q}} \\ &\quad + \sum_i \tilde{w}_{\alpha 0}(\boldsymbol{x} - \boldsymbol{R}_{\alpha m}) \tilde{w}_{\alpha i}(\boldsymbol{x} - \boldsymbol{R}_{\alpha m}) \rho_{\alpha i\boldsymbol{q}} + \cdots \Big] \\ &= \frac{1}{N_\alpha} \sum_{m\boldsymbol{q}} e^{i\boldsymbol{q}\cdot\boldsymbol{R}_{\alpha m}} \Big[\sum_{\mu} \tilde{w}^2_{\alpha\mu}(\boldsymbol{x} - \boldsymbol{R}_{\alpha m}) \rho_{\alpha d\mu\boldsymbol{q}} \\ &\quad - \sum_i \zeta_\alpha \nabla_i \tilde{w}^2_{\alpha 0}(\boldsymbol{x} - \boldsymbol{R}_{\alpha m}) \rho_{\alpha i\boldsymbol{q}} + \cdots \Big], \end{aligned} \tag{10.1.15}$$

where

$$\rho_{\alpha d\mu\boldsymbol{q}} = \sum_{\boldsymbol{p}} a^{\dagger}_{\alpha\mu\boldsymbol{p}} a_{\alpha\mu\boldsymbol{p}+\boldsymbol{q}}, \tag{10.1.16}$$

$$\rho_{\alpha i\boldsymbol{q}} = \sum_{\boldsymbol{p}} (a^{\dagger}_{\alpha i\boldsymbol{p}} a_{\alpha 0\boldsymbol{p}+\boldsymbol{q}} + a^{\dagger}_{\alpha 0\boldsymbol{p}} a_{\alpha i\boldsymbol{p}+\boldsymbol{q}}). \tag{10.1.17}$$

$\rho_{\alpha d\mu\boldsymbol{q}}$ and $\rho_{\alpha i\boldsymbol{q}}$ are intra-band and inter-band density fluctuation operators, respectively. Substituting Eq. (10.1.15) into the interaction Hamiltonian in Eq. (10.1.1) we obtain the interaction Hamiltonian for intra-band and inter-band density fluctuations:

$$H_I = \frac{1}{2}\sum_{\alpha\beta\mu\nu\boldsymbol{q}}{}' \frac{1}{\sqrt{N_\alpha N_\beta}} V^d_{\alpha\mu\beta\nu}(\boldsymbol{q})\rho^\dagger_{\alpha d\mu\boldsymbol{q}}\rho_{\beta d\nu\boldsymbol{q}} + \frac{1}{2}\sum_{\alpha\beta ij\boldsymbol{q}} \frac{1}{\sqrt{N_\alpha N_\beta}} V^{od}_{\alpha i\beta j}(\boldsymbol{q})\rho^\dagger_{\alpha i\boldsymbol{q}}\rho_{\beta j\boldsymbol{q}}, \tag{10.1.18}$$

$$V^d_{\alpha\mu\beta\nu}(\boldsymbol{q}) \equiv \sqrt{N_\alpha N_\beta}\, V_{\alpha\mu\mu,\beta\nu\nu}, \tag{10.1.19}$$

$$V^{od}_{\alpha i\beta j}(\boldsymbol{q}) \equiv \sqrt{N_\alpha N_\beta}\, V_{\alpha 0i,\beta 0j}, \tag{10.1.20}$$

$$V_{\alpha\mu\mu',\beta\nu\nu'}(\boldsymbol{q}) = \frac{1}{N_\alpha N_\beta}\sum_{m\neq n} e^{-\mathrm{i}\boldsymbol{q}\cdot(\boldsymbol{R}_{\alpha m}-\boldsymbol{R}_{\beta n})}\int d^3x d^3y\, \tilde{w}_{\alpha\mu}(\boldsymbol{x}-\boldsymbol{R}_{\alpha m}) \times \tilde{w}_{\alpha\mu'}(\boldsymbol{x}-\boldsymbol{R}_{\alpha m})V_{\alpha\beta}(\boldsymbol{x}-\boldsymbol{y})\tilde{w}_{\beta\nu}(\boldsymbol{y}-\boldsymbol{R}_{\beta n})\tilde{w}_{\beta\nu'}(\boldsymbol{y}-\boldsymbol{R}_{\beta n}), \tag{10.1.21}$$

where the prime on Σ means H_I excludes loop diagrams, which are taken into account in the potential $\Phi_{\boldsymbol{R}_{\alpha n}}$. In the classical limit [1,17]:

$$\tilde{w}^2_{\alpha 0}(\boldsymbol{x}-\boldsymbol{R}_{\alpha m}) \to \delta(\boldsymbol{x}-\boldsymbol{R}_{\alpha m})$$

and

$$n(x) \to \frac{1}{(2\pi)^3}\int d^3q[e^{\mathrm{i}\boldsymbol{q}\cdot\boldsymbol{x}}\rho_\alpha - \sum_{i\alpha}\zeta_\alpha\nabla_i e^{\mathrm{i}\boldsymbol{q}\cdot\boldsymbol{x}}\rho_{\alpha i}]$$

in Eq. (10.1.15), the first term in Eq. (10.1.18) tends to the interaction potentials between density fluctuations and the second term after partial integration of ∇_i tends to the harmonic potentials between the displaced positions of particles with the magnitude $\zeta_\alpha\rho_\alpha$ from the assigned positions $\boldsymbol{R}_{\alpha m}$ with the spring constants $\nabla_i\nabla_j V_{\alpha\beta}(\boldsymbol{R}_{\alpha m} - \boldsymbol{R}_{\beta n})$. Thus the first and second terms lead to sound (a density fluctuation wave) and phonons (elastic waves), respectively.

We employ the unperturbed Hamiltonian (10.1.12) and the interaction Hamiltonian (10.1.18) of intra-band density fluctuations $\rho_{\alpha 0\boldsymbol{q}}$ and inter-band density fluctuations $\rho_{\alpha i\boldsymbol{q}}$ to construct the dynamical equations for intra-band and inter-band density fluctuations, which describe the dynamical and thermodynamical properties of sound and phonons. Here since we are concerned with the low temperature $\beta\hbar\omega_\alpha < 1$, we only consider the lower band in the intra-band density fluctuations. Therefore we abbreviate $\rho_{\alpha d0\boldsymbol{q}}$ as $\rho_{\alpha 0\boldsymbol{q}}$. The dynamical equations also describe life time of sound and phonons, when these dispersion curves merge into the continuum of the intra-band and inter-band elementary excitations. However, the dynamical equations originating from Eqs. (10.1.12) and (10.1.18) cannot describe the relaxation and dissipative processes, because they dismiss the randomness of particle distributions. Therefore we must include the effect of the randomness.

10.1.2 The Random Scattering Interaction Hamiltonian

Here we include the effect of the randomness by taking into account the random scattering processes due to the random harmonic frequencies and the random hopping magnitudes. The original model Hamiltonian, Eq. (10.1.9) reflects the randomly distributed particles, while the configurationally averaged unperturbed Hamiltonian, Eqs. (10.1.11) and (10.1.12) reflect the free particle picture. Therefore we take into account the random scattering processes through the random scattering interaction Hamiltonian H_{RI} due to the random harmonic frequencies and the random hopping magnitudes:

$$H_{RI} = \sum_{\alpha m\mu} \hbar(\tilde{\omega}_{\alpha m\mu} - \tilde{\omega}_{\alpha\mu}) b^{\dagger}_{\alpha m\mu} b_{\alpha m\mu} + \sum_{\alpha mn\mu} \hbar\{J_{\alpha m\mu\alpha n\mu} - J_{\alpha\mu}(\boldsymbol{R}_{\alpha m} - \boldsymbol{R}_{\alpha n})\} b^{\dagger}_{\alpha m\mu} b_{\alpha n\mu}. \tag{10.1.22}$$

Note that since the term $J_{\alpha m\mu\alpha n\mu}$ involves operators $n(y)$, H_{RI} is essentially nonlinear. But at low temperatures a particle stays longer at the same site. Thus we make the approximation that the term $J_{\alpha m\mu\alpha n\mu}$ is a random c-number. Using Eq. (10.1.3), we rewrite Eq. (10.1.22) as

$$H_{RI} = \frac{1}{N_\alpha} \sum_{\alpha\mu \boldsymbol{p}\boldsymbol{p}'} \left[\sum_{m} \hbar(\tilde{\omega}_{\alpha\mu m} - \tilde{\omega}_{\alpha\mu}) e^{-\mathrm{i}(\boldsymbol{p}-\boldsymbol{p}')\cdot \boldsymbol{R}_{\alpha m}} + \sum_{mn} \hbar\{J_{\alpha m\mu\alpha n\mu} - J_{\alpha\mu}(\boldsymbol{R}_{\alpha m} - \boldsymbol{R}_{\alpha n})\} e^{-\mathrm{i}\boldsymbol{p}\cdot\boldsymbol{R}_{\alpha m} + \mathrm{i}\boldsymbol{p}'\cdot\boldsymbol{R}_{\alpha n}} \right] a^{\dagger}_{\alpha\mu\boldsymbol{p}} a_{\alpha\mu\boldsymbol{p}'}. \tag{10.1.23}$$

The relaxation and dissipative processes come from the simultaneously scattering of two particles by the same random scattering potentials due to the random harmonic frequencies and the random hopping magnitudes in Eq. (10.1.22) and (10.1.23). The sum of configurationally averaged simultaneously scattering processes by the same random scattering potentials over all sites due to random harmonic frequencies yields the configurationally averaged correlation of random harmonic frequencies, $U_{\omega}^{\alpha\mu\nu}$, and that due to random hopping magnitudes yields a correlation function of random hopping magnitudes, $U_{J}^{\alpha\mu\nu}$ as illustrated in Fig. 7.1,

$$U_{\omega}^{\alpha\mu\nu} = \frac{1}{N_\alpha^2} \sum_{m} \left(\langle \tilde{\omega}_{\alpha\mu m} \tilde{\omega}_{\alpha\nu n} \rangle_c - \tilde{\omega}_{\alpha\mu} \tilde{\omega}_{\alpha\nu} \right), \tag{10.1.24}$$

$$U_{J}^{\alpha\mu\nu}(q) = \frac{1}{N_\alpha} \int d^3R g(R) e^{-\mathrm{i}\boldsymbol{q}\cdot\boldsymbol{R}} \{ \langle J_{\alpha m\mu n\mu} J_{\alpha m\nu n\nu} \rangle_c - J_{\alpha\mu}(\boldsymbol{R}) J_{\alpha\nu}(\boldsymbol{R}) \}|_{\boldsymbol{R}=\boldsymbol{R}_{\alpha m}-\boldsymbol{R}_{\alpha n}}. \tag{10.1.25}$$

These elementary scattering processes $U_{\omega,J}^{\alpha\mu\nu}$ play an essential role in constructing the vertex corrected bubble diagrams of particle-hole pairs. The sum of configurationally averaged simultaneously scattering processes by the same random scattering potentials over all random positions, constrains the dynamical processes, which leads to dissipation. Eqs. (10.1.24), and (10.1.25) yield the relaxation times; the β- and α-relaxation, respectively. The elementary scattering processes, $U^{\alpha\mu\nu}_{\omega,J}$ correspond to the simultaneously scattering processes of two electrons due to the same random impurities in electric conductivity.

10.2 Sound and Diffusivity

Sound and diffusivity in multi-component systems also originate from the collective modes of intra-band density fluctuations. The correlation functions of intra-band density fluctuations are the sum of an infinite series of interacting bubbles of intra-band particle-hole pairs, where the bubbles are formed by the same kind of particles. We take into account the random scattering processes only to the vertex corrected bubbles, because the random scattering processes between different bubbles are negligible. We first develop the general theory of Green's functions of intra-band density fluctuations, then we apply the general theory to the two-component systems and show the concrete sound and diffusivity.

10.2.1 The General Theory of Green's Functions of Intra-Band Density Fluctuations

We first investigate Green's functions for intra-band density fluctuations at finite temperatures [1,6,15,16]. In the low temperature regime, we confine ourselves to the lower band.

$$F_{\alpha\beta\boldsymbol{q}}(\tau_1-\tau_2)\equiv-\frac{1}{\hbar\sqrt{N_\alpha N_\beta}}\left\langle T_\tau\rho_{\alpha 0\boldsymbol{q}}(\tau_1)\rho^\dagger_{\beta 0\boldsymbol{q}}(\tau_2)\right\rangle_c$$
$$\equiv\frac{1}{\beta\hbar}\sum_{\mathrm{i}\nu_n}e^{-\mathrm{i}\nu_n(\tau_1-\tau_2)}F_{\alpha\beta}(q), \tag{10.2.1}$$

where we should not confuse the temperature $\beta = 1/k_BT$ and the indices β; the temperature β appears as an independent parameter with the inverse energy dimension, while the indices β stand for particles. In $F_{\alpha\beta}(q)$, $q = (\mathrm{i}\nu_n, \mathbf{q})$, where $\mathrm{i}\nu_n$ is a Matsubara temperature frequency. In the random phase approximation, we obtain

$$F_{\alpha\beta}(q) = P_\alpha(q)\delta_{\alpha\beta} + \sum_\gamma P_\alpha(q)V_{\alpha\gamma}(\boldsymbol{q})F_{\gamma\beta}(q), \tag{10.2.2}$$

$$P_\alpha(q) \cong -\frac{\beta f'\omega^2_{\alpha\boldsymbol{q}}}{q_0(q_0+\frac{\mathrm{i}}{\tau_{\alpha 0}})-\omega^2_{\alpha\boldsymbol{q}}}, \tag{10.2.3}$$

$$V_{\alpha\beta}(\boldsymbol{q}) = \left\langle \frac{1}{\sqrt{N_\alpha N_\beta}}\sum_{m\neq n}e^{-\mathrm{i}\boldsymbol{q}\cdot(\boldsymbol{R}_m-\boldsymbol{R}_n)}V_{\alpha\beta}(\boldsymbol{R}_m-\boldsymbol{R}_n)\right\rangle_c$$
$$=\sqrt{\frac{N_\alpha}{N_\beta}}\int d^3R g_{\alpha\beta}(R)e^{\mathrm{i}\boldsymbol{q}\cdot\boldsymbol{R}}V_{\alpha 0\beta 0}(\boldsymbol{R}), \tag{10.2.4}$$

$$V_{\alpha0\beta0}(\boldsymbol{R}_{\alpha m}-\boldsymbol{R}_{\beta n})=\int d^3x d^3y \tilde{w}^2_{\alpha m}(\boldsymbol{x}-\boldsymbol{R}_{\alpha m}) \times V_{\alpha\beta}(\boldsymbol{x}-\boldsymbol{y})\tilde{w}^2_{\beta n}(\boldsymbol{y}-\boldsymbol{R}_{\beta n}), \quad (10.2.5)$$

$$\frac{1}{2\tau_{\alpha0}}=\sqrt{N_\alpha(U^{\alpha00}_{\omega}+U^{\alpha00}_{J})} \quad (10.2.6)$$

$$(\omega_{\alpha q})^2 \cong \frac{(v_{\alpha \mathrm{p}})^2}{3}q^2, \qquad v_{\alpha \mathrm{p}}=\frac{1}{N}\sum_p \frac{\partial \epsilon_{\alpha0p}}{\partial p}, \quad (10.2.7)$$

where we have analytically continued $i\nu_n \to q_0$ in Eq. (10.2.3). $\tau_{\alpha0}$ and $v_{\alpha\mathrm{p}}$ are the relaxation time and the mean velocity of the α-particle, respectively. The derivation of Eq. (10.2.3) is presented in Chapter 7. Hereafter, we omit the indices d in $V^d_{\alpha0\beta0}$.

Now we apply the general theory of Green's function of intra-band density fluctuations to that of a two-component liquid.

10.2.2 The Green's Function in a Two-Component Liquid

Restricting Eq. (10.2.2) to a two-component liquid composed of A and B particles, we obtain the dynamical equation:

$$\begin{pmatrix} 1-P_{\mathrm{A}}V_{\mathrm{AA}} & P_{\mathrm{A}}V_{\mathrm{AB}} \\ P_{\mathrm{B}}V_{\mathrm{BA}} & 1-P_{\mathrm{B}}V_{\mathrm{BB}} \end{pmatrix}\begin{pmatrix} F_{\mathrm{AA}} & F_{\mathrm{AB}} \\ F_{\mathrm{BA}} & F_{\mathrm{BB}} \end{pmatrix}=\begin{pmatrix} P_{\mathrm{A}} & 0 \\ 0 & P_{\mathrm{B}} \end{pmatrix} \quad (10.2.8)$$

with the solution:

$$\begin{pmatrix} F_{\mathrm{AA}} & F_{\mathrm{AB}} \\ F_{\mathrm{BA}} & F_{\mathrm{BB}} \end{pmatrix}=\frac{1}{\det||}\begin{pmatrix} (1-P_{\mathrm{B}}V_{\mathrm{BB}})P_{\mathrm{A}} & P_{\mathrm{A}}V_{\mathrm{AB}}P_{\mathrm{B}} \\ P_{\mathrm{B}}V_{\mathrm{BA}}P_{\mathrm{A}} & (1-P_{\mathrm{A}}V_{\mathrm{AA}})P_{\mathrm{B}} \end{pmatrix}, \quad (10.2.9)$$

$$\det||=(1-P_{\mathrm{A}}V_{\mathrm{AA}})(1-P_{\mathrm{B}}V_{\mathrm{BB}})-P_{\mathrm{A}}V_{\mathrm{AB}}P_{\mathrm{B}}V_{\mathrm{BA}}. \quad (10.2.10)$$

We can obtain the modes by solving the secular equation of Eq. (10.2.8):

$$\{1-P_{\mathrm{A}}(q)V_{\mathrm{AA}}(q)\}\{1-P_{\mathrm{B}}(q)V_{\mathrm{BB}}(q)\}-P_{\mathrm{A}}(q)P_{\mathrm{B}}(q)V_{\mathrm{AB}}(q)V_{\mathrm{BA}}(q)=0. \quad (10.2.11)$$

Before we solve Eq. (10.2.11), we investigate the secular equation for the sound in a one-component liquid.

10.2.3 The Secular Equation in an α-Component Liquid

The secular equation in an α-component liquid is given by

$$0=1-P_\alpha(q)V_{\alpha\alpha}(q) = \frac{q_0(q_0+\frac{\mathrm{i}}{\tau_{\alpha0}})-(1-\beta f' V_{\alpha\alpha}(q))\omega^2_{\alpha \boldsymbol{q}}}{q_0(q_0+\frac{\mathrm{i}}{\tau_{\alpha0}})-\omega^2_{\alpha \boldsymbol{q}}}. \quad (10.2.12)$$

We investigate the numerator of Eq. (10.2.12) = 0 in the two limiting cases:

1. $q_0\tau_{\alpha 0} \gg 1$
Eq. (10.2.12) leads to the sound frequency for α-particles:

$$q_0^2 = \omega_{\alpha S}^2(q) = (1 - \beta f' V_{\alpha\alpha}(q))\omega_{\alpha q}^2 = \frac{\omega_{\alpha q}^2}{S_\alpha(q)} = c_{\alpha T}^2 q^2, \tag{10.2.13}$$

where $S_\alpha(q)$ and $c_{\alpha T}$ are the static structure factor and the isothermal sound velocity of the α-particle:

$$S_\alpha(q) \equiv (1 - \beta f' V_{\alpha\alpha}(q))^{-1}, \tag{10.2.14}$$

$$c_{\alpha T} = \frac{v_{\alpha p}}{\sqrt{3 S_\alpha(0)}}. \tag{10.2.15}$$

In the classical limit [1,17] for $f' = -1 : v_p^2/2M = 3k_B T/2$, $S(0) = M\rho\chi_T$, the velocity of sound tends to the classical sound velocity: $c_{\alpha T}^2 = (\frac{\partial p}{\partial \rho})_T/M$, where ρ, χ_T, p are density, isothermal compressibility, and pressure, respectively. Next we investigate the sound disappearing. In a first approximation, if we consider a first peak for a pair distribution function, we can put

$$g_{\alpha\alpha}(R) = \rho_\alpha \delta(R - a_{\alpha\alpha}), \tag{10.2.16}$$

where $a_{\alpha\alpha}$ is the mean distance between the α-particles. Then we obtain

$$V_{\alpha\alpha}(q) = 4\pi a_{\alpha\alpha}^2 \rho_\alpha V_{\alpha\alpha}(a_{\alpha\alpha}) j_0(a_{\alpha\alpha} q), \quad j_0(x) \equiv \frac{\sin x}{x}, \tag{10.2.17}$$

where $j_0(x)$ is a zeroth spherical Bessel function, which is an oscillatory decreasing function of x, starting with $j_0(0) = 1$. Thus $V_{\alpha\alpha}$ has the minimum negative value at a reciprocal particle distance $\widetilde{K}_{\alpha\alpha} \cong 3\pi/2a_{\alpha\alpha}$. Since

$$\lim_{q_0 \to 0} P_\alpha(q) \cong \beta f', \tag{10.2.18}$$

the sound disappearing in the α-liquid occurs at the temperature $T_{\alpha 0}$ and the reciprocal particle distance $\widetilde{K}_{\alpha\alpha}$ in Eq. (10.2.12):

$$1 - \beta_{\alpha 0} f' V_{\alpha\alpha}(\widetilde{K}_{\alpha\alpha}) = 0. \tag{10.2.19}$$

2. $q_0\tau_{\alpha 0} \ll 1$
Eq. (10.2.12) leads to the diffusion mode

$$q_0 = -\mathrm{i}\tau_{\alpha 0} c_{\alpha T}^2 q^2, \tag{10.2.20}$$

where the diffusion coefficient is given by

$$D_\alpha = \tau_{\alpha 0} c_{\alpha T}^2. \tag{10.2.21}$$

Now we go back to the secular equation Eq. (10.2.11) for the two-component liquid.

10.2.4 The Secular Equation in a Two-Component Liquid

Substituting Eq. (10.2.3) into the secular equation Eq. (10.2.11) in the two-component liquid, we investigate the two limiting cases. Hereafter for simplicity we put $f' = -1$.

1. $q_0 \tau_{\alpha 0} \gg 1$
   We obtain the sound modes:

$$
\begin{aligned}
q_0^2 &= \omega_{s\pm}^2(q) \\
&= \frac{1}{2}\Big[\omega_{As}^2(q) + \omega_{Bs}^2(q) \\
&\quad \pm \sqrt{(\omega_{As}^2(q) - \omega_{Bs}^2(q))^2 + 4\beta^2 \omega_{Aq}^2 \omega_{Bq}^2 V_{AB}(q) V_{BA}(q)}\Big] \\
&= c_{T\pm}^2 q^2,
\end{aligned} \tag{10.2.22}
$$

where $c_{T\pm}$ are the sound velocities:

$$
c_{T\pm} = \sqrt{\frac{1}{2}\left\{c_{AT}^2 + c_{BT}^2 \pm \sqrt{(c_{AT}^2 - c_{BT}^2)^2 + \frac{4}{9}\beta^2 v_{Ap}^2 v_{Bp}^2 V_{AB}(q) V_{BA}(q)}\right\}}. \tag{10.2.23}
$$

If the term $V_{AB} V_{BA}$ dominates, ω_s- and c_T- disappear. Next we investigate the sound disappearing. In a first approximation, if we consider a peak for the pair distribution function, we can put

$$
g_{\alpha\beta}(R) = \rho_\beta \delta(R - a_{\alpha\beta}), \tag{10.2.24}
$$

where $a_{\alpha\beta}$ is the mean distance between the α and β particles. Then we obtain

$$
V_{\alpha\beta}(q) = 4\pi a_{\alpha\beta}^2 \rho_\beta V_{\alpha\beta}(a_{\alpha\beta}) j_0(a_{\alpha\beta} q), \tag{10.2.25}
$$

where $j_0(a_{\alpha\beta} q)$ has the following properties:

a. $j_0(x)$ is an oscillatory decreasing function of x, starting with $j_0(0) = 1$.
b. The larger, the value of $a_{\alpha\beta}$ takes, the more frequently $j_0(a_{\alpha\beta} q)$ oscillates and the more rapidly it decreases with respect to q.

In order to obtain the sound disappearing in a two-component liquid, we investigate the secular equation Eq. (10.2.11) with (10.2.18). We put

$$
\begin{aligned}
\alpha &\equiv a_{AB} q, \\
f(\alpha) &= \{1 + \beta V_{AA}(q)\}\{1 + \beta V_{BB}(q)\} - \beta^2 V_{BB}(q) V_{BA}(q).
\end{aligned}
$$

If we suppose $a_{AB} \cong a_{BA}$, the second term, $V_{AB}(q) V_{BA}(q) \propto j_0^2(\alpha)$, has the second maximum at $\alpha = 3\pi/2$. In the long wavelength regime, since the sound exists, we expect $f(\alpha) > 0$ for $0 < \alpha < \pi$. Thus the sound disappearing occurs at a reciprocal particle distance $\widetilde{K} \cong 3\pi/(2a_{AB})$ at the sound disappearing temperature T_0 satisfying:

$$
\{1 + \beta_0 V_{AA}(\widetilde{K})\}\{1 + \beta_0 V_{BB}(\widetilde{K})\} - \beta_0^2 V_{AB}(\widetilde{K}) V_{BA}(\widetilde{K}) = 0. \tag{10.2.26}
$$

T_0 is given by

$$T_0 = \frac{1}{2k_\mathrm{B}}\Big[- V_{\mathrm{AA}}(\widetilde{K}) - V_{\mathrm{BB}}(\widetilde{K}) + \sqrt{\{V_{\mathrm{AA}}(\widetilde{K}) - V_{\mathrm{BB}}(\widetilde{K})\}^2 + 4V_{\mathrm{AB}}(\widetilde{K})V_{\mathrm{BA}}(\widetilde{K})}\Big]. \qquad (10.2.27)$$

2. $q_0\tau_{\alpha 0} \ll 1$

We obtain the diffusion modes:

$$\begin{aligned} q_0 &= \mathrm{i}D^{\pm}q^2 \\ &= \frac{\mathrm{i}}{2}\Big[\tau_{\mathrm{A}0}\omega_{\mathrm{A}s}^2(q) + \tau_{\mathrm{B}0}\omega_{\mathrm{B}s}^2(q) \\ &\quad \pm\sqrt{\{\tau_{\mathrm{A}0}\omega_{\mathrm{A}s}^2(q) - \tau_{\mathrm{B}0}\omega_{\mathrm{B}s}^2(q)\}^2 + 4\beta^2\tau_{\mathrm{A}0}\omega_{\mathrm{A}q}^2\tau_{\mathrm{B}0}\omega_{\mathrm{B}q}^2 V_{\mathrm{AB}}(q)V_{\mathrm{BA}}(q)}\Big]q^2, \end{aligned} \qquad (10.2.28)$$

where $D^{\pm}$ are diffusion coefficients:

$$\begin{aligned} D^{\pm} &= \frac{1}{2}\Big[\tau_{\mathrm{A}0}v_{\mathrm{As}}^2 + \tau_{\mathrm{B}0}v_{\mathrm{Bs}}^2 \\ &\quad \pm\sqrt{(\tau_{\mathrm{A}0}v_{\mathrm{As}}^2 - \tau_{\mathrm{B}0}v_{\mathrm{Bs}}^2)^2 + \frac{4}{9}\beta^2\tau_{\mathrm{A}0}v_{\mathrm{Ap}}^2\tau_{\mathrm{B}0}v_{\mathrm{Bp}}^2 V_{\mathrm{AB}}(q)V_{\mathrm{BA}}(q)}\Big] \\ &= \frac{1}{2}\Big[D_\mathrm{A} + D_\mathrm{B} \pm\sqrt{(D_\mathrm{A} - D_\mathrm{B})^2 + \frac{4}{9}\beta^2\tau_{\mathrm{A}0}v_{\mathrm{Ap}}^2\tau_{\mathrm{B}0}v_{\mathrm{Bp}}^2 V_{\mathrm{AB}}(q)V_{\mathrm{BA}}(q)}\Big]. \end{aligned} \qquad (10.2.29)$$

If the term $V_{\mathrm{AB}}V_{\mathrm{BA}}$ dominates, D^{-} disappears.

10.3 Phonons, Boson Peaks, and Viscosities in Multi-Component Liquids

Phonons and viscosities in a multi-component glass are also the collective modes of inter-band density fluctuations. The correlation functions of inter-band density fluctuations are the sum of an infinite series of interacting bubbles of inter-band particle-hole pairs, where we consider only the bubbles from the same kind of particles. We take into account the random scattering processes only to the vertex corrected bubbles, because the random scattering processes between different bubbles are negligible. The secular equations of the Green's functions of inter-band density fluctuations associated with the gap equations, which are derived by the WT relations, lead to the dynamical equations of the collective modes. The particles in multi-component systems are polarized, so we include the effect of the Coulomb potential between particles by using the Evald's method. The dynamical matrix of the dynamical equations of the collective modes is constructed.

10.3.1 The Green's Function of Inter-Band Density Fluctuations

We investigate Green's functions of inter-band density fluctuations:

$$D_{\alpha i\beta j\boldsymbol{q}}(\tau_1-\tau_2)\equiv-\frac{1}{\hbar\sqrt{N_\alpha N_\beta}}\langle T_\tau\rho_{\alpha i\boldsymbol{q}}(\tau_1)\rho^\dagger_{\beta j\boldsymbol{q}}(\tau_2)\rangle_c$$
$$\equiv\frac{1}{\beta\hbar}\sum_{\mathrm{i}\nu_n}e^{-\mathrm{i}\nu_n(\tau_1-\tau_2)}D_{\alpha i\beta j}(q). \tag{10.3.1}$$

In the random phase approximation, we obtain the dynamical equations for phonons:

$$D_{\alpha i\beta j}(q)=Q_\alpha(q)\delta_{\alpha\beta}\delta_{ij}+Q_\alpha(q)\sum_{\gamma l}V^{od}_{\alpha i\gamma l}(\boldsymbol{q})D_{\gamma l\beta j}(q), \tag{10.3.2}$$

where

$$V^{od}_{\alpha i\beta j}(\boldsymbol{q})=\left\langle\frac{\zeta_\alpha\zeta_\beta}{\sqrt{N_\alpha N_\beta}}\sum_{m\neq n}e^{-\mathrm{i}\boldsymbol{q}\cdot(\boldsymbol{R}_{\alpha m}-\boldsymbol{R}_{\beta n})}V_{\alpha i\beta j}(\boldsymbol{R}_{\alpha m}-\boldsymbol{R}_{\beta n})\right\rangle_\epsilon \tag{10.3.3}$$

$$V_{\alpha i\beta j}(\boldsymbol{R}_{\alpha m}-\boldsymbol{R}_{\beta n})\equiv\int d^3xd^3y\nabla_i\tilde{w}^2_{\alpha m0}(\boldsymbol{x}-\boldsymbol{R}_{\alpha m})$$
$$\times V_{\alpha\beta}(\boldsymbol{x}-\boldsymbol{y})\nabla_j\tilde{w}^2_{\beta 0}(\boldsymbol{y}-\boldsymbol{R}_{\beta n}). \tag{10.3.4}$$

$$Q_\alpha(q)=\frac{2\omega_\alpha/\hbar}{(q_0+\mathrm{i}/2\tau_{\alpha\mathrm{M}})^2-\omega_\alpha^2}, \tag{10.3.5}$$

where $\tau_{\alpha\mathrm{M}}$ is called the Maxwell relaxation time and

$$\frac{1}{2\tau_{\alpha\mathrm{M}}}\equiv\frac{1}{2\tau_{\alpha0}}+\frac{1}{2\tau_{\alpha1}}-\frac{1}{2\tau^o_{\alpha0}}-\frac{1}{2\tau^o_{\alpha1}}. \tag{10.3.6}$$

$\tau_{\alpha0,1}$ are the relaxation times of the particles in the lower or upper bands, respectively. $\tau^o_{\alpha0,1}$ comes from the simultaneous scattering processes of particles between the lower and upper bands by the same random potentials. The latter term corresponds to the cosine-function term with the scattering angle of an electron in the electric conductivity.

10.3.2 The Gap Equations of Inter-Band Density Fluctuations

The broken symmetry of the system requires the gap equation. In order to obtain the gap equation, we establish the WT relations associated with the translational invariance. adding the infinitesimal symmetry breaking term H_ϵ to the invariant Hamiltonian H:

$$H_\epsilon=-\epsilon\int d^3xg_{\alpha\delta}(x)\psi^\dagger_\alpha(x)\psi_\alpha(x), \tag{10.3.7}$$

$$g_{\alpha\delta}=\langle n_\alpha(x)\rangle_c=\langle\psi^\dagger_\alpha(x)\psi_\alpha(x)\rangle_c, \tag{10.3.8}$$

and using the Noether's current we obtain a general Ward-Takahashi relation:

$$\sum_i \langle T_\tau \{\psi_1(x_1) \cdots \nabla \psi_i(x_i) \cdots \psi_n(x_n)\rangle_c$$
$$= -\frac{\epsilon}{\hbar} \int d^3x \langle T_\tau \psi_1(x_1) \cdots \psi_n(x_n) \nabla n_\alpha(x)\rangle_c g_{\alpha\delta}(x), \tag{10.3.9}$$

where $g_{\alpha\delta}(x)$ is the pair distribution function between atoms α and δ separated by a distance $|\boldsymbol{x}_\alpha - \boldsymbol{x}_\delta| \equiv x$. $\psi_i(x_i)$ stands for $\psi(x_i)$ or $\psi^\dagger(x_i)$. Here it should be noted that $\langle \cdots \rangle_\epsilon$ is simply denoted as $\langle \cdots \rangle_c$ as mentioned before. We obtain the following simple Ward-Takahashi relations:

$$\nabla \langle T_\tau n_\alpha(x)\rangle_c = \frac{\epsilon}{\hbar} \sum_\beta \int d^4y \langle T_\tau n_\alpha(x) n_\beta(y)\rangle_c \nabla g_{\beta\delta}(y), \tag{10.3.10}$$

Following the approximation procedure of Section 6.3, we obtain

$$\left\langle \cdots \int d^3x n_\alpha(x) \right\rangle_c \nabla_i g_{\alpha\delta}(x)$$
$$\cong -\left\langle \cdots \frac{\zeta_\alpha}{N_\alpha} \sum_{\alpha m, \delta n \boldsymbol{q}} \int_{\{\boldsymbol{R}_{\delta n}\}} d^3x e^{-\mathrm{i}\boldsymbol{q}\cdot\boldsymbol{R}_{\alpha m}} \nabla_i \tilde{w}^2_{\alpha m 0}(\boldsymbol{x} - \boldsymbol{R}_{\alpha m}) \rho^\dagger_{\alpha i \boldsymbol{q}} \right\rangle_c \nabla_i g_{\alpha\delta}(x)$$
$$\cong -\sum_{\boldsymbol{q}} \left\langle \cdots \rho^\dagger_{\alpha i \boldsymbol{q}} \right\rangle \delta_{\boldsymbol{q}0} \frac{\zeta_\alpha C_{\alpha\delta}}{N_\alpha}, \tag{10.3.11}$$

$$\int_{\{\boldsymbol{R}_{\delta n}\}} d^3x \nabla_i g_{\alpha\delta}(x) \nabla_j g_{\alpha\delta}(x) = \delta_{ij} C_{\alpha\delta}, \tag{10.3.12}$$

$$g_{\alpha\delta}(x) = \left\langle \sum_m \tilde{w}_0^2(\boldsymbol{x} - \boldsymbol{R}_{\alpha m}) \right\rangle_c. \tag{10.3.13}$$

In Eq. (10.3.11), the subscript $\{\boldsymbol{R}_{\delta n}\}$ means the summation of m is taken over the positions around n-particle. Since we can consider that the configurationally averaged distribution of particles $\langle \sum_m \nabla_i \tilde{w}_0^2(\boldsymbol{x} - \boldsymbol{R}_{\alpha m})\rangle_c$ around $\boldsymbol{R}_{\delta n}$ is of the same form, the summation of n leads to $\delta_{0,\boldsymbol{q}}$. Note that in Eq. (10.3.11) there does not appear the first term of $n(x)$, the intra-band density fluctuations, in Eq. (10.1.15). In Eq. (10.3.12) the subscript $\{\boldsymbol{R}_{\delta n}\}$ of the integral means we have regarded $g_{\alpha\delta}(x)$ as the actual pair distribution function with the origin at $\{\boldsymbol{R}_{\delta n}\}$ and the pair distribution function is spherically symmetric.

Substituting Eq. (10.1.15) into Eq. (10.3.10) and comparing the coefficients of $\tilde{w}_{\alpha 0}(\boldsymbol{x} - \boldsymbol{R}_{\alpha m})\tilde{w}_{\alpha i}(\boldsymbol{x} - \boldsymbol{R}_{\alpha m})$ on both the sides of Eq. (10.3.10), we obtain the left-hand side equation:

$$
\begin{aligned}
\nabla_i \langle T_\tau n_\alpha(x)\rangle_\epsilon &= \mp\frac{1}{N_\alpha \beta\hbar}\sum_{mp}\nabla_i\{\langle \tilde{w}_{\alpha 0}^2(\boldsymbol{x}-\boldsymbol{R}_{\alpha m})\rangle_\epsilon G_{\alpha 0}(p) \\
&\quad +\langle \tilde{w}_{\alpha i}^2(\boldsymbol{x}-\boldsymbol{R}_{\alpha m})\rangle_\epsilon G_{\alpha i}(p)\} \\
&\cong \pm\frac{1}{\zeta_\alpha N_\alpha \beta\hbar}\sum_{mp}\langle \tilde{w}_{\alpha 0}(\boldsymbol{x}-\boldsymbol{R}_{\alpha m})\tilde{w}_{\alpha i}(\boldsymbol{x}-\boldsymbol{R}_{\alpha m})\rangle_\epsilon \\
&\quad \times\{G_{\alpha 0}(p)-G_{\alpha i}(p)\} \\
&= \frac{1}{N_\alpha}\sum_m \langle \tilde{w}_{\alpha 0}(\boldsymbol{x}-\boldsymbol{R}_{\alpha m})\tilde{w}_{\alpha i}(\boldsymbol{x}-\boldsymbol{R}_{\alpha m})\rangle_\epsilon \widetilde{\Delta}_{\alpha i},
\end{aligned} \tag{10.3.14}
$$

where since the system is configurationally averaged, the wave vector $\boldsymbol{q}$ in $e^{\mathrm{i}\boldsymbol{q}\cdot\boldsymbol{R}_{\alpha m}}$ can be put $\boldsymbol{q} = 0$.

$$
\widetilde{\Delta}_{\alpha i} \equiv \pm\frac{1}{\zeta_\alpha \beta\hbar}\sum_p \{G_{\alpha 0}(p)-G_{\alpha i}(p)\}, \tag{10.3.15}
$$

and the right-hand side equation:

$$
\begin{aligned}
&\frac{\epsilon}{\hbar}\sum_\beta \int d^4y \langle T_\tau n_\alpha(x) n_\beta(y)\rangle_\epsilon \nabla_i g_{\beta\delta}(y) \\
&\quad \cong \frac{1}{N_\alpha}\langle \tilde{w}_{\alpha 0}(\boldsymbol{x}-\boldsymbol{R}_{\alpha m})\tilde{w}_{\alpha i}(\boldsymbol{x}-\boldsymbol{R}_{\alpha m})\rangle_\epsilon \\
&\qquad \times \sum_{\beta m j \boldsymbol{q}}\int d^4y \sum_j \langle \rho_{\alpha j \boldsymbol{q}}(\tau_x) n_\beta(y)\rangle_\epsilon \nabla_i g_{\beta\delta}(y) \\
&\quad = \frac{1}{N_\alpha}\langle \tilde{w}_{\alpha 0}(\boldsymbol{x}-\boldsymbol{R}_{\alpha m})\tilde{w}_{\alpha i}(\boldsymbol{x}-\boldsymbol{R}_{\alpha m})\rangle_\epsilon \widetilde{\Xi}_{\alpha i},
\end{aligned} \tag{10.3.16}
$$

$$
\begin{aligned}
\widetilde{\Xi}_{\alpha i} &= -\frac{\epsilon}{\hbar}\sum_{\beta j \boldsymbol{q}}\int d^4y \sum_{\boldsymbol{q}} \langle \rho_{\alpha j \boldsymbol{q}}(\tau_x)\rho^\dagger_{\beta i \boldsymbol{q}}(\tau_y)\rangle_\epsilon \delta_{\boldsymbol{q},0}\frac{\zeta_\beta C_{\beta\delta}}{N_\beta} \\
&= \frac{\epsilon}{\hbar}\int_0^{\beta\hbar} d\tau_y \hbar\sqrt{N_\alpha N_\beta} D_{\alpha j \beta i \boldsymbol{q}}(\tau_x-\tau_y)\delta_{\boldsymbol{q},0}\frac{\zeta_\beta C_{\beta\delta}}{N_\beta} \\
&= \epsilon\int_0^{\beta\hbar} d\tau_y \frac{1}{\beta\hbar}\sum_{\mathrm{i}\nu_n} e^{-\mathrm{i}\nu_n(\tau_x-\tau_y)} D_{\alpha j \beta i \boldsymbol{q}}(\mathrm{i}\nu_n)\delta_{\boldsymbol{q},0}\frac{\zeta_\beta C_{\beta\delta}\sqrt{N_\alpha}}{\sqrt{N_\beta}} \\
&= \sum\nolimits_{\beta j} \epsilon D_{\alpha j \beta i}(0)\frac{\zeta_\beta C_{\beta\delta}\sqrt{N_\alpha}}{\sqrt{N_\beta}}.
\end{aligned} \tag{10.3.17}
$$

Considering that $\widetilde{\Delta}_\alpha$ does not depend on the index i, because we are concerned with an isotropic system, we have

$$\widetilde{\Delta}_\alpha \delta_{ij} = \sum_\beta \epsilon D_{\alpha j\beta i}(0) \frac{\zeta_\beta C_{\beta\delta}\sqrt{N_\alpha}}{\sqrt{N_\beta}}. \tag{10.3.18}$$

Substituting Eq. (10.3.2) into $D_{\alpha j\beta i}(0)$, we obtain the gap equation:

$$\frac{\widetilde{\Delta}_\alpha}{\sqrt{N_\alpha}} \delta_{ij} = Q_\alpha(0) \sum_\beta \left\langle V^{od}_{\alpha i\beta j}(0) \right\rangle_c \frac{\widetilde{\Delta}_\beta}{\sqrt{N_\beta}}. \tag{10.3.19}$$

At low temperatures, from the definition (10.3.15), we obtain

$$\widetilde{\Delta}_\alpha = \frac{N_\alpha}{\zeta_\alpha}, \tag{10.3.20}$$

and from Eq. (10.3.3), we obtain

$$V^{od}_{\alpha i\beta j}(\boldsymbol{q}) = \zeta_\alpha \zeta_\beta \sqrt{\frac{N_\alpha}{N_\beta}} \sum_{n\neq 0} e^{\mathrm{i}\boldsymbol{q}\cdot\boldsymbol{R}_{\beta n}} V_{\alpha i\beta j}(\boldsymbol{R}_{\beta n}) \tag{10.3.21}$$

so that the gap Eq. (10.3.19) leads to

$$\delta_{ij} = \zeta_\alpha^2 Q_\alpha(0) \sum_{\beta,n\neq 0} \left\langle V_{\alpha i\beta j}(\boldsymbol{R}_{\beta n}) \right\rangle_c, \tag{10.3.22}$$

where $n \neq 0$ in the sum means we exclude $\boldsymbol{R}_{\beta n} = 0$ when $\alpha = \beta$.

This equation is alternatively obtained by considering

$$\nabla_i \langle n(x)\rangle_\epsilon \longrightarrow \nabla_i g_{\alpha\delta}(x), \tag{10.3.23}$$

$$\begin{aligned}
&\frac{\epsilon}{\hbar}\sum_\beta \int d^4y \langle T_\tau n_\alpha(x) n_\beta(y)\rangle_c \nabla_i g_{\beta\delta}(y) \longrightarrow \\
&-\frac{\epsilon}{\hbar}\sum_\beta \int_0^{\beta\hbar} d\tau_y \sum_j \sum_{\boldsymbol{q}} \frac{\zeta_\alpha}{N_\alpha} \nabla_j g_{\alpha\delta}(x) \langle \rho_{\alpha j\boldsymbol{q}}(\tau_x) \rho^\dagger_{\beta i\boldsymbol{q}} \tau_y\rangle_\epsilon \delta_{\boldsymbol{q},0} \frac{\zeta_\beta C_{\beta\delta}}{N_\beta} \\
&= \epsilon \sum_j \nabla_j g_{\alpha\delta}(x) D_{\alpha j\beta i}(0) \frac{\zeta_\alpha \zeta_\beta C_{\beta\delta}}{\sqrt{N_\alpha N_\beta}}.
\end{aligned} \tag{10.3.24}$$

Thus we obtain

$$\delta_{ij} = \epsilon D_{\alpha i\beta j}(0) \frac{\zeta_\alpha \zeta_\beta C_{\beta\delta}}{\sqrt{N_\alpha N_\beta}}. \tag{10.3.25}$$

This equation is just Eq. (10.3.18) with Eq. (10.3.20).

Considering $\zeta_\alpha = \langle \zeta_{\alpha n} \rangle_c = \sqrt{\frac{\hbar}{2M_\alpha \omega_\alpha}}$ and the condition $\omega_\alpha \tau_{\alpha \mathrm{M}} \ll 1$, from Eq. (10.3.22), we obtain

$$M_\alpha \omega_\alpha^2 = -\sum_{\beta n \neq 0} \langle V_{\alpha i \beta i}(\boldsymbol{R}_{\beta n}) \rangle_c, \tag{10.3.26}$$

which means that every α-particle feels the same statistical averaged potential made up by the surrounding particles.

10.3.3 The Dynamical Matrix for Phonons

In order to obtain the matrix elements of the dynamical Eq. (10.3.2), we rewrite the gap Eq. (10.3.19) with (10.3.20) as

$$\delta_{ij} = Q_\alpha(0) \sum_\beta V_{\alpha i \beta j}^{od}(0) \sqrt{\frac{N_\beta}{N_\alpha} \frac{\zeta_\alpha}{\zeta_\beta}} \tag{10.3.27}$$

and we use the property of the bubble diagrams (10.3.5):

$$Q_\alpha(q) - Q_\alpha(0) = -\frac{\hbar q_0 (q_0 + \frac{\mathrm{i}}{\tau_{\alpha M}})}{2\omega_\alpha} Q_\alpha(q) Q_\alpha(0), \tag{10.3.28}$$

The secular equation of inter-band density fluctuations in (10.3.2) is written as

$$\begin{aligned}
&Q_\alpha(q) V_{\alpha i \beta j}^{od}(\boldsymbol{q}) - \delta_{\alpha\beta}\delta_{ij} \\
&= Q_\alpha(q) V_{\alpha i \beta j}^{od}(\boldsymbol{q}) - \delta_{\alpha\beta}\delta_{ij} Q_\alpha(0) \sum_\gamma V_{\alpha i \gamma j}^{od}(0) \sqrt{\frac{N_\gamma}{N_\alpha} \frac{\zeta_\alpha}{\zeta_\gamma}} \\
&= Q_\alpha(q) \left\{ V_{\alpha i \beta j}^{od}(\boldsymbol{q}) - \delta_{\alpha\beta}\delta_{ij} \sum_\gamma V_{\alpha i \gamma j}^{od}(0) \sqrt{\frac{N_\gamma}{N_\alpha} \frac{\zeta_\alpha}{\zeta_\gamma}} \right\} \\
&\quad + \{Q_\alpha(q) - Q_\alpha(0)\} \delta_{\alpha\beta}\delta_{ij} \sum_\gamma V_{\alpha i \gamma j}^{od}(0) \sqrt{\frac{N_\gamma}{N_\alpha} \frac{\zeta_\alpha}{\zeta_\gamma}} \\
&= Q_\alpha(q) \left\{ V_{\alpha i \beta j}^{od}(\boldsymbol{q}) - \delta_{\alpha\beta}\delta_{ij} \sum_\gamma V_{\alpha i \gamma j}^{od}(0) \sqrt{\frac{N_\gamma}{N_\alpha} \frac{\zeta_\alpha}{\zeta_\gamma}} \right\} \\
&\quad - \delta_{\alpha\beta}\delta_{ij} \frac{\hbar q_0 \left(q_0 + \frac{\mathrm{i}}{\tau_{\alpha M}}\right)}{2\omega_\alpha} Q_\alpha(q) \\
&= -\zeta_\alpha^2 Q_\alpha(q) \left[\delta_{\alpha\beta}\delta_{ij} M_\alpha q_0 \left(q_0 + \frac{\mathrm{i}}{\tau_{\alpha M}}\right) \right. \\
&\quad \left. - \frac{1}{\zeta_\alpha^2} \left\{ V_{\alpha i \beta j}^{od}(\boldsymbol{q}) - \delta_{\alpha\beta}\delta_{ij} \sum_\gamma V_{\alpha i \gamma j}^{od}(0) \sqrt{\frac{N_\gamma}{N_\alpha} \frac{\zeta_\alpha}{\zeta_\gamma}} \right\} \right]
\end{aligned}$$

$$\equiv -\zeta_\alpha^2 Q_\alpha(q) M_\alpha \left[q_0 \left(q_0 + \frac{\mathrm{i}}{\tau_{\alpha\mathrm{M}}} \right) \delta_{\alpha\beta}\delta_{ij} - M_{\alpha i\beta j}(\boldsymbol{q}) \right] = 0,$$

Thus we obtain the secular equations of the collective modes

$$q_0(q_0 + \frac{\mathrm{i}}{\tau_{\alpha\mathrm{M}}})\delta_{\alpha\beta}\delta_{ij} - M_{\alpha i\beta j}(\boldsymbol{q}) = 0, \qquad (10.3.29)$$

where the dynamical matrix $M_{\alpha i\beta j}$ is written as

$$\begin{aligned} M_{\alpha i\beta j}(\boldsymbol{q}) &\equiv \frac{1}{M_\alpha}\frac{1}{\zeta_\alpha^2}\left\{ V_{\alpha i\beta j}^{od}(\boldsymbol{q}) - \delta_{\alpha\beta}\delta_{ij}\sum_\gamma V_{\alpha i\gamma j}^{od}(0)\sqrt{\frac{N_\gamma}{N_\alpha}\frac{\zeta_\alpha}{\zeta_\gamma}} \right\} \\ &= \frac{1}{M_\alpha}\frac{1}{\zeta_\alpha^2}\left\{ \zeta_\alpha\zeta_\beta\sqrt{\frac{N_\alpha}{N_\beta}}\sum_{n\neq 0} e^{\mathrm{i}\boldsymbol{q}\cdot\boldsymbol{R}_{\beta n}} V_{\alpha i\beta j}(\boldsymbol{R}_{\beta n}) \right. \\ &\quad \left. -\delta_{\alpha\beta}\delta_{ij}\sum_\gamma \zeta_\alpha^2 \sum_{n\neq 0} V_{\alpha i\gamma j}(\boldsymbol{R}_{\gamma n}) \right\} \\ &= \frac{1}{M_\alpha}\sum_{\gamma n}\left\{ \delta_{\beta\gamma}\frac{\zeta_\gamma}{\zeta_\alpha}\sqrt{\frac{N_\alpha}{N_\gamma}} e^{\mathrm{i}\boldsymbol{q}\cdot\boldsymbol{R}_{\gamma n}} - \delta_{\alpha\beta}\delta_{ij} \right\} V_{\alpha i\gamma j}(\boldsymbol{R}_{\gamma n}) \\ &= \frac{1}{M_\alpha}\sum_\gamma \int d^3R g_{\alpha\gamma}(R)\left\{ \delta_{\beta\gamma}\frac{\zeta_\gamma}{\zeta_\alpha}\sqrt{\frac{N_\alpha}{N_\gamma}} e^{\mathrm{i}\boldsymbol{q}\cdot\boldsymbol{R}} - \delta_{\alpha\beta}\delta_{ij} \right\} V_{\alpha i\gamma j}(\boldsymbol{R}). \end{aligned} \qquad (10.3.30)$$

where in the third equation $n = 0$ in the sum is included, because the first and the second term cancel when $\alpha = \gamma$.

10.3.4 The Interaction Potentials Between Particles

Now we investigate the interaction potentials. The interaction potentials between particles with no electric charge are short range ones. But particles in multi-component systems are electrically polarized. so the Coulomb potential acts between particles. The Coulomb potential is a long range one. So we separate the potential V into the short range one, V^N and the Coulombic one, V^C:

$$V_{\alpha i\beta j}(\boldsymbol{R}_{\alpha m} - \boldsymbol{R}_{\beta n}) = V_{\alpha i\beta j}^N(\boldsymbol{R}_{\alpha m} - \boldsymbol{R}_{\beta n}) + V_{\alpha i\beta j}^C(\boldsymbol{R}_{\alpha m} - \boldsymbol{R}_{\beta n}). \qquad (10.3.31)$$

It is hard task to include the effect of the Coulomb potentials from the quantum field theory. We take into account the effect of the Coulomb potential in the semi-classical way. When we take into account the polarization effect, we make the rigid ion approximation, the Ewald's method [12–14]. In the rigid ion approximation; $w_{\alpha 0}^2(\boldsymbol{R}) \longrightarrow \delta(\boldsymbol{R})$, we obtain

$$V_{\alpha i\beta j}^C(\boldsymbol{R}) = -\frac{\partial^2}{\partial x_i \partial x_j}\frac{e_\alpha e_\beta}{4\pi\epsilon_0 R}, \qquad (10.3.32)$$

where we have used the SI units, ϵ_0 is the dielectric constant in vacuum and e_α is the charge of an α-particle. We define the Fourier transformed potential as

$$\widehat{V}_{\alpha i\beta j}(\boldsymbol{q}) = \left\langle \sum_n e^{-\mathrm{i}\boldsymbol{q}\cdot(\boldsymbol{R}_{\alpha m}-\boldsymbol{R}_{\beta n})} V_{\alpha i\beta j}(\boldsymbol{R}_{\alpha m} - \boldsymbol{R}_{\beta n}) \right\rangle_c$$
$$= \widehat{V}^N_{\alpha i\beta j}(\boldsymbol{q}) + \widehat{V}^C_{\alpha i\beta j}(\boldsymbol{q}). \tag{10.3.33}$$

We first investigate the Coulomb potential V^C.

10.3.5 The Coulomb Potential and the Ewald's Method

We investigate the Coulomb potential:

$$\varphi^C_{\alpha\beta}(R) = \frac{e_\alpha e_\beta}{4\pi\epsilon_0 R}, \tag{10.3.34}$$

and its Fourier transformation:

$$\hat{\varphi}^C_{\alpha\beta}(q) = \frac{e_\alpha e_\beta}{\epsilon_0 q^2}, \tag{10.3.35}$$

which satisfy

$$\varphi^C_{\alpha\beta}(R) = \frac{1}{(2\pi)^3}\int d^3q e^{\mathrm{i}\boldsymbol{q}\cdot\boldsymbol{R}}\hat{\varphi}^C(q). \tag{10.3.36}$$

In the Ewald's method, we define a potential, $\varphi^G_{\alpha\beta}(R)$, which is the potential energy of a point charge e_α at a distance R from a three dimensional Gaussian distribution $\rho^G_\beta(R)$ with the total charge e_β:

$$\rho^G_\beta(R) = e_\beta\left(\frac{P}{\pi}\right)^{3/2} e^{-PR^2}, \tag{10.3.37}$$

and its Fourier transformation

$$\hat{\rho}^G_\beta(q) = e_\beta e^{-q^2/4P}, \tag{10.3.38}$$

which satisfies

$$\rho^G_\beta(R) = \frac{1}{(2\pi)^3}\int d^3q\hat{\rho}^G_\beta(q)e^{\mathrm{i}\boldsymbol{q}\cdot\boldsymbol{R}}, \tag{10.3.39a}$$

$$\int_0^\infty d^3R'\rho^G_\beta(R') = e_\beta. \tag{10.3.39b}$$

Thus we obtain

$$\varphi^G_{\alpha\beta}(R) = -\frac{e_\alpha}{4\pi\epsilon_0}\int_R^\infty \frac{1}{R^2}\int_0^R \rho^G_\beta(R')dR'dR$$

$$= \frac{e_\alpha}{4\pi\epsilon_0}\left[\frac{1}{R}\int_0^R \rho_\beta^G(R')d^3R' + \int_R^\infty \frac{\rho_\beta^G(R')}{R'}d^3R'\right], \qquad (10.3.40)$$

where we have used the partial integration. The Poisson equation:

$$\nabla^2\varphi_{\alpha\beta}^G(R) = \frac{1}{(2\pi)^3}\int d^3q(-q^2\hat{\varphi}_{\alpha\beta}^G(q))e^{\mathrm{i}\boldsymbol{q}\cdot\boldsymbol{R}} = -\frac{e_\alpha}{\epsilon_0}\rho_\beta^G(R), \qquad (10.3.41)$$

and Eq. (10.3.39a) leads to

$$\hat{\varphi}_{\alpha\beta}^G(q) = \frac{e_\alpha e_\beta}{\epsilon_0 q^2}e^{-q^2/4P}. \qquad (10.3.42)$$

Now we introduce $\varphi_{\alpha\beta}^H(R)$ as

$$\varphi_{\alpha\beta}^C(R) = \varphi_{\alpha\beta}^G(R) + \varphi_{\alpha\beta}^C(R) - \varphi_{\alpha\beta}^G(R) \equiv \varphi_{\alpha\beta}^G(R) + \varphi_{\alpha\beta}^H(R) \qquad (10.3.43)$$

Using Eqs. (10.3.34), (10.3.40), and (10.3.39b), we obtain

$$\begin{aligned}\varphi_{\alpha\beta}^H(R) &= \frac{e_\alpha}{4\pi\epsilon_0}\int_R^\infty\left(\frac{1}{R}-\frac{1}{R'}\right)\rho_\beta^G(R')d^3R' \\ &= \frac{e_\alpha e_\beta}{\epsilon_0}(\frac{P}{\pi})^{\frac{3}{2}}\int_R^\infty\left(\frac{R'}{R}-1\right)R'e^{-PR'^2}dR' \\ &= \frac{e_\alpha e_\beta}{2\epsilon_0\pi^{3/2}}\frac{1}{R}\int_{\sqrt{P}R}^\infty e^{-s^2}ds, \end{aligned} \qquad (10.3.44)$$

where from the second to the last equation, we have carried out the partial integration. Here we define

$$\begin{aligned}\widehat{V}_{\alpha i\beta j}^C(\boldsymbol{q}) &= -\left\langle\sum_n e^{-\mathrm{i}\boldsymbol{q}\cdot\boldsymbol{R}}\frac{\partial^2}{\partial x_i\partial x_j}\{\varphi_{\alpha\beta}^G(R)+\varphi_{\alpha\beta}^H(R)\}|_{\boldsymbol{R}=\boldsymbol{R}_{\alpha m}-\boldsymbol{R}_{\beta n}}\right\rangle_c \\ &\equiv \widehat{V}_{\alpha i\beta j}^G(\boldsymbol{q}) + \widehat{V}_{\alpha i\beta j}^H(\boldsymbol{q}). \end{aligned} \qquad (10.3.45)$$

First we investigate the term $\widehat{V}^G$, which is written as

$$\begin{aligned}&\widehat{V}_{\alpha i\beta j}^G(\boldsymbol{q}) \\ &= -\left\langle\sum_n e^{-\mathrm{i}\boldsymbol{q}\cdot\boldsymbol{R}}\frac{\partial^2}{\partial x_i\partial x_j}\left\{\frac{1}{(2\pi)^3}\int d^3q' e^{\mathrm{i}\boldsymbol{q}'\cdot\boldsymbol{R}}\hat{\varphi}^G(\boldsymbol{q}')\right\}\Big|_{\boldsymbol{R}=\boldsymbol{R}_{\alpha m}-\boldsymbol{R}_{\beta n}}\right\rangle_c \\ &= \left\langle\sum_n e^{\mathrm{i}(\boldsymbol{q}'-\boldsymbol{q})\cdot(\boldsymbol{R}_{\alpha m}-\boldsymbol{R}_{\beta n})}\frac{e_\alpha e_\beta}{(2\pi)^3\epsilon_0}\int d^3q'\frac{q_i'q_j'}{q'^2}e^{-q'^2/4P}\right\rangle_c \\ &= \left\langle\frac{e_\alpha e_\beta}{(2\pi)^3\epsilon_0}\sum_n\int d^3q'\left[\frac{q_i'q_j'}{q'^2}+\frac{q_i'q_j'}{q'^2}\left\{e^{-q'^2/4P}-1\right\}\right]e^{\mathrm{i}(\boldsymbol{q}'-\boldsymbol{q})\cdot(\boldsymbol{R}_{\alpha m}-\boldsymbol{R}_{\beta n})}\right\rangle_c\end{aligned}$$

$$= \frac{e_\alpha e_\beta}{(2\pi)^3 \epsilon_0 v_a} \int d^3 R g_{\alpha\beta}(R) \int d^3 q' \left[\frac{q'_i q'_j}{q'^2} + \frac{q'_i q'_j}{q'^2} \left\{ e^{-q'^2/4P} - 1 \right\} \right] e^{\mathrm{i}(\boldsymbol{q}'-\boldsymbol{q})\cdot \boldsymbol{R}}$$

$$\cong \frac{e_\alpha e_\beta \rho_\beta}{\epsilon_0 v_a} \left[\frac{q_i q_j}{q^2} + \frac{q_i q_j}{q^2} \left\{ e^{-q^2/4P} - 1 \right\} \right], \tag{10.3.46}$$

where we have made the approximation; $g_{\alpha\beta}(R) \cong \rho_\beta$ and v_a is a primitive unit cell. The term V^H is given by

$$V^H_{\alpha i \beta j}(\boldsymbol{q}) = -\left\langle \sum_n e^{-\mathrm{i}\boldsymbol{q}\cdot\boldsymbol{R}} \frac{\partial^2}{\partial x_i \partial x_j} \left\{ \frac{1}{(2\pi)^3} \int d^3 q' e^{\mathrm{i}\boldsymbol{q}'\cdot \boldsymbol{R}} \hat{\varphi}^H(\boldsymbol{q}') \right\} \Big|_{\boldsymbol{R}=\boldsymbol{R}_{\alpha m}-\boldsymbol{R}_{\beta n}} \right\rangle_c$$

$$= \left\langle \sum_n e^{-\mathrm{i}\boldsymbol{q}\cdot\boldsymbol{R}} \frac{\partial^2}{\partial x_i \partial x_j} \frac{e_\alpha e_\beta}{2\epsilon_0 \pi^{3/2}} \frac{1}{R} \int_{\sqrt{P}R}^{\infty} ds e^{-s^2} \Big|_{\boldsymbol{R}=\boldsymbol{R}_{\alpha m}-\boldsymbol{R}_{\beta n}} \right\rangle_c$$

$$= \frac{e_\alpha e_\beta}{4\pi \epsilon_0 v_a} P^{3/2} \int d^3 R g_{\alpha\beta}(R) H_{ij}\left(\sqrt{P}R\right) e^{-\mathrm{i}\boldsymbol{q}\cdot\boldsymbol{R}}$$

$$\cong \frac{e_\alpha e_\beta \rho_\beta}{4\pi \epsilon_0 v_a} P^{3/2} \int d^3 R H_{ij}\left(\sqrt{P}R\right) e^{-\mathrm{i}\boldsymbol{q}\cdot\boldsymbol{R}}, \tag{10.3.47}$$

$$H_{ij}(x) \equiv \frac{\partial^2}{\partial x_i \partial x_j} \frac{2}{\sqrt{\pi}} \frac{1}{x} \int_x^{\infty} ds e^{-s^2}. \tag{10.3.48}$$

It should be noted the function $H_{ij}(x)$ is singular at $x = 0$. When we treat the crystalline state [12–14], $H_{ij}(x)$ is replaced by $H^0_{ij} = \frac{\partial^2}{\partial x_i \partial x_j} \frac{2}{\sqrt{\pi}} \frac{1}{x} \int_0^x ds e^{-s^2}$, because we should exclude the singularity, since the dynamical matrix includes $\boldsymbol{R}_m = \boldsymbol{R}_n$. But in our case, taking into account the spherical symmetry $\sum_i H_{ii} = \frac{4}{\sqrt{\pi}} e^{-x^2}$ and the property of the argument of integration $d^3 R = 4\pi R^2 dR$ yields the integration finite. Thus Eqs. (10.3.46) and (10.3.47) lead to

$$V^C_{\alpha i \beta j}(q) = e_\alpha e_\beta \rho_\beta \left[\frac{q_i q_j}{v_a \epsilon_0 q^2} - Q_{ij}(q) \right], \tag{10.3.49}$$

$$Q_{ij}(q) = -\frac{q_i q_j}{\epsilon_0 v_a} \left[e^{-\frac{q^2}{4P}} - 1 \right] + \frac{P^{3/2}}{4\pi \epsilon_0 v_a} \int d^3 R H_{ij}(\sqrt{P}R) e^{-\mathrm{i}\boldsymbol{q}\cdot\boldsymbol{R}}, \tag{10.3.50}$$

where the first term and the second term in Eq. (10.3.49) relate to the macroscopic electric field and the Lorentz field, respectively. In the spherical symmetry, we obtain

$$Q_{ij}(q) = \delta_{ij} Q(q) \cong \delta_{ij} Q(0) = \delta_{ij} \frac{P^{\frac{1}{2}}}{4\pi v_a} \int d^3 R \frac{4}{3\sqrt{\pi}} e^{-PR^2}$$

$$= \delta_{ij} \frac{1}{3\epsilon_0 v_a}. \tag{10.3.51}$$

There is the electrical charge neutrality:

$$\sum_\alpha e_\alpha \rho_\alpha = 0. \tag{10.3.52}$$

The particles are polarized, but the total charge is zero.

10.3.6 The Short Range Potentials

Next we investigate the short range potentials $V^N_{\alpha i\beta j}(\boldsymbol{R})$, which are given by

$$V^N_{\alpha i\beta j}(\boldsymbol{R}) \equiv V_{\alpha i\beta j}(\boldsymbol{R}) = \int d^3x d^3y \nabla_i \tilde{w}^2_{\alpha 0}(\boldsymbol{x}) V(\boldsymbol{x}-\boldsymbol{y}) \nabla_j \tilde{w}^2(\boldsymbol{y}-\boldsymbol{R}). \tag{10.3.53}$$

In order to calculate $V_{\alpha i\beta j}(\boldsymbol{R})$, we introduce the couplings like the chemical bondings:

$$V_{\sigma\alpha\beta}(R) = -\int d^3x d^3y \nabla_{x_3} \tilde{w}^2_{\alpha 0}(\boldsymbol{x})\, V_{\alpha\beta}(\boldsymbol{x}-\boldsymbol{y})\, \nabla_{y_3} \tilde{w}^2_{\beta 0}(\boldsymbol{y}-R\boldsymbol{e}_3), \tag{10.3.54}$$

$$V_{\pi\alpha\beta}(R) = -\int d^3x d^3y \nabla_{x_1} \tilde{w}^2_{\alpha 0}(\boldsymbol{x}) V_{\alpha\beta}(\boldsymbol{x}-\boldsymbol{y})\, \nabla_{y_1} \tilde{w}^2_{\beta 0}(\boldsymbol{y}-R\boldsymbol{e}_3). \tag{10.3.55}$$

Here we take

$$\boldsymbol{q} = q\boldsymbol{e}_z, \tag{10.3.56}$$

and represent $\boldsymbol{R}$ by the polar coordinates:

$$\boldsymbol{R} = R(\sin\theta\cos\varphi, \sin\theta\sin\varphi, \cos\theta). \tag{10.3.57}$$

Then we have

$$V^N_{\alpha x\beta x}(\boldsymbol{R}) = -V_{\sigma\alpha\beta}(R)\sin^2\theta\cos^2\varphi - V_{\pi\alpha\beta}(R)(1-\sin^2\theta\cos^2\varphi), \tag{10.3.58}$$

$$V^N_{\alpha y\beta y}(\boldsymbol{R}) = -V_{\sigma\alpha\beta}(R)\sin^2\theta\sin^2\varphi - V_{\pi\alpha\beta}(R)(1-\sin^2\theta\sin^2\varphi), \tag{10.3.59}$$

$$V^N_{\alpha z\beta z}(\boldsymbol{R}) = -V_{\sigma\alpha\beta}(R)\cos^2\theta - V_{\pi\alpha\beta}(R)\sin^2\theta, \tag{10.3.60}$$

Otherwise, the integration of polar angles leads to zero. The integration of the polar angles, $d\Omega = \sin\theta d\theta d\varphi$ and $\xi = qR$ leads to

$$\begin{aligned}\int d\Omega e^{-\mathrm{i}\boldsymbol{q}\cdot\boldsymbol{R}} V^N_{\alpha x\beta x}(\boldsymbol{R}) &= \int d\Omega e^{-\mathrm{i}\boldsymbol{q}\cdot\boldsymbol{R}} V_{\alpha y\beta y}(\boldsymbol{R}) \\ &= -4\pi\left\{V_{\sigma\alpha\beta}(R)\left(-\frac{\cos\xi}{\xi^2}+\frac{\sin\xi}{\xi^3}\right)\right. \\ &\quad \left. +V_{\pi\alpha\beta}(R)\left(\frac{\sin\xi}{\xi}+\frac{\cos\xi}{\xi^2}-\frac{\sin\xi}{\xi^3}\right)\right\}\end{aligned} \tag{10.3.61}$$

$$\begin{aligned}\int d\Omega e^{-\mathrm{i}\boldsymbol{q}\cdot\boldsymbol{R}} V^N_{\alpha z\beta z}(\boldsymbol{R}) &= -4\pi\left\{V_{\sigma\alpha\beta}(R)\left(\frac{\sin\xi}{\xi}+\frac{2\cos\xi}{\xi^2}-\frac{2\sin\xi}{\xi^3}\right)\right. \\ &\quad \left. +V_{\pi\alpha\beta}(R)2\left(-\frac{\cos\xi}{\xi^2}+\frac{\sin\xi}{\xi^3}\right)\right\}\end{aligned} \tag{10.3.62}$$

$$\int d\Omega e^{-\mathrm{i}\boldsymbol{q}\cdot\boldsymbol{R}} V^N_{\alpha i\beta j}(\boldsymbol{R})\Big|_{i\neq j} = 0 \tag{10.3.63}$$

Thus we have obtained the short range potential.

10.3.7 The Calculation of the Dynamical Matrix

In order to calculate the dynamical matrix, we first investigate the total potential. Considering $q_i = q\delta_{i3}$ in $V^C_{\alpha i\beta j}$ in Eq. (10.3.49), we obtain

$$V_{\alpha x\beta x}(\boldsymbol{q}) = V_{\alpha y\beta y}(\boldsymbol{q}) = -\int d^3Rg_{\alpha\beta}(R)\left\{V_{\sigma\alpha\beta}(R)\left(-\frac{\cos\xi}{\xi^2}+\frac{\sin\xi}{\xi^3}\right)\right.$$
$$\left.+V_{\pi\alpha\beta}(R)\left(\frac{\sin\xi}{\xi}+\frac{\cos\xi}{\xi^2}-\frac{\sin\xi}{\xi^3}\right)\right\}$$
$$-e_\alpha e_\beta\rho_\beta\frac{1}{3\epsilon_0 v_a}, \tag{10.3.64}$$

$$V_{\alpha z\beta z}(\boldsymbol{q}) = -\int d^3Rg_{\alpha\beta}(R)\left\{V_{\sigma\alpha\beta}(R)\left(\frac{\sin\xi}{\xi}+\frac{2\cos\xi}{\xi^2}-\frac{2\sin\xi}{\xi^3}\right)\right.$$
$$\left.+V_{\pi\alpha\beta}(R)2\left(-\frac{\cos\xi}{\xi^2}+\frac{\sin\xi}{\xi^3}\right)\right\}+e_\alpha e_\beta\rho_\beta\frac{2}{3\epsilon_0 v_a}, \tag{10.3.65}$$

$$\sum_\beta V_{\alpha x\beta x}(0) = \sum_\beta V_{\alpha y\beta y}(0) = \sum_\beta V_{\alpha z\beta z}(0)$$
$$= -\sum_\beta\int d^3Rg_{\alpha\beta}(R)\left\{\frac{V_{\sigma\alpha\beta}(R)}{3}+\frac{2V_{\pi\alpha\beta}(R)}{3}\right\}, \tag{10.3.66}$$

where in the last equation, we have considered the electric charge neutrality.

Substituting Eqs. (10.3.64), (10.3.65), (10.3.66) into the dynamical matrix in Eq. (10.3.30), we obtain

$$M_{\alpha x\alpha x}(\boldsymbol{q}) = M_{\alpha y\alpha y}(\boldsymbol{q})$$
$$= -\frac{1}{M_\alpha}\int d^3Rg_{\alpha\alpha}(R)\left[V_{\sigma\alpha\alpha}(R)\left(-\frac{\cos\xi}{\xi^2}+\frac{\sin\xi}{\xi^3}\right)\right.$$
$$\left.+V_{\pi\alpha\alpha}(R)\left(\frac{\sin\xi}{\xi}+\frac{\cos\xi}{\xi^2}-\frac{\sin\xi}{\xi^3}\right)+e_\alpha^2\rho_\alpha\frac{1}{3\epsilon_0 v_a}\right]$$
$$+\frac{1}{M_\alpha}\sum_\gamma\int d^3Rg_{\alpha\gamma}(R)\left(\frac{V_{\sigma\alpha\gamma}(R)}{3}+\frac{2V_{\pi\alpha\gamma}(R)}{3}\right), \tag{10.3.67}$$

$$M_{\alpha z\alpha z}(\boldsymbol{q}) = -\frac{1}{M_\alpha}\int d^3Rg_{\alpha\alpha}(R)\left[V_{\sigma\alpha\alpha}(R)\left(\frac{\sin\xi}{\xi}+\frac{2\cos\xi}{\xi^2}-\frac{2\sin\xi}{\xi^3}\right)\right.$$
$$\left.+V_{\pi\alpha\alpha}(R)\left(-\frac{2\cos\xi}{\xi^2}+\frac{2\sin\xi}{\xi^3}\right)-e_\alpha^2\rho_\alpha\frac{2}{3\epsilon_0 v_a}\right]$$
$$+\frac{1}{M_\alpha}\sum_\gamma\int d^3Rg_{\alpha\gamma}(R)\left(\frac{V_{\sigma\alpha\gamma}(R)}{3}+\frac{2V_{\pi\alpha\gamma}(R)}{3}\right), \tag{10.3.68}$$

$$M_{\alpha x\beta x}(\boldsymbol{q}) = M_{\alpha y\beta y}(\boldsymbol{q})$$
$$= -\frac{1}{M_\alpha}\frac{\zeta_\beta}{\zeta_\alpha}\sqrt{\frac{N_\alpha}{N_\beta}}\int d^3Rg_{\alpha\beta}(R)\left[V_{\sigma\alpha\beta}(R)\left(-\frac{\cos\xi}{\xi^2}+\frac{\sin\xi}{\xi^3}\right)\right.$$

$$+V_{\pi\alpha\beta}(R)\left(\frac{\sin\xi}{\xi}+\frac{\cos\xi}{\xi^2}-\frac{\sin\xi}{\xi^3}\right)+e_\alpha e_\beta\rho_\beta\frac{2}{3\epsilon_0 v_a}\Bigg], \quad (10.3.69)$$

$$M_{\alpha z\beta z}(\boldsymbol{q}) = -\frac{1}{M_\alpha}\frac{\zeta_\beta}{\zeta_\alpha}\sqrt{\frac{N_\alpha}{N_\beta}}\int d^3R g_{\alpha\beta}(R)$$
$$\times\Bigg[V_{\sigma\alpha\beta}(R)\left(\frac{\sin\xi}{\xi}+\frac{2\cos\xi}{\xi^2}-\frac{2\sin\xi}{\xi^3}\right)$$
$$V_{\pi\alpha\beta}(R)\left(-\frac{2\cos\xi}{\xi^2}+\frac{2\sin\xi}{\xi^3}\right)-e_\alpha e_\beta\rho_\beta\frac{2}{3\epsilon_0 v_a}\Bigg], \quad (10.3.70)$$

where the other terms are zero. Since we have taken $\boldsymbol{q} = q\boldsymbol{e}_3$, the z-component of the dynamical matrix corresponds to the longitudinal mode and x- and y-components to the transverse modes.

10.4 Phonons, Boson Peaks, and Viscosities in Two-Component Liquids

In the previous section, we have derived the secular equation of the collective modes and dynamical matrix in multi-component systems in general. Now we concrete to calculate the collective modes by applying the general secular equation of the collective modes to the two-component systems composed of A and B particles. We obtain the phonon modes at high frequencies and the viscosities at low frequencies. In the short wavelength regime, we obtain the boson peaks in the phonon dispersion curves. We illustrate the acoustic phonon dispersion curves, a longitudinal, and two transverse modes [1, 3–5].

10.4.1 The Secular Equation of Inter-Band Density Fluctuations

Here, we solve the secular equation of Eq. (10.3.2), (10.3.29):

$$\mathrm{Det}\left|q_0\left(q_0+\frac{\mathrm{i}}{\tau_{\alpha\mathrm{M}}}\right)\delta_{\alpha\beta}\delta_{ij}-M_{\alpha i\beta j}(\boldsymbol{q})\right| = 0. \quad (10.4.1)$$

If we define

$$q_0\left(q_0+\frac{\mathrm{i}}{\tau_{\alpha\mathrm{M}}}\right)\delta_{\alpha\beta}\delta_{ij}-M_{\alpha i\beta j}(\boldsymbol{q}) \equiv \Phi_{\Lambda\Xi}, \quad (10.4.2)$$
$$\Lambda,\ \Xi = \mathrm{A1, A2, A3, B1, B2, B3}, \quad (10.4.3)$$

we can solve the secular equation:

$$\mathrm{Det}|\Phi| = \begin{vmatrix} \alpha & 0 & 0 & \gamma & 0 & 0 \\ 0 & \alpha & 0 & 0 & \gamma & 0 \\ 0 & 0 & \alpha' & 0 & 0 & \gamma' \\ \delta & 0 & 0 & \beta & 0 & 0 \\ 0 & \delta & 0 & 0 & \beta & 0 \\ 0 & 0 & \delta' & 0 & 0 & \beta' \end{vmatrix} = (\alpha\beta-\gamma\delta)^2(\alpha'\beta'-\gamma'\delta') = 0, \quad (10.4.4)$$

where $\alpha\beta - \gamma\delta = 0$ and $\alpha'\beta' - \gamma'\delta' = 0$ represent

$$\left\{q_0\left(q_0 + \frac{\mathrm{i}}{\tau_{\mathrm{AM}}}\right) - M_{\mathrm{A}i\mathrm{A}i}(\boldsymbol{q})\right\}\left\{q_0\left(q_0 + \frac{\mathrm{i}}{\tau_{\mathrm{BM}}}\right) - M_{\mathrm{B}i\mathrm{B}i}(\boldsymbol{q})\right\}$$
$$-M_{\mathrm{A}i\mathrm{B}i}(\boldsymbol{q})M_{\mathrm{B}i\mathrm{A}i}(\boldsymbol{q}) = 0, \tag{10.4.5}$$

for $i = x.y$ and $i = z$, respectively. $i = z$ corresponds to a longitudinal mode and $i = x, y$ to two transverse modes.

In order to obtain phonon modes and viscosities, we investigate the two limiting cases: (I) $q_0\tau_{\alpha\mathrm{M}} \gg 1$ for phonon modes, (II) $q_0\tau_{\alpha\mathrm{M}} \ll 1$ for viscosities. First, we investigate Phonon modes.

I. Phonon modes and boson peaks ; $q_0\tau_{\alpha\mathrm{M}} \gg 1$

We obtain phonon modes

$$q_0^2 = \omega_{i\pm}^2(\boldsymbol{q}) = \frac{1}{2}\Bigg[M_{\mathrm{A}i\mathrm{A}i}(\boldsymbol{q}) + M_{\mathrm{B}i\mathrm{B}i}(\boldsymbol{q})$$
$$\pm\sqrt{\{M_{\mathrm{A}i\mathrm{A}i}(\boldsymbol{q}) - M_{\mathrm{B}i\mathrm{B}i}(\boldsymbol{q})\}^2 + 4M_{\mathrm{A}i\mathrm{B}i}(\boldsymbol{q})M_{\mathrm{B}i\mathrm{A}i}(\boldsymbol{q})}\Bigg]. \tag{10.4.6}$$

Note that the term $M_{\alpha i\alpha i}$ consists of the αth individual phonon frequency and the frequency shift due to the coupling of particles with different particles, whereas $M_{\alpha i\beta i}$ plays a role in the mixing between the individual phonon frequencies.

Next we investigate the phonon frequencies in the two limiting cases: First we introduce

$$v_{\sigma\alpha}^{(n)} \equiv \frac{1}{M_\alpha}\int d^3R g_{\alpha\alpha}(R)\frac{V_{\sigma\alpha\alpha}}{3}\xi^n,$$
$$v_{\pi\alpha}^{(n)} \equiv \frac{1}{M_\alpha}\int d^3R g_{\alpha\alpha}(R)\frac{2V_{\pi\alpha\alpha}}{3}\xi^n,$$
$$v'^{(n)}{}_{\sigma\alpha} \equiv \frac{1}{M_\alpha}\int d^3R g_{\alpha\beta}(R)\frac{V_{\sigma\alpha\beta}}{3}\xi^n,$$
$$v'^{(n)}{}_{\pi\alpha} \equiv \frac{1}{M_\alpha}\int d^3R g_{\alpha\beta}(R)\frac{2V_{\pi\alpha\beta}}{3}\xi^n, \tag{10.4.7}$$
$$c_\alpha \equiv \frac{1}{M_\alpha}e_\alpha^2\rho_\alpha\frac{1}{3\epsilon_0 v_a}. \tag{10.4.8}$$

1. $qR = \xi \ll 1$; in the long wavelength regime:
   If we define as

$$A_{\alpha t} \equiv A_{\alpha x,y} \equiv v'^{(0)}{}_{\sigma\alpha} + v'^{(0)}{}_{\pi\alpha} - c_\alpha \equiv A_\alpha - c_\alpha,$$
$$B_{\alpha t} \equiv B_{\alpha x,y} \equiv \frac{v_{\sigma\alpha}^{(2)}}{10} + \frac{v_{\pi\alpha}^{(2)}}{5},$$
$$B'{}_{\alpha t} \equiv B'{}_{\alpha x} = B'{}_{\alpha y} \equiv \frac{v'^{(2)}{}_{\sigma\alpha}}{10} + \frac{v'^{(2)}{}_{\pi\alpha}}{5}, \tag{10.4.9}$$

$$A_{\alpha l} \equiv A_{\alpha z} \equiv v'^{(0)}{}_{\sigma\alpha} + v'^{(0)}{}_{\pi\alpha} + 2c_\alpha \equiv A_\alpha + 2c_\alpha,$$
$$B_{\alpha l} \equiv B_{\alpha z} \equiv \frac{3v^{(2)}_{\sigma\alpha}}{10} + \frac{v^{(2)}_{\pi\alpha}}{10},$$
$$B'_{\alpha l} \equiv B'_{\alpha z} \equiv \frac{3v'^{(2)}{}_{\sigma\alpha}}{10} + \frac{v'^{(2)}{}_{\pi\alpha}}{10}, \tag{10.4.10}$$

we obtain

$$M_{\alpha x,y\alpha x,y} = A_{\alpha t} + B_{\alpha t} = A_\alpha + B_{\alpha t} - c_\alpha$$
$$M_{\alpha z\alpha z} = A_{\alpha l} + B_{\alpha l} = A_\alpha + B_{\alpha l} + 2c_\alpha,$$
$$M_{\alpha x,y\beta x,y} = \frac{\zeta_\beta}{\zeta_\alpha}\sqrt{\frac{N_\alpha}{N_\beta}}\{-A_{\alpha t} + B'_{\alpha t}\}$$
$$= \frac{\zeta_\beta}{\zeta_\alpha}\sqrt{\frac{N_\alpha}{N_\beta}}\{-A_\alpha + B'_{\alpha t} + c_\alpha\},$$
$$M_{\alpha z\beta z} = \frac{\zeta_\beta}{\zeta_\alpha}\sqrt{\frac{N_\alpha}{N_\beta}}\{-A_{\alpha l} + B'_{\alpha l}\}$$
$$= \frac{\zeta_\beta}{\zeta_\alpha}\sqrt{\frac{N_\alpha}{N_\beta}}\{-A_\alpha + B'_{\alpha t} - 2c_\alpha\}. \tag{10.4.11}$$

In the long wavelength limit, we obtain the phonon frequencies:

$$q_0^2 = \frac{1}{2}\Bigg[A_{Ai} + A_{Bi} + B_{Ai} + B_{Bi} \pm \Bigg(|A_{Ai} + A_{Bi}| + \frac{A_{Ai}(B_{Ai} - B_{Bi} - 2B'_{Bi}) + A_{Bi}(B_{Bi} - B_{Ai} - 2B'_{Ai})}{|A_{Ai} + A_{Bi}|}\Bigg)\Bigg]. \tag{10.4.12}$$

For $A_{\alpha i} > 0$,

$$\omega_{t\pm}^2(\boldsymbol{q}) = \begin{cases} A_A + A_B - c_A - c_B & + \frac{A_A - c_A}{A_A + A_B - c_A - c_B}(B_{At} - B'_{Bt}) \\ & + \frac{A_B - c_B}{A_A + A_B - c_A - c_B}(B_{Bt} - B'_{At}), \\ \frac{A_A - c_A}{A_A + A_B - c_A - c_B}(B_{Bt} + B'_{Bt}) & + \frac{A_B - c_B}{A_A + A_B - c_A - c_B}(B_{At} + B'_{At}). \end{cases} \tag{10.4.12a}$$

$$\omega_{l\pm}^2(\boldsymbol{q}) = \begin{cases} A_A + A_B + 2c_A + 2c_B & + \frac{A_A + 2c_A}{A_A + A_B + 2c_A + 2c_B}(B_{Al} - B'_{Bl}) \\ & + \frac{A_B + 2c_B}{A_A + A_B + 2c_A + 2c_B}(B_{Bl} - B'_{Al}), \\ \frac{A_A + 2c_A}{A_A + A_B + 2c_A + 2c_B}(B_{Bl} + B'_{Bl}) & + \frac{A_B + 2c_B}{A_A + A_B + 2c_A + 2c_B}(B_{Al} + B'_{Al}). \end{cases} \tag{10.4.13a}$$

$\omega_{\lambda q+}$ are the optical modes and $\omega_{\lambda q-}$ the acoustic modes. The velocities of the acoustic phonons are given by

$$v_t = \frac{\omega_{t-}(\boldsymbol{q})}{q}$$
$$= \sqrt{\frac{2A_A - 2c_A}{A_A + A_B - c_A - c_B}(B_{Bt} + B'_{Bt}) + \frac{2A_B - 2c_B}{A_A + A_B - c_A - c_B}(B_{At} + B'_{At})\frac{1}{q}}, \tag{10.4.14a}$$

$$v_l = \frac{\omega_{l-}(\boldsymbol{q})}{q}$$
$$= \sqrt{\frac{2A_A + 4c_A}{A_A + A_B + 2c_A + 2c_B}(B_{Bl} + B'_{Bl}) + \frac{2A_B + 4c_B}{A_A + A_B + 2c_A + 2c_B}(B_{Al} + B'_{Al})\frac{1}{q}}. \tag{10.4.15a}$$

For $A_{\alpha i} < 0$,

$$\omega^2_{t\pm}(\boldsymbol{q}) = \begin{cases} \dfrac{A_A - c_A}{A_A + A_B - c_A - c_B}(B_{Bt} + B'_{Bt}) + \dfrac{A_B - c_B}{A_A + A_B - c_A - c_B}(B_{At} + B'_{At}), \\ A_A + A_B - c_A - c_B + \dfrac{A_A - c_A}{A_A + A_B - c_A - c_B}(B_{At} - B'_{Bt}) \\ \quad + \dfrac{A_B - c_B}{A_A + A_B - c_A - c_B}(B_{Bt} - B'_{At}), \end{cases} \tag{10.4.12b}$$

$$\omega^2_{l\pm}(\boldsymbol{q}) = \begin{cases} \dfrac{A_A + 2c_A}{A_A + A_B + 2c_A + 2c_B}(B_{Bl} + B'_{Bl}) + \dfrac{A_B + 2c_B}{A_A + A_B + 2c_A + 2c_B}(B_{Al} + B'_{Al}), \\ A_A + A_B + 2c_A + 2c_B + \dfrac{A_A + 2c_A}{A_A + A_B + 2c_A + 2c_B}(B_{Al} - B'_{Bl}) \\ \quad + \dfrac{A_B + 2c_B}{A_A + A_B + 2c_A + 2c_B}(B_{Bl} - B'_{Al}). \end{cases} \tag{10.4.13b}$$

where $\omega_{t+}(\boldsymbol{q})$ and $\omega_{l+}(\boldsymbol{q})$ are the acoustic modes, while $\omega_{t-}(\boldsymbol{q})$ and $\omega_{l-}(\boldsymbol{q})$ are imaginary numbers so disappear. These aspects originate from the mixing of the individual phonons. The velocities of the acoustic phonons are given by

$$v_t = \frac{\omega_{t+}(\boldsymbol{q})}{q}$$
$$= \sqrt{\frac{2A_A - 2c_A}{A_A + A_B - c_A - c_B}(B_{Bt} + B'_{Bt}) + \frac{2A_B - 2c_B}{A_A + A_B - c_A - c_B}(B_{At} + B'_{At})\frac{1}{q}}, \tag{10.4.14b}$$

$$
\begin{aligned}
vl &= \frac{\omega_{l+}(\boldsymbol{q})}{q} \\
&= \sqrt{\frac{2A_{\mathrm{A}}+4c_{\mathrm{A}}}{A_{\mathrm{A}}+A_{\mathrm{B}}+2c_{\mathrm{A}}+2c_{\mathrm{B}}}(B_{\mathrm{B}l}+B'_{\mathrm{B}l})+\frac{2A_{\mathrm{B}}+4c_{\mathrm{B}}}{A_{\mathrm{A}}+A_{\mathrm{B}}+2c_{\mathrm{A}}+2c_{\mathrm{B}}}(B_{\mathrm{A}l}+B'_{\mathrm{A}l})\frac{1}{q}}.
\end{aligned} \tag{10.4.15b}
$$

2. $qR \gg 1$; in the short wavelength regime:
   In the short wavelength limit, we obtain

$$
\begin{aligned}
M_{\alpha x\alpha x} &= M_{\alpha y\alpha y} = v_{\sigma\alpha}^{(0)} + v_{\pi\alpha}^{(0)} + v'^{(0)}{}_{\sigma\alpha} + v'^{(0)}{}_{\pi\alpha} - c_\alpha \\
&\equiv S_\alpha - c_\alpha, \\
M_{\alpha z\alpha z} &= v_{\sigma\alpha}^{(0)} + v_{\pi\alpha}^{(0)} + v'^{(0)}{}_{\sigma\alpha} + v'^{(0)}{}_{\pi\alpha} + 2c_\alpha \\
&\equiv S_\alpha + 2c_\alpha \\
M_{\alpha x\beta x} &= M_{\alpha y\beta y} = \frac{\zeta_\beta}{\zeta_\alpha}\sqrt{\frac{N_\alpha}{N_\beta}}c_\alpha \\
M_{\alpha z\beta z} &= \frac{\zeta_\beta}{\zeta_\alpha}\sqrt{\frac{N_\alpha}{N_\beta}}(-2c_\alpha),
\end{aligned} \tag{10.4.16}
$$

Then we obtain

$$
\begin{aligned}
\omega_{t\pm}^2(\boldsymbol{q}) &= \frac{1}{2}\{S_{\mathrm{A}} + S_{\mathrm{B}} - c_{\mathrm{A}} - c_{\mathrm{B}} \\
&\quad \pm\sqrt{(S_{\mathrm{A}} - S_{\mathrm{B}} - c_{\mathrm{A}} + c_{\mathrm{B}})^2 + 4c_{\mathrm{A}}c_{\mathrm{B}}}\},
\end{aligned} \tag{10.4.17}
$$

$$
\begin{aligned}
\omega_{l\pm}^2(\boldsymbol{q}) &= \frac{1}{2}\{S_{\mathrm{A}} + S_{\mathrm{B}} + 2c_{\mathrm{A}} + 2c_{\mathrm{B}} \\
&\quad \pm\sqrt{(S_{\mathrm{A}} - S_{\mathrm{B}} + 2c_{\mathrm{A}} - 2c_{\mathrm{B}})^2 + 16c_{\mathrm{A}}c_{\mathrm{B}}}\},
\end{aligned} \tag{10.4.18}
$$

These frequencies correspond to the boson peaks, where the density of states have peaks. In nonpolarized liquids, because of $c_\alpha = 0$ the corresponding longitudinal and transverse frequencies and the boson peaks are of the same.

10.4.2 Phonon Dispersion Curves of the Longitudinal Modes

Here we show the figures for three specific cases of phonon dispersion curves as illustrated in Figures 10.1, 10.2, 10.3. We restrict ourselves to the longitudinal modes, $V_\pi = 0$ and $c_\alpha = 0$. To the first order, we put $g_{\alpha\alpha}(R) = \rho_\alpha\delta(R-a)$ and $g_{\alpha\beta}(R) = \rho'_\alpha\delta(R-c)$, where a and c are the particle distances between α particles and between unlike particles, respectively. According to the restrictions, in this paragraph, we abbreviate the parameters as:

$$
\begin{aligned}
v_\alpha &\equiv v_{\sigma\alpha}^{(0)}, \\
v'_\alpha &\equiv v'^{(0)}{}_{\sigma\alpha}.
\end{aligned}
$$

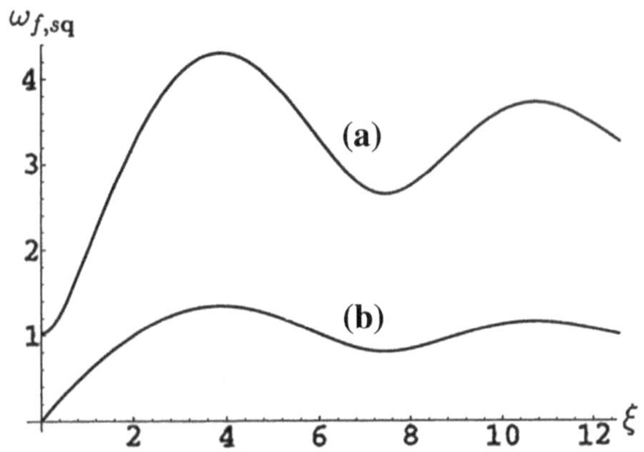

Figure 10.1 Phonon dispersion curves versus $\xi = aq$ for a binary mixture with large mass difference between components A and B; $M_B/M_A = 20$: (a) ω_{fq}, (b) ω_{sq}. Potential parameters: $v_A = 10,\ v_B = 1;\ v'_A = 1,\ v'_B = 0.05$.

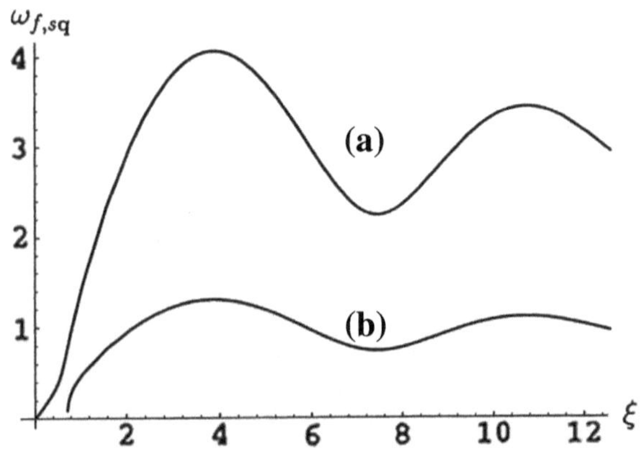

Figure 10.2 Same as Fig. 10.1 expect that here $v'_A = -1,\ v'_B = -0.05$.

The figures show the results for three specific cases of phonon dispersion curves with $N_A = N_B$ and $a/c = 2$:

1. $M_B/M_A = 20,\quad v_A = 10,\quad v_B = 1,\quad v'_A = 1,\quad v'_B = 0.05,$ in Figure 10.1.
2. $M_B/M_A = 20,\quad v_A = 10,\quad v_B = 1,\quad v'_A = -1,\quad v'_B = -0.05,$ in Figure 10.2.
3. $M_B/M_A = 1,\quad v_A = v_B = 0,\quad v'_A = v'_B = 1,$ in Figure 10.3.

We have illustrated the longitudinal phonons in two-component liquids with large mass difference in Figures 10.1, 10.2 and with the same mass in Figure 10.3. When the mass difference is large, given that the wave function of a light mass particle spreads out,

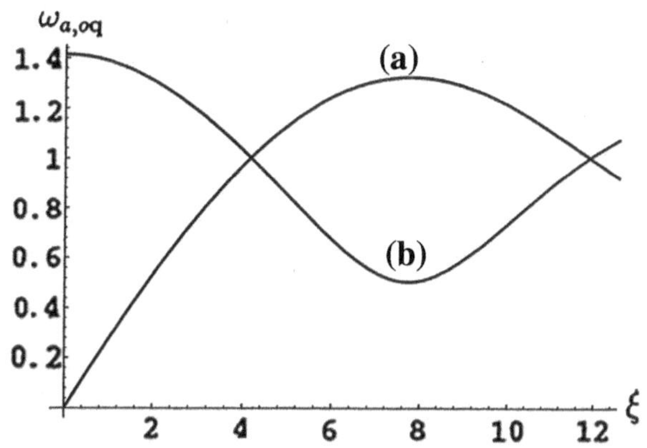

Figure 10.3 Phonon dispersion curves versus $\xi - aq$ for a binary mixture with the same mass. $M_B/M_A = 1$: (a) ω_{aq}, (b) ω_{oq}. Potential parameter $v_A = v_B = 0$, $v'_A = v'_B = 1$.

the coupling between light mass particles dominates the other couplings. The coupling between the different mass atoms causes a shift in the frequencies of individual phonons and the mixing between individual phonons.

10.4.3 The Characteristic Features of the Large Mass Difference, (i) and (ii)

In the cases of the large mass difference, (i) and (ii), we denote fast and slow frequency modes as ω_{fq} and ω_{sq}, respectively. The characteristic features of these modes are as follows [18–25]:

1. In the short wavelength regime:
   Eq. (10.4.18) leads to

$$\left.\begin{array}{l}\omega_{fq}\\ \omega_{sq}\end{array}\right\} \cong \left\{\begin{array}{l}\sqrt{S_{\rm A}} = \sqrt{M_{{\rm A}z{\rm A}z}(q)} = \sqrt{v_A l(\xi) + v'_A},\\ \sqrt{S_{\rm B}} = \sqrt{M_{{\rm B}z{\rm B}z}(q)} = \sqrt{v_B l(\xi) + v'_B},\end{array}\right. \tag{10.4.19}$$

   where

$$l(\xi) = 1 - 3\left(\frac{\sin\xi}{\xi} + \frac{2\cos\xi}{\xi^2} - \frac{2\sin\xi}{\xi^3}\right), \qquad \xi = aq. \tag{10.4.20}$$

   ω_{fq} is a frequency mode mainly involves the light particles, while ω_{sq} mainly involves the heavier particles. This behavior is similar to that found for fast and slow phonons, although there are important difference in the hydrodynamic limit when compared to the kinetic theory calculation carried out by Compa and Cohen [22–24]. It should be noted that the individual phonon frequencies $\sqrt{v_A l(\xi)}$ and $\sqrt{v_B l(\xi)}$ are shifted upward or downward corresponding to the $\pm$ sign of the terms v'_α , respectively. For the negative value of the term v'_B, if the absolute value of v'_B exceeds the minimum value of $v_B l(\xi)$ near $\xi = 7.5$, ω_{sq} becomes imaginary and the phonon mode disappears [19]. This case is expected for not so much large mass difference. In the short wavelength limit,

$$\lim_{q\to\infty} \omega_{f,s,\boldsymbol{q}} = \omega_{A,B}. \tag{10.4.21}$$

These frequencies correspond to the Boson peaks.

2. In the long wavelength regime:
In this regime, the hybridization effect is most important. We investigate the phonon frequencies for the two cases (i) $v'_\alpha > 0$ and (ii) $v'_\alpha < 0$.

a. $v'_\alpha > 0$. Eq. (10.4.13a) leads to

$$\left.\begin{array}{l}\omega_{f\boldsymbol{q}}\\ \omega_{s\boldsymbol{q}}\end{array}\right\} \cong \left\{\begin{array}{l}\sqrt{v'_{\mathrm{A}} + \frac{3}{10}v_{\mathrm{A}}\xi^2}\\ \sqrt{\frac{3}{10}v_{\mathrm{B}}\xi}\end{array}\right. \tag{10.4.22a}$$

$\omega_{f\boldsymbol{q}}$ pushed the individual mode $\omega_{\mathrm{A}\boldsymbol{q}} = \sqrt{\frac{3}{10}v_{\mathrm{A}}\xi}$ upward, while $\omega_{s\boldsymbol{q}}$ is the individual longitudinal acoustic phonon mode $\sqrt{\frac{3}{10}v_B}\,\xi$.

b. $v'_\alpha < 0$. The phonon dispersion curves change drastically. Eq. (10.4.13b) leads to

$$\left.\begin{array}{l}\omega_{f\boldsymbol{q}}\\ \omega_{s\boldsymbol{q}}\end{array}\right\} \cong \left\{\begin{array}{l}\sqrt{\frac{3}{10}v_{\mathrm{B}}\xi}\\ \sqrt{v'_{\mathrm{A}} + \frac{3}{10}v_{\mathrm{A}}\xi^2}\end{array}\right. . \tag{10.4.22b}$$

The hybridization effect causes the individual phonon modes $\omega_{\mathrm{A}\boldsymbol{q}}$ and $\omega_{\mathrm{B}\boldsymbol{q}}$ transfer to the frequencies $\omega_{f\boldsymbol{q}} = \omega_{\mathrm{B}\boldsymbol{q}}$ and $\omega_{s\boldsymbol{q}} = \sqrt{v'_{\mathrm{A}} + \frac{3}{10}v_{\mathrm{A}}\xi^2}$ pushed $\omega_{\mathrm{A}\boldsymbol{q}}$ downward, respectively. So, $\omega_{f\boldsymbol{q}}$ is the longitudinal acoustic phonon mode, while $\omega_{s\boldsymbol{q}}$ is imaginary and disappears in the regions, $0 < \xi < \frac{10|v'_{\mathrm{A}}|}{3v_{\mathrm{A}}}$. It is interesting to note that the experimental results for $Mg_{70}Zn_{30}$ glass [25] show longitudinal acoustic and optical dispersion curves similar to those shown in Figure10.2.

10.4.4 The Characteristic Features of the Same Mass, (iii)

In glasses of the case (iii), we can consider that each primitive cell contains two particles like Silicon. Eq. (10.4.6) yields

$$q_0^2 = M_{AiAi}(\boldsymbol{q}) \mp M_{AiBi}(\boldsymbol{q}). \tag{2.4.23}$$

In the glasses, the nearest neighbor interaction between different particles are most dominant. Thus we can put $v_{\sigma A} = v_{\sigma B} \cong 0$. We obtain a longitudinal acoustic phonon $\omega_{a\boldsymbol{q}}$ and a longitudinal optical phonon $\omega_{o\boldsymbol{q}}$ [3–5]

$$\left.\begin{array}{l}\omega_{a\boldsymbol{q}}\\ \omega_{o\boldsymbol{q}}\end{array}\right\} \cong \left\{\begin{array}{l}\sqrt{v'_{\mathrm{A}}l(\xi)}\\ \sqrt{v'_{\mathrm{A}}\{1+3m(\xi)\}}\end{array}\right. , \tag{10.4.24}$$

where

$$m(\xi) = \frac{\sin\xi}{\xi} + \frac{2\cos\xi}{\xi^2} - \frac{2\sin\xi}{\xi^3}, \tag{10.4.25}$$

and $l(\xi) = 1 - 3m(\xi)$ in Eq. (10.4.20).

Now we investigate the characteristic feature of Eq. (10.4.24).

1. In the short wavelength regime: ω_{aq} behaves the same phonon dispersion curve as the single particle system so that it has a roton type minimum at a reciprocal nearest to neighbor particle distance. On the other hand ω_{op} behaves an opposite magnitude of ω_{aq}.
   In the short wavelength limit, $q \to \infty$, both the modes tend to
   $$\lim_{q\to\infty} \omega_{aq},\ \omega_{oq} \to v'_{\mathrm{A}}. \tag{10.4.26}$$
   This corresponds to the boson peak.
2. In the long wavelength regime:
   $$\left.\begin{matrix}\omega_{aq} \\ \omega_{oq}\end{matrix}\right\} \cong \left\{\begin{matrix}\sqrt{v'_{\mathrm{A}}\xi} \\ \sqrt{v'_{\mathrm{A}}(2 - \frac{3}{10}\xi^2)}\end{matrix}\right.$$
   ω_{aq} behaves as the NG boson, while ω_{oq} as an optical phonon.

II. Viscosities ; $q_0\tau_{\alpha\mathrm{M}} \ll 1$

Next we investigate the secular Eq. (10.4.5) in the low frequency regime: Eq. (10.4.5) leads to

$$(q_0 + \mathrm{i}\tau_{\alpha\mathrm{M}}M_{\alpha i\alpha i})(q_0 + \mathrm{i}\tau_{\beta\mathrm{M}}M_{\beta i\beta i}) - \mathrm{i}\tau_{\alpha\mathrm{M}}M_{\alpha i\beta i}\mathrm{i}\tau_{\beta\mathrm{M}}M_{\beta i\alpha i} = 0. \tag{10.4.27}$$

Thus we obtain the dissipative modes due to viscosity

$$\begin{aligned} q_0 &= \frac{\mathrm{i}}{2}\Big[\tau_{\mathrm{AM}}M_{\mathrm{A}i\mathrm{A}i} + \tau_{\mathrm{BM}}M_{\mathrm{B}i\mathrm{B}i} \\ &\quad -\sqrt{(\tau_{\mathrm{AM}}M_{\mathrm{A}i\mathrm{A}i} - \tau_{\mathrm{BM}}M_{\mathrm{B}i\mathrm{B}i})^2 + 4\tau_{\mathrm{AM}}M_{\mathrm{A}i\mathrm{B}i}\tau_{\mathrm{BM}}M_{\mathrm{B}i\mathrm{A}i}}\Big], \\ &\cong \mathrm{i}\frac{2\tau_{\mathrm{AM}}A_{\mathrm{A}i}}{\tau_{\mathrm{AM}}A_{\mathrm{A}i} + \tau_{\mathrm{BM}}A_{\mathrm{B}i}}\tau_{\mathrm{BM}}(B_{\mathrm{B}i} + B'_{\mathrm{B}i}) \\ &\quad + \mathrm{i}\frac{2\tau_{\mathrm{BM}}A_{\mathrm{B}i}}{\tau_{\mathrm{AM}}A_{\mathrm{A}i} + \tau_{\mathrm{BM}}A_{\mathrm{B}i}}\tau_{\mathrm{AM}}(B_{\mathrm{A}i} + B'_{\mathrm{A}i}), \end{aligned} \tag{10.4.28}$$

where the modes corresponding to the optical modes disappear because of the large imaginary value. We should not confuse the imaginary number i and the indices i; the i in front of the terms means the imaginary number, but i in the indices means the component of the coordinates. Thus viscosities are given by

$$\eta_t = \left\{\frac{2\tau_{\mathrm{AM}}(A_{\mathrm{A}} - c_{\mathrm{A}})}{\tau_{\mathrm{AM}}(A_{\mathrm{A}} - c_{\mathrm{A}}) + \tau_{\mathrm{BM}}(A_{\mathrm{B}} - c_{\mathrm{B}})}\tau_{\mathrm{BM}}(B_{\mathrm{B}t} + B'_{\mathrm{B}t})\right.$$

$$+\frac{2\tau_{\rm BM}(A_{\rm B}-c_{\rm B})}{\tau_{\rm AM}(A_{\rm A}-c_{\rm A})+\tau_{\rm BM}(A_{\rm B}-c_{\rm B})}\tau_{\rm AM}(B_{{\rm A}t}+B'_{{\rm A}t})\Bigg\}^{\frac{1}{2}}\frac{1}{q}, \tag{10.4.29}$$

$$\eta_l=\Bigg\{\frac{2\tau_{\rm AM}(A_{\rm A}+2c_{\rm A})}{\tau_{\rm AM}(A_{\rm A}+2c_{\rm A})+\tau_{\rm BM}(A_{\rm B}+2c_{\rm B})}\tau_{\rm BM}(B_{{\rm B}l}+B'_{{\rm B}l})$$

$$+\frac{2\tau_{\rm BM}(A_{\rm B}+2c_{\rm B})}{\tau_{\rm AM}(A_{\rm A}+2c_{\rm A})+\tau_{\rm BM}(A_{\rm B}+2c_{\rm B})}\tau_{\rm AM}(B_{{\rm A}l}+B'_{{\rm A}l})\Bigg\}^{\frac{1}{2}}\frac{1}{q}, \tag{10.4.30}$$

where η_l and η_t correspond to longitudinal and transverse acoustic phonons, respectively.

10.5 The Kauzmann Entropy Crisis and the VTF Law; Specific Heat, Relaxation Times, and Transport Coefficients

The sum of an infinite series of interacting bubbles of intra-band particle-hole pairs leads to sound modes in multi-component system as well as one-component systems. The extra bubbles of intra-band particle-hole pairs not constituting the sound yield the fluctuation entropy of sound, which always lower the entropy. In the supercooling process, enormous bubbles not constituting the sound mode are created. The fluctuation entropy mixes with the entropy of the sound mode. This mixing entropy is the Kauzmann entropy. From the Einstein relation on entropy and probability, the Kauzmann entropy yields the VTF law. The random scattering processes due to random harmonic frequencies and random hopping magnitudes lead to the relaxation and transport processes.

10.5.1 The Kauzmann Entropy and the VTF Law

First, we investigate the entropy due to intra-band density fluctuations, the correlation function of which yields sound. Since sound depends on the temperature, in deriving the entropy, we must carefully treat the calculation process. Therefore we calculate the thermodynamical function by gradually increasing the interaction Hamiltonian. We introduce the Hamiltonian with a parameter λ [1,15,16]:

$$H(\lambda)=H_0+\lambda H_I. \tag{10.5.1}$$

The thermodynamical function Ω_λ for the above Hamiltonian is given by

$$\Omega_\lambda=-\frac{1}{\beta}\ln\{{\rm Tr}e^{-\beta(H_0-\mu N+\lambda H_I)}\}. \tag{10.5.2}$$

Here we calculate the thermodynamical function due to intra-band density fluctuations denoted by $\Omega_{{\rm intra}\lambda}$. From Eq. (10.1.18), we obtain

$$\frac{\partial\Omega_{{\rm intra}\lambda}}{\partial\lambda}=\frac{1}{\lambda}\langle\lambda H_\lambda\rangle_\lambda$$

$$\cong \frac{1}{2}\sum_{\alpha\beta\boldsymbol{q}} \frac{1}{\sqrt{N_\alpha N_\beta}} V_{\alpha\beta}(q)\langle \rho^{\dagger}_{\alpha 0\boldsymbol{q}} \rho_{\beta 0\boldsymbol{q}}\rangle_\lambda,$$

$$= -\frac{1}{2\beta}\sum_{\alpha\beta\boldsymbol{q}} V_{\alpha\beta}(q) F^{\lambda}_{\beta\alpha}(q), \tag{10.5.3}$$

where $\langle\cdots\rangle_\lambda$ and $F^\lambda(q)$ are calculated by H_I replaced by λH_I in $\langle\cdots\rangle_c$and $F(q)$. Eq. (10.2.2) leads to

$$\sum_{\gamma}\{\delta_{\alpha\gamma} - P_\alpha(q)\lambda V_{\alpha\gamma}(q)\}F^{\lambda}_{\gamma\beta}(q) = P_\alpha(q)\delta_{\alpha\beta}, \tag{10.5.4}$$

where we have put

$$P^{\lambda}_{\alpha}(q) \cong P_\alpha(q) = \frac{\beta\omega^2_{\alpha\boldsymbol{q}}}{q_0(q_0 + \frac{\mathrm{i}}{\tau_{\alpha 0}}) - \omega^2_{\alpha\boldsymbol{q}}}, \tag{10.5.5}$$

because of the condition $\beta\hbar|J_\mu| \ll 1$ in Eq. (10.2.3).

Hereafter we confine ourselves to two-component systems. Then Eq. (10.5.4) leads to

$$\begin{pmatrix} F^{\lambda}_{\mathrm{AA}} & F^{\lambda}_{\mathrm{AB}} \\ F^{\lambda}_{\mathrm{BA}} & F^{\lambda}_{\mathrm{BB}} \end{pmatrix} = \frac{1}{\det||}\begin{pmatrix} (1-\lambda P_{\mathrm{B}}V_{\mathrm{BB}})P_{\mathrm{A}} & \lambda P_{\mathrm{A}}V_{\mathrm{AB}}P_{\mathrm{B}} \\ \lambda P_{\mathrm{B}}V_{\mathrm{BA}}P_{\mathrm{B}} & (1-\lambda P_{\mathrm{A}}V_{\mathrm{AA}})P_{\mathrm{B}} \end{pmatrix}, \tag{10.5.6}$$

$$\det|| = (1-\lambda P_{\mathrm{A}}V_{\mathrm{AA}})(1-\lambda P_{\mathrm{B}}V_{\mathrm{BB}}) - \lambda^2 P_{\mathrm{A}}V_{\mathrm{AB}}P_{\mathrm{B}}V_{\mathrm{BA}}. \tag{10.5.7}$$

Thus Eq. (10.5.3) yields

$$\frac{\partial\Omega_{\mathrm{intra}\lambda}}{\partial\lambda}$$

$$= -\frac{1}{2\beta}\sum_{q}\frac{(1-\lambda P_{\mathrm{B}}V_{\mathrm{BB}})P_{\mathrm{A}}V_{\mathrm{AA}} + (1-\lambda P_{\mathrm{A}}V_{\mathrm{AA}})P_{\mathrm{B}}V_{\mathrm{BB}} + 2\lambda P_{\mathrm{A}}P_{\mathrm{B}}V_{\mathrm{AB}}V_{\mathrm{BA}}}{(1-\lambda P_{\mathrm{A}}V_{\mathrm{AA}})(1-\lambda P_{\mathrm{B}}V_{\mathrm{BB}}) - \lambda^2 P_{\mathrm{A}}P_{\mathrm{B}}V_{\mathrm{AB}}V_{\mathrm{BA}}}$$

$$= \frac{1}{2\beta}\sum_{q}\frac{\partial}{\partial\lambda}\ln\left\{(1-\lambda P_{\mathrm{A}}V_{\mathrm{AA}})(1-\lambda P_{\mathrm{B}}V_{\mathrm{BB}}) - \lambda^2 P_{\mathrm{A}}P_{\mathrm{B}}V_{\mathrm{AB}}V_{\mathrm{BA}}\right\}. \tag{10.5.8}$$

Integrating Eq. (10.5.8) in λ and putting $\lambda \to 1$, we obtain the thermodynamical function due to the intra-band density fluctuations, Ω_{intra}:

$$\Omega_{\mathrm{intra}} = \frac{1}{2\beta}\sum_{\mathrm{i}\nu_n\boldsymbol{q}}\ln\{(1-P_{\mathrm{A}}V_{\mathrm{AA}})(1-P_{\mathrm{B}}V_{\mathrm{BB}}) - P_{\mathrm{A}}P_{\mathrm{B}}V_{\mathrm{AB}}V_{\mathrm{BA}}\} + \Omega_{\mathrm{intra}\,0}, \tag{10.5.9}$$

where the term $\Omega_{\mathrm{intra}\,0}$ comes from the initial condition. In order to calculate the entropy, we neglect $1/\tau_0$ terms under the condition $\beta\hbar/\tau_0 \ll 1$ and we put $q_0 \longrightarrow \mathrm{i}\nu_n$. From Eq. (10.5.5),

$$P_\alpha(q) = \frac{\beta\omega^2_{\alpha q}}{\mathrm{i}\nu_n^2 - \omega^2_{\alpha q}},$$

we obtain

$$\begin{aligned}\Omega_{\text{intra}} &= \frac{1}{2\beta}\sum_q \ln \frac{(i\nu_n^2-\omega_{\text{A}s}^2(q))(i\nu_n^2-\omega_{\text{B}s}^2(q))-\beta^2\omega_{\text{A}q}^2\omega_{\text{B}q}^2 V_{\text{AB}}V_{\text{BA}}}{(i\nu_n^2-\omega_{\text{A}q}^2)(i\nu_n^2-\omega_{\text{B}q}^2)}\\ &\quad +\Omega_{\text{intra0}}\\ &= \frac{1}{2\beta}\sum_q \ln \frac{(i\nu_n^2-\omega_{\text{s+}}^2(q))(i\nu_n^2-\omega_{\text{s-}}^2(q))}{(i\nu_n^2-\omega_{\text{A}q}^2)(i\nu_n^2-\omega_{\text{B}q}^2)}+\Omega_{\text{intra0}}. \end{aligned} \tag{10.5.10}$$

The Contour integration of Eq. (10.5.10) gives

$$\begin{aligned}\Omega_{\text{intra}} &= \frac{1}{\beta}\sum_{\boldsymbol{q}}\Big[\ln(1-e^{-\beta\hbar\omega_{\text{s+}}(q)})(1-e^{-\beta\hbar\omega_{\text{s-}}(q)})\\ &\quad -\ln(1-e^{-\beta\hbar\omega_{\text{A}q}})(1-e^{-\beta\hbar\omega_{\text{B}q}})\Big]+\Omega_{\text{intra}\,0}. \end{aligned} \tag{10.5.11}$$

$$\Omega_{\text{intra0}} = \frac{1}{\beta}\sum_{\alpha\boldsymbol{q}}\ln(1-e^{-\beta\hbar\omega_{\alpha q}}), \tag{10.5.12}$$

where we have neglected the term $\frac{1}{2}\sum_q(\hbar\omega_{\text{s}q}-\hbar\omega_q^0)$. The first, the second, and the third terms in Eq. (10.5.11) correspond to the thermodynamical function for the sound, the fluctuations due to bubbles of intra-band particle-hole pairs and the dissipation due to the diffusion, respectively. $\Omega_{\text{intra}\,0}$ in Eq. (10.5.12) is determined such that in an equilibrium state, there remains the thermodynamical function of sound, i.e., the first term; the second and third terms must cancel. This means that the third term due to diffusivity compensates the second term due to the fluctuations of bubbles of intra-band particle-hole pairs and the system becomes a local equilibrium.

Now we investigate the dependence of Ω_{intra} on temperature.

1. Using $\beta\hbar\omega_{\text{s}\pm}(q)\ll 1$ and $\beta\hbar\omega_{\alpha q}^0 \ll 1$ under the condition $\beta\hbar|J_\mu|\ll 1$, we expand the first and second terms in Eq. (10.5.11).
2. From the condition of the disappearing of sound, Eq. (10.2.26), the dominant contribution of the temperature dependence to the thermodynamical function in the first and the second terms comes from the regions $q\sim K$.
3. We divide the number of states in the phase space in $\boldsymbol{q}$ into the number of states N_0 around K denoted as $\boldsymbol{q}\in N_0$ and the remaining part denoted as $\boldsymbol{q}\notin N_0$.

Thus we obtain

$$\begin{aligned}\Omega_{\text{intra}} &\cong \frac{N_0}{\beta}\ln\left(\frac{\omega_{\text{s+}}(q)\omega_{\text{s}-}(q)}{\omega_{\text{A}q}\omega_{\text{B}q}}\right)\Bigg|_{q\sim K}+\Omega'_{\text{sound}}\\ &= \frac{N_0}{\beta}\ln\sqrt{\frac{\omega_{\text{A}}^2(q)\omega_{\text{B}}^2(q)-\beta^2\omega_{\text{A}q}^2\omega_{\text{B}q}^2V_{\text{AB}}(q)V_{\text{BA}}(q)}{\omega_{\text{A}q}^2\omega_{\text{B}q}^2}}\Bigg|_{q\sim K}+\Omega'_{\text{sound}}\\ &= \frac{N_0}{2\beta}\ln\left\{(1+\beta V_{\text{AA}}(K)(1+\beta V_{\text{BB}}(K))-\beta^2V_{\text{AB}}(K)V_{\text{BA}}(K)\right\}\end{aligned}$$

$$+\Omega'_{\text{sound}}, \tag{10.5.13}$$

$$\Omega'_{\text{sound}} = \frac{1}{\beta}\sum_{q}{}' \left[\ln(1 - e^{-\beta\hbar\omega_{s+}(q)}) + \ln(1 - e^{-\beta\hbar\omega_{s-}(q)})\right], \tag{10.5.14}$$

where the prime on Ω_{sound} and Σ means that the region of the wave vector $\boldsymbol{q}$ is limited to the remaining region $\boldsymbol{q} \notin N_0$ with the number of states $N - N_0$, excluded the region $\boldsymbol{q} \in N_0$ with the number of states N_0. Note that the dissipation part compensates the intra-band fluctuation part in the remaining regions $\boldsymbol{q} \notin N_0$ with the number of states $N - N_0$.

Now we investigate the first term of Eq. (10.5.13) using Eq. (10.2.26).

$$\begin{aligned}
&(1 + \beta V_{\text{AA}}(K))(1 + \beta V_{\text{BB}}(K)) - \beta^2 V_{\text{AB}}(K) V_{\text{BA}}(K) \\
&\quad = 1 + \{V_{\text{AA}}(K) + V_{\text{BB}}(K)\}\beta + \{V_{\text{AA}}(K)V_{\text{BB}}(K) - V_{\text{AB}}(K)V_{\text{BA}}(K)\}\beta^2 \\
&\quad = \{V_{\text{AA}}(K)V_{\text{BB}}(K) - V_{\text{AB}}(K)V_{\text{BA}}(K)\}(\beta_0 - \beta)(\beta' - \beta) \\
&\quad = \{V_{\text{AA}}(K)V_{\text{BB}}(K) - V_{\text{AB}}(K)V_{\text{BA}}(K)\}\left\{\beta_0\beta' - (\beta_0 + \beta')\beta + \beta^2\right\} \\
&\quad = \left(1 - \frac{\beta}{\beta_0}\right)\left(1 - \frac{\beta_0}{\beta'}\right) \\
&\quad = \left(1 - \frac{T_0}{T}\right)\left(1 - \frac{\beta_0}{\beta'}\right),
\end{aligned} \tag{10.5.15}$$

where β' is another solution of the disappearing sound equation. Comparing the second equation with the fourth one, we have got

$$\{V_{\text{AA}}(K)V_{\text{BB}}(K) - V_{\text{AB}}(K)V_{\text{BA}}(K)\}\beta_0\beta' = 1, \tag{10.5.16}$$

Thus we obtain

$$\Omega_{\text{intra}} \cong \frac{N_0}{2\beta}\ln\left\{1 - \frac{T_0}{T}\right\} + \Omega'_{\text{sound}}, \tag{10.5.17}$$

where we have neglected the constant term, which does not contribute to entropy. The entropy due to the intra-band density fluctuations, S_{intra} is given by

$$S_{\text{intra}} = S_{\text{K}} + S'_{\text{sound}}, \tag{10.5.18}$$

$$S_{\text{K}} \cong -\frac{N_0 k_{\text{B}}}{2}\frac{T_0}{T - T_0}, \tag{10.5.19}$$

$$\begin{aligned}
S'_{\text{sound}} &\cong -k_{\text{B}}\sum_{q\notin N_0}\left\{\ln(1 - e^{-\beta\hbar\omega_{s+}(q)}) + \ln(1 - e^{-\beta\hbar\omega_{s-}(q)}\right\} \\
&\quad + \frac{1}{T}\sum_{q\notin N_0}\left\{\frac{\hbar\omega_{s+}(q)}{e^{\beta\hbar\omega_{s+}(q)} - 1} + \frac{\hbar\omega_{s-}(q)}{e^{\beta\hbar\omega_{s-}(q)} - 1}\right\}.
\end{aligned} \tag{10.5.20}$$

S_{K} manifests the Kauzmann crisis. We call S_{K} the Kauzmann entropy.

A state of N-particles distributed randomly in a space corresponds to a minimum of N-particle potential in a configuration space. A hopping of a particle from a site to a vacancy corresponds to a jump from a deep valley to another deep valley in the multi-dimensional configuration space in the energy landscape model (ELM) as the α-relaxation process. The successive hoppings constitute a configuration space. A hopping generates intra-band elementary excitations so that the successive hoppings yield the intra-band fluctuation entropy. The hopping probability is proportional to the configuration number, which is e^{S_K/k_B} from the Einstein relation on entropy and probability. The hopping probability per particle is given uniquely by the hopping amplitude:

$$J = e^{\frac{zS_K}{Nk_B}} = e^{-\frac{E}{T-T_0}}, \tag{10.5.21}$$

where $E = zN_0T_0/(2N)$ and z is of the order of the number of the surrounding molecules composed of A and B particles. This equation manifests the Vogel-Tamman-Fulcher (VTF) law. The sound $\omega_{s\pm}(q)$ consists of the individual sound frequencies $\omega_A(q)$ and $\omega_B(q)$. As seen in the first terms on the right-hand side of Eq. (10.5.13), the fact that

$$\left.\frac{\omega_{s+}(q)\omega_{s-}(q)}{\omega_{Aq}\omega_{Bq}}\right|_{q\sim K} \cong (1+\beta V_{AA}(K))(1+\beta V_{BB}(K)) - \beta^2 V_{AB}(K)V_{BA}(K)$$

determines a unique hopping amplitude J. This originates from the temperature-dependence of sound. Thus J governs all individual quantities v_α and $J_{\alpha\mu}$. Then we can put

$$|J_{\alpha\mu}| \propto J, \tag{10.5.22}$$

$$\sqrt{U_J^{\alpha\mu\nu}} \propto J. \tag{10.5.23}$$

Since all $J_{\alpha\mu}$ are governed by a unique J, we can regard z as the order of the number of the surrounding molecules composed of A and B particles. It should be noted that the hopping J generates the intra-band particle-hole pairs, the sound $\omega_{s\pm}(q)$, and the Kauzmann entropy S_K, while S_K determines J in Eq. (10.5.21). Thus $\omega_{s\pm}(q)$ in Eq. (10.2.22) is self-consistently determined.

10.5.2 The Phonon Fluctuation Entropy and the Randomness of Harmonic Frequencies

Next we investigate the entropy due to inter-band density fluctuations. The thermodynamical function due to the inter-band density fluctuations is given in a similar manner to the intra-band density fluctuations:

$$\Omega_{\text{inter}} = \frac{1}{2\beta}\sum_{\lambda \boldsymbol{q}} \ln \frac{(i\nu_n^2 - \omega_{\lambda+}^2(\boldsymbol{q}))(i\nu_n^2 - \omega_{\lambda-}^2(\boldsymbol{q}))}{(i\nu_n^2 - \omega_A^2)(i\nu_n^2 - \omega_B^2)} + \Omega_{\text{inter}\,0},$$

$$= \frac{1}{\beta}\sum_{\lambda \boldsymbol{q}} \Big[\ln(1 - e^{-\beta\hbar\omega_{\lambda+}(\boldsymbol{q})})(1 - e^{-\beta\hbar\omega_{\lambda-}(\boldsymbol{q})})$$

$$-\ln(1-e^{-\beta\hbar\omega_{\mathrm{A}}})(1-e^{-\beta\hbar\omega_{\mathrm{B}}})\Big] + \Omega_{\mathrm{inter}\,0}, \tag{10.5.24}$$

$$\Omega_{\mathrm{inter}\,0} = \frac{1}{\beta}\sum_{\lambda \boldsymbol{q}} \ln(1-e^{-\beta\hbar\omega_{\mathrm{A}}})(1-e^{-\beta\hbar\omega_{\mathrm{B}}}) \tag{10.5.25}$$

where we have neglected the term $\frac{1}{2}\sum_{\lambda \boldsymbol{q}}(\hbar\omega_{\lambda q}-\hbar\omega)$. The entropy due to the inter-band density fluctuations is given by

$$S_{\mathrm{inter}} = S_{\mathrm{phonon}} + S_{\Omega} - S_{\Omega}, \qquad S_{\Omega} = S_{\Omega_{\mathrm{A}}} + S_{\Omega_{\mathrm{B}}}, \tag{10.5.26}$$

$$S_{\mathrm{phonon}} = -k_{\mathrm{B}}\sum_{\lambda \boldsymbol{q}}\left[\ln(1-e^{-\beta\hbar\omega_{\lambda+}(\boldsymbol{q})}) + \ln(1-e^{-\beta\hbar\omega_{\lambda-}(\boldsymbol{q})})\right]$$
$$+\frac{1}{T}\sum_{\lambda \boldsymbol{q}}\left[\frac{\hbar\omega_{\lambda+}(\boldsymbol{q})}{e^{\beta\hbar\omega_{\lambda+}(\boldsymbol{q})}-1} + \frac{\hbar\omega_{\lambda-}(\boldsymbol{q})}{e^{\beta\hbar\omega_{\lambda-}(\boldsymbol{q})}-1}\right]. \tag{10.5.27}$$

$$S_{\Omega_\alpha} = k_{\mathrm{B}}\sum_{\lambda \boldsymbol{q}}\ln(1-e^{-\beta\hbar\omega_\alpha}) - \frac{1}{T}\sum_{\lambda \boldsymbol{q}}\frac{\hbar\omega_\alpha}{e^{\beta\hbar\omega_\alpha}-1}. \tag{10.5.28}$$

In Eq. (10.5.26), the first, the second, and the third terms correspond to the phonon entropy, the fluctuation entropy due to bubbles of intra-band particle-hole pairs and the dissipation entropy due to the viscosity, respectively. Here we also reconfirm that in an equilibrium state, there only remains the phonon entropy; the inter-band fluctuation entropy and the dissipative entropy cancel. In this case, the phonon entropy and the inter-band fluctuation entropy do not mix. This originates from the temperature-independence of phonons. The inter-band fluctuation entropy lowers the equilibrium entropy, but the dissipative entropy compensates the inter-band fluctuation entropy and the system becomes a local equilibrium.

An up and down transition of a particle at a site corresponds to a jump of a shallow valley to another shallow valley in the multi-dimensional configuration space in the ELM as the β-relaxation process. The successive up and down transitions constitute another configuration space different from hoppings. The probability of the magnitude of the randomness of harmonic frequencies of α-particles is proportional to the configuration number $e^{S_{\Omega_\alpha}/k_{\mathrm{B}}}$. The probability of the magnitude of randomness of the harmonic frequency of an α-particle is proportional to

$$\Omega_\alpha = e^{\frac{z_\alpha S_{\Omega_\alpha}}{N_\alpha k_{\mathrm{B}}}} = \exp\left\{-\frac{3z_\alpha\beta\hbar\omega_\alpha}{e^{\beta\hbar\omega_\alpha}-1}\right\}, \tag{10.5.29}$$

where z_α is of the order of the α-surrounding particles of an α-particles. Thus we obtain

$$\sqrt{U_\omega^{\alpha\mu\nu}} \propto \Omega_\alpha. \tag{10.5.30}$$

10.5.3 The Liquid-Glass Transition

The Kauzmann entropy S_{K} diverges negatively at T_0 so that the entropy crisis seems to occur. But the inter-band fluctuation entropy S_{Ω} increases as the temperature decreases toward T_0 and eventually overwhelms the Kauzmann entropy S_{K} at the liquid-glass transition temperature T_g above T_0:

$$S_{\mathrm{K}} = S_{\Omega}|_{T=T_g}, \tag{10.5.31}$$

so that S_{Ω} prevents the crisis. The Kauzmann entropy S_{K} dominates above T_g, while the fluctuation entropy S_{Ω} dominates below T_g. But below T_g, the dissipative entropy compensates the S_{Ω} completely. Sound is density fluctuation waves essential in a fluid, while phonons are elastic waves essential in a solid. The glass transition is a sort of dynamical transition.

10.5.4 The Specific Heats

Next we investigate the specific heat. First, we calculate the specific heat due to the entropy of the intra-band density fluctuations, which consists of the Kauzmann entropy S_{K} and the entropy due to the conventional sound, S'_{sound}. The specific heat due to S_{K}, C_{K} is given by

$$C_{\mathrm{K}} = \begin{cases} \frac{N_0 k_{\mathrm{B}} T}{2} \frac{T_0}{(T-T_0)^2} & \text{for } T_g < T \\ 0 & \text{for } T < T_g \end{cases}. \tag{10.5.32}$$

The specific heat due to the sound, C'_{sound} is given by

$$C'_{\mathrm{sound}} = \frac{k_{\mathrm{B}}}{4} \sum_{q \notin N_0} \left[\frac{(\beta\hbar\omega_{\mathrm{s}+}(q))^2}{\sinh^2 \frac{\beta\hbar\omega_{\mathrm{s}+}(q)}{2}} + \frac{(\beta\hbar\omega_{\mathrm{s}-}(q))^2}{\sinh^2 \frac{\beta\hbar\omega_{\mathrm{s}-}(q)}{2}} \right] \cong k_{\mathrm{B}}(N - N_0), \tag{10.5.33}$$

where we have used the relation $\beta\hbar\omega_{\mathrm{s}\pm}(q) \ll 1$. C'_{sound} remains $k_{\mathrm{B}}(N - N_0) =$ constant around the glass transition temperature T_g.

Next we investigate the specific heat due to the inter-band density fluctuations. In this case, since the inter-band fluctuation entropy cancels with the dissipative entropy, there remains only the specific heat of phonons. This is because the structure of the phonons does change little above and below the glass transition, while the structure of the sound essentially depends on the temperature. We obtain the conventional specific heat of phonons, C_{phonon} as

$$C_{\mathrm{phonon}} = \frac{k_{\mathrm{B}}}{4} \sum_{\lambda \boldsymbol{q}} \left[\frac{(\beta\hbar\omega_{\lambda+}(\boldsymbol{q}))^2}{\sinh^2 \frac{\beta\hbar\omega_{\lambda+}(\boldsymbol{q})}{2}} + \frac{(\beta\hbar\omega_{\lambda-}(\boldsymbol{q}))^2}{\sinh^2 \frac{\beta\hbar\omega_{\lambda-}(\boldsymbol{q})}{2}} \right]. \tag{10.5.34}$$

At low temperatures, $C_{\mathrm{phonon}} \propto T^3$.

10.5.5 The VTF Law

Now, we investigate the VTF law on the transport coefficients, the relaxation times. and the velocities of the modes.

1. $T_0 < T$:
   The velocities of the particles $v_{\alpha p}$ and the sound velocity $c_{T\pm}$ only depend on $J_{\alpha 0}$, so we obtain

$$v_{\alpha p},\ c_{T\pm} \propto J, \tag{10.5.35}$$

$$c_{\lambda\pm} \cong \text{constant}. \tag{10.5.36}$$

   Here it should be noted the phonon velocities $c_{\lambda\pm}$ is constant.
2. $T_0 < T_g < T$:
   The term $\sqrt{U^{\alpha\mu\nu}} \propto J_{\alpha\mu}$ dominates; the α-relaxation. We obtain

$$\tau_{\alpha 0}^{-1},\ \tau_{\alpha \mathrm{M}}^{-1} \propto J \tag{10.5.37}$$

   If we consider diffusivity $D^{\pm}$ in (10.2.29) and viscosity η_λ in Eqs. (10.4.25) and (10.4.26), we obtain

$$D^{\pm} \propto J \tag{10.5.38}$$

$$\eta_\lambda^{-1} \propto J. \tag{10.5.39}$$

   It should be noted that the sound velocity $c_{T\pm} \propto J$ in Eq. (10.5.35) plays an essential role in the diffusivity $D^{\pm}$. Thus $D^{\pm}$ and η_λ satisfies the Stokes law.
3. $T_0 < T < T_g$:
   The term $\sqrt{U_\omega^{\alpha\mu\nu}} \propto \Omega_\alpha$ dominates; the β-relaxation. Thus we obtain

$$\tau_{\alpha 0}^{-1},\quad \tau_{\alpha \mathrm{M}}^{-1} \propto \Omega, \tag{10.5.40}$$

$$D^{\pm} \propto J^2, \tag{10.5.41}$$

$$\eta_\lambda \propto \Omega^{-1}. \tag{10.5.42}$$

   In this regime there remains little diffusion.

10.6 Concluding Remarks

We have developed a general theory of the liquid-glass transition in multi-component liquids in the mean field approximation taking into account the electric polarization effect. Applying the general theory to two-component liquids, we have elucidated the universal features of the liquid-glass transition such as the Kauzmann paradox, the VTF law on the relaxation and transport processes, the jump of the specific heat, and the boson peaks. Deducing the theoretical development in the two-component liquids, we can elucidate the universal features of the glass transition in any multi-component liquids. Although the approximation procedure is simple, we have been able to establish a unified theory of the liquid-glass transition in any multi-component liquid, even in polymerization. The theoretical formulae: $S_{\mathrm{K}} \cong -\frac{N_0 k_{\mathrm{B}}}{2} \frac{T_0}{T-T_0}$, $J = e^{\frac{zS_{\mathrm{K}}}{Nk_{\mathrm{B}}}}$, $S_\Omega = -k_{\mathrm{B}} \frac{3N\beta\hbar\omega}{e^{\beta\hbar\omega}-1}$, $S_{\mathrm{K}} = S_\Omega|_{T=T_g}$, yield experimental parameters: N, T_0, N_0, z, ω (an average harmonic frequency), and T_g. These parameters can be obtained by experimental results:

$$
\begin{aligned}
C_{\mathrm{K}} &= \frac{1}{T}\frac{\partial S_{\mathrm{K}}}{\partial T} \longrightarrow S_{\mathrm{K}} \longrightarrow N_0,\ T_0, \\
c_T, \tau_0^{-1}, \tau_M^{-1},\ D,\ \eta_\lambda{}^{-1} &\propto J \longrightarrow S_{\mathrm{K}},\ z, \\
\lim_{q\to\infty} \omega_\lambda &= \omega \longrightarrow S_\Omega, \\
S_{\mathrm{K}} &= S_\Omega \longrightarrow T_g,
\end{aligned}
$$

where N is given.

Once a unified theory has been established, the unified theory plays a crucial role in arranging experimental results and developing a more precise theory. The VTF law is constructed from the Einstein relation on entropy and probability so that the VTF law includes naturally the Kauzmann entropy in exponent; an inverse relaxation time is given by

$$1/\tau \propto J = \exp\{zS_{\mathrm{K}}/(Nk_{\mathrm{B}})\} : \quad \text{The unified theory.}$$

The Kauzmann entropy explains the Kauzmann paradox and the jump of the specific heat so that the unified theory elucidates the universal features consistently.

On the other hand, the well-known Adam-Gibbs formula for the VTF law includes inversely the configuration entropy S_c corresponding to the Kauzmann entropy in exponent;

$$1/\tau \propto \exp\{-F/(TS_c)\} : \quad \text{The Adam-Gibbs formula,}$$

where F is a constant number. The Adam-Gibbs formula has been usually used in arranging experimental data, the transport coefficients, and the relaxation times. However, the configuration entropy S_c cannot explain the Kauzmann paradox and the jump of the specific heat well so that the Adam-Gibbs formula faces some conflicting points.

The Kauzmann entropy S_{K} originates from the mixing between the sound and the intra-band fluctuation entropies, and determines a unique hopping amplitude J. J governs the hopping magnitudes $J_{\mu\alpha}$ for the individual particles, the individual particle velocities v_α and sound velocity c_T. On the other hand, since the mixing between the phonon and inter-band fluctuation entropies does not occur, the inter-band fluctuation entropies for the individual particles S_{Ω_α} are independent. The inter-band fluctuation entropy $S_\Omega = \sum_\alpha S_{\Omega_\alpha}$ prevents the Kauzmann entropy crisis at T_g and determines the glass transition temperature T_g. Below T_g the remaining intra-band fluctuation entropy free from the mixing and the inter-band fluctuation entropy are compensated by the dissipative entropies due to diffusion and viscosity, respectively; there remains the sound entropy only with the remaining number of states $N - N_0$ free from the mixing, and the complete phonon entropy.

We have elucidated the VTF law on the relaxation times and the transport coefficients taking into account the random scattering processes due to the random hopping magnitudes and the random harmonic frequencies. The random hopping magnitudes determine the VTF law and the α-relaxation, while the random harmonic frequencies determine the β-relaxation.

We have calculated phonons and viscosity taking into account the electric polarization. Since the electric polarization relates to the p-states of particles, the electric polarization connects the inter-band density fluctuations and does not connect the intra-band density fluctuations. The effect of the electric polarization on phonon modes is to separate the longitudinal and transverse modes constant values according to the parameters c_α in Eq. (10.4.8). We have also elucidated the boson peaks Eqs. (10.4.17), (10.4.18).

References

[1] T. Kitamura, Phys. Rep. 383 (2003) 1.

[2] T. Kitamura, AIP Conference Proceedings 708 (2004) 631.

[3] T. Kitamura, Phys. Lett. A 147 (1990) 511.

[4] T. Kitamura, Physica A 174 (1991) 578.

[5] T. Kitamura, M. Silbert, Phys. Lett. A 215 (1996) 69.

[6] T. Kitamura, Physica A 383 (2007) 232.

[7] E. F. Tombari, C. Ferrari, G. Salvetti, G. P. Johari, J. Phys. Condens. Matter 9 (1997) 7017.

[8] S. Correzzi, D. Fioretto, P. Rolla, Nature 420 (2002) 653.

[9] T. Kitamura, M. Peyrard, S. C. Lopez, Phys. Biol. 2 (2005) 200.

[10] J. P. Hansen and I. R. MacDonald, Theory of Simple Liquid, 2nd Edition, Academic Press, London, 1990.

[11] D. M. Heyes, The Liquid State, Wiley, Chichester, 1997.

[12] T. Kitamura, Il Nuovo Cimento D 11 (1989) 1441.

[13] A.A. Maradudin, E.W. Montroll, G.H. Weiss, I.P. Ipatova, Solid State Phys. 2nd Edition, Supple. 3, Academic Press, New York, 1971.

[14] P. P. Ewald, Ann. Phys. 54 (1917) 519, 557; 64 (1921) 253.

[15] A. A. Abrikosov, L. P. Gorkov, I. E. Dzyloshinski, Methods of Quantum Field Theory in Statistical Physics, Prentice-Hall, NJ, 1963..

[16] A. L. Fetter, J.D. Walecka, Quantum Theory of Many-Particle Systems, MacGrow-Hill, New York, 1971.

[17] T. Kitamura, Phys. Lett. A 282 (2001) 59.

[18] W. Montfrooij, P. Westerhuijs, V.O. de Haan and I.M. de Schepper, Phys. Rev. Lett. 63 (1989) 544.

[19] G.H. Wegdam, A. Bot, R.P.C. Schram and H.M. Schaink, Phys. Rev. Lett. 63 (1989) 2697.

[20] P. Westerhuijs, W. Montfrooij, L.A. de Graaf and I.M. Schepper, Phys. Rev. A45 (1992) 3749.

[21] J. Bosse, G. Jaccucci, M. Ronchetti and W. Schirmacher, Phys. Rev. Lett. 57 (1986) 3277.

[22] A. Campa and E.G.D. Cohen, Phys. Rev. A39 (1989) 4909.

[23] A. Campa and E.G.D. Cohen, Phys. Rev. A41 (1990) 5451.

[24] A. Campa and E.G.D. Cohen, Physica A174 (1991) 214.

[25] C.J. Benmore, B.J. Olives, J.B. Suck, R.A. Robinson and P.A. Egelstaff, J. Phys.: Condens. Matter 7 (1995) 4775.

11 Extension of the Two Band Model

In the condensed state, the liquid and the glassy states, a particle stays in a randomly localized potential made up by the surrounding particles. The potential is approximated as a harmonic one. At the lower temperature, we can take two levels. The hopping of the particle makes the two levels bands. A particle makes up and down transition at the potential for a while, then hops to a neighboring vacancy: this picture leads to the two band model, where intra-band and inter-band density fluctuations play an essential role in the condensed state. The two band model elucidates the universal features such as the Kauzmann entropy crisis and the VTF law in the liquid-glass transition.

Random nature in the condensed state has been treated at the two stages:

1. At the first stage, we have developed the dynamical and thermodynamical processes in the configurationally averaged Hamiltonian. We have clarified the glassy state and liquid state.

   a. *The glassy state*: The glassy state has a kind of long range order, the pair distribution function looks the same to each observer at his own particle. The pair distribution function holds constant every time so that it spontaneously breaks the translational symmetry of the system, which yields the existence of the NG bosons, phonons, where up and down transition of particles is essential. The phonons are formed as the sum of an infinite series of interacting bubble diagrams of inter-band particle-hole pairs. The extra bubble diagrams of inter-band particle-hole pairs not constituting the phonons yield the inter-band fluctuation entropy.

   b. *The liquid state*: The liquid state has no long range order. The pair distribution function holds at instant time, but changes with time. The hopping of particles is essential. The sound is formed as the sum of an infinite series of interacting bubble diagrams of intra-band particle-hole pairs. The extra bubble diagrams of intra-band particle-hole pairs not constituting the sound yield the intra-band fluctuation entropy. Near the liquid-glass transition temperature T_g, since the intra-band elementary excitations become very narrow because of very narrow bands, the sound mode keeps alive over the long and short wavelength regimes. The sound mode disappears at the sound disappearing temperature T_0 and at a reciprocal particle distance K.

2. At the second stage, we have developed the relaxation and dissipative processes by taking into account the random scattering processes by the random scattering potentials due to the random harmonic frequencies and random hopping magnitudes. The configuration average of the simultaneously scattering processes of the two

Liquid Glass Transition. DOI: http://dx.doi.org/10.1016/B978-0-12-407177-3.00011-8

particles by the same random scattering potentials constrains the dynamics of the system so that this process leads to the relaxation and dissipative processes.

3. Finally, we have developed the liquid-glass transition. In a supercooling process in a liquid, enormous bubble diagrams of intra-band particle-hole pairs are created. The extra bubbles of intra-band particle-hole pairs not constituting the sound form the intra-band fluctuation entropy, which mixes the sound entropy. This mixing entropy yields the Kauzmann entropy with a Curie law with a critical temperature T_0 and negative sign. As the temperature decreases toward T_0, the Kauzmann entropy seems to drop to a negative infinity, but the inter-band fluctuation entropy due to the extra bubbles of inter-band particle-hole pairs not constituting the phonons increases and eventually overwhelms the Kauzmann entropy at the glass transition temperature T_g so that the inter-band fluctuation entropy prevents the Kauzmann entropy crisis. The hopping amplitude is the probability corresponding to the Kauzmann entropy per particle, from the Einstein relation on probability and entropy. The hopping amplitude just manifests the VTF law in the relaxation and dissipative processes.

In this chapter, we extend the two band model to three cases. Eventually, we clarify that the two band model elucidates the mechanism of panics; a panic occurs under a supercooling process in a two level system.

First, we consider a quantum liquid. In a normal liquid, the pair distribution function changes with time so that the pair distribution function does not play a role of the long range order; sound is not the NG boson. However, in the bose-condensed state of a quantum liquid, there appears another long range order: every annihilation or creation operator has the constant phase. The constant phase spontaneously breaks the phase symmetry of the system so that there is the existence of the Nambu–Goldstone boson, phason. In Section 11.1, we extend the dynamical equation of the intra-band density fluctuations in liquids to that in a quantum liquid, taking into account the bose-condensed state.

Second, we extend the unified theory of the liquid-glass transition to the origin of the RNA, the polymerization of RNA in a pre-biotic age. In Section 11.2, we develop the polymerization as the liquid-glass transition in a multi-component liquids. Furthermore, we investigate how the primitive RNA appears around volcanoes in an ancient time.

Third, we extend the two band model to the panics, especially, the financial panic. Panics are ubiquitous; panics occur in financial markets, psychology, etc., and even in nature. This is because the two level systems are ubiquitous. The hopping of constituent matters yields the two bands. The two band model covers very wide regions. Regarding constituent matters in the system as constituent particles in a liquid, we can develop the dynamical and dissipative processes in a parallel way to those of intra-band and inter-band particle-hole pairs in a liquid. Whenever a two level system is under a supercooling process, a panic occurs in the system. We call each supercooling process to avoid completing own cycles a panic. We can regard the liquid-glass transition as a kind of panic in the liquid. In the panics, the dynamical and dissipative processes are governed by the universal features such as the Kauzmann entropy crisis and the VTF law. In Section 11.3, regarding securities in the financial markets as particles in a liquid, we clarify the dynamical and dissipative processes in the financial market and the financial panic in the supercooling process. We formulate the criteria of the financial panic. Finally, we formulate the criteria of panics.

11.1 Excitations in a Bose-Condensed Liquid

The experiments and theories on excitations in a bose-condensed liquid have been reviewed by many authors. Griffin and Glyde have proposed the dielectric formulation in explaining the excitations in a bose-condensed liquid [1, 2]: The density correlation function and the particle Green's function have the same dielectric function in their denominators. The dielectric function determines the excitations in the condensed liquid; the excitations are expressed as the hybridization of the sound and particle excitations in the normal state through the condensate state.

We extend the two band model in normal liquids to the dielectric formulation in a bose-condensed liquid to investigate the bose-condensed excitations [3]. The particle excitations are essentially determined by the hopping of particles in normal liquids and the Ward-Takahashi relation associated with the spontaneous breakdown of the phase symmetry [3, 4] in the bose-condensed liquid; the particle excitations are the Nambu-Goldstone boson, which relates to the Hugenholtz-Pines relation between the chemical potential and the self-energy parts of the particle Green's functions [5].

We construct sound mode with the relaxation time in the long wavelength regime taking into account the random scattering processes and then extend the sound mode to the short wavelength regime under some assumptions. The excitations in the bose-condensed liquid are the hybridization between the Bogoluibov excitations and sound excitations. In the long wavelength regime, the excitations are essentially determined by the hybridizations. In the short wavelength regime, the excitations inherit the Bogoluibov excitations, where the lower band excitations of particles almost determine the Bogoluibov excitations.

11.1.1 *The Ward-Takahashi Relations Associated with the Phase Broken Symmetry*

The Ward-Takahashi relations play an important role in determining individual and collective excitations associated with the spontaneous breakdown of the phase symmetry [4]. In order to obtain the Ward-Takahashi relations at finite temperatures, we add an infinitesimal symmetry breaking term H_ϵ to the invariant Hamiltonian H:

$$H_\epsilon = -\epsilon \int d^3x\{\psi(x) + \psi^\dagger(x)\}, \tag{11.1.1}$$

$$v \equiv \langle \psi(x) + \psi^\dagger(x)\rangle_\epsilon, \tag{11.1.2}$$

where $\psi(x)$ is a field operator and any operator O satisfies

$$O(x) = e^{\frac{(H+H_\epsilon-\mu N)\tau}{\hbar}} O(\boldsymbol{x}) e^{-\frac{(H+H_\epsilon-\mu N)\tau}{\hbar}}. \qquad x \equiv (\boldsymbol{x}, \tau), \tag{11.1.3}$$

The ensemble average is defined as

$$\langle \cdots \rangle_\epsilon = \frac{\langle \mathrm{Tr} e^{-\beta(H+H_\epsilon-\mu N)} \cdots \rangle_c}{\langle \mathrm{Tr} e^{-\beta(H+H_\epsilon-\mu N)} \rangle_c}, \tag{11.1.4}$$

where $\beta = \frac{1}{k_B T}$ and $\langle \cdots \rangle_c$ means the ensemble average of the position of atoms.

Now we introduce the generator of the phase transformation:

$$N(\tau) = \int d^3x N_0(x), \qquad N_0(x) = -\hbar\psi^\dagger(x)\psi(x) \tag{11.1.5}$$

satisfies

$$[\psi(x), N(x)] = -\hbar\psi(x), \tag{11.1.6}$$

$$\frac{\partial N(\tau)}{\partial \tau} = \epsilon \int d^3x \{\psi(x) - \psi^\dagger(x)\}, \tag{11.1.7}$$

which leads to $e^{i\theta N/\hbar}\psi(x)e^{-i\theta N/\hbar} = e^{i\theta}\psi(x)$.

Next we consider the derivative of the product of operators in τ:

$$\begin{aligned}
&\frac{\partial}{\partial \tau} T_\tau\{N(\tau)\psi_1(x_1)\cdots\psi_n(x_n)\} \\
&= \sum_{i=1}^{n} \delta(\tau - \tau_i) T_\tau\{\psi(x_1)\cdots[N(\tau), \psi_i(x_i)]\cdots\psi_n(x_n)\} \\
&+ T_\tau\{\frac{\partial}{\partial \tau} N(\tau)\psi_1(x_1)\cdots\psi_n(x_n)\},
\end{aligned} \tag{11.1.8}$$

where $\psi_i(x_i)$ stands for $\psi(x_i)$ or $\psi^\dagger(x_i)$. After we integrate both the sides of Eq. (11.1.8) in τ from 0 to $\beta\hbar$, we take the ensemble average of the equation. Then using Eqs. (11.1.6) and (11.1.7), and the properties of Tr operator, we obtain a general Ward-Takahashi relation:

$$\begin{aligned}
&\sum\nolimits_i \langle T_\tau \psi_1(x_1)\cdots(-1)^{\alpha_i}\psi_i(x_i)\cdots\psi_n(x_n)\rangle_\epsilon = \\
&-\frac{\epsilon}{\hbar}\int d^4x \langle T_\tau\{\psi(x) - \psi^\dagger(x)\}\psi_1(x_1)\cdots\psi_n(x_n)\rangle_\epsilon,
\end{aligned} \tag{11.1.9}$$

where

$$(-1)^{\alpha_i} = \begin{cases} -1 & \text{for } \psi_i = \psi \\ 1 & \text{for } \psi_i = \psi^\dagger \end{cases}.$$

Here we investigate a simple Ward-Takahashi relation:

$$\langle \psi(x) + \psi^\dagger(x)\rangle_\epsilon = \frac{\epsilon}{\hbar}\int d^4x' \langle T_\tau\{\psi(x) - \psi^\dagger(x)\}\{\psi(x') - \psi^\dagger(x')\}\rangle_\epsilon, \tag{11.1.10}$$

$\psi(x) - \psi^\dagger(x)$ has a free boson field [4]:

$$\mathrm{i}\{\psi(x) - \psi^\dagger(x)\} = Z^{\frac{1}{2}}(\partial)\chi^0(x) + \cdots, \tag{11.1.11}$$

where $\chi^0(x)$ satisfies $D(\partial)\chi^0(x) = 0$ and $Z^{\frac{1}{2}}(\partial)$ can contain derivatives. We can write

$$-\langle T_\tau \chi^0(x)\chi^0(x')\rangle_\epsilon = \frac{1}{\beta\hbar}\sum_p e^{\mathrm{i}p\cdot(x-x')-\mathrm{i}\nu_n(\tau-\tau')}\frac{Z(p)}{(\mathrm{i}\nu_n)^2 - E_p^2 + \epsilon A_p}, \tag{11.1.12}$$

where ϵA_p is the shift of energy due to H_ϵ. Substituting Eq. (11.1.12) into Eq. (11.1.10), we obtain

$$E_{p=0} = 0, \qquad v = \frac{Z(0)}{\hbar A_0}. \tag{11.1.13}$$

This manifests the free boson field is a Nambu-Goldstone boson.

11.1.2 The Hamiltonian

In Chapter 6, we have established the Hamiltonian for the dynamical equations of intra- and inter-band density fluctuations: We start with the following Hamiltonian:

$$H = \int d^3x \psi^\dagger(x)\hbar\epsilon^0(-\mathrm{i}\nabla)\psi(x) + \frac{1}{2}\int d^3x d^3y n(x)V(\boldsymbol{x}-\boldsymbol{y})n(y), \tag{11.1.14}$$

where ψ is an annihilation field operator of a particle, $\hbar\epsilon^0(-i\nabla)$ is the energy operator of a free particle, $n(x)$ is a density operator of the particle, $n(x) = \psi^\dagger(x)\psi(x)$, and V is the interaction potential between particles.

First we obtain the unpertubed Hamiltonian. To do so, we consider the following model Hamiltonian

$$H = \sum_{m\mu} \hbar\tilde{\omega}_{m\mu} b^\dagger_{m\mu} b_{m\mu} + \sum_{m\mu n\nu} \hbar J_{m\mu n\nu} b^\dagger_{m\mu} b_{n\nu}. \tag{11.1.15}$$

If we introduce an annihilation operator of the particle in a localized harmonic potential at the site $\boldsymbol{R}_m$, $b_{m\mu}$, we can write

$$\psi(x) = \sum_{m\mu} \tilde{w}_{m\mu}(\boldsymbol{x}-\boldsymbol{R}_m) b_{m\mu}. \tag{11.1.16a}$$

Here $\tilde{w}_{m\mu}(\boldsymbol{x}-\boldsymbol{R}_m)$ is the wave function of the μth state, $\tilde{\omega}_{m\mu}$ at the site $\boldsymbol{R}_m$.

Next we take the configuration average of Eq. (11.1.15). Here we define the configurationally averaged harmonic frequencies and hopping magnitudes:

$$\tilde{\omega}_\mu \equiv \langle\tilde{\omega}_{m\mu}\rangle_c, \qquad J_\mu(\boldsymbol{R}_m - \boldsymbol{R}_n) \equiv \langle J_{m\mu n\nu}\rangle_c,$$

where we only consider the hopping between the same states. Then we obtain

$$H_0 = \langle H\rangle_c = \sum_{m\mu} \hbar\tilde{\omega}_\mu b^\dagger_{m\mu} b_{m\mu} + \sum_{mn\mu} \hbar J_\mu(\boldsymbol{R}_m - \boldsymbol{R}_n) b_{m\mu} b_{n\nu}. \tag{11.1.17a}$$

Now we introduce the following extended operators:

$$b_{m\mu} = \frac{1}{\sqrt{N}} \sum_{p} e^{\mathrm{i}\boldsymbol{p}\cdot\boldsymbol{R}_m} a_{\mu\boldsymbol{p}}, \qquad \mu = 0, 1, 2, 3, \tag{11.1.18}$$

Then the field operator is written down as

$$\psi(x) = \frac{1}{\sqrt{N}} \sum_{\mu p m} e^{\mathrm{i}\boldsymbol{p}\cdot\boldsymbol{R}_m} \tilde{w}_{m\mu}(\boldsymbol{x} - \boldsymbol{R}_m) a_{\mu\boldsymbol{p}}, \tag{11.1.16b}$$

and the unperturbed Hamiltonian

$$H = \sum_{\mu\boldsymbol{p}} \hbar\epsilon_{\mu\boldsymbol{p}} a^{\dagger}_{\mu\boldsymbol{p}} a_{\mu\boldsymbol{p}}, \tag{11.1.17b}$$

where

$$\epsilon_{\mu\boldsymbol{p}} = \tilde{\omega}_{\mu} + J_{\mu}(\boldsymbol{p}), \tag{11.1.19}$$

$$J_{\mu}(\boldsymbol{p}) = \frac{1}{N} \int d^3R g(R) e^{\mathrm{i}\boldsymbol{p}\cdot\boldsymbol{R}} J_{\mu}(\boldsymbol{R}). \tag{11.1.20}$$

In the low temperature liquids, we can restrict ourselves to the lower band, so we obtain

$$\begin{aligned} n(x) &= \sum_{m} \left\{ \tilde{w}^2_{m0}(\boldsymbol{x} - \boldsymbol{R}_m) b^{\dagger}_{m0} b_{m0} + \cdots \right\} \\ &= \frac{1}{N} \sum_{m\boldsymbol{q}} e^{-\mathrm{i}\boldsymbol{q}\cdot\boldsymbol{R}_m} \left[\tilde{w}^2_{m0}(\boldsymbol{x} - \boldsymbol{R}_m) \rho^{\dagger}_{0\boldsymbol{q}} + \cdots \right], \end{aligned} \tag{11.1.21}$$

where

$$\rho^{\dagger}_{0\boldsymbol{q}} = \sum_{\boldsymbol{p}} a^{\dagger}_{0\boldsymbol{p}} a_{0\boldsymbol{p}-\boldsymbol{q}}, \tag{11.1.22}$$

$\rho^{\dagger}_{0\boldsymbol{q}}$ are intra-band density fluctuation operators.

Substitution of Eq. (1.1.21) into Eq. (11.1.14) yields

$$H = \sum_{\mu\boldsymbol{p}} \hbar\epsilon_{0\boldsymbol{p}} a^{\dagger}_{0\boldsymbol{p}} a_{0\boldsymbol{p}} + \frac{1}{2N} \sum_{\mu\nu\boldsymbol{q}}{}' V(\boldsymbol{q}) \rho^{\dagger}_{0\boldsymbol{q}} \rho_{0\boldsymbol{q}}, \tag{11.1.23}$$

$$V(\boldsymbol{q}) = \frac{1}{N} \sum_{m \neq n} e^{-\mathrm{i}\boldsymbol{q}\cdot(\boldsymbol{R}_m - \boldsymbol{R}_n)} \int d^3x d^3y \tilde{w}^2_{m0}(\boldsymbol{x} - \boldsymbol{R}_m) V(\boldsymbol{x} - \boldsymbol{y}) \tilde{w}^2_{n0}(\boldsymbol{y} - \boldsymbol{R}_n), \tag{11.1.24}$$

where we have only considered intra-band density fluctuations. Here it should be noted that since the unperturbed Hamiltonian involves the term $\int d^3xd^3yn(x)V(\boldsymbol{x}-\boldsymbol{y})\langle n(y)\rangle_c$, the interaction Hamiltonian H_I must exclude the term. Thus the prime on $\sum$ in Eq. (11.1.23) means that H_I excludes one loop diagrams.

So far we have surveyed the Hamiltonian in the normal state. Now we extend the Hamiltonian to the bose-condensed state. The condensate exists at $\boldsymbol{p}=0$ in the lowest band of the particles $\epsilon_{0\boldsymbol{p}}$. To do so, we abbreviate the notation as follows:

$$a_{0\boldsymbol{p}} \longrightarrow a_{\boldsymbol{p}},$$
$$\rho_{0\boldsymbol{q}} \longrightarrow \rho_{\boldsymbol{q}} = \sum_{\boldsymbol{p}} a_{\boldsymbol{q}}^{\dagger} a_{\boldsymbol{p}+\boldsymbol{q}},$$
$$\epsilon_{0\boldsymbol{p}} \longrightarrow \epsilon_{\boldsymbol{p}} - \mu,$$

where we introduce the chemical potential μ explicitly, because we can compare with the conventional theories.

Here we consider the condensed state:

$$a_0^{\dagger} = a_0 = \sqrt{N_0}, \tag{11.1.25}$$

where N_0 is the number of the condensed particles. Therefore, the number of the normal state particle N' is given by

$$N' = N - N_0. \tag{11.1.26}$$

If we introduce the notation:

$$A_{\boldsymbol{q}} = a_{\boldsymbol{q}} + a_{-\boldsymbol{q}}^{\dagger}, \tag{11.1.27}$$

we have

$$\rho_{\boldsymbol{q}}^{\dagger} = \sqrt{N_0}(a_{-\boldsymbol{q}} + a_{\boldsymbol{q}}^{\dagger}) + \sum_{\boldsymbol{p}}{}' a_{\boldsymbol{p}}^{\dagger} a_{\boldsymbol{p}-\boldsymbol{q}} = \sqrt{N_0}A_{\boldsymbol{q}}^{\dagger} + \sum_{\boldsymbol{p}}{}' a_{\boldsymbol{p}}^{\dagger} a_{\boldsymbol{p}-\boldsymbol{q}}, \tag{11.1.28}$$

$$\rho_{\boldsymbol{q}} = \sqrt{N_0}(a_{\boldsymbol{q}} + a_{-\boldsymbol{q}}^{\dagger}) + \sum_{\boldsymbol{p}}{}' a_{\boldsymbol{p}}^{\dagger} a_{\boldsymbol{p}+\boldsymbol{q}} = \sqrt{N_0}A_{\boldsymbol{q}} + \sum_{\boldsymbol{p}}{}' a_{\boldsymbol{p}}^{\dagger} a_{\boldsymbol{p}+\boldsymbol{q}}, \tag{11.1.29}$$

where the prime on $\sum$ means the exclusion of zero momentum. Thus we take the following model Hamiltonian in the condensed state:

$$H = \sum_{\boldsymbol{p}}{}' \hbar(\epsilon_{\boldsymbol{p}} - \mu) a_{\boldsymbol{p}}^{\dagger} a_{\boldsymbol{p}} + \hbar(\epsilon_0 - \mu)N_0 + \frac{N_0^2}{2N'}V(0)$$
$$+ \frac{1}{2N'}\sum_{\boldsymbol{q}}{}' V(\boldsymbol{q})\left[N_0 A_{\boldsymbol{q}}^{\dagger} A_{\boldsymbol{q}} + \sqrt{N_0}(A_{\boldsymbol{q}}^{\dagger}\rho_{\boldsymbol{q}} + \rho_{\boldsymbol{q}}^{\dagger}A_{\boldsymbol{q}}) + \rho_{\boldsymbol{q}}^{\dagger}\rho_{\boldsymbol{q}}\right], \tag{11.1.30}$$

where the configurationally averaged diagonal potential function is given by

$$V(q) \equiv \langle V_{00}(\boldsymbol{q})\rangle_c = \int d^3Rg(R)e^{-\mathrm{i}\boldsymbol{q}\cdot\boldsymbol{R}}V_{00}(\boldsymbol{R}), \tag{11.1.31}$$

$$V_{00}(\boldsymbol{R}) \equiv \int d^3xd^3y\tilde{w}_0^2(\boldsymbol{x})V(\boldsymbol{x}-\boldsymbol{y})\tilde{w}_0^2(\boldsymbol{y}-\boldsymbol{R}), \tag{11.1.32}$$

where $\tilde{w}_0(\boldsymbol{x})$ is configurationally averaged. In summing the position of particles $\boldsymbol{R}_m$ in Eq. (11.1.31), we take into account only the position of the normal state particles; $N' = N - N_0$. The pair distribution function $g(R)$ in Eq. (11.1.31) is defined as $\lim_{R\to\infty} g(R) = \frac{N'}{V}$, where V is the volume of the system. This Hamiltonian resembles the traditional one in the condensed state. However, it should be noted that $\epsilon_{\boldsymbol{p}}$ is the lowest particle energy band and $\rho_{\boldsymbol{q}}$ is the inter-band density fluctuations in the lowest band.

11.1.3 The Dielectric Formulation

Next we reformulate the dielectric formulation [1,2] in our scheme. First we introduce particle Green's functions in the bose-condensed state:

$$G_{\alpha\beta}(\boldsymbol{p},\tau) \equiv -\langle T_\tau a_{\boldsymbol{p}\alpha}(\tau)a^\dagger_{\boldsymbol{p}\beta}(0)\rangle_\epsilon = \frac{1}{\beta\hbar}\sum_p e^{-i\omega_n\tau}G_{\alpha\beta}(p), \tag{11.1.33}$$

where

$$a_{\boldsymbol{p}\alpha} = \begin{cases} a_{\boldsymbol{p}} & \text{for } \alpha = 1 \text{ or} + \\ a^\dagger_{-\boldsymbol{p}} & \text{for } \alpha = 2 \text{ or} - \end{cases}. \tag{11.1.34}$$

The matrix form for $G_{\alpha\beta}$ is given by

$$G(\boldsymbol{p},\tau) = -\begin{pmatrix} \langle T_\tau a_{\boldsymbol{p}}(\tau)a^\dagger_{\boldsymbol{p}}(0)\rangle_\epsilon & \langle T_\tau a_{\boldsymbol{p}}(\tau)a_{-\boldsymbol{p}}(0)\rangle_\epsilon \\ \langle T_\tau a^\dagger_{-\boldsymbol{p}}(\tau)a^\dagger_{\boldsymbol{p}}(0)\rangle_\epsilon & \langle T_\tau a^\dagger_{-\boldsymbol{p}}(\tau)a_{-\boldsymbol{p}}(0)\rangle_\epsilon \end{pmatrix} \tag{11.1.35}$$

If we introduce the bare Green's functions:

$$G^0_{\alpha\beta}(p) = \frac{\delta_{\alpha\beta}}{\alpha i\omega_n - (\epsilon_{\boldsymbol{p}} - \mu)}, \tag{11.1.36}$$

the Dyson equation is given by in the matrix form:

$$G(p) = G^0(p) + G^0(p)\Sigma(p)G(p). \tag{11.1.37}$$

Thus we obtain

$$G(p) \equiv \frac{N(p)}{D(p)} \equiv \frac{1}{D(p)} \times \begin{pmatrix} i\omega_n + \epsilon_{\boldsymbol{p}} - \mu + \Sigma_{22}(p) & -\Sigma_{12}(p) \\ -\Sigma_{21}(p) & -i\omega_n + \epsilon_{\boldsymbol{p}} - \mu + \Sigma_{11}(p) \end{pmatrix}, \tag{11.1.38}$$

$$D(p) \equiv \{i\omega_n - \epsilon_{\boldsymbol{p}} + \mu - \Sigma_{11}(p)\}\{i\omega_n + \epsilon_{\boldsymbol{p}} - \mu + \Sigma_{22}(p)\} + \Sigma_{12}(p)\Sigma_{21}(p). \tag{11.1.39}$$

Using the rotational invariance, the time-reversal symmetry and the space-inversion symmetry and the fact $a_{\boldsymbol{p}\alpha} = a^\dagger_{-\boldsymbol{p}-\alpha}$ [1,2], we find

$$G_{\alpha\beta}(\boldsymbol{p}, i\omega_n) = G_{\alpha\beta}(|\boldsymbol{p}|, i\omega_n) = G_{\beta\alpha}(\boldsymbol{p}, i\omega_n) = G_{-\alpha-\beta}(\boldsymbol{p}, -i\omega_n). \tag{11.1.40}$$

The self-energy parts satisfy the same relations. In particular

$$\Sigma_{11}(\boldsymbol{p}, \mathrm{i}\omega_n) = \Sigma_{22}(\boldsymbol{p}, -\mathrm{i}\omega_n), \qquad \Sigma_{12}(p) = \Sigma_{21}(p). \tag{11.1.41}$$

Using Eqs. (11.1.39), (11.1.40), (11.1.41), analytically continuing $\mathrm{i}\omega_n \to p_0 + \mathrm{i}\delta$ and putting

$$D(\boldsymbol{p}, p_0 + \mathrm{i}\delta) \equiv p_0^2 - E_{\boldsymbol{p}}^2 + \mathrm{i}\delta, \tag{11.1.42}$$

from the Nambu-Goldstone theorem (11.1.13), we obtain

$$E_{\boldsymbol{p}=0}^2 = \{\epsilon_0 - \mu + \Sigma_{11}(0) - \Sigma_{12}(0)\}\{\epsilon_0 - \mu + \Sigma_{11}(0) + \Sigma_{12}(0)\} = 0.$$

Here we investigate the weak coupling limit:

$$\Sigma_{11}(p) = \Sigma_{22}(p) = n_0 V(0) + n_0 V(p), \qquad \Sigma_{12}(p) = \Sigma_{21}(p) = n_0 V(p),$$

as illustrated in Figure 11.1, where

$$n_0 \equiv \frac{N_0}{N}. \tag{11.1.43}$$

Then we have

$$E_0^2 = \{\epsilon_0 - \mu + n_0 V(0)\}\{\epsilon_0 - \mu + 3n_0 V(0)\} = 0.$$

If we take $\epsilon_0 - \mu + 3n_0 V(0) = 0$, then $\epsilon_0 - \mu + n_0 V(0) < 0$. Thus because of $E_0^2 > 0$, we must take

$$\epsilon_{\boldsymbol{p}=0} - \mu + \Sigma_{11}(0) - \Sigma_{12}(0) = 0. \tag{11.1.44}$$

This is nothing but the Hugenholtz-Pines relation [5]. Note that in our theory $\epsilon_{\boldsymbol{p}=0} = \tilde{\omega}_0 + \tilde{J}_0$. If we put $\epsilon_0 = 0$, we obtain the ordinary Hugenholtz-Pines relation.

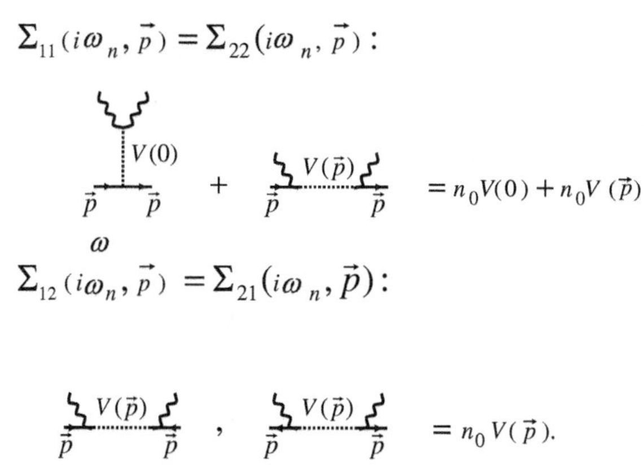

Figure 11.1 The Feynman diagrams of the self-energy parts in the weak coupling approximation (the Bogoluibov approximation) are shown. The wavy line indicates $\sqrt{n_0}$.

Next we introduce the density–density correlation function:

$$F(\boldsymbol{q},\tau) = -\frac{1}{\hbar N'}\langle T_\tau \rho_{\boldsymbol{q}}(\tau)\rho_{\boldsymbol{q}}^\dagger(0)\rangle_\epsilon = \frac{1}{\beta\hbar}\sum_{\mathrm{i}\omega_n} e^{-\mathrm{i}\omega_n\tau} F(q). \tag{11.1.45}$$

We introduce the irreducible diagrams, which cannot be separated into two parts by cutting a single interaction line. Hereafter we will denote an irreducible diagram as a tilded one. Then we have

$$F(q) = \widetilde{F}(q) + \widetilde{F}(q)V(q)F(q) = \frac{\widetilde{F}(q)}{\epsilon(q)}, \tag{11.1.46}$$

where

$$\epsilon(q) = 1 - V(q)\widetilde{F}(q). \tag{11.1.47}$$

Considering Eqs. (11.1.28) and (11.1.29), we can also write $F(q)$ as

$$F(q) = \Lambda^T(q)G(q)\Lambda(q) + F_R(q), \tag{11.1.48}$$

$$\Lambda(q) = \begin{pmatrix} \Lambda_1 \\ \Lambda_2 \end{pmatrix}, \tag{11.1.49}$$

$$\Lambda_\alpha(q) = \widetilde{\Lambda}_\alpha(q) + \widetilde{F}_R(q)V(q)\Lambda_\alpha(q) = \frac{\widetilde{\Lambda}_\alpha}{\epsilon_R(q)}, \tag{11.1.50}$$

$$\epsilon_R(q) = 1 - V(q)\widetilde{F}_R(q), \tag{11.1.51}$$

$$\widetilde{F}(q) = \widetilde{F}_R(q) + \widetilde{\Lambda}^T\widetilde{G}(q)\widetilde{\Lambda}(q), \tag{11.1.52}$$

$$F_R(q) = \widetilde{F}_R(q) + \widetilde{F}_R(q)V(q)F_R(q) = \frac{\widetilde{F}_R(q)}{\epsilon_R(q)}, \tag{11.1.53}$$

where Λ^T is the transposed matrix of Λ. $F_R(q)$ has no term with a single particle Green's function as the intermediate state. $\widetilde{F}_R(q)$ is the irreducible diagram of $F_R(q)$.

Here we introduce the self-energy part of the irreducible $\widetilde{G}(q)$ as

$$\widetilde{G}^{-1}(q) = \{G^0(q)\}^{-1} - \widetilde{\Sigma}(q), \tag{11.1.54}$$

$$\widetilde{G}(p) \equiv \frac{\widetilde{N}(p)}{\widetilde{D}(p)} \equiv \frac{1}{\widetilde{D}(p)} \times \begin{pmatrix} \mathrm{i}\omega_n + \epsilon_{\boldsymbol{p}} - \mu + \widetilde{\Sigma}_{22}(p) & -\widetilde{\Sigma}_{12}(p) \\ -\widetilde{\Sigma}_{21}(p) & -\mathrm{i}\omega_n + \epsilon_{\boldsymbol{p}} - \mu + \widetilde{\Sigma}_{11}(p) \end{pmatrix}, \tag{11.1.55}$$

$$\widetilde{D}(p) \equiv \{\mathrm{i}\omega_n - \epsilon_{\boldsymbol{p}} + \mu - \widetilde{\Sigma}_{11}(p)\}\{\mathrm{i}\omega_n + \epsilon_{\boldsymbol{p}} - \mu + \widetilde{\Sigma}_{22}(p)\} + \widetilde{\Sigma}_{12}(p)\widetilde{\Sigma}_{21}(p). \tag{11.1.56}$$

Then we have

$$\Sigma_{\alpha\beta}(q) = \widetilde{\Sigma}_{\alpha\beta}(q) + \Sigma^c_{\alpha\beta}(q) \equiv \widetilde{\Sigma}_{\alpha\beta}(q) + \widetilde{\Lambda}_\alpha(q)\frac{V(q)}{\epsilon_R}\widetilde{\Lambda}_\beta(q), \tag{11.1.57}$$

$$N(q) = \widetilde{N}(q) + \widetilde{\Lambda}'(q)\frac{V(q)}{\epsilon_R}\widetilde{\Lambda}'^T(q), \tag{11.1.58}$$

$$\Lambda'(q) \equiv \begin{pmatrix} \Lambda_2 \\ -\Lambda_1 \end{pmatrix}. \tag{11.1.59}$$

Using the relation $\Sigma^c_{11}\Sigma^c_{22} = \Sigma^c_{12}\Sigma^c_{21}$, we obtain

$$\begin{aligned} D &= (-N_{22})N_{11} + N_{12}N_{21} \\ &= (-\widetilde{N}_{22} - \Sigma^c_{11})(\widetilde{N}_{11} + \Sigma^c_{22}) + (\widetilde{N}_{12} - \Sigma^c_{12})(\widetilde{N}_{21} - \Sigma^c_{21}) \\ &= \widetilde{D} - \sum_{\alpha\beta} \widetilde{N}_{\alpha\beta}\Sigma^c_{\beta\alpha} \\ &= \widetilde{D} - \frac{V(q)}{\epsilon_R}\widetilde{\Lambda}^T\widetilde{N}\widetilde{\Lambda} \\ &= \frac{\widetilde{D}}{\epsilon_R}\left[\epsilon_R - V(q)\widetilde{\Lambda}^T\widetilde{G}\widetilde{\Lambda}\right]. \end{aligned} \tag{11.1.60}$$

Considering

$$\epsilon = \epsilon_R - V(q)\widetilde{\Lambda}^T\widetilde{G}\widetilde{\Lambda} = 1 - V(q)\widetilde{F}_R(q) - V(q)\widetilde{\Lambda}^T\widetilde{G}\widetilde{\Lambda}. \tag{11.1.61}$$

we have

$$\epsilon_R\widetilde{D} = \epsilon D \equiv C. \tag{11.1.62}$$

Considering $\widetilde{\Lambda}^T N\widetilde{\Lambda} = \widetilde{\Lambda}^T\widetilde{N}\widetilde{\Lambda}$ because of $\widetilde{\Lambda}'^T\widetilde{\Lambda}' = 0$, we can write F and G with the common denominator C:

$$F = \frac{\widetilde{D}\widetilde{F}_R + \widetilde{\Lambda}^T\widetilde{\Lambda}}{C}, \tag{11.1.63}$$

$$G = \frac{\epsilon_R\widetilde{N} + V\widetilde{\Lambda}'\widetilde{\Lambda}'^T}{C}. \tag{11.1.64}$$

The Green's functions F and G have the same excitations in the bose-condensed state.

11.1.4 The Calculation of the Particle Green's Function in the Random Phase Approximation

In order to calculate the particle Green's function, we make the Bogoliubov approximation $n_0 \ll 1$:

$$\widetilde{\Lambda}_\alpha \cong n_0^{\frac{1}{2}}, \tag{11.1.65}$$

$$\Sigma_{11}(p) = \Sigma_{22}(p) \cong n_0 V(0) + n_0\frac{V(p)}{\epsilon_R}, \tag{11.1.66}$$

$$\Sigma_{12}(p) = \Sigma_{21}(p) = n_0\frac{V(p)}{\epsilon_R}. \tag{11.1.67}$$

Thus we obtain

$$G(p) = \frac{1}{D}\begin{pmatrix} i\omega_n + \tilde{\epsilon}_p + n_0\frac{V(p)}{\epsilon_R} & -n_0\frac{V(p)}{\epsilon_R} \\ -n_0\frac{V(p)}{\epsilon_R} & -i\omega_n + \tilde{\epsilon}_p + n_0\frac{V(p)}{\epsilon_R} \end{pmatrix}, \tag{11.1.68}$$

$$D(p) = \{i\omega_n\}^2 - \left\{\tilde{\epsilon}_p^2 + 2\tilde{\epsilon}_p n_0 \frac{V(p)}{\epsilon_R}\right\}, \tag{11.1.69}$$

where we have put

$$\tilde{\epsilon}_p = \epsilon_p - \mu + n_0 V(0). \tag{11.1.70}$$

From the Nambu-Goldstone theorem (11.1.44), we have

$$\tilde{\epsilon}_0 = \epsilon_0 - \mu + n_0 V(0) = 0. \tag{11.1.71}$$

Noting the relation

$$\frac{V(p)}{\epsilon_R(p)} = V(p) + V^2(p)\frac{\widetilde{F}_R(p)}{1 - V(p)\widetilde{F}_R(p)} = V(p) + V^2(p)\widetilde{F}(p), \tag{11.1.72}$$

and under the weak coupling limit, i.e. $\frac{V(p)}{\epsilon_R} \to V(p)$, we obtain the Bogoluibov approximation. We denote the dispersion relation as

$$\epsilon_{Bp}^2 = \tilde{\epsilon}_p^2 + \Delta_p, \qquad \Delta_p \equiv 2V(p)n_0\tilde{\epsilon}_p. \tag{11.1.73}$$

Here we take the irreducible particle Green's function as

$$\widetilde{G}(p) = \frac{\widetilde{N}}{\widetilde{D}} = \frac{1}{(i\omega_n)^2 - \tilde{\epsilon}_p^2}\begin{pmatrix} i\omega_n + \tilde{\epsilon}_p & 0 \\ 0 & -i\omega_n + \tilde{\epsilon}_p \end{pmatrix}. \tag{11.1.74}$$

11.1.5 The Calculation of the Sound Green's Function

Next we investigate the sound Green's function, $F_R(p)$. In order to calculate the vertex corrected bubble diagrams, $\widetilde{F}_R(p)$, we employ the calculation procedure of the Bethe–Salpeter equation in Section 7.1. We put

$$\widetilde{F}_R(q) = -\frac{1}{\beta\hbar^2 N'}\sum_p{}' \phi(p, q), \tag{11.1.75}$$

where

$$\phi(p, q) = \sum_{p'} \phi_{pp'}(i\omega_n, q), \tag{11.1.76}$$

$$\phi_{pp'}(i\omega_n, q) \equiv \langle G_{p,p'}(i\omega_n) G_{p-q,p'-q}(i\omega_n - i\nu_n)\rangle_c. \tag{11.1.77}$$

Then the Bethe-Salpeter equation is given by

$$\phi_{pp'}(i\omega_n, q) = G(p)G(p-q)\left[\delta_{pp'} + \sum_{p_1} U_{pp_1}(q)\phi_{p_1p'}(i\omega_n, q)\right]. \tag{11.1.78}$$

Introducing the self-energy parts of the Green's function as

$$G^{-1}(p) \equiv \mathrm{i}\omega_n - \tilde{\epsilon}_{\boldsymbol{p}} - \Sigma(p), \tag{11.1.79}$$

and dividing the Bethe-Salpeter equation by $G(p)G(p-q)$, we obtain the alternative Bethe-Salpeter equation:

$$\begin{aligned}\left\{-\mathrm{i}\nu_n + \tilde{\epsilon}_{\boldsymbol{p}} - \tilde{\epsilon}_{\boldsymbol{p-q}} + \Delta\Sigma(p,q)\right\}\phi(p,q)\\ = \Delta G(p, p-q)\left[1 + \sum_{p_1} U_{\boldsymbol{p}\boldsymbol{p}_1}(q)\phi(p_1 q)\right],\end{aligned} \tag{11.1.80}$$

where

$$\Delta G(p,q) = G(p) - G(p-q), \tag{11.1.81}$$

$$\Delta\Sigma(p,q) = \Sigma(p) - \Sigma(p-q). \tag{11.1.82}$$

Now we calculate the vertex function $\phi(p,q)$ in the long wavelength regime, so we make the following approximation:

1. We retain only the following terms in the Bethe-Salpeter equation:

$$\sum_p{}' \phi(p,q) \equiv \phi, \qquad \sum_p{}' \boldsymbol{q} \cdot \nabla\tilde{\epsilon}_{\boldsymbol{p}}\phi(p,q) = \boldsymbol{q} \cdot \Phi, \tag{11.1.83}$$

2. We make the ladder approximation to the vertex part:

$$U^{00}_{\boldsymbol{pp}'}(q) \cong U^{00}_{\omega}(0) + U^{00}_{J}(0), \tag{11.1.84}$$

as schematically illustrated in Figure 7.1. The corresponding self-energy part is given by

$$\begin{aligned}\Sigma(p_0 + \mathrm{i}\delta) &\cong (U^{00}_{\omega} + U^{00}_{J}(0))\sum_{\boldsymbol{p}}{}' G(\boldsymbol{p}, p_0 + \mathrm{i}\delta)\\ &= (U^{00}_{\omega} + U^{00}_{J}(0))\sum_{\boldsymbol{p}}{}' \frac{1}{p_0 + \mathrm{i}\delta - \tilde{\epsilon}_{0\boldsymbol{p}} - \Sigma^0(p_0 + \mathrm{i}\delta)},\end{aligned} \tag{11.1.85}$$

which is schematically illustrated in Figure7.5a.

3. Since the imaginary part of the Green's function is a Lorentzian, we obtain

$$\Sigma(p_0 + i\delta) \cong -\frac{\mathrm{i}}{2\tau_0}, \qquad \frac{1}{2\tau_0} \cong \sqrt{N'(U^{00}_{\omega} + U^{00}_{J})}, \tag{11.1.86}$$

where we have analytically continued, $\mathrm{i}\omega_n \to q_0 + \mathrm{i}\delta$.

The approximation procedure (ii) satisfies the WT relation:

$$\Delta\Sigma(p,q)\phi(p,q) = \sum_{p'} U_{pp'}(q)\Delta G(p',q), \tag{11.1.87}$$

therefore, from Eq. (11.1.80) we obtain

$$-q_0\phi + \boldsymbol{q}\cdot\boldsymbol{\Phi} = \sum_p{}' \Delta G(p,q). \tag{11.1.88}$$

Here we have

$$\begin{aligned}\sum\nolimits_{\mathrm{i}\omega_n} \Delta G(p,q) &= -\beta\hbar\left[f(\tilde{\epsilon}_p - \frac{\mathrm{i}}{2\tau_0}) - f(\tilde{\epsilon}_{p-q} + \frac{\mathrm{i}}{2\tau_0})\right]\\ &= (\beta\hbar)^2\left[\boldsymbol{q}\cdot\nabla\tilde{\epsilon}_p + \frac{\mathrm{i}}{\tau_0}\right]\frac{df(\tilde{\epsilon}_p)}{d(\beta\hbar\tilde{\epsilon}_p)}\\ &= (\beta\hbar)^2\left[\boldsymbol{q}\cdot\nabla\tilde{\epsilon}_p + \frac{\mathrm{i}}{\tau_0}\right]\frac{1}{4\sinh^2(\frac{\beta\hbar\tilde{\epsilon}_p}{2})}.\end{aligned} \tag{11.1.89}$$

Considering

$$\frac{1}{\tau_0} \cong 0, \tag{11.1.90}$$

we can put

$$-q_0\phi + \boldsymbol{q}\cdot\boldsymbol{\Phi} \cong 0. \tag{11.1.91}$$

In order to obtain another equation, we multiply $\boldsymbol{q}\cdot\nabla\tilde{\epsilon}_p$ to (11.1.80) and summing over $\boldsymbol{p}$, then we obtain

$$\begin{aligned}&\sum_p{}'\left[-q_0\boldsymbol{q}\cdot\nabla\tilde{\epsilon}_p + (\boldsymbol{q}\cdot\nabla\tilde{\epsilon}_p)^2 + \boldsymbol{q}\cdot\nabla\tilde{\epsilon}_p\Delta\Sigma(p,q)\right]\phi(p,q)\\ &\quad= \sum_p{}' \boldsymbol{q}\cdot\nabla\tilde{\epsilon}_p\Delta G(p,q),\end{aligned} \tag{11.1.92}$$

where we have neglected the term corresponding to the last term of the right-hand side equation. First we estimate the second term of the left-hand side of Eq. (11.1.92):

$$\frac{1}{N'}\sum_p{}' \left[\boldsymbol{q}\cdot\nabla\tilde{\epsilon}_p\right]^2 \equiv \omega_q^2 \cong c_{0s}^2 q^2, \tag{11.1.93}$$

$$c_{0s} \equiv \frac{v_p}{\sqrt{3}}, \qquad v_p \equiv \frac{1}{N'}\sum_p{}' \frac{\partial\tilde{\epsilon}_p}{\partial p}, \tag{11.1.94}$$

where v_p is the averaged velocity of particles. The term of the right-hand side is estimated as

$$\begin{aligned}\sum_p{}' \boldsymbol{q}\cdot\nabla\tilde{\epsilon}_p\Delta G(p,q) &\cong (\beta\hbar)^2\sum_p{}' (\boldsymbol{q}\cdot\nabla\tilde{\epsilon}_p)^2\frac{1}{4\sinh^2(\frac{\beta\hbar\tilde{\epsilon}_p}{2})}\\ &\cong (\beta\hbar\omega_q)^2\sum_p{}' \frac{1}{4\sinh^2(\frac{\beta\hbar\tilde{\epsilon}_p}{2})}.\end{aligned} \tag{11.1.95}$$

Since we are concerned with the long wavelength regime and the low temperatures, we can make the following approximation:

$$\tilde{\epsilon}_p \cong E_p \cong cp, \qquad \beta\hbar c p_{max} \gg 1, \tag{11.1.96}$$

where we have replaced $\tilde{\epsilon}_p$ by the true atomic excitations E_p. c is the sound velocity in the condensed state and p_{max} is the wavelength at the maximum value of $\tilde{\epsilon}_p$. Then we obtain

$$\sum_p{}' \boldsymbol{q} \cdot \nabla\tilde{\epsilon}_p \Delta G(p,q) \cong \frac{VA\omega_q^2}{\beta\hbar c^3}, \qquad A = \frac{1}{\pi^2}\int_0^\infty \frac{x^2 dx}{\sinh^2 x}, \tag{11.1.97}$$

where V is the volume of the system. Thus we have another equation

$$-q_0\boldsymbol{q}\cdot\Phi + \omega_q^2\phi - \frac{i}{\tau_0}\boldsymbol{q}\cdot\Phi = \frac{VA\omega_q^2}{\beta\hbar c^3}. \tag{11.1.98}$$

From Eqs. (11.1.91), (11.1.98), and (11.1.75), we obtain

$$\tilde{F}_R(q) = \frac{\kappa\omega_q^2}{q_0(q_0+\frac{i}{\tau_0}) - \omega_q^2}, \qquad \kappa = \frac{VA}{\beta^2\hbar^3 N' c^3} = \frac{A}{\beta^2\hbar^3\rho c^3}, \tag{11.1.99}$$

where $\rho = \frac{N'}{V}$ is the density of normal particles. Thus, we obtain

$$F_R(q) = \frac{\kappa\omega_q^2}{q_0(q_0+\frac{i}{\tau_0}) - \omega_{sq}^2}, \tag{11.1.100}$$

where ω_{sq} is the sound frequency.

$$\omega_{sq}^2 \equiv \omega_q^2 + \alpha_q = (1+\kappa V(q))\omega_q^2, \qquad \alpha_q \equiv \kappa V(q)\omega_q^2, \tag{11.1.101}$$

Sound velocity is given by

$$c_s = (1+\kappa V(0))c_{0s}. \tag{11.1.102}$$

We should note that ω_q and ω_{sq} are calculated in the long wavelength regime. However, if we consider higher order Legendre terms in ϕ, we can extend ω_q and ω_{sq} to the whole regime.

11.1.6 Excitations in a Bose-Condensed Liquid

Substitution of Eqs. (11.1.65), (11.1.74), and (11.1.99) into Eq. (11.1.61) leads to

$$\epsilon(p) = 1 - \frac{\alpha_p}{p_0(p_0+\frac{i}{\tau_0}) - \omega_p^2} - \frac{\Delta_p}{p_0^2 - \tilde{\epsilon}_p^2}. \tag{11.1.103}$$

$\epsilon(q)$ corresponds to the model dielectric function taken by Glyde and Griffin. However, $\epsilon(p)$ is valid over the whole regime within the random phase approximation, if we extend the sound mode calculated in the long wavelength regime to that in the short wavelength regime. The hybridized mode are obtained by putting $\epsilon(p) = 0$ with $1/\tau_0 \cong 0$

$$p_0^2 = \frac{1}{2}\left[\omega_{sp}^2 + \epsilon_{Bp}^2 \pm \sqrt{(\omega_{sp}^2 - \epsilon_{Bp}^2)^2 + 4\Delta_p \alpha_p}\right] \equiv \Omega_{\pm p}^2. \tag{11.1.104}$$

In the long wavelength regime, Since $\epsilon_{Bp}^2 \sim \Delta_p$,

$$\begin{aligned}\sqrt{(\omega_{sp}^2 - \epsilon_{Bp})^2 + 4\Delta_p \alpha_p} &\sim \sqrt{(\omega_{sp}^2 - \Delta_p)^2 + 4\Delta_p(\omega_{sp}^2 - \omega_p^2)} \\ &\sim \sqrt{(\omega_{sp}^2 + \epsilon_{Bp}^2)^2 - 4\Delta_p \omega_p^2},\end{aligned}$$

Ω_{-p} becomes an imaginary number, so it disappears. Thus Ω_{+p} is identified with the Nambu-Goldstone mode;

$$E_p = \Omega_{+p} = \sqrt{\frac{1}{2}\left[\omega_{sp}^2 + \epsilon_{Bp}^2 + \sqrt{(\omega_{sp}^2 - \epsilon_{Bp}^2)^2 + 4\Delta_p \alpha_p}\right]}. \tag{11.1.105}$$

Frequencies $E_p, \omega_{sp}, \epsilon_{Bp}, \Delta_p, \alpha_p$ are all spherical symmetry, so we do not distinguish the arguement $\boldsymbol{p}$ with scalar p.

Next we estimate the dipersion curves, ω_p and ϵ_p. In order to estimate them we put the radial distribution function as $g(R) = \rho\delta(R - a)$, where $\rho = N'/V$, V is the volume, and a is the averaged particle distance. From Eqs. (11.1.19, 11.1.20) and (11.1.31, 11.1.32), we can estimate

$$\epsilon_p = \tilde{\omega}_0 + J\frac{\sin\alpha}{\alpha}, \qquad \alpha = pa, \tag{11.1.106}$$

$$J = \widetilde{J}_0 = \frac{4\pi a^2}{N'}\rho J_0(a), \tag{11.1.107}$$

$$V(p) = V(0)\frac{\sin\alpha}{\alpha}, \qquad V(0) = 4\pi a^2 \rho V_{00}(a). \tag{11.1.108}$$

Combining Eqs. (11.1.70) and (11.1.71) leads to

$$\tilde{\epsilon}_p = |J|\left(1 - \frac{\sin\alpha}{\alpha}\right), \tag{11.1.109}$$

where $\lim_{p\to 0}\tilde{\epsilon}_p \cong \frac{|J|}{6}a^2 p^2$. This dispersion curve, ϵ_{Bq} in the long wavelength regime, behaves in the same form as that of the Bogoluibov approximation. Its sound velocity is given by

$$c_B = \frac{V(0)n_0}{m} = \frac{V(0)n_0|J|a^2}{3}. \tag{11.1.110}$$

Schematic dispersion curves of $\tilde{\epsilon}_p$ and ϵ_{Bp} are shown in Figure 11.2, where $|J| = 1$, $V(0) = 2$, $n_0 = 0.1$;

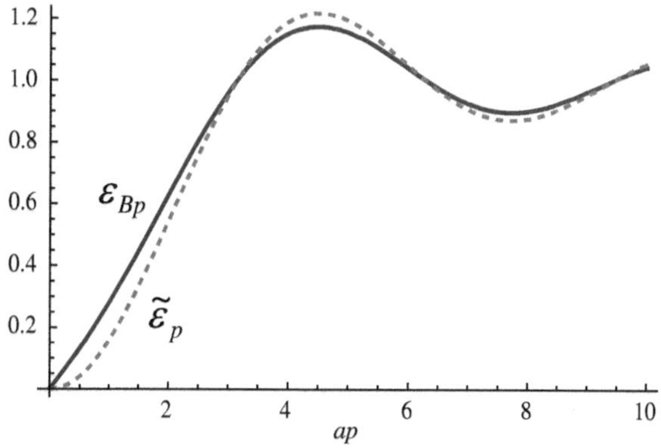

Figure 11.2 Schematic dispersion curves of $\tilde{\epsilon}_p$ and ϵ_{Bp} are illustrated, where $|J| = 1$, $V(0) = 2, \quad n_0 = 0.1$.

$$\epsilon_{Bp} = \sqrt{\tilde{\epsilon}_p^2 + \Delta_p}, \qquad \tilde{\epsilon}_p = 1 - \frac{\sin\alpha}{\alpha}, \qquad \Delta_p = 0.4\frac{\sin\alpha}{\alpha}(1 - \frac{\sin\alpha}{\alpha}).$$

Next we investigate the sound dispersion curves, $\omega_{\boldsymbol{q}}$ and $\omega_{s\boldsymbol{q}}$. In calculating these terms, we have restricted ourselves to the long wavelength regime. But we need dispersion curves in the maxon and roton regions. We can expect the frequencies of ω_p are depressed in the short wavelength regions, so we replace $\omega_q^2 = c_{0q}^2 q^2 = \frac{c_{0q}^2}{a^2}\alpha^2$ by

$$\omega_q^2 = \frac{6c_{0s}^2}{a^2}(1 - \frac{\sin\alpha}{\alpha}), \tag{11.1.111}$$

where $\lim_{\alpha\to 0} \frac{6c_{0s}^2}{a^2}(1 - \frac{\sin\alpha}{\alpha}) = c_{0s}^2 q^2$. Here we can expect the sound frequencies are lower than the particle frequencies in the maxon and roton regions, so we take

$$\omega_q^2 = 0.4(1 - \frac{\sin\alpha}{\alpha}), \qquad \frac{6c_{0s}^2}{a^2} = 0.4,$$

where $\lim_{\alpha\to 0} \omega_q^2 = \lim_{\alpha\to 0} \Delta_q = \lim_{\alpha\to 0} \epsilon_{Bq}^2$. Since the sound $\omega_{sq}^2 = (1+\kappa V(0))\omega_q^2$ does not soften, we put

$$\alpha_q = \kappa V(q)\omega_q = 0.16\frac{\sin\alpha}{\alpha}(1 - \frac{\sin\alpha}{\alpha}), \qquad \kappa = 0.2.$$

Schematic dispersion curves of ω_q and ω_{sq} are shown in Figure 11.3., where we have taken $\frac{6c_{0s}^2}{a^2} = 0.4$, and $\kappa = 0.2$.

Now we can obtain the excitations in a bose-condensed liquid E_p in Eq. (11.1.105). In the long wavelength regime, we have

$$\epsilon_{Bp}^2 \cong c_B^2 p^2, \qquad \omega_{sp}^2 \cong c_s^2 p^2, \qquad \Delta_p \cong \epsilon_{Bq}^2 \cong c_B^2 p^2,$$
$$\alpha_p = \omega_{sp}^2 - \omega_p^2 \cong (c_s^2 - c_{0s}^2)p^2.$$

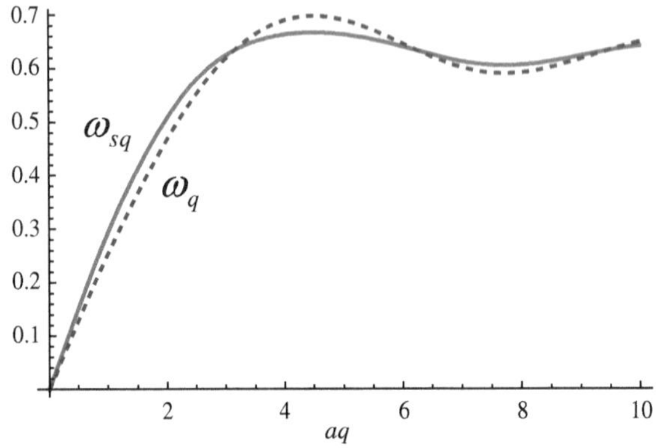

Figure 11.3 Schematic dispersion curves of ω_q and ω_{sq} are illustrated, where $\frac{6c_{0s}^2}{a^2} = 0.4$ and $\kappa = 0.2$.

Thus we obtain the velocity of E_p, c in the long wavelength regime is given by

$$c = \sqrt{\frac{1}{2}\left\{c_s^2 + c_B^2 + \sqrt{(c_s^2 + c_B^2)^2 - 4c_{0s}^2 c_B^2}\right\}}, \tag{11.1.112}$$

Taking the parameters: $|J| = 1$, $V(0) - 2$, $n_0 = 0.1$, $\frac{6c_{0s}^2}{a^2} = 0.4$, $\kappa = 0.2$, we show schematic dispersion curves of excitations in a bose-condensed liquid, $E(p)$ with ϵ_{Bp} and ω_{sp} in Figure 11.4.

E_p is determined as the upper dispersion curves Ω_{+p} due to the hybridization between the particle excitations ϵ_{Bp} and the sound ω_{sp}. The upper or the lower value of dispersion curves $\Omega_{\pm p}$ are essentially the larger or smaller value of ϵ_{Bp} and ω_{sp}. When the dispersion curves of ϵ_{Bp} and ω_{sp} cross over, the dispersion curves around

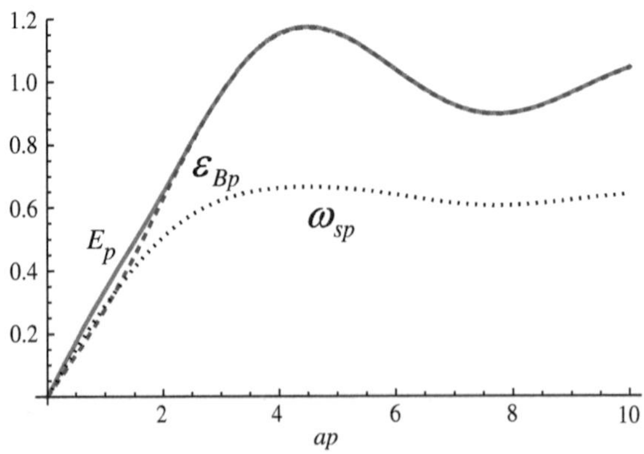

Figure 11.4 Schematic dispersion curves of $E(p)$ with ϵ_{Bp}, and ω_{sp} are illustrated, where $|J| = 1$, $V(0) - 2$, $n_0 = 0.1$, $\frac{6c_{0s}^2}{a^2} = 0.4$, and $\kappa = 0.2$.

the crossover of two curves are split into the upper and lower dispersion so that $\Omega_{\pm p}$ inherits the upper or lower part. The lower dispersion Ω_{-p} around the crossover point $p = 0$ has negative values, so Ω_{-p} disappears around $p = 0$. Therefore when the sound frequencies ω_{sp} are lower than the particle frequencies ϵ_{Bp}, the excitations E_p are determined essentially by the particle frequencies ϵ_{Bp}.

The experimental frequencies of the excitations in the short wavelength regime are higher than those of E_p illustrated in Figure 11.4. The dispersion of E_p is essentially determined by the hopping frequencies $\epsilon_p = \tilde{\omega}_0 + J\frac{\sin\alpha}{\alpha}$ in Eq. (11.1.106). Since the hopping frequencies are bound to the harmonic frequency $\tilde{\omega}_0$, in the long wavelength regime, $ap \gg 1, \epsilon_p \longrightarrow \tilde{\omega}_0$, E_p cannot increase as the wave vector increases. However, in actual liquids, as the wavelength increases over a reciprocal particle distance, particles split from the binding so that E_p increases as free particles. Therefore the actual dispersion curves are higher than E_p.

11.1.7 The Dynamical and Static Structure Factors

The dynamical and static structure factors are closely related with experimental results [6,7]. We present the dynamical and static structure factors in a bose-condensed liquid and explain the experimental result of the temperature dependence of the static structure factor.

The dynamical structure factor is defined as

$$S(\boldsymbol{q}, q_0) = \frac{1}{2\pi}\int_{-\infty}^{\infty} dt e^{\mathrm{i}q_0 t}\frac{1}{N'}\langle\rho_{\boldsymbol{q}}(t)\rho_{\boldsymbol{q}}^{\dagger}(0)\rangle_{\epsilon}, \tag{11.1.113}$$

which satisfies

$$S(\boldsymbol{q}, -q_0) = e^{-\beta\hbar q_0}S(\boldsymbol{q}, q_0), \tag{11.1.114}$$

$$\begin{aligned}-\frac{\mathrm{i}}{N'}\int_{-\infty}^{\infty} dt\langle[\rho_{\boldsymbol{q}}(t), \rho_{\boldsymbol{q}}^{\dagger}(0)]\rangle_{\epsilon}e^{\mathrm{i}q_0 t} &\equiv \int_{-\infty}^{\infty} dt K(\boldsymbol{q}, t)e^{\mathrm{i}q_0 t}\\ &= -2\pi\mathrm{i}(1 - e^{-\beta\hbar q_0})S(\boldsymbol{q}, q_0),\end{aligned} \tag{11.1.115}$$

where in the first line we have defined the function $K(\boldsymbol{q}, \tau)$ and in the second line we have used the symmetry of the space inversion and the relation $\langle\rho_{\boldsymbol{q}}(0)\rho_{\boldsymbol{q}}^{\dagger}(t)\rangle_{\epsilon} = \langle\rho_{\boldsymbol{q}}(t - \mathrm{i}\beta\hbar)\rho_{\boldsymbol{q}}^{\dagger}(0)\rangle_{\epsilon}$.

Here we introduce the retarded Green's function for the density fluctuation operators as

$$\begin{aligned}F_{\boldsymbol{q}}^{r}(t) &\equiv -\frac{\mathrm{i}}{\hbar N'}\langle\rho_{\boldsymbol{q}}(t), \rho_{\boldsymbol{q}}^{\dagger}(0)]\rangle_{\epsilon}\theta(t)\\ &= \frac{1}{2\pi}\int_{-\infty}^{\infty} dt e^{-\mathrm{i}q_0 t}F^{r}(\boldsymbol{q}, q_0).\end{aligned} \tag{11.1.116}$$

Using the properties of the function $K(\boldsymbol{q}, t)$:

1. real: $K^*(\boldsymbol{q},t) = K(\boldsymbol{q},t)$,
2. odd with respect $t: K(\boldsymbol{q},-t) = -K(\boldsymbol{q},t)$,

we obtain

$$\int_{-\infty}^{\infty} dt e^{iq_0 t} K(\boldsymbol{q},t) = 2i\hbar \Im F^r(\boldsymbol{q},q_0). \tag{11.1.117}$$

Equations (11.1.115) and (11.1.117) lead to the fluctuation-dissipation theorem:

$$\begin{aligned} S(\boldsymbol{q},q_0) &= -\frac{1}{1-e^{-\beta\hbar q_0}}\frac{\hbar}{\pi}\mathrm{Im}F^r(\boldsymbol{q},q_0) \\ &= -\frac{1}{1-e^{-\beta\hbar q_0}}\frac{\hbar}{\pi}\Im F(\boldsymbol{q},q_0+i\delta), \end{aligned} \tag{11.1.118}$$

where we have used the relation of the retarded Green's function and the temperature Green's function: $F^r(\boldsymbol{q},q_0) = F(\boldsymbol{q},q_0+i\delta)$ in the last equation. Since from Eq. (11.1.46)

$$F(q) = \frac{\widetilde{F}(q)}{\epsilon} = \frac{\widetilde{F}(q)}{1-V(q)\widetilde{F}(q)} = \frac{1}{V(q)}\frac{V(q)\widetilde{F}(q)-1+1}{1-V(q)\widetilde{F}(q)} = \frac{1}{V(q)}(\frac{1}{\epsilon}-1),$$

we obtain

$$S(\boldsymbol{q},q_0) = -\frac{1}{1-e^{-\beta\hbar q_0}}\frac{\hbar}{\pi V(q)}\Im\frac{1}{\epsilon}. \tag{11.1.119}$$

In order to obtain $\Im\frac{1}{\epsilon}$, we first investigate ϵ

$$\begin{aligned} \epsilon &= 1 - V(q)\widetilde{F}_R(q) - V(q)\widetilde{\Lambda}^T\widetilde{G}(q)\widetilde{\Lambda} \\ &= 1 - V(q)\frac{\frac{\kappa}{\hbar}\omega_q^2}{q_0^2+i\Gamma q_0-\omega_q^2} - V(q)n_0\frac{2\tilde{\epsilon}_q}{q_0^2-\tilde{\epsilon}_q^2} \\ &= \frac{q_0^2+i\Gamma q_0-\omega_{sq}^2}{q_0^2+i\Gamma q_0-\omega_q^2} - \frac{\Delta_q}{q_0^2-\tilde{\epsilon}_q^2} \\ &= \frac{(q_0^2-\omega_{sq}^2)(q_0^2-\epsilon_{Bq}^2)-\Delta_q\alpha_q+i\Gamma q_0(q_0^2-\epsilon_{Bq}^2)}{(q_0^2-\omega_q^2)(q_0^2-\tilde{\epsilon}_q^2)+i\Gamma q_0(q_0^2-\tilde{\epsilon}_q^2)}, \end{aligned}$$

where $\Gamma = 1/\tau_0$, then we have

$$\begin{aligned} \Im\frac{1}{\epsilon} &= \Im\frac{(q_0^2-\omega_q^2)(q_0^2-\tilde{\epsilon}_q^2)+i\Gamma q_0(q_0^2-\tilde{\epsilon}_q^2)}{(q_0^2-\omega_{sq}^2)(q_0^2-\epsilon_{Bq}^2)-\Delta_q\alpha_q+i\Gamma q_0(q_0^2-\epsilon_{Bq}^2)} \\ &= -\frac{\alpha\Gamma q_0(q_0^2-\tilde{\epsilon}_q^2)^2}{[(q_0^2-\epsilon_{sq}^2)(q_0^2-\epsilon_{Bq}^2)-\Delta_q\alpha_q]^2+[\Gamma q_0(q_0^2-\epsilon_{Bq}^2)]^2}. \end{aligned}$$

Considering the limiting case, $\Gamma \to 0$, and using $\lim_{\Gamma\to 0}\frac{\Gamma}{x^2+\Gamma^2} = \pi\delta(x)$, we obtain

$$
\begin{aligned}
\Im\frac{1}{\epsilon} &= \frac{-\Gamma q_0(q_0^2-\epsilon_{Bq}^2)\frac{\alpha\Gamma q_0((q_0^2-\tilde{\epsilon}_q)^2}{\Gamma q_0(q_0^2-\epsilon_{Bq}^2)}}{[(q_0^2-\epsilon_{sq}^2)(q_0^2-\epsilon_{Bq}^2)-\Delta_q\alpha_q]^2+[\Gamma q_0(q_0^2-\epsilon_{Bq}^2)]^2} \\
&\to -\pi\alpha\mathrm{sign}(q_0)\frac{(q_0^2-\epsilon_{Bq}^2+\Delta_q)^2}{q_0^2-\epsilon_{Bq}^2}\delta[(q_0^2-\epsilon_{sq}^2)(q_0^2-\epsilon_{Bq}^2)-\Delta_q\alpha_q] \\
&= -\pi\alpha\mathrm{sign}(q_0)\frac{(q_0^2-\epsilon_{Bq}^2+\Delta_q)^2}{q_0^2-\epsilon_{Bq}^2}\delta[(q_0^2-\Omega_{+q}^2)(q_0^2-\Omega_{-q}^2)].
\end{aligned}
$$

Consider the $E_q = \Omega_{+q}$ branch, using the relation:

$$
\mathrm{sign}(x)\delta(x^2-a^2) = \frac{1}{2a}\{\delta(x-a)-\delta(x+a)\},
$$

we obtain

$$
\begin{aligned}
\Im\frac{1}{\epsilon} &\longrightarrow -\pi\,\mathrm{sign}(q_0)\alpha\frac{(E_q^2-\epsilon_{Bq}^2+\Delta_q)^2}{E_q^2-\epsilon_{Bq}^2}\frac{1}{E_q^2-\Omega_{-q}^2}\delta(q_0^2-E_q^2) \\
&= -\pi\alpha\frac{(E_q^2-\epsilon_{Bq}^2+\Delta_q)^2}{E_q^2-\epsilon_{Bq}^2}\frac{1}{2E_q\sqrt{(\omega_{sq}^2-\epsilon_{Bq}^2)^2+4\Delta_q\alpha_q}} \\
&\quad\times[\delta(q_0-E_q)-\delta(q_0+E_q)].
\end{aligned}
\tag{11.1.120}
$$

Thus we obtain the dynamic structure factor in the $\Gamma \to 0$ limit,

$$
\begin{aligned}
S(\boldsymbol{q},q_0) &= \frac{1}{1-e^{-\beta\hbar q_0}}\frac{\hbar\kappa\omega_q^2(E_q^2-\epsilon_{Bq}^2+\Delta_q)^2}{2E_q(E_q-\epsilon_{Bq}^2)\sqrt{(\omega_{sq}^2-\epsilon_{Bq}^2)^2+4\Delta_q\alpha_q}} \\
&\quad\times\{\delta(q_0-E_q)-\delta(q_0+E_q)\},
\end{aligned}
\tag{11.1.121}
$$

where we have considered $\alpha = \kappa V(q)\omega_q^2$. The static structure factor is given by

$$
\begin{aligned}
S(\boldsymbol{q}) &= \int_{-\infty}^{\infty} dq_0 S(\boldsymbol{q},q_0) \\
&= \frac{\hbar\kappa\omega_q^2(E_q^2-\epsilon_{Bq}^2+\Delta_q)^2}{2E_q(E_q^2-\epsilon_{Bq}^2)\sqrt{(\omega_{sq}^2-\epsilon_{Bq}^2)^2+4\Delta_q\alpha_q}}\coth\frac{\beta\hbar E_q}{2}.
\end{aligned}
\tag{11.1.122}
$$

It should be noted that if we consider the actual values of κ and $\omega_{\boldsymbol{q}}$, Eqs. (11.1.121) and (11.1.122) are valid over the whole regimes of the wavelength. In the long wavelength regime, we obtain

$$
S(0) = \frac{\kappa}{\beta}\frac{c^2c_{0s}^2}{(c^2-c_B^2)\sqrt{(c_s^2+c_B^2)^2-4c_{0s}^2c_B^2}},
\tag{11.1.123}
$$

where we have considered $\epsilon_{Bq} \cong \Delta_q$ in the short wavelength regime. Since $\kappa = \frac{A}{\beta^2\hbar^3\rho c^3} \propto T^2$, we have

$$S(0) \propto T^3. \tag{11.1.124}$$

This explains the experimental results [7].

11.1.8 Concluding Remarks

The two band model in the normal state is extended to the bose-condensed state. In the normal state, there are two particle excitations, upper and lower bands, and collective excitations, sound. In the bose-condensed state, there is an excitation due to the hybridization between the Bogoluibov excitation $\epsilon_{Bp} = \sqrt{\tilde{\epsilon}_p^2 + \Delta_p}$ and sound excitation, ω_{sq}. The hybridization of two modes are essential in the long wavelength regime, but the hybridization inherits the Bogoluibov excitations in the short wavelength regime, where $\epsilon_{Bp} \sim \tilde{\epsilon}_p = |J|\left(1 - \frac{\sin\alpha}{\alpha}\right)$; the NG boson originate from the hybridization in the long wavelength regime, where both the Bogoluibov and sound excitations are gapless, while the maxion and the roton from the the lower particle excitation $\tilde{\epsilon}_p = |J|\left(1 - \frac{\sin\alpha}{\alpha}\right)$.

Conventionally, this excitations in the bose-condensed state is considered as a phonon mode. This is not correct from two meanings: first, in the bose-condensed state the NG boson comes from the phase broken symmetry, while the phonons in the glassy state from the translational broken symmetry. Second, in the liquid state, there is no translational broken symmetry like in the glassy state; sound is not a NG boson.

The dynamical and static structure factors are calculated from the correlation functions of the particle density fluctuations. Since conventional theories do not distinguish between sound and phonons, the intra-band and inter-band density fluctuations, they cannot correctly explain the behaviors of the dynamical and static structure factors.

However, in the present theory, we have not investigated the sound excitations in the short wavelength regime, but we have assumed the dispersion curves there, because the hybridization does not reflect the sound excitation in the short wavelength regime. In normal liquids near the glass transition, the bandwidth of particles is very narrow, $\beta\hbar|J| \ll 1$, we can extend the sound excitation in the long wavelength regime to that in the short one so that we can develop the liquid-glass transition, but this condition is not satisfied in the bose-condensed state, because the bose-condensed state is a liquid one.

11.2 A Model on the Origin of RNA

The RNA hypothesis advocates that the origin of life started with the primitive RNA. A RNA is a kind of polymer, which is formed by physical or chemical vitrification. In both physical and chemical vitrification, the slowing down of the dynamics due to the reduction of the configuration entropy corresponding to the Kauzmann entropy occurs so that the polymerization closely relates with the glass transition. The RNA is in a glassy state. Therefore the primitive RNA could be formed under the occasions of the supercooling process to create enormous bubbles of nucleotide-hole pairs not to complete the nucleotide-hole cycles (to avoid crystalization) in the nucleotide-condensed state.

Around volcanoes there could exist a lot of different nucleotides, which were creating bubble of nucleotide-hole pairs. The temperature could be fluctuating sometimes higher and sometimes lower. Under some appropriate conditions, the glass transition, the polymerization of RNA in a nucleotide-condense state, could occur. In the glassy state the ribose subunits are joined by a $2' - 5'$ or $3' - 5'$ phosphodiester linkage to form the ribose-phosphate backbone similar to oligosaccharides and polysaccharides. The occurrence of the glass transition requires two conditions:

1. A supercooled state in a nucleotide-condensed state should exist below the temperature at which the whole RNA hydrolyzes.
2. The Gibbs free energy due to the Kauzmann entropy, which obeys a Curie law with a negative sign, must be larger than the height of the potential barrier for nucleotides to overcome to form the binding.

We first briefly survey a history of theories on the origin of life and propose a model on the origin of RNA.

11.2.1 A Brief Historic Survey of Theories on the Origin of Life

The origin of life is one of the most challenging subjects. The earth was formed about 4.6×10^9 years ago. The fossil record strongly supports the existence of cellular life by 3.6×10^9 years ago. It is certain that some primitive organic compounds should be formed at the earlier stage of the cellular life. By 1930s Oparin[8–10] and Haldane[11–13] had proposed that the atmosphere of the pre-biotic earth was strongly reducing like that of outer planets so that it contained very little oxygen and was rich in hydrogen and compounds such as methane (CH_4) and ammonia (NH_3). In an experiment under such a pre-biotic atmosphere condition, in 1953 Miller and Urey [14,15] produced a suite of simple organic compounds including hydrogen cyanide (HCN) and aldehydes (compounds containing the group CHO), which produce amino acids and ammonia. In an experiment to obtain amino acids by simpler chemistry than that in Miller-Urey experiment, in 1961 Oró [16] found that the mixture of hydrogen cyanide and ammonia in an aqueous solution produced amino acids, and unexpectedly discovered that the mixture produced abundant adenine. Adenine is one of the four nitrogen-containing bases present in RNA and DNA.

Life must possess genetic information and function of replication. In contemporary organisms, catalytic properties of proteins play an essential role in replicating genetic information and synthesizing proteins. On the other hand, the production of proteins is governed by genetic information through the order of amino acids. In late 1960s, Woese [17], Crick [18,19], and Orgel [20–22] independently proposed that certain RNAs might have catalytic activity. Cech [23–26] and Altman [27–29] independently discovered that RNA can cut and splice as an enzyme. Cech and Szostak carried out stringing together nucleotides or oligonucleotides. Szostak et al. [30,31] also found that random sequences of RNA derive structurally complex and highly active RNA ligases.

RNA possesses both genetic information and function of replication indispensable to life, which is now called the RNA world [32]. RNA must have been pre-biotically synthesized with nucleotides. However, there are several obstacles to confirm that life began with RNA. Without enzymes, synthesis of ribose in adequate quantity and purity and an efficient synthesis of pyrimidine nucleosides are impossible. The greatest problem with pyrimidines is not their synthesis but the difficulty in attaching them to ribose to form pyrimidine nucleotides. Cairns-Smith [33] proposed that life began with self-replacing clays, with the genetic information stored as a distribution of charges on a clay and replicated by ionic interactions between existing and newly formed surface layers. Minerals might have catalyzed both the synthesis of properly structured nucleotides and their polymerization to a random family of oligonucleotides. Szostak [34] proposed that one of the random polymers produced on minerals happened to be a ribozyme that could copy its complement and thus duplicate itself. However, although structural irregularities in clay were complicated enough to set the stage to the emergence of RNA, the structure probably would not be amenable to accurate self-replication. As an alternative genetic materials instead of ribose in RNA, Eschenmoser [35] has proposed pyranosyl RNA, where ribose in RNA is replaced by a six-member ring. Nielsen [36] has also proposed peptide nucleic acid, which has nucleic acid bases but a protein-like backbone. But any RNA-like system faces several obstacles [37]: (i) the lack for end-to-end copying of the template before the development of a mechanism to ensure that replication begins at a discrete initiation site. (ii) The tendency of molecules with regions of self-complementarity to form intermolecular duplex structures. (iii) The problem of accurate copying of the genetic information in the absence of a sophisticated replicase enzyme.

Therefore, although it is plausible, the RNA hypothesis, "genome first" scenario, is still disputed. A serious obstacle is that it is unlikely that sufficient quantities of all the necessary nucleotides have been produced on the primitive Earth [38, 39]. Following "metabolism first" scenario from Operin [10], Dyson , and Kauffman [42, 43] subsequently proposed alternative models for homeostasis and potential self-replication in mutually catalitic molecular assemblies without genome. In Dyson's model, highly organized catalytic interactions develop over time in a manner dependent on assembly size and the diversity of the components involved. In Kauffman's model, "catalytic closure" is achieved through molecular interactions that gradually increase molecular density. Extending the Dyson's and Kauffman's models, Lancet and co-workers [44–46] proposed the Graded Autocatalysis Replication Domain (GARD) model based on a set of plausible chemical kinetic assumptions concerning mutually catalytic growth. But as it is unlikely that any specific scenario attempting to explain how life started can be proven, it is important to evaluate the limitations of all the possible pathways that lead to the emergence of life. It is here that the present work intends to contribute, by examining one of the crucial steps of the emergence of life through RNA, the polymerization, and showing how it could be achieved through a mechanism analogous to a liquid-glass transition.

So far, synthesis of nucleotides and RNA has been considered at a biological temperature. But nucleotides, even pyrimidine, and RNA could be synthesized at a higher temperature such as around volcanoes where water was evaporated and nucleotides were condensed. We call this state a nucleotide-condensed state.

Here we present a model on the origin of RNA by applying the theory of the liquid-glass transition based on the two band model [47,48] to the transition in a nucleotide-condensed state assuming that there are adequate quantity of nucleotides [49].

A glass is normally formed by supercooling the viscous liquid fast enough to avoid crystallization. However, a glass may also be formed by chemical vitrification, the process of which involves progressive polymerization of the constituent molecules via the formation of irreversible chemical bonds. In both physical and chemical vitrification, the slowing down of the dynamics due to the reduction of the configuration entropy, corresponding to the Kauzmann entropy, is commonly observed [50,51]. The polymerization of liquids is a process that modifies molecular motions in a reacting system, introducing configuration restriction which progressively reduces the accessible phase space, leading to a glassy behavior. This leads to a strong decrease of the configuration entropy. Here we consider that the Gibbs free energy associated to the Kauzmann entropy of nucleotides is released to overcome the barrier that opposes to their binding in the synthesis of RNA. In a biological system, $5'$-triphosphate nucleotides provide the energy to overcome the potential barrier, but another pathway may exist, through a glass transition, if the proper conditions are present.

Let T_{M1} be the temperature above which the whole RNA hydrolyzes to nucleotides. Below T_{M1} a condensed state of nucleotides, which are not bound to form RNA, behaves like a supercooled state in a liquid-glass transition. The Kauzmann entropy has a very large negative value, and evolves with temperature with a Curie law with a negative sign. There exists a critical temperature (the sound disappearing temperature), which is essentially determined by the interactions between hydroxyls connected with phosphate and $3'$ or $2'$ of ribose by taking account of the intermediate state: the interactions between phosphate and oxygen, and between oxygen and $3'$ or $2'$. At the glass transition temperature above the sound disappearing temperature, the negative divergence of the Kauzmann entropy (the so-called Kauzmann entropy crisis) is prevented by the entropy of the bubbles of inter-band particle-hole pairs. In the context of RNA synthesis, this "glass transition" is actually the polymerization of the nucleotides, i.e. the binding of the ribose subunits by a $2' - 5'$ or $3' - 5'$ phosphodiester linkage to form the ribose-phosphate backbone similar to oligosaccharides and polysaccharides. The system after the transition can be viewed as similar to a cotton candy. The binding is made possible by the Gibbs free energy due to the Kauzmann entropy.

The occurrence of such a "glass transition" for nucleotides requires two conditions:

1. A supercooled state should exist below T_{M1}; the sound disappearing temperature must be lower than T_{M1}.
2. The Gibbs free energy due to the Kauzmann entropy must be larger than the height of the potential barrier to overcome to form the binding.

The temperature around a volcano could be fluctuating sometimes higher and sometimes lower. The glass transition in a nucleotide-condensed state could occur at the glass transition temperature. After the formation of the glass, when the temperature increased, the RNAs in the glassy state could break up. Thus there could appear single-strand RNAs.

At a temperature lower than T_{M1}, there exists another melting temperature T_{M2}, which is defined by the temperature above which all hydrogen bonds in RNA split up. Below T_{M2}, if two parts of a single-strand RNA in a random sequence contain the appropriate pairs according to the pairing rule, they can join and establish double-strand segments. Thus there appears complex three-dimensional structure. At this stage RNAs possess catalytic properties.

Here we confine ourselves to the formation of RNAs in the glassy state above T_{M2}.

11.2.2 The Glass Transition in a Nucleotide-Condensed State

Now we extend the theory of the liquid-glass transition in multi-component liquids to that in a nucleotide-condensed state. Hereafter, we assume that there are adequate quantity of nucleotides. We consider an averaged base, though bases are different. The number of nucleotides N. In order to investigate the liquid-glass transition in multi-component liquids, we first clarify the components of nucleotide-condensed matter, and then calculate the correlation functions of intra-band and inter-band density fluctuations, which yield sound and phonons, respectively. Bubble diagrams non-constituting their respective collective modes yield their respective fluctuation entropies.

Sound plays an essential role in the liquid-glass transition. Sound disappears at a reciprocal particle distance K and at sound disappearing temperature T_0. The intra-band fluctuation entropy mixes with sound entropy around K. This mixing entropy yields the Kauzmann entropy, which obeys a Curie law with a negative sign and the critical temperature T_0. The Kauzmann entropy and the inter-band fluctuation entropy determine the liquid-glass transition at the glass transition temperature T_g above T_0.

Therefore we first clarify the components of nucleotide-condensed matter. A nucleotide consists of nucleoside, phosphate (PO^{2-}), a hydroxyl (OH) bound to the phosphate, and hydroxyls connected with $2'$ and $3'$ of ribose, which we regard as the components of the nucleotide-condensed matter and denote as R, P, O1, and O2, respectively. Nucleotides are schematically shown in Figure 11.5(a). Above the glass transition temperature, T_g, the interaction between O1 and O2 plays the most important role in the glass transition.

Next we investigate the correlation functions $F_{\alpha\beta}(q)$ for intra-band density fluctuations with the components, P, R, O1, and O2, which lead to the dynamical equations with 4×4 matrices:

$$F_{\alpha\beta}(q) = \delta_{\alpha\beta} P_\alpha(q) + \sum_\gamma P_\alpha(q) V_{\alpha\gamma}(\boldsymbol{q}) F_{\gamma\beta}(q),$$

$$P_\alpha(q) \cong -\frac{\beta f' \omega_{\alpha\boldsymbol{q}}^2}{q_0(q_0 + \frac{\mathrm{i}}{\tau_{\alpha 0}}) - \omega_{\alpha\boldsymbol{q}}^2},$$

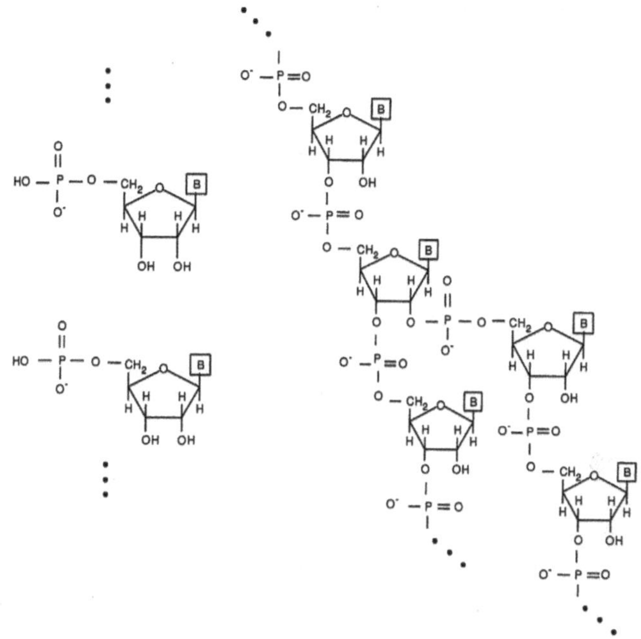

(a) (b)

Figure 11.5 Nucleotides in a nucleotide-condensed state (a) and nucleotides in a glassy state (b) are schematically shown. (a) A nucleotide consists of nucleoside, phosphate (PO^{2-}), a hydroxyle (OH) bound to the phosphate, and hydroxyles connected with $2'$ and $3'$ of ribose. (b) A nucleotide consists of nucleoside, phosphate, and an oxgen connected with $2'$ or $3'$ of ribose and phosphate PO^{2-}. B in a square denotes a base.

where $P_\alpha(q)$ is a bubble diagram of intra-band particle-hole pairs and the argument $q = (q_0, \boldsymbol{q}), q_0$ and $\boldsymbol{q}$ are a frequency parameter and a wave vector, respectively. $\beta = \frac{1}{k_B T}$, where k_B is the Boltzmann constant. f is a distribution function and f' its derivative. Suffices, α, β, and γ are P, R, O1, and O2. Here we should not confuse the temperature $\beta = \frac{1}{k_B T}$ and the suffix β. Above the glass transition temperature T_g, there are potential functions $V_{\alpha\beta}(\boldsymbol{q})$ between O1–P, P–R, R–O2, and O2–O1. In the term $P_\alpha(q)$, $\omega_{\alpha\boldsymbol{q}} \cong \frac{v_{\alpha p}}{\sqrt{3}} q$, where $v_{\alpha p}$ is an average velocity of α-particle. $\tau_{\alpha 0}$ is the relaxation time of the sound of α-particles. In deriving sound and entropy, we neglect the term $\frac{i}{\tau_{\alpha 0}}$. The sound modes are obtained from the secular equation of the dynamical equation of $F_{\alpha\beta}(q)$:

$$\det \left| \delta_{\alpha\beta} - P_\alpha(q) V_{\alpha\beta}(\boldsymbol{q}) \right| = 0,$$

We obtain four-component sound modes.

Below T_g a nucleotide condenses into the macromolecule RNA which is a nucleotide in the glassy state, i.e. the polymerized state and a water molecule H_2O. A nucleotide in the glassy state consists of a phosphate (PO^{2-}), a nucleoside, and an oxygen connected with both $2'$ or $3'$ of ribose and phosphate PO^{2-}, which we denote as P, R, and O, respectively. Nucleotides are schematically shown in Figure 11.5(b). The correlation functions $F_{\alpha\beta}(q)$ for the intra-band density fluctuations due to P, R, and O lead to the dynamical equations with 3×3 matrices so that we obtain three-component sound modes; there are potential functions $V_{\alpha\beta}(\boldsymbol{q})$ between O–P, P–R, and R–O. And the correlation function for intra-band density fluctuations due to H_2O leads to a sound mode of water.

The intra-band density fluctuations of nucleotides play an essential role in the glass forming process, where we are concerned with the potential functions $V_{\alpha\beta}(\boldsymbol{q})$ between O1–P, P–R, R–O2, and O2–O1. The sound disappearing occurs at a sound disappearing temperature T_0 and a reciprocal particle distance K, when the secular equation of the correlation functions for intra-band density fluctuations $F_{\alpha\beta}(q)$ with $\lim_{q_0 \to 0} P_\alpha(q) \cong +\beta f'$:

$$\det\left|\delta_{\alpha\beta} - \beta_0 f' V_{\alpha\beta}(K)\right| = 0,$$

where we choose the highest temperature as T_0 among four solutions [52]. The mixing entropy of the sound and intra-band fluctuation entropies leads to the Kauzmann entropy. However, since there are a lot of unknown parameters such as potential functions $V_{\alpha\beta}(\boldsymbol{q})$ and the pair distribution functions $g_{\alpha\beta}(R)$, we only investigate the condition for the occurrence of the glass transition. The Kauzmann entropy is given by:

$$S_K \cong -\frac{N_0 k_B}{2} \frac{T_0}{T - T_0}, \tag{11.2.1}$$

where N_0 is the number of states around the reciprocal particle distance K. In this case we can estimate the mean distance between R and O1 as the particle distance.

Now we investigate the glass forming process in the context of RNA synthesis. Since a supercooled state in a nucleotide-condensed state should exist below T_{M1}, the following condition is required:

$$T_0 < T_{M1}. \tag{11.2.2}$$

If T_{M1} were smaller than T_0, the glass transition would not occur, because the system would remain a nucleotide-condensed state.

Since the glass is formed by $2' - 5'$ or $3' - 5'$ phosphodiester linkage, P–O–R, the Kauzmann entropy should come from this interaction P–O–R. But for a supercooling process in a nucleotide-condensed state the interaction P–O1–O2–R dominates. However, at such a high temperature the interaction O1–O2 has an intermediate state: O1–O2 $\rightleftarrows$ $\{O + H_2O\}$. This intermediate state $\{O + H_2O\}$ plays an important role in forming the glassy state, because the P–O–R process comes from P–O1–O2–R through the intermediate state $\{O + H_2O\}$: P–O1–O2–R $\rightleftarrows$ $\{P - O - R + H_2O\}$. We can include this intermediate state introducing the interaction Hamiltonian between O1, O2, and O. If we consider the potential functions $V_{\alpha\beta}(\boldsymbol{q})$ taking account of the intermediate state; $V_{PO1} P_{O1} V_{O1O2} P_{O2} V_{O2R} \to V_{PO} P_O V_{OR}$, we can expect that the condition (11.2.2) is satisfied for a supercooling process in a nucleotide-condensed state.

Next we investigate the Gibbs free energy of the nucleotides. The synthesis of RNA, i.e. the binding of nucleotides, needs a large Gibbs free energy ΔG, where $G = H - TS$, H being the enthalpy. If the Gibbs free energy ΔG exceeds the height of a potential barrier h, the synthesis of RNA occurs. Free energy for a nucleotide-condensed state, a glassy state, and a potential barrier is schematically illustrated in Figure 11.6. In a biological system triphosphate nucleotides are released to overcome the potential barrier so that a $5'$-nucleotide links to another $3'$-nucleotide. The magnitude of the free energy for ATP is nearly 7.5 kcal/mol $\cong 3.8 \times 10^3$ K$k_{\rm B}$/molecule. In a biological system catalysis plays an essential role in lowering the height of the potential barrier. Even in a pre-biotic system such as around volcanoes, there could be some catalytic materials such as minerals, which could lower the height of the potential barrier.

We consider a nucleotide-condensed system at a temperature T around T_{M1}. If the height of the potential barrier h had been lower than the $k_{\rm B}T_{M1}$, RNA could have been synthesized at $T = T_{M1}$. But the height h is expected to be much larger than $k_{\rm B}T_{M1}$. The Gibbs free energy due to the Kauzmann entropy $S_{\rm K}$ is given by $\Delta G = -TS_{\rm K}$. If ΔG in the nucleotide-condensed state is larger than the height of the potential barrier h above the glass transition temperature T_g:

$$h < \Delta G = -TS_{\rm K}, \tag{11.2.3}$$

ΔG is released to the system so that the glass transition occurs.

The correlation functions for inter-band density fluctuations $D_{\alpha i\beta j}$ due to P, R, O1, and O2 for nucleotide-condensed state lead to the dynamical equations with 12×12 matrices so that we obtain 12-component phonon modes:

$$D_{\alpha i\beta j}(q) = \delta_{\alpha\beta}\delta_{ij}Q_\alpha(q) + Q_\alpha(q)\sum_{\gamma l} V_{\alpha i\gamma l}(\boldsymbol{q})D_{\gamma l\beta j}(q),$$

$$Q_\alpha(q) = \frac{2\omega_\alpha/\hbar}{(q_0 + \frac{\mathrm{i}}{2\tau_{\alpha M}})^2 - \omega_\alpha^2},$$

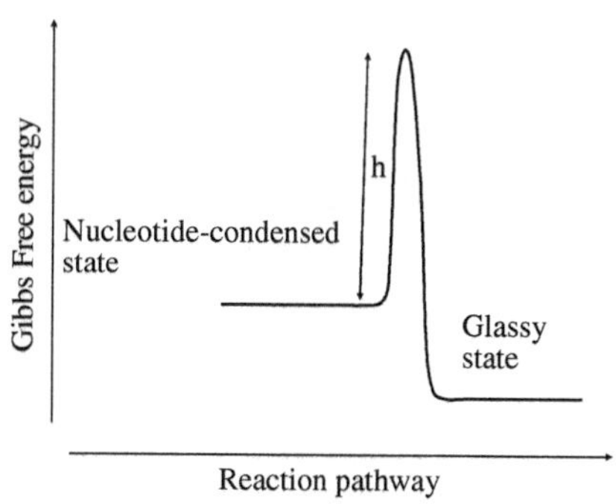

Figure 11.6 Free energy for a nucleotide-condensed state, a glassy state, and a potential barrier is schematically illustrated. The height of the potential barrier h is also shown. The Free energy for nucleotide-condensed state corresponds to the horizontal part on the left-hand side of the potential barrier and that for glassy state corresponds to the horizontal part on the right-hand side of the potential barrier.

where i and j denote the Cartesian coordinates [47,52]. In the term $Q_\alpha(q)$, ω_α is the harmonic frequency of α-particle and $\tau_{\alpha M}$ the Maxwell relaxation time of α-particle. In deriving phonon modes and entropies, we neglect the term $\frac{\mathrm{i}}{\tau_{\alpha M}}$.

The inter-band fluctuation entropy S_{Ω_α} is given by

$$S_{\Omega_\alpha} \cong -k_{\mathrm{B}} \frac{3N\beta\hbar}{e^{\beta\hbar\omega_\alpha} - 1}$$

If we know the harmonic frequencies of P, R, O1, and O2, we obtain the inter-band fluctuation entropy S_Ω for the nucleotide-condensed state summing up the inter-band fluctuation entropies due to P, R, O1, and O2.

The correlation functions for inter-band density fluctuations $D_{\alpha i\beta j}$ due to P, R, and O, and H_2O for RNAs in a glassy state and water molecules lead to the dynamical equations with 9–9 and 3–3 matrices. But if we know the harmonic frequencies of P, R, and O, and H_2O, we obtain the inter-band fluctuation entropy S_Ω for RNAs and water molecules summing up the inter-band fluctuation entropies due to P, R, and O, and H_2O.

In summing up the inter-band fluctuation entropy, we assume that the inter-band fluctuation entropies for both condensed nucleotides and RNAs are of the same magnitude and we further make the following approximation:

$$S_\Omega = 4S_{\Omega_\alpha}\big|_{\omega=\tilde{\omega}} = -k_{\mathrm{B}} \frac{12N\beta\hbar\tilde{\omega}}{e^{\beta\hbar\tilde{\omega}} - 1}, \tag{11.2.4}$$

where $\tilde{\omega}$ is a harmonic frequency for a nucleotide.

Eventually, if the conditions (11.2.2) and (11.2.3) are satisfied, the glass transition occurs at the glass transition temperature T_g:

$$S_{\mathrm{K}} = S_\Omega \mid_{T=T_g}. \tag{11.2.5}$$

Next we investigate the glass transition in a nucleotide-condensed state by plotting

$$\frac{S_{\mathrm{K}}}{Nk_{\mathrm{B}}} = -\frac{N_0}{2N}\frac{T_0}{T - T_0}, \qquad \frac{S_\Omega}{Nk_{\mathrm{B}}} = -\frac{12\beta\hbar\tilde{\omega}}{e^{\beta\hbar\tilde{\omega}} - 1} = -\frac{12/x}{e^{1/x} - 1}$$

and

$$-\frac{h}{Nk_{\mathrm{B}}T} \equiv -\frac{T_h}{T}$$

with respect to $x \equiv \frac{k_{\mathrm{B}}T}{\hbar\tilde{\omega}} = \frac{T}{T_{\tilde{\omega}}}$, where $T_{\tilde{\omega}} = \hbar\tilde{\omega}/k_{\mathrm{B}}$, and $T_{\tilde{\omega}}$ is the order of the boiling temperature, which is defined by the temperature above which all nucleotides split up [47]. If the conditions of the occurrence of the glass transition, $T_0 < T_{M1}$ in Eq. (11.2.2) and

$$\frac{S_{\mathrm{K}}(T_g)}{Nk_{\mathrm{B}}} < -\frac{T_h}{T_g}$$

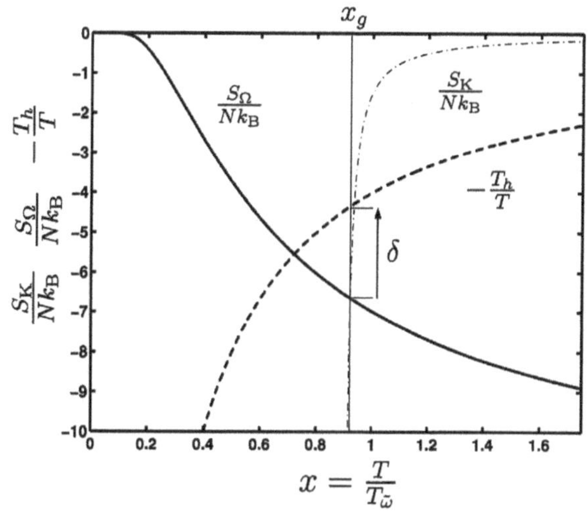

Figure 11.7 We illustrate $\frac{S_K}{Nk_B} = -\frac{N_0}{2N}\frac{T_0}{T-T_0}$, $\frac{S_\Omega}{Nk_B} = -\frac{12\beta\hbar\tilde{\omega}}{e^{\beta\hbar\tilde{\omega}}-1} = -\frac{12/x}{e^{1/x}-1}$ and $-\frac{h}{Nk_BT} \equiv -\frac{T_h}{T}$ with respect to $x \equiv \frac{k_BT}{\hbar\tilde{\omega}} = \frac{T}{T_{\tilde{\omega}}}$. We show an example satisfying the conditions of the occurrence of the glass transition, $T_0 < T_{M1}$ in Eq. (11.2.2) and $\frac{S_K(T_g)}{Nk_B} < -\frac{T_h}{T_g}$ in Eq. (11.2.3) for $N_0/N = 1/3$, $T_0 = 450$K, $T_{\tilde{\omega}} = 500$K, $T_h = 2000$K. The glass transition temperature is $T_g = 461$K. A measure on the occurrence of the glass transition $\delta \equiv -\frac{T_h}{T_g} - \frac{S_K(T_g)}{Nk_B}$ is shown. $x_g \equiv \frac{T_g}{T_{\tilde{\omega}}}$ is also shown.

in Eq. (11.2.3) are satisfied, the glass transition temperature is determined by $S_K = S_\Omega|_{T=T_g}$. Since we do not know the actual values of N_0, T_0, ω, T_h, and T_{M1}, we show an example in Figure 11.7, satisfying the conditions (11.2.2) and (11.2.3) for $N_0/N = 1/3$, $T_0 = 450$K, $T_\omega = 500$K, $T_h = 2000$K. The glass transition temperature is $T_g = 461$K.

Here we introduce a measure on the occurrence of the glass transition by

$$\delta \equiv -\frac{T_h}{T_g} - \frac{S_K(T_g)}{Nk_B}.$$

The larger value δ takes, the more easily the glass transition occurs. We investigate T_h, $\widetilde{T_0}$, and $T_{\tilde{\omega}}$-dependence of δ under the conditions (11.2.2) and (11.2.3).

(i) T_h, T_0 =constant: As $T_{\tilde{\omega}}$ becomes higher, δ becomes smaller.
(ii) T_h, $T_{\tilde{\omega}}$ =constant: As T_0becomes higher, δbecomes larger.
(iii) As $T_h$becomes higher, δbecomes smaller.

We have shown an example in Figure 11.7 with $T_h = 2000$K, $T_0 = 450$K, and $T_{\tilde{\omega}} = 500$K. But even if $T_h = 4000$K, the glass transition occurs when $T_{\tilde{\omega}} < 157$K for $T_0 = 400$K and when $T_{\tilde{\omega}} < 284$K for $T_0 = 450$K because of the conditions (i) and (ii).

11.2.3 Concluding Remarks

Before the appearance of RNAs in evolutionary history, there could be many intermediate processes such as clay model and pyranosyl RNA model. We have proposed a model on the origin of RNA at higher temperatures than in a biological system. We have assumed that there had been adequate quantity of nucleotides at a certain pre-biotic time. That is because unless there had been adequate quantity of nucleotides, RNAs could not appear. At such higher temperatures even pyrimidine nucleotides could be produced, while in a biological system, it is difficult to attach pyrimidines to ribose. Synthesis of RNA, i.e. binding nucleotides, needs a large Gibbs free energy because the nucleotides have to overcome a high potential barrier. Standard chemical kinetics would conclude that it is impossible in a reasonable time. We have considered that the Gibbs free energy due to the Kauzmann entropy of nucleotides in the glass forming process is released to overcome it. T_h, T_0, and the $T_{\tilde{\omega}}$ dependence of $\delta \equiv -\frac{T_h}{T_g} - \frac{S_K(T_g)}{Nk_B}$ under the conditions, Eqs. (11.2.2) and (11.2.3) show that the lower T_h, the higher T_0, and the lower $T_{\tilde{\omega}}$ becomes, the larger δ becomes, i.e. the more easily the glass transition occurs. We have shown an example in Figure 11.7. We could expect the glass transition for reasonable values T_0 and $T_{\tilde{\omega}}$, if $T_h \cong 2000$K. However, the value of T_h is possibly rather high. In a biological system, 5′-triphosphate nucleotides are released to overcome the potential barrier. The released free energy for an ATP corresponds to $k_B T_h \cong k_B 4000$K. In the case of $T_h = 4000$K, the glass transition occurs if $T_{\tilde{\omega}} < 157$K for $T_0 = 400$K and if $T_{\tilde{\omega}} < 284$K for $T_0 = 450$K, but the values of $T_{\tilde{\omega}}$ are rather low. In a biological system, catalysis plays an important role in lowering the height of the potential barrier. Even in a pre-biotic system such as around volcanoes, there could be some catalytic materials such as minerals, which could lower the height of the potential barrier. Thus considering such catalytic materials, even if T_h is high, we could expect the glass transition in a pre-biotic system.

Under the conditions of Eqs. (11.2.2) and (11.2.3), the glass transition occurs. In the glassy state, the ribose subunits are joined by a $2'-5'$ or $3'-5'$ phosphodiester linkage to form the ribose-phosphate backbone similar to oligosaccharides and polysaccharides. The condition defined by Eq. (11.2.2) requires that the sound disappearing temperature is lower than the temperature above which RNA hydrolyzes. This is guaranteed by taking account of the intermediate state in the interaction between O1 and O2: O1–O2 $\rightleftarrows$ {O+H_2O}. The intermediate state plays an essential role in forming the glassy state. Thus the potential function $V_{\alpha\beta}(\boldsymbol{q})$ should be taken account of the intermediated state. The condition, Eq. (11.2.3) requires that the Gibbs free energy due to the Kauzmann entropy is large enough to overcome the potential barrier.

Although it is hard to give specific conditions for an experiment to test the ideas of this work, the polymerization of RNA by the liquid-glass transition could perhaps be attempted by starting from condensed nucleotides at high temperature, below T_{M1}, and reducing temperature rather quickly, possibly in the presence of minerals as catalysts.

In this study we showed a possible pathway to the synthesis of RNA at the early stage of life. Of course it is not sufficient to validate the hypothesis of the RNA world, because the difficulty of the polymerization is not the only possible limitation of this hypothesis. For instance our study cannot address the question of the availability of the necessary nucleotides, but, at least it shows that the barrier to polymerization may be surmounted in favorable conditions.

11.3 A Model on the Financial Panic

We extend the two band model on the unified theory of the liquid-glass transition to that of the financial panic in a capitalist society [53]. In order to extend the two band model, we must identify the system of the capitalist society with the two band system. We consider a simplified capitalist society consisting of workers and enterprises. A worker produces the two values, his wage for his life and savings (securities), and a surplus value for his enterprise. The averaged wage and surplus value make two levels, the lower and the upper levels. The lower and upper bands are formed by the hopping (transfer) of securities of workers and enterprises in financial markets, respectively. Sound-like mode is constructed as the sum of an infinite series of interacting bubbles of buy–sell pairs of securities in financial markets through the interaction potentials of buy–sell pairs; the securities in the financial markets play the same role as particles in liquids. Phonon-like modes are constructed as the sum of an infinite series of interacting bubbles of consumption-production and production-consumption pairs in enterprises through the interaction potentials of the consumption-production and production-consumption pairs, the processes of which play the same role as up and down transitions of the particles between the two bands in glasses.

Thus a financial panic occurs on an occasion of the supercooling process to avoid completing the financial (buy-sell) cycles under the circumstances: an extremely high consumption causes enormous bubbles of buy-sell pairs of securities in financial markets, which form the configuration entropy (the Kauzmann entropy) and trigger a chain recession. As the society cools down, the dynamical processes of financial panic are governed by the universal features such as the Kauzmann entropy crisis and the drastic drop of the dynamical motions (the VTF law), which also cause further recessions so that a chain recession occurs. And the financial markets eventually freeze at the freezing point at which the Kauzmann entropy and the fluctuation entropy of production-consumption modes cross over, where the financial markets hit the bottom.

11.3.1 *The Criteria of the Financial Panic*

The Wall Street Crash of 1929 is well known as the great financial panic. Historically, the first panic is known as tulip mania in the United Provinces, the peak of which was in 1637 [54]. The first worldwide panic occurred in 1857. Since the first panic, panics or great depressions occurred almost every decade except the periods shifting to monopolistic capitalism in the late 19th century and the periods of the World Wars I and II. In the regulated capitalism, since the governments had intervened the financial markets, such drastic financial crisis as in 1929 had not occurred, but instead so-called "great depressions" had appeared. However, in the deregulated capitalism, panic or great depression occurs as a natural law. The financial crisis of 2007–2009 originating from the subprime mortgages was a typical example [55, 56]. Although the governments bailed out banks and enterprises, a great panic like the Wall Street Crash in 1929 occurred. Deregulated financial markets conceive potentially financial panic.

The conventional economic theories have tried to explain the cause of the financial panic, but no theory has succeeded to explain the dynamical processes of financial panic [57, 58]. This is because they have limits in treating the dynamical processes of financial panic that occurs under the supercooling process to avoid completing the financial (buy-sell) cycles. First, a financial panic occurs as collective motions so that we must treat the dynamical processes through the collective modes such as the securities density waves and the production waves. Second, the fluctuation entropy due to the bubbles of buy-sell pairs of securities not completing the financial cycles mixes with the entropy of the securities density waves. The mixing entropy forms the configuration entropy (the Kauzmann entropy), which leads to the entropy crisis. This means that decision-makers cannot act optimally in the financial markets under the supercooling process. Third, the dynamical processes are very sensitive to the activity of the society (the temperature of the society). As the society cools down, the dynamical motions drastically drop (the VTF law).

Agent-based models may seem to explain the dynamical processes of financial panic [57, 58]. An agent-based model is a computerized simulation of a number of decision-makers (agents) and enterprises, which interact through prescribed rules [58]. If ones want to elucidate the universal features in the dynamical processes of the financial panic such as the Kauzmann entropy crisis and the VTF law, they must take the above mentioned three points into account. Agent-based models lack these three points.

To elucidate the universal features of the dynamical processes of financial panic under the supercooling process, taking these three points into account, we propose a model on the financial panic from the viewpoint of statistical mechanics. In our model we consider the following simplified capitalist society: The society consists of workers and enterprises; each worker belongs to his own enterprise. All social products are produced by workers. In an equilibrium society, the distribution of social products is governed by the statistical mechanics as a natural law [59–61]. The gross social products can be regarded as the social energy E. In the equilibrium state of the society, the entropy of the society $S(E)$ is defined by $S(E) = k_B \log W(E)$, where $W(E)$ is the number of states which is defined by the maximum value among the number of ways of the distribution of the gross social products and k_B is the Boltzmann constant. In a normal society, the activity of the society can be regarded as the temperature of the society T, the inverse of which is defined by the partial derivative of $S(E)$ with respect to E. The temperature is expected to be an increasing function of E [60]. The number of way of the distribution fluctuates from that of the equilibrium distribution. Such fluctuations lower the entropy as the fluctuation entropy. In the normal society the dissipation of the losses due to the fluctuations smoothly compensate for the fluctuation entropy so that the society reaches an equilibrium. Thus the normal society is regulated to keep the system in an equilibrium state.

In normal markets, individuals make decision through buy-sell processes [62]. A financial cycle starts with selling one product in the social products, it goes through a series of buy-sell processes of securities in financial markets, and it finally ends with buying another product in the social products. We call a buy-sell process of securities in financial markets, a bubble of buy-sell pairs of securities. In the financial markets, the bubbles of buy-sell pairs are always created and they are interacting with each other through dealers; securities hop around in financial markets like particles in a liquid. The number of a series of interacting bubbles of buy-sell pairs can be any integer. In the normal markets, the sum of an infinite series of interacting bubbles of buy-sell pairs yields a mode, securities density waves, like a sound mode (particle density waves) in a liquid. The buy-sell processes propagate with the velocity of the securities density waves like the sound velocity. The securities density waves constitute the maximum entropy of the society. The bubbles of the buy-sell pairs which do not constitute the securities density waves are just those that do not complete the financial (buy-sell) cycles so that they make fluctuations. Such fluctuations lower the entropy from the maximum entropy as the fluctuation entropy. In the normal markets since the dissipation due to diffusion of the securities losses compensate for such negative fluctuation entropy, the society reaches an equilibrium. The normal markets play a role in regulating the gross social products and their distribution to keep a balance [63–66].

However, in financial markets dealers create enormous bubbles of buy-sell pairs of securities not completing the financial (buy-sell) cycles for speculation. The fluctuation entropy due to such bubbles of buy-sell pairs mixes the entropy of securities density waves. The mixing entropy essentially has a form of the Curie law with a negative sign, $S_K \propto -\frac{1}{T-T_0}$, where the critical temperature T_0 indicates the fragility of the society. We call S_K the Kauzmann entropy. If an extremely high consumption causes enormous bubbles of buy-sell pairs not completing the financial cycles, which form the Kauzmann entropy and trigger a chain recession, then the supercooling process to avoid completing the financial cycles appears. The activity of the society cools down; the temperature of the society T decreases. With the temperature of the society decreasing toward T_0, S_K tends to a negative infinity so that the total entropy of the society seems to become negative. This is called the Kauzmann entropy crisis. But, the financial markets eventually freeze at the freezing temperature of the society T_f above T_0, where S_K crosses over the fluctuation entropy of the production waves (the recovering modes) so that the Kauzmann entropy crisis is prevented, where the financial markets hit bottom.

The transfer (the hopping) of securities from one dealer to another is governed by the hopping amplitude, J, which is determined by the Einstein relation on the Kauzmann entropy, $S_K \propto -\frac{1}{T-T_0}$, and probability, $J = \exp\left[-\frac{A}{T-T_0}\right]$, which is called the Vogel-Tamman-Fulcher (VTF) law, where A is a numerical value. With the temperature decreasing, the dynamical motions of the financial markets drastically drop according to $J = \exp\left[-\frac{A}{T-T_0}\right]$; the velocity of the securities density waves, the inverse relaxation time of the financial markets, the diffusivity of the loss of securities, the inverse viscosity of the propagation of the production processes, drastically drop toward zero according to $J = \exp\left[-\frac{A}{T-T_0}\right]$. Since these processes also cause further recessions, a chain recession occurs: a recession (a T-decrease) $\rightarrow$ the VTF law $\rightarrow$ further recessions (further T-decreases) $\rightarrow$ the VTF law $\rightarrow$ .... Eventually the financial markets freeze at the freezing temperature $T_f(> T_0)$.

Thus we obtain the criteria of the financial panic as follows [66–72]:

1. A financial panic occurs on an occasion of the supercooling process to avoid completing the financial (buy-sell) cycles under the circumstances: an extremely high consumption causes enormous bubbles of buy-sell pairs of securities in financial markets, which form the Kauzmann entropy and trigger a chain recession (the temperature of the society T decreases).
2. The Kauzmann entropy S_K drops according to the Curie law with a negative sign, $S_K \propto -\frac{1}{T-T_0}$. The dynamical motions of financial markets drastically drop according to the VTF law, $J = \exp\left[-\frac{A}{T-T_0}\right]$, which also cause further recessions so that a chain recession occurs. The financial markets eventually freeze at the freezing point $T_f(> T_0)$, where the Kauzmann entropy and the fluctuation entropy of the production waves cross over.

In the financial crisis of 2007–2009, 1: The supercooling process to avoid completing the financial cycles was created under the circumstances: the extremely high consumption through subprime mortgages caused enormous bubbles of the buy-sell pairs of securities, which formed the Kauzmann entropy and then frozen subprime mortgages triggered a chain recession. 2: The Kauzmann entropy dropped, while the dynamical motions of the financial markets drastically dropped (the VTF law), which also caused recessions so that a chain recession occurred. Although the government bailed out to banks and enterprises relaxed the dynamical processes of the financial panic, the dynamical motions of the financial markets drastically dropped so that a lot of banks and enterprises went bankrupt.

Panics occur in financial markets, psychology, etc., and even in nature. Each panic occurs on an occasion of the supercooling process to avoid completing its own cycle, while the dynamical processes of those panics are governed by the universal features such as the Kauzmann entropy crisis and the VTF law. In order to elucidate the universal features in the dynamical processes of the panics, we propose the two band model from statistical mechanics. We can regard the liquid-glass transition as a kind of panic in the liquid state; the liquid-glass transition occurs under the supercooling process to avoid completing particle-hole cycles (to avoid the crystallization). We have established the unified theory of the liquid-glass transition on the basis of the two band model to elucidate the universal features in the liquid-glass transition [47, 48]. Therefore, we extend the unified theory of the liquid-glass transition to the financial panic. To do so, we first construct the two band model for the financial markets.

11.3.2 The Two Band Model

In order to construct the two band model in a capitalist society, we consider a simplified capitalist society consisting of N workers and M enterprises. Each worker belongs to his own enterprise. Hereafter, M enterprises are not distinguished. All products are produced by workers. An m-worker's work force p^m in his enterprise produces the two values, his wage p_w^m for his life and a surplus value p_s^m for his enterprise; $p^m = p_w^m + p_s^m$. The wage p_w^m consists of his living cost p_{wl}^m and his savings p_{ws}^m; $p_w^m = p_{wl}^m + p_{ws}^m$. Here we assume that the savings p_{ws}^m are thrown into financial markets as securities, because his living cost p^m_{wl} is consumed for his life, but his savings p^m_{ws} can be freely spent. Thus we get the important results: the first relation $p^m = p^m_w + p^m_s$ shows that the simplified capitalist society is a two level system. Also through the transfer of the securities p^m_w between dealers the securities hop around in the financial markets, which yields the lower level a lower band. Similarly, we obtain an upper band. Thus it is shown that a simplified capitalist society has the two band structure.

However, in order to develop the dynamical processes in such a simplified capitalist society, we must clarify the Hamiltonian of the society; the states of the individual workers and the interaction potentials between workers. The interactions between workers are divided into two cases: the interactions between workers in the same enterprise and those in different enterprises. For the former case, the interactions work between workers who randomly locate in a two-dimensional space of an enterprise (an enterprise space). For the latter case, interactions work between workers in different enterprise spaces, which can be thought to randomly locate along another axis perpendicular to an enterprise space. If we assume the interaction strength between workers in different enterprises is the same as that in the same enterprise, considering the two cases we can regard that the interactions work between workers who randomly locate in an isotropic three-dimensional social space. Therefore, we assume that an m-worker locates at a random social position $\boldsymbol{R}_m$ in the three-dimensional social space. We also assume that the workers are interacting with each other with the interaction potential $V_{00}(\boldsymbol{R}_m - \boldsymbol{R}_n)$, which has a hard core and a valley with a minimum near a mean worker distance a as illustrated in Figure 11.8a, where $V_{00}(\boldsymbol{R})$ is in a spherical symmetry; $V_{00}(\boldsymbol{R}) = V_{00}(R)$ and $|\boldsymbol{R}| = R$. This assumption is reasonable, because workers become repulsive as they get closer to each other, but they feel comfortable at an appropriate distance a. The actual interaction potentials for a capitalist society are very complicated, but we take such a simple interaction potential as that in a simple liquid, because the human economic activities depend on so many factors that they are governed by simple laws as a statistical ensemble.

The random distribution of workers in the social space yields a pair distribution function of workers $g(R)$, which is defined by the configurationally averaged number density of workers at the distance R from a worker at an original point. $g(R)$ has a peak near the mean worker distance a as shown in Figure 11.8b. An m-worker feels a potential well which is made up by the surrounding workers. We approximate the well as a harmonic potential. An m-harmonic potential has a harmonic frequency ω_m. We consider the two levels for the harmonic potential. The two levels for the ith direction ($i = x, y$ and z) are $\omega_{m0} = \omega_m/2$ and $\omega_{mi} = 3\omega_m/2$, as illustrated in Figure 11.8b. The two levels for the three-dimensional harmonic potential are given as

$$\tilde{\omega}_{m0} = 3\omega_m/2, \qquad \tilde{\omega}_{mi} = 5\omega_m/2,$$

where $\tilde{\omega}_{m0}$ means that all components are ω_{m0}, while $\tilde{\omega}_{mi}$ means that i-component is ω_{mi} and the others are ω_{m0}. The higher levels $\tilde{\omega}_i$ are degenerated, so we can put $\tilde{\omega}_{mi} = \tilde{\omega}_{m1}$. Thus we can identify the two levels:

$$kp^m_w = \hbar\tilde{\omega}_{m0}, \qquad kp^m_s = \hbar\tilde{\omega}_{m1},$$

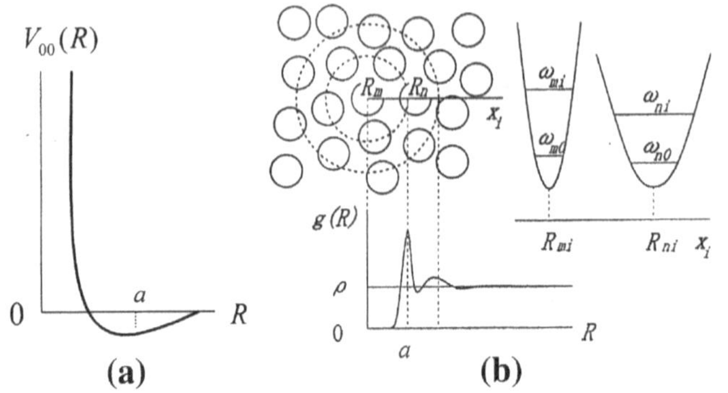

Figure 11.8 The interaction potentials between workers and the random distribution of workers in the respective harmonic potentials are schematically illustrated. (a) The interaction potential $V_{00}(R)$ with a hard core and a valley with a minimum near a mean worker distance a are shown. (b) The random distribution of workers and a pair distribution of workers $g(R)$ with a peak near a worker distance a are shown. The random harmonic potentials with two levels in the ith direction are also shown [47].

where k is a rate of energy to money and $\hbar$ is a rate of energy to frequency, the Plank constant. This means that in our simplified capitalist society we assume that workers produce products in three ways corresponding to $\tilde{\omega}_{mi}$. The harmonic frequency ω_m represents the elasticity for the harmonic potential and also represents the worker's ability.

Here we investigate how to treat the dynamical processes in such a random system. It is a hard task to solve directly the dynamical processes. So we divide the calculation processes of the random system into two steps. At the first step, we derive the dynamical motions in the configurationally averaged system. At the next step we take the randomness of the system into account as the scattering processes by the random potentials.

Therefore, at the first step we take the configuration average of random two level system:

$$\hbar\tilde{\omega}_0 = 3\hbar\omega/2 = kp_w, \qquad \hbar\tilde{\omega}_1 = \hbar\tilde{\omega}_i = 5\hbar\omega/2 = k(p_w + p_s),$$

where $\tilde{\omega}_\mu = \langle\tilde{\omega}^m_\mu\rangle_c$ $(\mu = 0, i)$, $\omega = \langle\omega_m\rangle_c$, $p_{w.s} = \langle p^m_{w,s}\rangle_c$. $\langle\cdots\rangle_c$ means the configuration average. Here we pay our attention to the savings $\{p^m_{ws}\}$ in $\{p^m_w\}$. We assume that the savings are thrown into financial markets as securities. Each security is assigned to its own worker, but they are bundled in financial markets by dealers. The transfer of securities through dealers determines a lower band. We introduce the transfer of securities from an m-dealer to an n-dealer, $J_0(\boldsymbol{R}_m - \boldsymbol{R}_n)$, as the hopping magnitude of the securities. If we denote the Fourier transformation of the configuration average of $J_0(\boldsymbol{R}_m - \boldsymbol{R}_n)$ as $J_0(\boldsymbol{p})$ we obtain the energy dispersion of the lower band,

$$\epsilon_0(\boldsymbol{p}) = \tilde{\omega}_0 + J_0(\boldsymbol{p}),$$

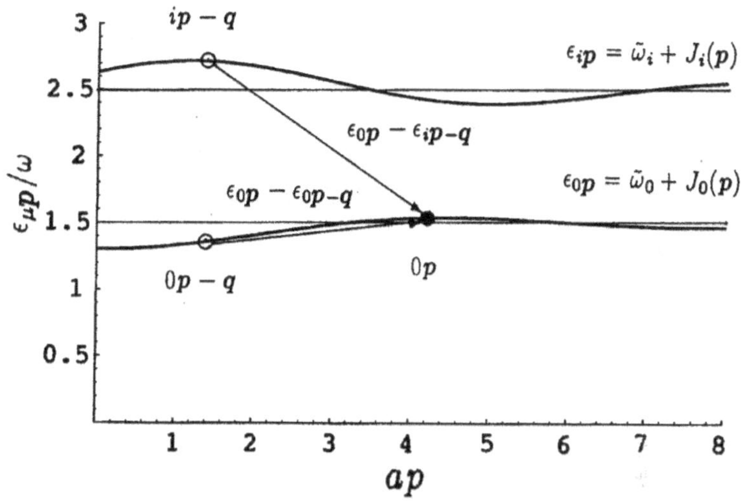

Figure 11.9 The two bands are schematically illustrated. Energy dispersion curves normalized by the configuration averaged harmonic frequency $\omega =< \omega_m >_c$ as a function of ap are shown, where a is a mean worker distance, $\boldsymbol{p}$ is a wave vector, $p = |\boldsymbol{p}|$, $\tilde{\omega}_0 = 3\omega/2$, $\tilde{\omega}_i = \omega_1 = 5\omega/2$. $J_0(\boldsymbol{p})$ is isotropic, while $J_i(\boldsymbol{p})$ depends on the direction, so that we show the averaged value over the directions. An intra-band elementary excitation $\epsilon_{0\boldsymbol{p}} - \epsilon_{0\boldsymbol{p}-\boldsymbol{q}}$ and an inter-band elementary excitation $\epsilon_{0\boldsymbol{p}} - \epsilon_{i\,\boldsymbol{p}-\boldsymbol{q}}$ are shown.

as illustrated in Figure 11.9. $J_0(\boldsymbol{p})$ has a maximum value near the reciprocal worker distance $K \sim 2\pi/a$. Similarly, we can construct the energy dispersion of the upper band

$$\epsilon_i(\boldsymbol{p}) = \tilde{\omega}_i + J_i(\boldsymbol{p})$$

in Figure 11.9. Thus we reach the two band model. Using this two band model, we can solve the dynamical equations of motions through interactions potentials.

At the next step, the randomness is taken into account by the scattering processes by the scattering potentials due to the random hopping magnitudes $J_\mu(\boldsymbol{R}_m - \boldsymbol{R}_n)$ and those due to the random frequencies ω_m, where $\mu = 0, i$. Calculating the random scattering processes due to $J_\mu(\boldsymbol{R}_m - \boldsymbol{R}_n)$ and ω_m, we can derive the relaxation processes for the securities density waves and the production waves, and the transport processes for the securities and the products.

Before we develop the dynamical processes in the random system from the statistical mechanics, we define the thermodynamical quantities. The gross social wage is $P_w = \sum_m p_w^m$. On the other hand, the gross social surplus value is $P_s = \sum_m p_s^m$ which consists of the constant capital (raw materials, machines for product, etc.) and the accumulation of the capital in the whole enterprises in the society. The gross social product is $P = \sum_m p^m = P_w + P_s$. We define the total energy of the society as

$$E = k(P_w + P_s) = \sum_m 5\hbar\omega_m/2 = 5N\hbar\omega/2.$$

The total entropy of the society, $S(E)$ is defined as

$$S(E) \equiv k_B \log W(E),$$

where $W(E)$= [*the maximum value among the number of ways of distribution of the total energy of the society*, E]. k_B is the Boltzmann constant. The temperature of the society T is regarded as the activity of the society and is defined by

$$\frac{1}{T} = \frac{\partial S(E)}{\partial E}.$$

The temperature T is expected to be an increasing function of the energy E. In an ideal society, we can expect $E = 3Nk_B T$.

In such a scheme, the magnitude of two bands is governed by the hopping amplitude $J : J_0(\boldsymbol{p}),\ J_i(\boldsymbol{p}) \propto J$. J is later shown to obey the VTF law, $J = \exp\left[-\frac{A}{T-T_0}\right]$. Near the financial panic, the transfer of the securities is slowed down so that these bands are very narrow according to the VTF law; $\beta\hbar J_\mu \ll 1$, where $\beta = 1/k_B T$ and $\mu = 0, i$.

11.3.3 Dynamical Processes of the Financial Panic

Here we survey the dynamical processes in the random two level system, extending the unified theory of the liquid-glass transition to the financial panic. First we investigate the dynamical motions of the simplified capitalist society. Next we take the randomness of the system into account as the scattering processes, which yield dissipations and relaxations. Finally, we investigate the dynamical processes of financial panic from statistical mechanics.

(1) The elementary excitations in the two band model can be decomposed into intra-band and inter-band transitions, which are shown in Figure 11.9. Intra-band transitions occur when securities are hopping between dealers so that they are associated with securities density waves corresponding to sound waves in a liquid [73–75]. In financial markets, the hopping of securities yields at a lower level $\tilde{\omega}_0$ a band $\epsilon_{0\boldsymbol{p}} = \tilde{\omega}_0 + J_0(\boldsymbol{p})$; securities hop around in financial markets like particles in a liquid. An elementary excitation, $\epsilon_{0\boldsymbol{p}} - \epsilon_{0\boldsymbol{p}-\boldsymbol{q}}$, corresponds to an elementary buy-sell process of a security that a dealer buys a security with a value $\epsilon_{0\boldsymbol{p}} - \epsilon_{0\boldsymbol{p}-\boldsymbol{q}}$, while another dealer sells it. This process is represented by a propagator of the elementary excitation $\epsilon_{0\boldsymbol{p}} - \epsilon_{0\boldsymbol{p}-\boldsymbol{q}}$, which we call an elementary bubble of buy-sell pair of securities (securities-antisecurities pair). Thus, the bubble of buy-sell pairs of securities with a wave vector difference $\boldsymbol{q}$ is defined as the average of the sum of the buy-sell pairs of securities with respect to $\boldsymbol{p}$ as follows:

$$P(q_0, \boldsymbol{q}) = \frac{1}{N\hbar}\sum_{\boldsymbol{p}} \frac{f(\epsilon_{0\boldsymbol{p}-\boldsymbol{q}}) - f(\epsilon_{0\boldsymbol{p}})}{q_0 - (\epsilon_{0\boldsymbol{p}} - \epsilon_{0\boldsymbol{p}-\boldsymbol{q}})} \cong -\frac{\beta f' {\omega_q}^2}{{q_0}^2 - {\omega_q}^2},$$

as illustrated in Figure 11.10a, where q_0 is a frequency, f is a distribution function, f' is its derivative, and we have considered the inversion symmetry with respect to $\boldsymbol{q}$ under the summation of $\boldsymbol{p}$; $\omega_{\boldsymbol{q}} = \omega_q$ and $P(q_0, \boldsymbol{q}) = P(q_0, q)$. The second equation is obtained under the condition $\beta\hbar J_0(\boldsymbol{p}) \ll 1$. The frequency of the bubble of the buy-sell

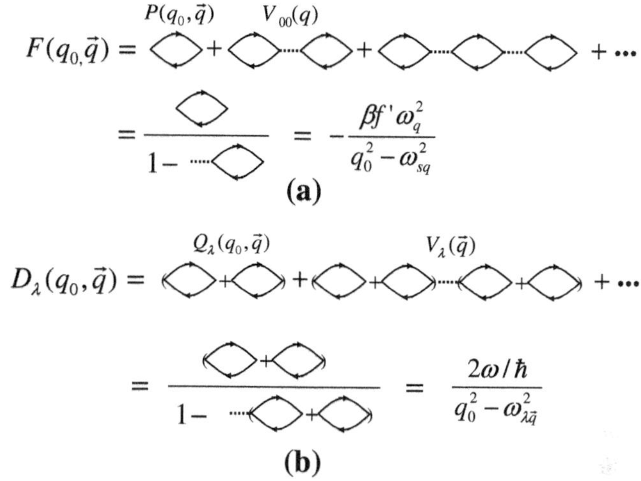

Figure 11.10 How to construct securities density waves and production waves with the respective bubbles is illustrated. (a) $F(q_0, q)$: the propagator of securities density waves. $P(q_0, q)$ represents the bubble of buy-sell pairs of securities. A dotted line indicates the interaction potential $V_{00}(\boldsymbol{q})$. (b) $D_\lambda(q_0, \boldsymbol{q})$: the propagator of production waves with a mode λ. $Q(q_0, \boldsymbol{q})$ represents the pair of bubbles of prod-cons or cons-prod pairs of goods. A dotted line indicates the interaction potential $V_\lambda(\boldsymbol{q})$.

pairs of securities is given by ω_q, because the dominant contribution of the frequency q_0 to the bubble comes from the denominator $q_0{}^2 - \omega_q{}^2 = 0$. Here $\omega_q \cong v_p q/\sqrt{3}$, where v_p is an average velocity of the securities in financial markets.

In financial markets, successive buy-sell processes between dealers are held through the interaction potentials $V_{00}(\boldsymbol{R}_m - \boldsymbol{R}_n)$. n successive buy-sell processes are represented by n bubbles of buy-sell pairs of securities connecting through $V_{00}(\boldsymbol{q})$ as illustrated in Figure 11.10(a), where $V_{00}(\boldsymbol{q})$ is the Fourier transformation of $V_{00}(\boldsymbol{R}_m - \boldsymbol{R}_n)$. The propagator of the securities density waves $F(q_0, \boldsymbol{q})$ is formed as the sum of an infinite series of the interacting bubbles of buy-sell pairs of securities, i.e. a collective mode-like particle density waves (sound waves), as shown in Figure 11.10a:

$$F(q_0, q) = P(q_0, q) + P(q_0, q)V_{00}(q)F(q_0, q)$$
$$= \frac{1}{1/P(q_0, q) - V_{00}(q)} = -\frac{\beta f' \omega_q{}^2}{q_0{}^2 - \omega_{sq}{}^2}.$$

The frequencies of securities density waves are given by

$$\omega_{sq}{}^2 = (1 - \beta f' V_{00}(q))\omega_q{}^2,$$

where $\omega_{sq} = c_s q$ and

$$c_s = \sqrt{1 - \beta f' V_{00}(q)}\, v_p/\sqrt{3}$$

is the velocity of securities density waves. v_p and c_s are proportional to the hopping amplitude J; $v_p,\ c_s \propto J$. The securities density waves are fluctuation waves of securities density in financial markets; the buy-sell processes propagate with the velocity of the securities density waves, c_s, in the financial markets like the velocity of sound in a liquid.

Here we note that the disappearing of securities density waves occurs at a reciprocal worker distance $K \sim 2\pi/a$ and at the disappearing temperature of securities density waves T_0, where the securities density waves disappear;

$$1 - \beta_0 f' V_{00}(K) = 0, \qquad \beta_0 = \frac{1}{k_B T_0}. \tag{11.3.1}$$

$V_{00}(q)$ has the negative minimum value at $q = K$.

(2) On the other hand, inter-band transitions occur when workers produce for their enterprises. Therefore inter-band transitions are associated with the production waves like elastic waves in a liquid [76–79]. In an enterprise, an elementary excitation, $\epsilon_{i\boldsymbol{p}} - \epsilon_{0\boldsymbol{p}-\boldsymbol{q}} \cong \omega$ or $\epsilon_{0\boldsymbol{p}} - \epsilon_{i\boldsymbol{p}-\boldsymbol{q}} \cong -\omega$, corresponds to an elementary production-consumption (prod-cons) or consumption-production (cons-prod) process that a worker produces or consumes a goods with a value ω, while another worker consumes or produces it, respectively. This process is represented by a propagator of the elementary excitation ω or $-\omega$, which we call an elementary bubble of prod-cons or cons-prod pair of goods. The bubble of prod-cons or cons-prod pairs of goods with a wave vector $\boldsymbol{q}$ is defined by the average of the sum of the bubbles of prod-cons or cons-prod pairs of goods with respect to $\boldsymbol{p}$ as follows:

$$Q(q_0, \boldsymbol{q}) = \frac{1}{N\hbar} \sum_{\boldsymbol{p}} \left[\frac{1}{q_0 - (\epsilon_{0\boldsymbol{p}} - \epsilon_{i\boldsymbol{p}-\boldsymbol{q}})} - \frac{1}{q_0 + (\epsilon_{0\boldsymbol{p}} - \epsilon_{i\boldsymbol{p}-\boldsymbol{q}})} \right]$$
$$\times [f(\epsilon_{i\boldsymbol{p}-\boldsymbol{q}}) - f(\epsilon_{0\boldsymbol{p}})] \cong \frac{2\omega/\hbar}{q_0{}^2 - \omega^2}$$

as illustrated in Figure 11.10(b). In enterprises, successive prod-cons or cons-prod processes between workers are held through the interaction potentials $V_{ij}(\boldsymbol{R}_m - \boldsymbol{R}_n)$, where $V_{ij}(\boldsymbol{R})$ is essentially written as

$$V_{ij}(\boldsymbol{R}) = \zeta^2 \frac{\partial^2 V_{00}(\boldsymbol{R})}{\partial R_i \partial R_j}, \qquad \zeta^2 = \frac{\hbar}{2M\omega},$$

where ζ is the averaged range of a worker's activity in the ith direction in the social space and ω is the ability of the workers. Then the worker's mass in the society is given by $M \sim \frac{\hbar}{2\zeta^2\omega}$ corresponding to a mass of particle in a liquid. If we put the Fourier transformation of $V_{ij}(\boldsymbol{R})$ as $V_{ij}(\boldsymbol{q})$ and $\boldsymbol{q} = q\mathbf{e_z}$ ($\boldsymbol{e}_z$ is the unit vector of the z-axis), we can diagonalize $V_{ij}(\boldsymbol{q})$ as $V_\lambda(\boldsymbol{q}) = V_{ii}(\boldsymbol{q})$, where λ is longitudinal for $i = z$ and transverse for $i = x, y$. Production waves are formed as the sum of an infinite series of interacting bubbles of prod-cons or cons-prod pairs of goods, i.e. three collective in Figure 11.10(b):

$$\begin{aligned} D_\lambda(q_0, \boldsymbol{q}) &= Q(q_0, \boldsymbol{q}) + Q(q_0, \boldsymbol{q}) V_\lambda(\boldsymbol{q}) D(q_0, \boldsymbol{q}) \\ &= \frac{1}{1/Q(q_0, \boldsymbol{q}) - V_\lambda(\boldsymbol{q})} = \frac{2\omega/\hbar}{q_0{}^2 - \omega_{\lambda \boldsymbol{q}}{}^2}. \end{aligned}$$

The frequencies of the production waves are given by

$$\omega_{\lambda \boldsymbol{q}}{}^2 = \frac{2\omega}{\hbar}[V_\lambda(\boldsymbol{q}) - V_\lambda(0)], \qquad \hbar\omega = -2V_\lambda(0),$$

where ω is the elasticity of the society. For $aq \ll 1$, $\omega_{\lambda \boldsymbol{q}} = c_\lambda q$, where c_λ is the velocity of the production waves with a mode λ. c_λ is constant, independent of the temperature of the society. The production waves are elastic waves in enterprises; the production processes propagate with the velocities of the production waves, c_λ, in the enterprises like the velocities of elastic waves in a liquid.

Production waves are the modes associated with the spontaneous breakdown of the spatially translational symmetry [80–83]; production waves play a role of the modes to recover from the symmetry broken society (the frozen society).

(3) Next we take the randomness of the two level systems into account, which yields the dissipation and relaxation processes in the society. Dissipative processes come from the simultaneous scattering of two securities by the same random potentials. Random potentials for the scattering processes in the lower band come from the random hopping magnitudes $J_0(\boldsymbol{R}_m - \boldsymbol{R}_n)$ and the random harmonic frequencies ω_m. The sum of the configurationally averaged simultaneously scattering processes over all positions yields the random scattering potentials between securities:

$$\begin{aligned} &U_J \sim \langle \hat{J}_0(\boldsymbol{R}_m - \boldsymbol{R}_n) \hat{J}_0(\boldsymbol{R}_m - \boldsymbol{R}_n) \rangle_c \quad \text{due to random hopping magnitudes,} \\ &U_\omega \sim \langle \hat{\omega}_m \hat{\omega}_n \rangle_c \quad \text{due to random harmonic frequencies,} \end{aligned}$$

where $\hat{}$ means the deviation from the average value. U_J is proportional to J^2, $U_J \propto J^2$, while U_ω is almost constant. Above the freezing point of the society T_f, U_J dominates the scattering processes. The statistical average on a random system constrains the dynamical processes, which leads to dissipation. In the dissipative processes, since two securities simultaneously participate in the same random scattering potentials $U_J + U_\omega$, above T_f the relaxation time τ_0 is given by

$$\frac{1}{\tau_0} \propto \sqrt{(U_J + U_\omega)} \cong \sqrt{U_J} \propto J. \tag{11.3.2}$$

Similarly, we obtain the relaxation time of the production waves τ_M, where τ_M is called the Maxwell relaxation time and

$$\frac{1}{\tau_M} \propto \frac{1}{\tau_0} \propto J.$$

When the relaxation time of securities density waves τ_0 and that of production waves τ_M are shorter than their respective periods; $q_0\tau_0$, $q_0\tau_M \ll 1$ and the wave number q is smaller than a reciprocal worker distance $K \sim 2\pi/a$; $aq \ll 1$, diffusivity D and viscosity η_λ appear, respectively, where q_0 is a frequency. The dynamical equations associated with the dissipation for securities density waves and production waves are given by

$$q_0{}^2 - \frac{\mathrm{i}}{\tau_\mu} q_0 - \omega_{\mu \boldsymbol{q}}{}^2 \cong -\frac{\mathrm{i}}{\tau_\mu} q_0 - \omega_{\mu \boldsymbol{q}}{}^2$$

$$= \frac{1}{\tau_\mu}(-\mathrm{i}q_0 - \tau_\mu c_\mu{}^2 q^2) \equiv \frac{1}{\tau_\mu}(-\mathrm{i}q_0 - D_\mu q^2) = 0,$$

where $\boldsymbol{q} = q\mathbf{e}_z$ and i is the imaginary number. Here $(q_0, \boldsymbol{q})$ corresponds to $(\mathrm{i}\frac{\partial}{\partial t}, -\mathrm{i}\nabla)$. For securities density waves, $\mu = 0$; $\omega_{0\boldsymbol{q}} \equiv \omega_{s\boldsymbol{q}}, c_0 \equiv c_s, D_0 \equiv D$. For production waves, $\mu = i \equiv \lambda$; $\tau_i \equiv \tau_M, D_\lambda \equiv \eta_\lambda$. Thus diffusivity D is given by

$$D = \tau_0 c_s{}^2 \propto J,$$

where the velocity of the securities density waves $c_s \propto J$. The diffusivity comes from the gradient of the securities density in the financial markets and represents the dissipation of the securities loss. Viscosity $\eta_\lambda$is given by

$$\eta_\lambda = \tau_M c_\lambda{}^2 \propto J^{-1},$$

where the velocity of the production waves c_λ=constant. The viscosity comes from the gradient of the worker's production magnitudes in the enterprises and represents the resistivity of the flow of the production processes of goods.

(4) Now we investigate the dynamical processes of the financial panic from the viewpoint of statistical mechanics. In the equilibrium state, the securities density waves constitute the maximum entropy for the society. The bubbles of buy-sell pairs of securities are always created. The securities density waves are constructed by the sum of an infinite series of interacting bubbles of buy-sell pairs of securities to completing the financial cycles. Completing the financial cycles corresponds to completing the buy-sell cycles, where assets and liabilities on balance sheets are extinguished. In normal markets, the bubbles of the buy-sell pairs not completing the financial (buy-sell) cycles lower the entropy as the fluctuation entropy. but the dissipative entropy due to the diffusion of securities loss compensates for the fluctuation entropy and the society reaches a local equilibrium so that the society holds the maximum entropy. Bubbles of buy-sell pairs not completing the financial cycles just correspond to the concept of bubbles in financial economics, which means a self-sustaining excursion of the price of an asset from its fundamental value.

However, in financial markets dealers create enormous bubbles of buy-sell pairs of securities not completing the financial cycles for speculation. The fluctuation entropy due to such bubbles of the buy-sell pairs mixes with the entropy of the securities density waves. To calculate the mixing entropy, we first calculate the mixing Free energy (denoted as F_K) between the securities density waves ω_{sq} and the bubbles of the buy-sell pairs ω_q. The most dominant contribution to the mixing Free energy F_K comes from the regions of the disappearing of the securities density waves; $T \sim T_0$ and $q \sim K$, where T_0 and K satisfy $1 - \beta_0 f' V_{00}(K)$ in Eq. (11.3.1). The mixing Free energy F_K for $\beta\hbar\omega_{sq},\ \beta\hbar\omega_q \ll 1$ is given by

$$F_\mathrm{K} = k_\mathrm{B} T \sum_{\boldsymbol{q}} [\ln(1 - e^{-\beta\hbar\omega_{sq}}) - \ln(1 - e^{-\beta\hbar\omega_q})]$$

$$\cong k_B T \sum_{q \sim K} \ln \frac{\omega_{sq}}{\omega_q} = k_B T \sum_{q \sim K} \ln[1 - \beta f' V_{00}(q)] \cong \frac{N_0 k_B T}{2} \ln\left[1 - \frac{T_0}{T}\right],$$

where N_0 is a number of states around the reciprocal worker distance K in the wave vector space. The disappearing temperature T_0 indicates "fragility" of the society. From the thermodynamical relation $S = -\frac{\partial F}{\partial T}$, we obtain the mixing entropy (denoted as S_K):

$$S_K \cong -\frac{N_0 k_B}{2} \frac{T_0}{T - T_0}. \tag{11.3.3}$$

This is called the Kauzmann entropy. If an extremely high consumption causes enormous bubbles of the buy-sell pairs not to complete the financial cycles, which form the Kauzmann entropy and trigger a chain recession, then the supercooling process occurs. As the temperature decreases toward T_0, the magnitude of $|S_K|$ overwhelms the entropy of the society so that the total entropy of the society becomes a negative value as illustrated in Figure 11.11. This situation is called the Kauzmann entropy crisis. However, the financial markets eventually freeze at the freezing temperature $T_f (> T_0)$ as discussed below.

The hopping amplitude relates to the configuration number for the Kauzmann entropy in the multi-dimensional configuration space, e^{S_K/k_B}, from the Einstein relation on entropy and probability [59–61]; the hopping amplitude per security is given by

$$J \equiv \exp\left[\frac{z S_K}{N k_B}\right] = \exp\left[-\frac{A}{T - T_0}\right], \qquad A = z N_0 T_0 / 2N, \tag{11.3.4}$$

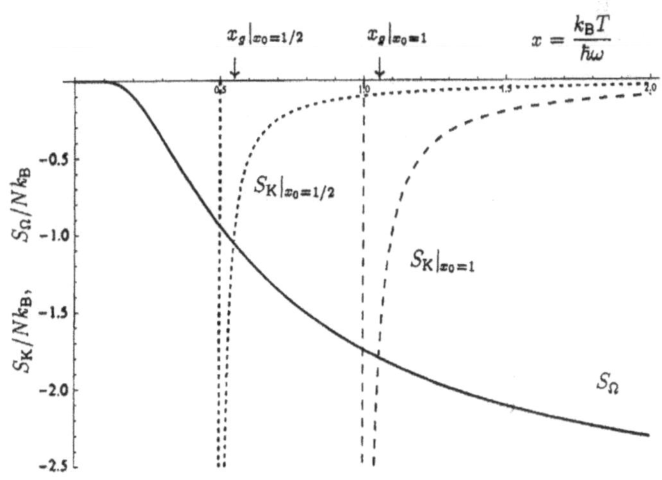

Figure 11.11 The Kauzmann entropy S_K and the fluctuation entropy of the production waves S_Ω normalized by Nk_B are illustrated in $x = k_B T/\hbar\omega$ for $N_0/N = 1/5$ for a fragile society $x_0 = k_B T_0/\hbar\omega = 1/2$ and a more fragile one $x_0 = 1$. The Kauzmann entropy S_K diverges at the instability temperature x_0, but crossovers S_Ω at the social freezing temperature $x_f = k_B T_f/\hbar\omega$ above x_0.

where z is the number of the surrounding workers. It should be noted that each security is assigned to its own worker, although they are bundled in financial markets. The hopping amplitude J manifests the VTF law. The velocities of securities and securities density waves v_p, c_s, the inverse relaxation times τ_0^{-1}, τ_M^{-1}, the diffusivity D, and the inverse viscosity η_λ^{-1}, are proportional to $J = \exp\left[-\frac{A}{T-T_0}\right]$;

$$v_p, c_s, \tau_0^{-1}, \tau_M{}^{-1}, D, \eta_\lambda^{-1} \propto J = \exp\left[-\frac{A}{T-T_0}\right],$$

which also cause further recessions so that a chain recession occurs. Note that the velocities of production waves, c_λ = constant.

(5) The freezing temperature T_f is related to the magnitude of the randomness of harmonic frequencies. In the equilibrium state, production waves constitute the maximum entropy for the society. The bubbles of prod-cons or cons-prod pairs of goods are always created. Production waves are formed by the sum of an infinite series of interacting bubbles of prod-cons or cons-prod pairs. The bubbles which do not constitute production waves lower the entropy as the fluctuation entropy, but the dissipative entropy due to viscosity compensates for the fluctuation entropy with a negative value and the system reaches a local equilibrium.

Since production waves do not depend on temperature, the mixing of entropies does not occur. The Free energy for the fluctuations of the production waves (denoted as F_Ω) is given as: $F_\Omega = 3Nk_BT\ln(1-e^{-\beta\hbar\omega})$. Thus the fluctuation entropy of production waves is given by

$$S_\Omega = -3Nk_B\frac{\beta\hbar\omega}{e^{\beta\hbar\omega}-1}, \tag{11.3.5}$$

where ω is the configuration averaged harmonic frequency, $\omega = \langle\omega_m\rangle_c$, ω represents the elasticity of the society. The amplitude of randomness of harmonic frequencies is related to the configuration number for the fluctuation entropy in the multi-dimensional configuration space, e^{S_Ω/k_B}; the amplitude of the randomness of harmonic frequencies per worker is given by $\Omega = \exp[zS_\Omega/(Nk_B)]$. As the temperature T decreases toward T_0, the fluctuation entropy of the production waves S_Ω increases so that S_Ω crosses over S_K at the freezing temperature of the society T_f as illustrated in Figure 11.11:

$$S_K = S_\Omega|_{T=T_f}. \tag{11.3.6}$$

Thus the Kauzmann entropy crisis is prevented by the fluctuation entropy of the production waves, because the production waves are the recovering modes from the frozen society.

The Kauzmann entropy S_K and the fluctuation entropy of the production waves S_Ω normalized by Nk_B are illustrated in Figure 11.11, in $x = k_BT/\hbar\omega$ for $N_0/N = 1/5$, for a fragile society $x_0 = k_BT_0/\hbar\omega = 1/2$ and a more fragile one $x_0 = 1$.

With the temperature (the activity of the society) decreasing, the Kauzmann entropy S_K drops according to the Curie law with a negative sign, $S_K \cong -\frac{N_0k_B}{2}\frac{T_0}{T-T_0}$, while the dynamical motions of financial markets drastically drop according to the VTF law $J = \exp\left[-\frac{A}{T-T_0}\right]$, i.e. the velocities of securities and securities density waves v_p, c_s, the inverse relaxation times τ_0^{-1}, τ_M^{-1}, the diffusivity D, and the inverse viscosity η_λ^{-1} drastically drop according to $J = \exp\left[-\frac{A}{T-T_0}\right]$; Drastically, the velocities and the diffusivity drop, the relaxation times lengthen and the viscosities strengthen. These processes cause further recessions so that a chain recession occurs. And eventually the financial markets freeze at the social freezing temperature T_f ($> T_0$).

11.3.4 Applications of the Theory

First, we apply our theory to the financial panic of 2007–2009:
The US bear market of 2007–2009 was declared in June 2008 when the Dow Jones Industrial Average (DJIA) had fallen down 20% from its October 11, 2007 high [84]. The DJIA peaked on October 9, 2007 with a closing price of 14,164.35. On October 11, 2007, the DJIA hits an intra-day peak of 14,198.10 before starting its decline. The decline of 20% by mid-2008 was followed by other stock exchanges across the globe. On September 29, 2008, the DJIA had a record-breaking drop of 777.68 with a close at 10,365.45. The DJIA hits a market low of 6,443.27 on March 6, 2009. The bear market reversed course on March 9, 2009, as the DJIA rebounded more than 20% from its low to 7924.56 after a mere 3 weeks of gains. These data say that the supercooling process appears from October 11, 2007 (the highest consumption) to March 6, 2009 (the freezing of the society).

Since the DJIA reflects the frequency ω_q in the bubbles of buy-sell pairs of securities, $P(q_0, \mathbf{q})$ ($\omega_q \propto J$), the DJIA is proportional to the hopping amplitude J; DJIA $= BJ$, where B is a coefficient. In order to estimate J and S_K, we take the parameters as follows: $N_0/N = 1/5$, $z = 20$, and $x_0 = \frac{k_B T_0}{\hbar\omega} = 1/10$. Then $\frac{S_K(x)}{k_B N} = -\frac{1}{10}\frac{0.1}{x-0.1}$, $\frac{S_\Omega(x)}{k_B N} = -\frac{3/x}{e^{\frac{1}{x}}-1}$, and $J(x) = e^{\frac{0.20}{x-0.1}}$, where $x = \frac{k_B T}{\hbar\omega}$. From Eq. (11.3.6) the freezing temperature $x_f = \frac{k_B T_f}{\hbar\omega}$ is determined as $x_f = 0.199425$. Since the freezing occurred on March 6, 2009, the coefficient is given as $B = 6443.27/J(0.199425) = 48163.6$, where $J(0.199425) = 0.133779$. Once B has been determined, from DJIA$=$ BJ we can determine J, x, and S_K successively. Thus we obtain the results:

Date	DJIA	$\frac{S_K}{k_B N}$	J	$x = \frac{k_B T}{\hbar\omega}$
Jan 3, 2007	12,474.52	−0.0675457	0.259004	0.248048
Oct 9, 2007	14,164.53	−0.0611931	0.294092	0.263417
Sep 29, 2008	10.365.45	−0.0768061	0.215214	0.230198
Mar 6, 2009	6,443.27	−0.100578	0.133779	x_f =0.199425

In this table, it should be noted that the DJIA is regulated by the temperature of the society x. We can see how the DJIA deviates from a normal DJIA on January 3, 2007 under the supercooling process.

In order to see the numerical situations, we show another example:
In 1945, just after the World War II, the worst inflation in history, a super-inflation occurred in Hungary [85]. The Hungarian super-inflation is one of the panics on an

occasion of the supercooling processes to avoid completing the financial cycles so that we can apply the criteria of panic to it. The price drastically increased 10^{30} times for about a year. Meanwhile, a thousand million billion (10^{21}) Pengo paper currency was issued [86]. This situation is explained by the Kauzmann entropy and the VTF law as follows: the velocity of currency flow is proportional to the hopping amplitude J so that the number of times of circulations of the currency is proportional to J. Here we assume that (the total amount of issued currency) $\times$(the number of times of the circulations) is constant, because even if the product changes a lot, its change can be neglected compared to the drastic change of the VTF law, J. Therefore, the total amount of issued currency is proportional to J^{-1}. Here we consider the case that Pengo paper currency was issued e^{50}/e^{1} times as high as usual, where $10^{21} = e^{(\ln 10)\times 21} \cong e^{48.3541}$. For simplicity we investigate the case $x_0 = \frac{k_B T_0}{\hbar\omega} = 1$ in Figure 11.11. In this case, $J^{-1} = \exp[\frac{z}{10}\frac{1}{x-1}]$, $\frac{S_K}{Nk_B} = -\frac{1}{10}\frac{1}{x-1}$, and $\frac{S_\Omega}{Nk_B} = -\frac{3\frac{1}{x}}{e^{\frac{1}{x}}-1}$, where $x = \frac{k_B T}{\hbar\omega}$. The social freezing temperature is $x_f \cong 1.05555$, where $\frac{S_K}{Nk_B} = \frac{S_\Omega}{Nk_B} \cong -1.80018$. Since $\frac{z}{10}\frac{1}{x_f-1} > 50$ is required, we obtain $z > 27.775$. Here we take $z = 28$. Then we obtain:

1. $J^{-1} = e^{1}$ and $\frac{S_K}{Nk_B} = -1/28 \cong -0.357143$ for $x = 3.8$.
2. $J^{-1} = e^{50}$ and $\frac{S_K}{Nk_B} = -50/28 \cong -1.78571$ for $x = 528/500 \cong 1.0560$.

Thus we could explain that the Pengo paper was issued e^{50}/e^{1} times at $x \cong 1.0560$ very close $x_f = 1.05555$ than at $x = 3.8$. It should be noted that as $k_B T_0$ becomes larger than $\hbar\omega$, z becomes smaller than 28.

Since ω represents the elasticity of a society and T_0 the fragility of the society, Figure 11.11 shows the following properties in the panic:

1. For $\omega =$ constant, as T_0 increases, the freezing temperature of the society T_f increases and the Kauzmann entropy S_K drops more sharply. This means that a less fragile society is less damaged than a more fragile one.
2. For $T_0 =$ constant, as ω increases, T_f decreases, and S_K drops less sharply. This means that a more elastic society is less damaged than a less elastic one.

11.3.5 Concluding Remarks

We have been concerned with a simplified capitalist society consisting of workers and enterprises. An m-worker produces his wage p_w^m and a surplus value p_s^m; $kp_w^m = 3\hbar\omega_m/2$ and $kp_s^m = \hbar\omega_m$, where ω_m represents his ability. We have paid our attention to his saving p_{ws}^m in the p_w^m, which is thrown into a financial market as securities. Through the transfer of securities between dealers, the securities hop around in the financial markets. The securities density waves are formed from the sum of an infinite series of interacting bubbles of buy-sell pairs of securities. In financial markets, dealers create enormous bubbles of buy-sell pairs of securities not completing the financial cycles for speculation. Such bubbles of the buy-sell pairs form the fluctuation entropy, which mixes with the entropy of the securities density waves. The mixing entropy with $q \sim K$ constitutes the Kauzmann entropy $S_K \cong -\frac{N_0 k_B}{2}\frac{T_0}{T-T_0}$ in Eq. (11.3.3), which plays an essential role in the panic in the supercooling process. The Einstein relation on entropy and probability leads to the VTF law for the hopping amplitude $J = \exp\left[-\frac{A}{T-T_0}\right]$ in Eq. (11.3.4).

On the other hand, production waves are formed from the sum of an infinite series of interacting bubbles of prod-cons or cons-prod pairs of goods. The production waves are the modes to recover from the symmetry broken (frozen) society. The quantity ω represents the elasticity of the society. Thus the fluctuation entropy of the production waves S_Ω in Eq. (11.3.5) prevents the Kauzmann entropy crisis at T_f in Eq. (11.3.6).

We have assumed that the interaction potential, $V_{00}(\boldsymbol{R})$ between workers in the social space, has a hard core and a valley with a minimum near the mean worker distance a, similar to that in a simple liquid. This assumption is reasonable, because workers become repulsive as they get closer to each other, but they feel comfortable at an appropriate distance a. The interaction potential $V_{ij}(\boldsymbol{R}_m - \boldsymbol{R}_n)$ between workers in the production processes is essentially written as $V_{ij}(\boldsymbol{R}) = \zeta^2 \frac{\partial^2 V_{00}(\boldsymbol{R})}{\partial R_i \partial R_j}$. The actual interaction potentials for a capitalism society are very complicated, but we have taken such a simple interaction potential as that in a simple liquid. This is because the human economic activities depend on so many factors that they are governed by simple laws as a statistical ensemble. This is why the current financial crisis occurs in such a highly information-oriented society.

Deregulated financial markets conceive potentially financial panic. In the markets a great number of bubbles of buy-sell pairs are created exclusively for the sake of speculation, not for the sake of intermediation of the buying or selling social products. In the case of the intermediation, the bubbles of buy-sell pairs play a role in constructing securities density waves, which constitute the maximum entropy of the society. But, in the case of the speculation, enormous bubbles of buy-sell pairs do not complete the financial cycles so that the fluctuation entropy due to the bubbles of buy-sell pairs of securities together with the entropy of the securities density waves with $q \sim K \sim 2\pi/a$ form the Kauzmann entropy S_K. If an extremely high consumption causes enormous bubbles of the buy-sell pairs not completing the financial cycles, which create the supercooling process; the Kauzmann entropy drops according to $S_K \cong -\frac{N_0 k_B}{2} \frac{T_0}{T-T_0}$ and the dynamical motions of the financial markets drastically drops according to the VTF law; $v_p, v_s, \tau_0^{-1}, \tau_M{}^{-1}, \eta_\lambda^{-1}, D \propto J = \exp\left[-\frac{A}{T-T_0}\right]$. These processes cause further recessions so that a chain recession occurs. The markets eventually freeze at the freezing point of the society T_f.

In the optimum society in the equilibrium state, there are the most diverse freedoms; there is no regulation. The entropy of the optimum society is larger than that of a regulated society in the equilibrium state. The current deregulated society seems to be an optimum society, but it produces a great number of bubbles of the buy-sell pairs not completing the financial cycles which form the Kauzmann entropy so that it makes the social gap and the poverty, and eventually induces a panic. The current deregulated society is not an optimum society but a premature society. Some regulations are required to prevent the financial market from reckless driving.

The two level systems are ubiquitous and the two band model elucidates the universal features in the dynamical processes of the panics such as the Kauzmann entropy crisis and the VTF law. Here we consider a system composed of randomly distributed matter in the matter space; we put tildes on quantities in the matter space corresponding to the quantities in the capitalist society. The interaction potential $\widetilde{V}(\widetilde{R})$ between matters is assumed to have a hard core and a valley with a minimum near a mean matter distance $\tilde{a}$. We can construct the two level systems. The difference of the two levels $\tilde{\omega}$ represents the elasticity of the system. If we consider that matters can hop from one position to another, we can construct the two band model so that we can extend the present theory on the financial panic to such a system, where the instability temperature for the matter density waves $\widetilde{T}_0$ is determined as $1 - \tilde{\beta}_0 \tilde{f}' \widetilde{V}(\widetilde{K}) = 0$, $\tilde{\beta}_0 = 1/k_B \widetilde{T}_0$, $\widetilde{K} \sim 2\pi/\tilde{a}$. For such system, the panic occurs as follows:

1. A panic occurs on an occasion of the supercooling processes to avoid completing matter-antimatter cycles under the circumstances: an extremely big happening causes enormous bubbles of matter-antimatter pairs, which form the Kauzmann entropy $\widetilde{S}_K$ and trigger a chain recession (the temperature $\widetilde{T}$ decreases).
2. The Kauzmann entropy $\widetilde{S}_K$ drops according to the Curie law with a negative sign, $\widetilde{S}_K \propto -\frac{\widetilde{N}_0 k_B}{2} \frac{\widetilde{T}_0}{\widetilde{T} - \widetilde{T}_0}$. The dynamical motions of the system drastically drop according to the VTF law, $\widetilde{J} \propto \exp\left[-\frac{\widetilde{A}}{\widetilde{T} - \widetilde{T}_0}\right]$, which cause further recessions so that a chain recession occurs. The system eventually freezes at the freezing point $\widetilde{T}_f (> \widetilde{T}_0)$, where the Kauzmann entropy and the fluctuation entropy of matter production waves cross over.

For a panic in personal psychology, the matter is information via neurons in the sufferer's brain. For a panic in mass psychology, the matter is information via humans en masse.

We can recognize that the two level systems are ubiquitous and the two band model elucidates the dynamical processes of their respective panics in the systems, where the Kauzmann entropy crisis and the VTF law play a universal role in the dynamical processes.

References

[1] A. Griffin, Excitations in a Bose-Condensed Liquid Cambridge University Press, London, 1993.
[2] H.R. Glyde, Excitations in Liquid and Solid Helium Oxford Universuty Press, New York, 1994.
[3] T. Kitamura, Physica A 290 (2001) 39.
[4] H. Umezawa, H. Matsumoto, M. Tachiki, Thermo Field Dynamics Condensed States North-Holland, Amsterdam, 1982.
[5] N. Hugenholtz, D. Pines, Phys. Rev. 116 (1959) 489.
[6] J.P. Hansen and I.R. MacDonald, Theory of Simple Liquids Academic Press, London, 1990.
[7] E.C. Svensson, V.F. Sears, A.D.B. Woods, P. Martel, Phys. Rev B 23 (1980) 3638.
[8] A.I. Oparin, Proiskhozhdenie Zhizny, Moskovskii Rabochii, Moskva (1924).

[9] A.I. Oparin, The Origin of Life, MacMillan 1938.
[10] A.I. Oparin, The Origin of Life, New York, Dover (1952).
[11] J.B.S. Haldane, Daedalus or Science and the Future, Cambrigde (1923).
[12] J.B.S. Haldane, Possible World, New York, Hugh & Bres 1928.
[13] J.B.S. Haldane, The Origin of Life, New Biology 16 12 (1954).
[14] S.L. Miller, Science 117 528 (1953).
[15] S.L. Miller, H.C. Urey, Science 130 245 (1959).
[16] J. Oró, Nature 19 389-390 (1961).
[17] C.R. Woese, The genetic Code: The Molecular Basis for Genetic Expression Harper & Row, New York, 1967.
[18] F.H.C. Crick, J. Mol. Biol. 38 (1968) 367-380.
[19] F.H.C. Crick, Science 204 (1979) 264-271.
[20] L.E. Orgel, J. Mol. Biol. 38 (1968) 381-393.
[21] J.E. Sulston, L.E. Orgel, De la Physique Théorique à la Biologie,in: M. Marois (Ed.), Editions du Centre National de la Recherche Scientifique, Paris, 1971.
[22] L.E. Orgel, J.E. Sulston, in: A.P. Kimball, J. Oró (Eds.), Prebiotic and Biochemical Evolution, North-Holland, Amsterdam, 1971 89-94.
[23] T.R. Cech, A.Z. Zaug, P.J. Grabowski, Cell 27 (1981) 487-496.
[24] K. Kruger, P.J. Grabowski, P.J. Zaug, J. Sands, D.E. Gottschling, T.R. Cech, Cell 31 (1982) 147-157 .
[25] A.J. Zaug, T.R. Cech, Science 231 (1986) 470-475.
[26] T.R. Cech, Proc. Natl. Acad. Sci. USA 83 (1983).
[27] S. Altman, J.D. Smith, Nature New Biol. 233 35-39 (1971) 4360-4363.
[28] C. Guerrier-Takada, K. Gardiner, T. Marsh, N. Pace, S. Altman, Cell 35 (1983) 849-857.
[29] C. Guerrier-Takada, K. Hydock, L. Allen, S. Altman, Biochemistry 25 (1986) 1509-1515.
[30] D.P. Bartel, J.W. Szostak, Science 261 (1993) 1411-1418.
[31] E.H. Ekland, J.W. Szostak, D.P. Bartel, Science 269 (1995) 364-370.
[32] W. Gilbert, Nature 319 618 (1986).
[33] A, G. Cairns-Smith, J. Theor. Biol. 10 (1966) 53-83.
[34] J.A. Doudna, J.W. Szostak, Nature 339 (1989) 519-522.
[35] A. Eschenmoser, D. Felix, G. Ohloff, Helv. Chim. Acta 50 (1967) 708-713.
[36] P.E. Nielsen, M. Egholm, R.H. Berg, O Buchardt, Science 254 (1991) 1497-1500.
[37] G.F. Joyce, Nature 338 (1989) 217-224.
[38] R. Shapilo, Origins of Life 14 (1984) 565-570.
[39] R. Shapilo, Proc. Natl. Acad. Sci. USA 96 (1999) 4396-4401.
[40] F.J. Dyson, J. Mol. Evol. 18 (1982) 344-350.
[41] F. Dyson, Origin of Life. Cambridge University Press, Cambridge, UK, 1999.
[42] S.A. Kauffman, J. Theor. Biol. 119 (1986) 1-24.
[43] S.A. Kauffman, The origin of Order - Self-organization and Selection in Evolution. Oxford University Press. Oxford, UK, 1993.
[44] D. Segré, D. Lancet, O. Kedem, Y. Pilpel, Orig. Life Evol. Biosphere 28 (1998) 501-514.
[45] D. Segré, Y. Pilpel, D. Lancet, Physica A 249 (1998) 558-564.
[46] D. Segré, D. Lancet, Biochem. Mol. Biol. 12 (1999) 382-397.
[47] T. Kitamura, Phys. Rep. 383 (2003) 1-94.
[48] T. Kitamura, AIP Conference Proceedings 708 (2004) 631-634.
[49] T. Kitamura, M. Peyrard, S.C. Lopez, Phys. Biol. 2 (2005)1.
[50] E.F. Tombari, C. Ferrari, G. Salvetti, G.P. Johari, J. Phys. Condens. Matter 9 (1997) 7017-7037.
[51] S. Correzzi, D. Fioretto, P. Rolla, Nature 420 (2002) 653-656.

[52] T. Kitamura, Physica A 383 (2007) 232.
[53] T. Kitamura, Phyica A 389 (2010) 2586.
[54] R.J. Schiller, International Exuberance, 2nd edn, Princeton University Press, Princeton, 2005.
[55] M. Kaneko, A. Dewit, Global Crisis, Iwanami Shuppan, Tokyo, 2008 (in Japanese).
[56] M.K. Brunnermeier, J. Econ. Perspect. 23 (2009) 77.
[57] M. Buchanan, Nature 680 (2009) 460.
[58] J.D. Farmer, D. Foley, Nature 685 (2009) 460.
[59] L.D. Landau, Statistical Physics, Pergamon Press, London, 1958.
[60] H.B. Callen, Thermodynamics, John Wiley & Sons, Inc., New York, 1960.
[61] D. Kondepudi, I. Prigogine, Modern Thermodynamics, John Wiley and Sons, New York, 1998.
[62] H. Chernoff, L.E. Moses, Elementary Decision Theory, Dover Publications, New York, 1987.
[63] F. Quesnay, Tableau économimique, L'ami des hommes: Pt. 6, Paris, 1760.
[64] A. Smith, An Inquiry into the Nature and Causes of the Wealth of Natations, 5th edn. Methuen & Co., Ltd., London, 1904.
[65] D. Ricardo, On the principles of Political Economy and Taxation, 3rd edn. John Murray. Publications, London, 1817.
[66] K. Marx, Das Kapital, Kritik der politischen Ökonomie, Erster Band, Dietz Verlag, Berlin, 1973.
[67] J. Robinson, The Accumulation of Capital, Macmillan, London, 1956.
[68] M. Kalecki, Social Science Information 6 (1968) 73.
[69] T.B. Veblen, The Theory of Business Enterprise, Charles Scribner's Sons, New York, 1904.
[70] J.M. Keynes, The General Theory of Employment, Interest, and Money, Macmillan Cambridge University Press, Cambridge, 1936.
[71] T. Fuwa, Shihonron zen 3-bu wo yomu, Shin Nihon Shuppan, Tokyo, 2003 (in Japanese).
[72] S. Strange, Casino Capitalism, Blackwell, Oxford, 1986.
[73] T. Schneider, R. Brout, H. Thomas, J. Feder, Phys. Rev. Lett. 25 (1970) 1423.
[74] K.S. Singwi, K. Sköld, M.P. Tosi, Phys. Rev. A 1, (1970) 454.
[75] L.D. Landau, E.M. Lifshitz, Fluid Mechanics, Pergamon Press, London, 1963.
[76] M. Born, K. Haung, Dynamical Theory of Crystal Lattices, Oxford Univ. Press, Oxford, 1954.
[77] A.A. Maradudin, E.W. Montroll, G.H. Weiss, I.P. Ipatova, Theory of Lattice Dynamics in the Harmonic Approximation, 2nd edn. Academic Press, New York, 1971.
[78] G.K. Horton, A.A. Maradudin, Dynamical Properties of Solids Vol. 1, North-Holland, New York, 1974.
[79] L.D. Landau, E.M. Lifshitz, Theory of Elasticity, Pergamon Press, London, 1970.
[80] Y. Nambu, G. Jona-Lasinio, Phys. Rev. 122 (1961) 345.
[81] Y. Nambu, G. Jona-Lasinio, Phys. Rev. 124 (1961) 246.
[82] J. Goldstone, Nuovo Cim. 19 (1961) 154.
[83] J. Goldstone, A. Salam, S. Weinberg, Phys. Rev. 127 (1962) 965.
[84] Dow Hits Bear-Market Territory, Signalling Woe for Economy, Wall Street Journal, June 28, 2008.
[85] W.A. Bomberger, G.E. Makinen, The Hungarian Hyperinflation and Stabilization of 1945-1946, J. Political Economy 91 (1983) 801.
[86] An issued "thousand million billion Pengo paper currency is preserved in The Bank of Japan Currency Museum.

Kolibri 360 GmbH
Ferdinand-Jühlke-Straße 7
99095 Erfurt, Deutschland
produktsicherheit@kolibri360.de